Near-Surface Applied Geophysics

Just a few meters below the Earth's surface lie features of great importance, from geological faults which can produce devastating earthquakes, or migrating contaminant plumes of hazardous industrial waste, to lost archaeological treasures!

Fueled by the growing need for careful management of the Earth's subsurface resources, traditional geophysical methods designed for deep-exploration targets have been complemented by significant advances in near-surface applied geophysics techniques and interpretation theory.

This book provides a refreshing, up-to-date exploration of the theoretical foundations and the latest developments in near-surface techniques. It is clear but rigorous, explaining theory and practice in simple physical terms, supported by intermediate-level mathematics. Techniques covered include magnetics, resistivity, seismic reflection and refraction, surface waves, induced polarization, self-potential, electromagnetic induction, ground-penetrating radar, magnetic resonance, interferometry, seismoelectric methods, and more. Sections on data analysis and inverse theory are provided and chapters are amply illustrated by case studies, providing a resource that gives students and professionals the necessary tools to plan, conduct, and analyze a near-surface geophysical survey.

This is an important textbook for advanced undergraduate and graduate students in geophysics, and a valuable reference for practicing geophysicists, geologists, hydrologists, archaeologists, civil and geotechnical engineers, and others who use geophysics in their professional activities.

Mark E. Everett is currently the Howard Karren Endowed Professor and Graduate Director in the Department of Geology and Geophysics at Texas A&M University, and was a Guest Professor in 2010 at the Institut fur Geophysik, ETH Zurich. He operates the Near-Surface Applied Geophysics laboratory at Texas A&M and has conducted geophysical field work in many locations including the Normandy D-Day landing site in France and on the island of Alcatraz. Dr Everett is on the editorial boards of *Geophysics* and *Geophysical Journal International* and has received the Texas A&M College of Geoscience Dean's Research Achievement Award. He is a fellow of the Royal Astronomical Society and a member of the American Geophysical Union and Society of Exploration Geophysicists. Dr Everett is a frequent consultant to the oil and gas, environmental, and geotechnical engineering sectors of industry, holds a professional license to practice geophysics in the State of Texas, and enjoys contributing his time and expertise to local historical archaeological projects.

Praise for this book:

"This book provides an excellent introduction to the rapidly emerging field of near-surface geophysics. The state-of-the-art material covered in the text will not only be helpful for undergraduate and graduate students, but it will also serve as a valuable reference for practitioners."

– **Professor Hansruedi Maurer**, *ETH Zürich*

"A great book for teaching undergraduates the essence of all geophysical techniques used in near surface exploration; the mix of theory, practice and case studies is just right for students."

– **Professor Graham Heinson**, *School of Earth and Environmental Sciences, The University of Adelaide*

"This is an excellent text for advanced geophysics undergraduate and graduate students, a valuable resource for scientists and engineers involved in characterizing the earth's near surface, and the material fills a void in the bookshelves of many geoscientists."

– **Professor Doug Oldenburg**, *University of British Columbia*

Near-Surface Applied Geophysics

MARK E. EVERETT

Texas A&M University

CAMBRIDGE
UNIVERSITY PRESS

University Printing House, Cambridge CB2 8BS, United Kingdom

One Liberty Plaza, 20th Floor, New York, NY 10006, USA

477 Williamstown Road, Port Melbourne, VIC 3207, Australia

4843/24, 2nd Floor, Ansari Road, Daryaganj, Delhi - 110002, India

79 Anson Road, #06-04/06, Singapore 079906

Cambridge University Press is part of the University of Cambridge.

It furthers the University's mission by disseminating knowledge in the pursuit of education, learning and research at the highest international levels of excellence.

www.cambridge.org
Information on this title: www.cambridge.org/9781107018778

First published 2013
Reprinted 2014

A catalogue record for this publication is available from the British Library

Library of Congress Cataloging in Publication data
Everett, Mark E., 1961–
Near-surface applied geophysics / Mark E. Everett, Texas A&M University.
pages cm
Includes bibliographical references and index.
ISBN 978-1-107-01877-8 (Hardback)
1. Geophysics–Methodology. I. Title.
QC808.5.E94 2013
550.72′4–dc23
2012036494

ISBN 978-1-107-01877-8 Hardback

To Elaine, Taylor, and Laura

Contents

The color plates can be found between pages 242 and 243.

Preface

Historically, geophysics has been used to characterize deep exploration targets, such as economic mineralization, oil and gas deposits, or new groundwater resources, in frontier environments that are relatively free of human impact. At the same time, civil engineers, archaeologists, soil scientists, and others have applied the traditional geophysical methods with long-trusted but simple interpretation schemes to detect, classify, and describe buried geological or anthropogenic targets in the shallow subsurface. In recent years however, as the amount of Earth's land area untouched by human impact has decreased and as the importance of responsible stewardship of Earth's subsurface resources has increased, a significant body of advances has been made in near-surface applied geophysics techniques and interpretation theory that have caused existing textbooks and monographs on the subject to become outdated.

The present book is designed to bring senior undergraduate and graduate students in geophysics and related disciplines up to date in terms of the recent advances in near-surface applied geophysics, while at the same time retaining material that provides a firm theoretical foundation on the traditional basis of the exploration methods. The plan of the book is to explain the new developments in simple physical terms, using intermediate-level mathematics to bring rigor to the discussion. The sections on data analysis and inverse theory enable the student to appreciate the full execution of applied geophysics, from data acquisition to data processing and interpretation. The material is amply illustrated by case histories sampled from the current, peer-reviewed scientific literature. This is a textbook that students will find challenging but should be able to master with diligent effort. The book will also serve as a valuable reference for geoscientists, engineers, and others engaged in academic, government, or industrial pursuits that call for near-surface geophysical investigation.

Acknowledgments

Many persons have provided warm friendship, steady guidance, unwavering support and valuable advice throughout the years; the following are those who have particularly stood out as being instrumental to the successful completion of this book: Chris Everett, Nigel Edwards, Steven Constable, Brann Johnson, Chris Mathewson, Eugene Badea, Zdenek Martinec, Dax Soule, Chester Weiss, Alfonso Benavides, Souvik Mukherjee, Rungroj Arjwech, and Tim De Smet.

1 Introduction

The purpose of this textbook is to introduce undergraduate and graduate students in geophysics, geology, geotechnical engineering, archaeology, and related disciplines to the application of geophysical methods for studying the uppermost tens of meters beneath Earth's surface. This portion of Earth both affects and is impacted by human activities such as building, excavating, tunneling, and storing or accidentally releasing hazardous materials. Many of the planet's mineral and groundwater resources are located in the uppermost subsurface layers. The techniques of near-surface geophysics are being increasingly used to benefit society in activities such as nuclear-waste storage, carbon sequestration, precision agriculture, archaeology, crime-scene investigations, and cultural-resources management. Near-surface geophysicists conduct a huge range of scientific and engineering investigations in service of society (Doll *et al.*, 2012). Some of these are shown in Table 1.1.

Moreover, near-surface geophysicists address basic scientific questions relevant to a spectrum of natural processes that include biogeochemistry, coastal processes, climate change, ecology, hydrology, tectonics, volcanology, and glaciology (Slater *et al.*, 2006a). Near-surface geophysical data are now routinely used, for example, in basic geoscience investigations such as environmental remediation, natural-hazards risk assessment, water-resource management, gas hydrates and permafrost studies, glacier and ice-sheet mass transport, watershed-scale and coastal hydrology, fault-zone characterization, and the reconstruction of Earth's tectonic, volcanic, and extraterrestrial impact history.

Near-surface geophysicists also develop new instrumentation, data acquisition and processing strategies, and explore fundamental aspects of Earth imaging, forward

Table 1.1 Near-surface applied geophysics in the service of society

Engineering and environmental	Shallow resources	Archaeology, forensics, military	Groundwater
Building foundations	Geological mapping	UXO/landmine detection	Water-well site location
Top of bedrock	Ore body delineation	Historic preservation	Water-table mapping
Non-destructive evaluation	Aggregate prospecting	Cultural-resources management	Fracture-zone delineation
Tunnel detection	Offshore gas hydrates	Gravesite location	Salinization mapping
Underground storage tanks	Coal-seam mapping	Crime-scene investigations	Aquifer characterization
Slope stability	Geothermal mapping	Nautical archaeology	Contaminant mapping

modeling, and inversion. Also of critical interest to near-surface geophysicists is the petrophysical interpretation step which provides the link from the geophysical image to the geological properties of the subsurface. The latter, which include such properties as porosity, salinity, and rock strength, are of broad scientific or engineering concern.

1.1 Workflow

The following list outlines a typical series of activities that a near-surface geophysicist, or a team of geophysicists, or a multi-disciplinary team including one or more geophysicists, might undertake in order to complete a scientific or engineering project. The order and extent of the individual activities varies from project to project.

A. Front-end

- define the scientific or engineering problem to be solved
- gather the available prior information
- formulate possible hypotheses
- initial site reconnaissance
- decide on relevant geophysical techniques
- design the experiment (subject to logistical and budget contraints, etc.)
- purchase/build/modify/test and sequester instrumentation
- assemble field crew and analysis team
- contingency planning/crew scheduling
- safety plan/logistics/shipping/permissions

B. In the field

- prepare daily checklist
- mobilization
- prepare field site for experiment (clear vegetation, etc.)
- determine navigation and acquisition tactics
- survey the site and lay out the grid
- test equipment, perform quality-control procedures
- acquire the data
- store/archive dataset
- demobilization (stow equipment for next use, etc.)

C. In the office

- notebook/data reconciliation
- visualization of raw data
- preliminary data processing (removal of bad data points, etc.)
- prepare initial field report
- advanced data processing (spectral analysis, filtering, etc.)
- modeling and inversion to construct Earth model

- geological interpretation
- formulate tentative conclusions

D. Back-end

- place results in broader context
- prepare presentation/draft report
- get feedback from colleagues
- refine hypotheses/acquire additional data/perform additional analysis
- finalize conclusions
- provide recommendations for further study
- prepare and submit technical report/scientific publication
- rigorous peer review
- acquire additional data/perform additional analysis/refine hypotheses
- prepare and submit revised technical report/scientific publication

There are numerous introductory field guides which describe the various geophysical techniques in simple terms and explain how to select the appropriate equipment, plan and conduct a geophysical experiment, and display and provide a preliminary interpretation of the data. An example is the chapter entitled "Geophysical Techniques" from the Field Sampling Procedures Manual produced by the New Jersey Department of Environmental Remediation and available on the web at www.nj.gov/dep/srp/guidance/fspm/. Practical guidelines for applying geophysics to engineering problems, with a special emphasis on transportation projects, is provided in NCHRP (2006). The American Society for Testing and Materials has produced a standard guide for selecting surface geophysical methods (ASTM, 2011).

1.2 Some applications of near-surface geophysics

Near-surface applied geophysics (NSG) differs from traditional exploration geophysics, as practiced by the oil and gas and mineral industries, in several key respects. First, there is a demand for sub-meter-scale depth and lateral resolution. Second, there often exists an opportunity for confirmation, or ground-truthing, of the geophysical interpretation via excavation or drilling. Third, near-surface geophysics involves certain specialized techniques not found in traditional geophysics such as ground-penetrating radar (GPR), metal detection, surface nuclear magnetic resonance, high-frequency seismology, and microgravity. Finally, a near-surface geophysics project is often motivated and constrained not by profit concerns but by public health and safety concerns, and the work often must be carried out subject to rigorous legal or regulatory requirements (Butler, 2005).

Over the past 30 years, the field of near-surface geophysics has grown rapidly, as indicated by the following analysis. The number of papers in the Science Citation Index (SCI) (www.wokinfo.com) published between 1980 and 2010 that contain the keyword "ground penetrating radar" in the title or abstract is shown in Figure 1.1.

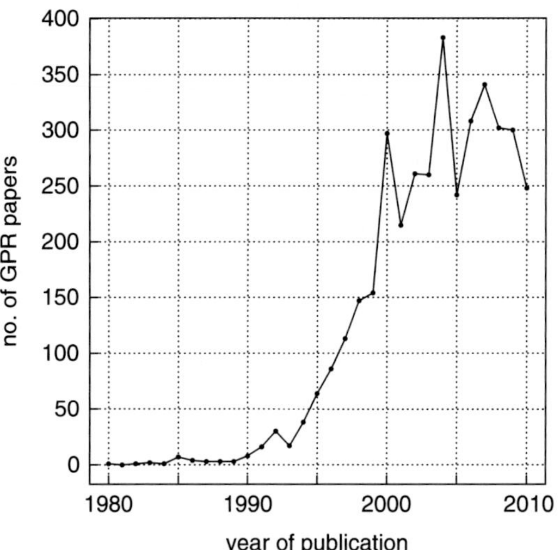

Figure 1.1 Numbers of papers in Science Citation Index between 1980 and 2010 with "ground penetrating radar" as keyword.

The figure shows that GPR literature is virtually non-existent in the SCI database prior to the early 1990s. The activity level in all aspects of NSG prior to 1991 is not well described by these data however since GPR is a relatively new technique compared to electromagnetic (EM), magnetic, seismic, and resistivity techniques. The number of GPR papers grew rapidly in the 1990s and since 2000 it has stabilized at ~ 300 per year. The variation in the number of GPR papers since the early 1990s serves as a rough proxy for the overall level of activity in near-surface applied geophysics over the same period of time.

In order to provide a rough estimate of how often the various techniques of near-surface applied geophysics are used, the abstracts of articles from the European Association of Geoscientists and Engineers (EAGE) journal *Near Surface Geophysics* during the years 2002–2010 were analyzed. The result of an analysis of the 299 articles is shown in Figure 1.2. Clearly, GPR and resistivity are the two most widely used techniques in this database, being mentioned in more than 50% of all abstracts. The next most prominent methods are the various EM and seismic techniques, the latter including surface-wave, reflection, and refraction analysis. Following these, in order of popularity, are magnetics, magnetic resonance, gravity, borehole seismic/radar, self-potential (SP), and induced-polarization (IP) methods. The "other" category includes specialized techniques such as seismoelectrics, ultrasonics, infrared imaging, microwave tomography, γ-ray spectrometry, cone-penetration testing, microseismic and ambient seismic noise studies, time-domain reflectometry, acoustics, and temperature mapping.

A number of case histories are now briefly described to provide the reader with an early glimpse of the range of science and engineering problems that can be addressed using the techniques of near-surface geophysics.

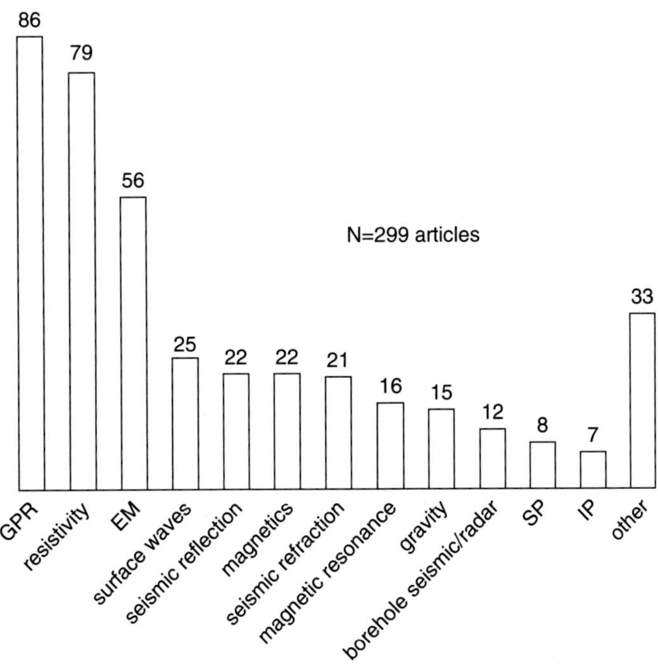

Figure 1.2 Number of times that the use of a specific technique is mentioned in articles published in *Near Surface Geophysics*, 2002–2010.

Example. UXO clearance.

An excellent example of the special characteristics of near-surface geophysics is provided by the problem of unexploded ordnance (UXO) clearance. The US Congress periodically directs the Department of Defense to close military bases that are no longer needed to fulfill mission requirements (MacDonald *et al.*, 2004). A major impediment to the transfer of these lands to civilian purposes is the environmental clean-up of potentially dangerous UXO resulting from military training and weapons testing. The primary technology used for UXO detection and its discrimination from harmless metal clutter is electromagnetic induction geophysics.

Two sample UXO electromagnetic induction geophysics datasets are shown in Figure 1.3. The quantity plotted is Q_{SUM} [ppm], the quadrature response summed over multiple frequencies, which for the present purpose may be regarded simply as a measure of the metal content in the ground. Figure 1.3a shows anomalies due to purposely buried metal objects, namely steel, copper, and aluminum pipes, at a test site in North Carolina. Figure 1.3b shows anomalies due to putative UXO at a live site in Kaho'olawe, Hawaii. The background response is noisy at the Hawaii site due to the high magnetic mineral content of the basaltic soils.

Data in UXO or landmine surveys are routinely acquired at 10 cm station spacing and 25 cm line spacing, both of which are a factor of ~ 1000 times less than the corresponding spacings in traditional oil and gas exploration geophysical surveys. The goal of a UXO/landmine geophysical survey is to maximize the probability of detection (no misses) while

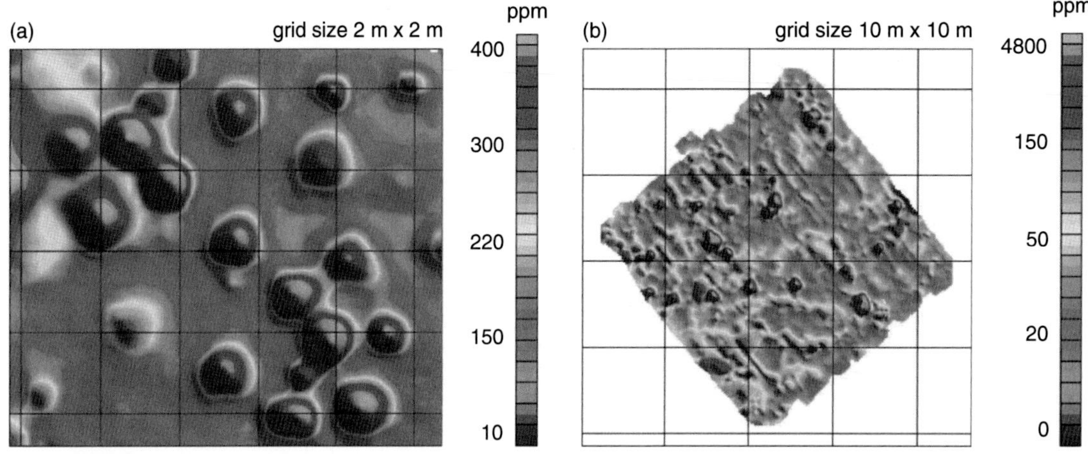

Figure 1.3 (a) EM induction response of metal pipes buried at a test site. (b) EM induction response of unexploded ordnance (UXO) at a live site with highly magnetic soil, Kaho'olawe, Hawaii. From Huang and Won (2003). See plate section for color version.

minimizing the false-alarm rate (no costly false positives) and to generate a reliable dig/no-dig decision for each anomaly. For the clearance operation, a priority dig list is constructed by the geophysicist and ground-truthing is performed by the site operator.

Example. GPR stratigraphy.

An example of a near-surface geophysics survey that uses GPR is shown in Figure 1.4. In this example, a GPR image is presented of the stratigraphy of an active dune on Parengarenga sandspit in New Zealand. The spit is an important source of high-quality (~ 94% quartz) fine sand used for glass manufacture. Besides economic factors, the possible environmental effects on the Parengarenga harbor and the regional coastal hydraulics of continued sand extraction has motivated the GPR investigation. The 200 MHz GPR section reveals the lateral continuity of the underlying "coffee rock," a semi-consolidated Quaternary paleosol. Cross-bedding within the dune can also be seen in the GPR section. The internal structure revealed by the GPR data provides valuable constraints on long-term evolution of the spit at this important coastal site.

Example. Mars radar sounding.

Shallow subsurface radar is finding new and exciting applications in planetary science. Recently, data from onboard the Mars Reconnaissance Orbiter has returned high-resolution images of the Martian north polar layered deposits (NPLD) and the underlying basal unit (BU), as shown in Figure 1.5. The orbiting SHARAD instrument is an 85-μs chirp radar with a central frequency of 20 MHz and a bandwidth of 10 MHz. The depth

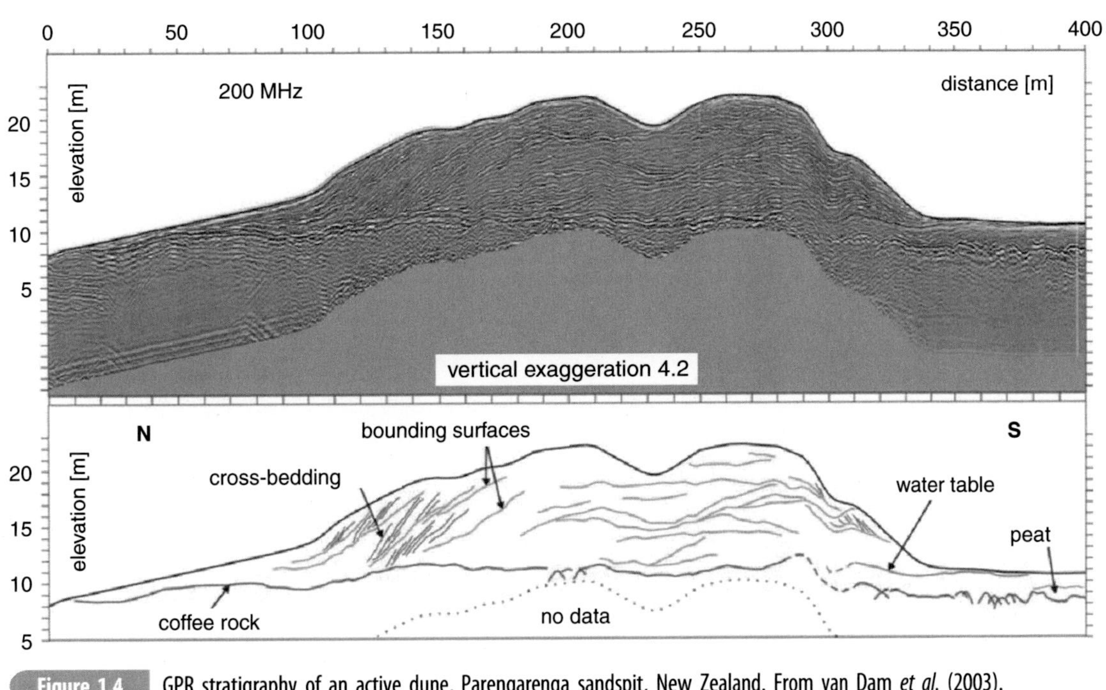

Figure 1.4 GPR stratigraphy of an active dune, Parengarenga sandspit, New Zealand. From van Dam *et al.* (2003).

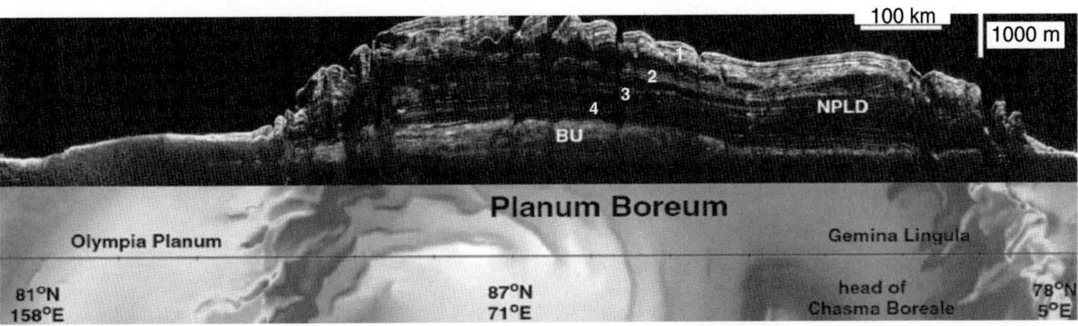

Figure 1.5 Orbiting 85-μs chirp radar image of stratigraphy at the Mars north polar ice cap. From Phillips *et al.* (2008). See plate section for color version.

resolution is ~ 15 m and the along-track horizontal resolution, after data processing, is ~ 1 km. The radar imagery indicates a laterally continuous deposition of high-reflectivity ice, dust, and lava underlain by a basal interbedded sequence of lower reflectivity. The NPLD vertical structure consists of distinct sediment packages characterized by different ice/dust fractions (see Figure 1.5). The stratigraphy may be explained in terms of Mars climatic variations that are governed by the ~ 10^6-year periodicities in planetary obliquity and orbital eccentricity (Phillips *et al.*, 2008). Reprinted with permission from AAAS.

Figure 1.6 Approximate locations of three seismic reflection profiles in the city of Barcelona, Spain (after Martí *et al*. 2008). The route of the proposed subway tunnel is also shown. Only the southernmost ~150 m of profile 1 is indicated in this figure; it continues for an additional ~300 m in the north-northwest direction.

Example. High-resolution urban seismic reflection.

Performing a geophysical survey within an urbanized area is challenging due to the presence of abundant cultural noise and above-ground obstacles. Geophysicists are

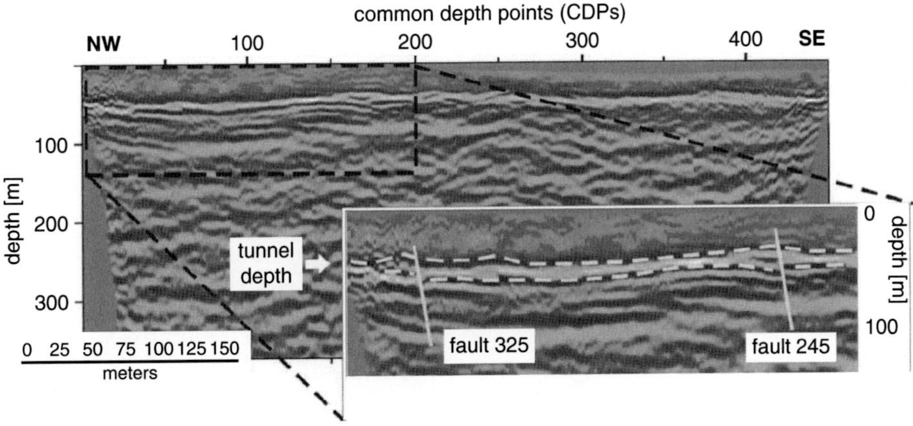

Figure 1.7 Seismic depth section from Martí *et al.* (2008), profile no.1 shown in the previous figure. The inset shows two mapped faults and the contact zone between the near-surface weathered layer and underlying intact bedrock.

increasingly tasked to work in urban settings, industrial brownfields, or other built-up areas since these locations contain a disproportionate number of the world's environmental hazards or require site characterization for further development or land reclamation. A high-resolution seismic-reflection survey was recently carried out within the city limits of Barcelona, Spain to support the overall engineering design for a new subway transportation tunnel. Due to abrupt lateral changes in hydraulic and mechanical rock properties, tunnel-boring conditions often become problematic or hazardous at intersections with subsurface permeable fracture zones. A series of seismic-reflection profiles, shown in Figure 1.6, were acquired along city streets during the relatively quiet overnight hours using a vibroseis truck with a frequency sweep of 14–120 Hz as the source (Martí *et al.*, 2008). The stacked seismic section from profile 1 is shown in Figure 1.7. Two previously mapped faults, labeled 245 and 325, have been identified. These faults may be associated with subsurface fracture zones and could cause engineering difficulties during the tunnel construction.

1.3 Communication of uncertainties

It is well known that geophysical data are insufficient to uniquely determine the distribution of subsurface properties, to any level of precision. There are always ambiguities in the interpretation of geophysical data. A major challenge for the near-surface geophysicist is to decide how the uncertainty associated with a given subsurface image should be communicated to stakeholders. Clearly, in all cases it is important that the geophysicist should understand the stakeholders' objectives. Given that, the geophysicist should explain how

Figure 1.8 Excavation of a leaking 30 kL steel tank from a former retail petroleum site. Photo courtesy of Joshua Gowan.

the geophysical result is relevant to the problem objectives, and carefully describe the limitations inherent in the geophysical interpretation.

If the geophysicist is acting purely in a scientific role, this might be the extent of the geophysicist's contribution. A scientist acts primarily as a provider of facts, an interpreter of evidence, or as an advisor and is not obliged, nor oftentimes expected, to contribute in a direct way to decision-making either in the public or a corporation's best interest.

On the other hand, if the geophysicist is acting in a professional role, an expert opinion is normally required. The professional geophysicist bears the responsibility to provide stakeholders with a definite statement in the face of many unknowns. The stakeholder, who has employed the geophysicist, is prepared to tolerate only a certain amount of uncertainty. The professional geophysicist must carefully state the inherent uncertainty in the interpretation, but then provide a definite professional opinion that contains *as much uncertainty as the stakeholder is willing to tolerate*. The professional geophysicist should always strive to act in the best interest of the stakeholders without any compromise of scientific integrity.

Example. Superfund.

A primary example of US federal legislation that has driven near-surface geophysical investigations is the 1980 Comprehensive Environmental Response, Compensation, and Liability Act (CERCLA), commonly known as Superfund (www.epa.gov/superfund/policy/cercla.htm). This law created a tax on the chemical and petroleum industries in order to maintain a public trust fund managed by the Environmental Protection Agency (EPA). The fund is used to enable a response to releases of hazardous substances that may endanger public health or the environment and to clean up waste sites. Near-surface geophysics is often used for subsurface characterization and remediation (Figure 1.8), and to perform tasks such as monitoring the subsurface transport and fate of a contamination plume.

1.4 Outline of the book

This book is intended to provide readers with an overview of the techniques and applications of near-surface geophysics. The mathematics is kept at an intermediate level with the result that the equations should be readily accessible to readers who have completed two semesters of physics and three semesters of calculus. Readers with not as much preparation in math and physics will be pleasantly surprised to find the text rich in conceptual detail and illustrated with abundant case histories. My experience has told me that most, if not all, scientific concepts, however difficult, can be explained in three different yet complementary ways: (a) in words; (b) in pictures; (c) in equations. These three presentation techniques provide three different perspectives on the topic at hand, but ultimately they should all convey the same essential information about the topic. Everyone has a different learning style that responds best to certain combinations of words, pictures, and equations. Some readers learn most effectively from words and pictures, others from words and equations, still others from pictures and equations, and so forth. Therefore, throughout the text, I try to use all three presentation techniques.

The book is arranged as follows. First a brief review of elementary data-analysis concepts is presented. The next set of chapters of the book are organized by geophysical technique. There are separate chapters on magnetics, resistivity, induced polarization and self-potential, seismic body and surface waves, EM induction, and ground-penetrating radar. Within each chapter, there are discussions on theory, instrumentation, data acquisition and processing, and modeling and interpretation methods. Each chapter includes a selection of illustrative case histories and most conclude with student exercises. Selected techniques that are either new or resurgent in popularity, such as nuclear magnetic resonance, time-lapse microgravity, and the seismoelectric technique, are treated in a separate chapter. The book does not discuss borehole methods, which have been reviewed by Paillet and Ellefson (2005). Other topics such as radiometrics and thermal methods are also left out so that the book maintains a manageable length. The book concludes with chapters on inversion, both linear and non-linear, treating local and global methods separately.

2 Data analysis

This chapter provides a general overview of some elementary concepts in geophysical data analysis such as information, sampling, aliasing, convolution, filtering, Fourier transforms, and wavelet analysis. Good introductions to many of these topics may be found in Kanasewich (1981) and Gubbins (2004). The material presented in this chapter is aimed to help the near-surface geophysicist to process and interpret data acquired in the field using the techniques that are discussed in the subsequent chapters of this book.

2.1 Information

Geophysicists gather data from which they attempt to extract information about the Earth, in order to better understand its subsurface structure and the wide variety of geological processes that shapes its evolution. The assumption is that geophysical datasets contain information about subsurface geology. It is useful to establish what is meant by the term *information*. There are many ways to define information but, at the most fundamental level, I regard it as *a quantum, or bit, of new knowledge*. Special emphasis should be placed on the word *new* in the foregoing definition, in order to reinforce the concept that previously existing knowledge is not information.

A simple example illustrates this concept of information. Consider a pathological ant that moves forward in the same direction exactly one unit of distance x at the stroke of midnight each night. The change in the ant's position $\Delta x(t)$ may be represented as a comb function, or a unit impulse signal that repeats at a given interval of time, in this case $\Delta t = 24$ h. The unit comb function is shown in Figure 2.1a. For an observer seeing the ant for the first time, there are just two bits of information in this graph: the length of the jump and the time interval between jumps. However, for the experienced observer who has learned the ant's behavior and has no reason to believe it should change, there is no longer any information contained in the ant's motion. Now suppose, unexpectedly, there happens to be an unusual day in which the ant moves ahead two units at precisely midnight, instead of the customary one unit, as shown in Figure 2.1b. Suddenly, even for the experienced observer, there is information here. In fact, there are two bits of information: the time t_0 at which the unusual jump took place, and the length $\Delta x(t_0) = 2$ of the unusual jump.

The above example illustrates that information is tied to expectation and that information, like beauty, is in the eye of the beholder. The information content of a data stream should be regarded as the number of bits of new, or unexpected, knowledge that it contains. A highly predictable data stream, such as an ocean tidal signal, has very little or no

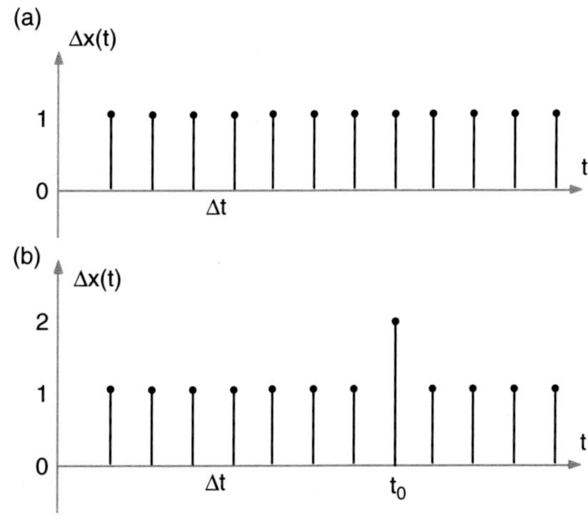

Figure 2.1 (a) The unit comb function consisting of regularly spaced unit impulses. This signal captures the regular movement of a pathological ant. (b) A signal which captures an unexpected movement of the ant.

information content for someone who is quite familiar with ocean tides. On the other hand, for all of us there is a great deal of information contained in a seismic data stream that records the ground motion associated with a large destructive earthquake, since it is a highly unpredictable event. Scientists cannot say they understand such earthquakes until the relevant seismic datasets carry no information, a situation which is not likely to happen any time soon. A general rule is that the information content of a geophysical time series decreases with increasing grasp of the underlying processes that generated the time series. In the extreme case, as Lindley (1956) points out: "*if the state of nature is known, then no experiment can be informative.*"

Information is oftentimes deeply embedded in datasets and it can take considerable physical insight on the part of the geophysicist to reveal it. For example, geophysical measurements may be affected by Earth parameters that are not always recognized. The effect of these parameters on a geophysical dataset is typically interpreted as uncertainty, or *scatter*, in the data. Fortunately, the uncertainty can be greatly reduced if the unrecognized parameter is discovered and taken into account. To illustrate this point in an elementary way, consider an artificial example of a scattered dataset shown in Figure 2.2, left. These synthetic data were generated using the relationship $q = p_1^3 p_2$ for the range $0 < p_1 < 1$. To create the plot in Figure 2.2, left, the variable p_2 was considered to be an unrecognized parameter and it was treated as a random variate drawn uniformly from the unit interval [0,1]. There are many geophysical scenarios to which this type of analysis is relevant. For example, the variable q might be a measure of the electrical conductivity of a fluid-saturated rock sample, while p_1 could be its porosity and p_2 its clay content.

It is the presence of the variable p_2 that has generated the scatter observed in the dataset. A geophysicist unaware of p_2 might be misled into believing that the scatter is due instead to measurement error in q. The plot in Figure 2.2, right, shows the effect of recognizing that a second parameter plays a role in shaping the dataset. The dataset in this case is partitioned into three subsets depending on the magnitude of p_2, as shown. Now, it is easy to see that

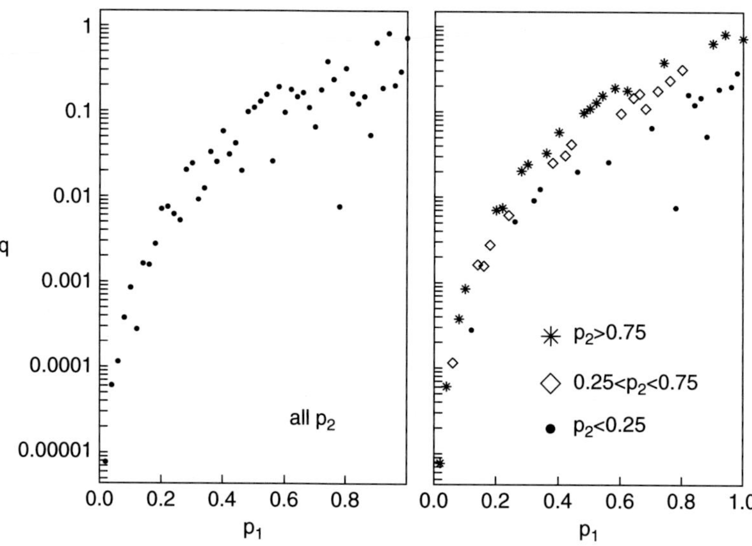

Figure 2.2 (Left) Scatter in data based on one explanatory variable. (Right) Partition of data into three discernible subsets based on the value of a second explanatory variable.

most of the original scatter in the data is not due to measurement error in q, but rather resulted from a failure to recognize the role of parameter p_2. In our hypothetical example in which q represents electrical conductivity, the rock samples are partitioned by their clay content, with the samples having the largest clay content (stars) generating a $q(p_1)$ relationship that is easily discernible from the ones generated by the samples with moderate (diamonds) and low (solid dots) clay content.

2.2 Sensors

A *sensor* is a device that converts a physical input signal, such as ground motion, into a voltage. For example, *geophones* (Figure 2.3, left), which will be discussed more fully in Chapter 6, use electromechanical coupling (transduction) to convert ground motion into a transient voltage output. An ideal sensor produces an output signal that is in direct proportion to the physical input signal (Gubbins, 2004). Owing to factors such as inertia and damping, however, a practical sensor has a finite frequency response, which means that only certain frequency components of the input signal are detected. Moreover, some frequency components are detected with more or less fidelity than others. A geophone, for example, senses components of the ground motion only within a certain band of frequencies. Since different types of seismic waves (see Figure 2.3, right) have different frequency contents and amplitudes, a seismometer with a large dynamic range (defined below) and a capability to sense signal components across a broad swathe of frequencies is often required to capture as many as possible of the essential details contained within the

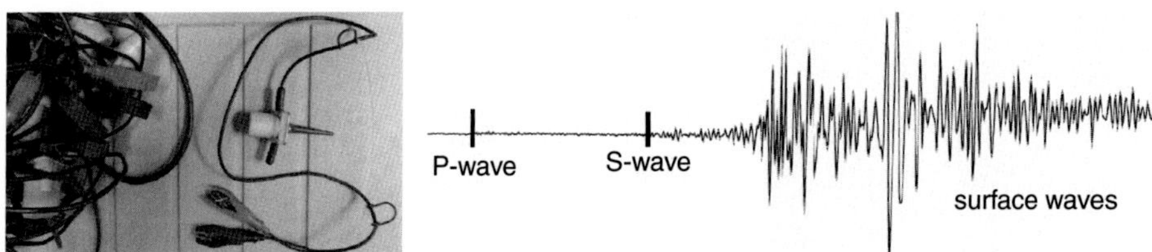

Figure 2.3 (Left) Geophones. (Right) Earthquake seismogram showing P-, S-, and surface waves (after Stein and Wysession, 2003).

entire seismic waveform. Physical signals, such as the ground motion due to an earthquake or an explosive source, generate a rich spectrum of frequencies.

A *digital recorder* typically samples and records the sensor output voltage $a(t)$ at some regular interval Δt in time. If the sampling interval is small, and the number of sampled points is large, a recorder can capture a broad range of the frequencies brought to it by the sensor. However, a fine sampling interval also implies a large data storage requirement. Digitized data are typically stored as a sequence of N real numbers

$$\{a\} = \{a_0, a_1, \ldots, a_{N-1}\} = \{a(n\,\Delta\,t)\} \quad \text{for} \quad n = 0, 1, 2, \ldots, N - 1. \tag{2.1}$$

A recording system is necessarily *band-limited* in the sense that it cannot track fluctuations in the sensor voltage that are slower than the entire measurement duration $(N - 1)\Delta t$ nor those faster than the sampling interval Δt. Keeping in mind that storage requirements increase as Δt decreases, it is worthwhile to capture as much information as practical.

The performance of a recorder is often expressed in terms of its *dynamic range*, g. This is the ratio of the maximum possible to minimum possible recorded signal. The dynamic range g of a digital recorder, expressed in decibels [dB],

$$g = 20\log\frac{a_{MAX}}{a_{MIN}}, \tag{2.2}$$

is determined by the number of bits used to store each data point. A 16-bit digital recorder can store integers ranging from $a_{MIN} = 1$ to $a_{MAX} = 2^{16} - 1 = 65535$ so it has a dynamic range of $g = 20\ \log_{10}(65535) = 96$ dB. Many modern seismometers are equipped with 24-bit recording systems, such that $g = 144$ dB.

2.3 Frequency response

The analysis of *frequency response* is important in evaluating the performance of a sensor. The concept is readily understood in terms of the reproduction of recorded music by a loudspeaker. An ideal loudspeaker should be able to faithfully reproduce audio signals at each frequency across the human audible range 20 Hz–20 kHz. The ideal frequency response therefore should be *flat* across much of the audible range and smoothly *roll off* toward zero at either end of the spectrum (see Figure 2.4). If the frequency response is irregular across the spectrum, as is the case for all realistic loudspeakers, some frequencies

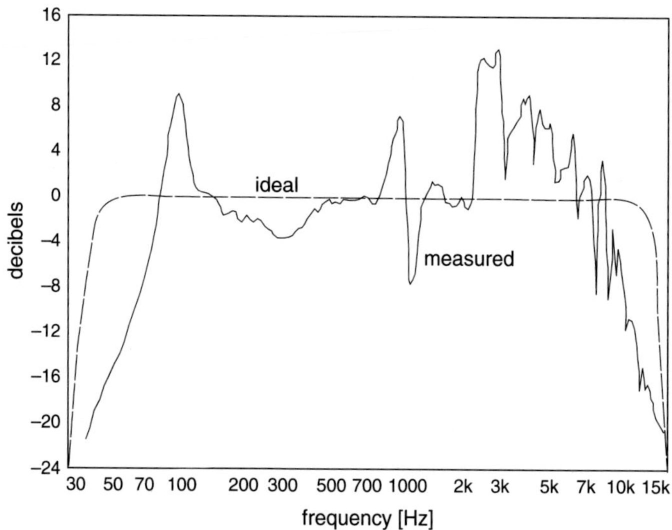

Figure 2.4 Measured and ideal frequency response of a loudspeaker. Reprinted from Corrington and Kidd (1951) with permission from IEEE. © IEEE 1951.

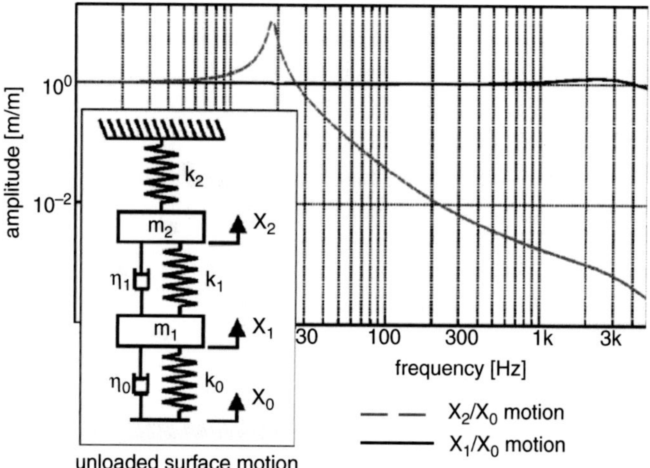

Figure 2.5 Modeled frequency response of a dual-mass accelerometer for landmine detection (after Martin *et al.*, 2006).

in the music will be reproduced at a louder or softer volume than was recorded and the music may sound distorted or unnatural (Corrington and Kidd, 1951). Of course, in audio engineering, many heuristic factors affect loudspeaker design including the architecture of the listening room, the locations of the speaker, the listener, and other objects within the room, and inherent listener preferences (Toole, 1986).

The frequency response of a geophysical sensor does not have to be flat; it can be shaped to suit a particular application. For instance, the modeled frequency response of a ground-contacting accelerometer developed for seismic detection of landmines (Martin et al., 2006) is shown in Figure 2.5. The lower mass m_1 is designed to have a flat amplitude spectrum which indicates it will move coherently with the ground motion. The upper mass m_2, on the other hand, has a *resonance peak* at ~ 20 Hz which indicates that the accelerometer will respond strongly and preferentially to ground vibrations at that frequency. Most landmines are hollow and cylindrical with a construction that produces a mechanical resonance close to ~ 20 Hz. The designed accelerometer will therefore be highly sensitive to the mechanical excitation of a buried landmine by the seismic source.

2.4 Discrete Fourier transform

Suppose we have recorded digital voltage readings from a geophysical sensor and have stored them as the sequence of N numbers $\{a\}$ as in Equation (2.1). Assume that the readings were acquired with a uniform time sampling interval, say Δt, as would be the case for a seismic trace, or at evenly spaced stations Δx, as in an idealized gravity or magnetics survey. Regardless of whether time or space is the independent variable, the sequence of numbers in Equation (2.1) shall be called a *time sequence* or a *time series*. It is well known that the techniques of continuous Fourier transforms can be applied to analyze the frequency components of a continuous function $f(t)$. An excellent reference to Fourier transforms and their properties is found in Bracewell (2000). As shown in Figure 2.6, a time series $\{a\}$ can be regarded as the discretely sampled version of a continuous function. A useful tool for analyzing the frequency content of a time series is the discrete Fourier transform (DFT).

A continuous Fourier representation of the data sequence $\{a\}$ in Equation (2.1) is *defined* by the complex function $A(\omega)$

$$A(\omega) = \frac{1}{N}\sum_{k=0}^{N-1} a_k \exp(-i\omega k\Delta t). \tag{2.3}$$

where $i = \sqrt{-1}$. The discrete Fourier transform (DFT) is developed from the Fourier representation in Equation (2.3) by simply discretizing the continuous variable ω. Let us select N values of the frequency ω_n (with $n = 0, 1, 2, \ldots, N-1$) such that, for each n, the total data acquisition time $T = N\Delta t$ is filled with an exact integer numbers of wavelengths, i.e.

$$\omega_n = \frac{2\pi n}{T} = \frac{2\pi n}{N\Delta t}. \tag{2.4}$$

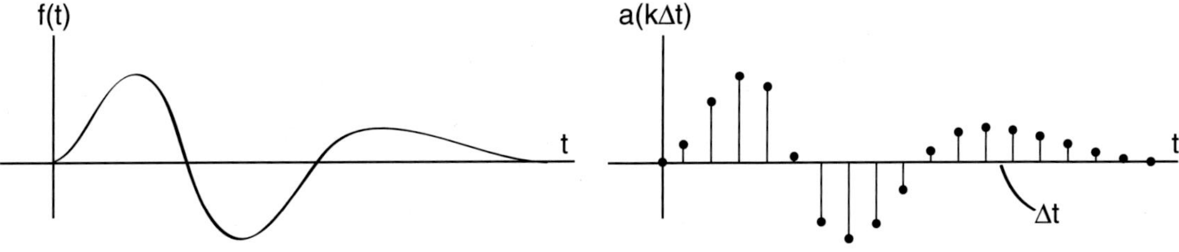

Figure 2.6 (Left) A continuous function $f(t)$ and (right) its discretely sampled representation $\{a(k\Delta t)\}$.

Substitution of the discrete frequency ω_n into the complex Fourier representation (2.3) gives the n-th coefficient A_n of the discrete Fourier transform

$$A_n = A(\omega_n) = \frac{1}{N}\sum_{k=0}^{N-1} a_k \exp\left(-\frac{2\pi ink}{N}\right). \tag{2.5}$$

The DFT therefore transforms the original data sequence $\{a\}$ of real numbers into a new sequence $\{A\}$ of complex numbers, namely the Fourier coefficients $\{A_0, A_1, \ldots, A_{N-1}\}$.

The utility of the DFT is that it enables the geophysicist to resolve an input data sequence into coefficients of a *harmonic series*, i.e. the set of frequencies which are integral multiples of the fundamental frequency $\omega_1 = 2\pi/T$ associated with acquisition time T. Each coefficient A_n describes the contribution of the particular frequency $\omega_n = 2\pi n/T$ to the original time series.

An inverse DFT may be used to recover the elements of a data sequence $\{a\}$ from its set of Fourier coefficients (Gubbins, 2004). The formula is

$$a_k = \sum_{n=0}^{N-1} A_n \exp\left(+\frac{2\pi ink}{N}\right). \tag{2.6}$$

Each complex Fourier coefficient A_n can be written in polar form as $A_n = R_n \exp(i\Phi_n)$ where R_n is the amplitude and Φ_n is the angle that describes the phase of the ω_n frequency contribution to the time series. A plot of $R_n(n)$ is called the amplitude spectrum; $R_n^2(n)$ is the power spectrum; and $\Phi_n(n)$ is the phase spectrum.

Following Gubbins (2004), it is instructive to examine the DFT of some elementary sequences that have wide applicability in time-series analysis. The first is the *spike*, or impulse signal which is a sequence $\{a\}$ of N numbers such that $a_k = \delta_{kj}$ where

$$\delta_{kj} = \begin{cases} 1 & \text{if } k = j \\ 0 & \text{otherwise} \end{cases}. \tag{2.7}$$

The impulse occurs at the time $j\Delta t$. The Fourier coefficients corresponding to the impulse signal are, using Equation (2.5),

$$A_n = \frac{1}{N}\sum_{k=0}^{N-1} \delta_{kj} \exp\left(-\frac{2\pi ink}{N}\right) = \frac{1}{N}\exp\left(-\frac{2\pi inj}{N}\right). \tag{2.8}$$

An important special case occurs if the spike is located at origin of the time series. In that case $j = 0$, such that the amplitude spectrum is constant and equal to $A_n = 1/N$. All frequencies play an equal role in shaping this transient impulse signal. The phase spectrum is zero since the amplitude spectrum is real. The phase spectrum changes, however, if the spike occurs at some other instant $j \neq 0$ of the time series, instead of the origin; see Figure 2.7. The amplitude spectrum remains unchanged regardless of the location of the spike within the time series.

The *boxcar* function (Figure 2.8, top) is important in time-series analysis for removal of bad data or otherwise isolating segments of a long data sequence. Multiplication of a time

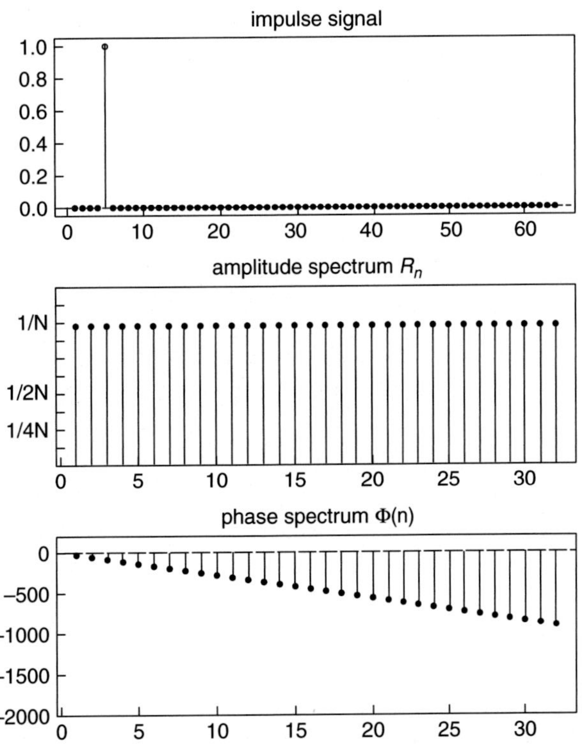

Figure 2.7 (Top) Impulse signal; (middle) amplitude spectrum R_n and; (bottom) phase spectrum $\Phi(n)$ of the discrete Fourier series.

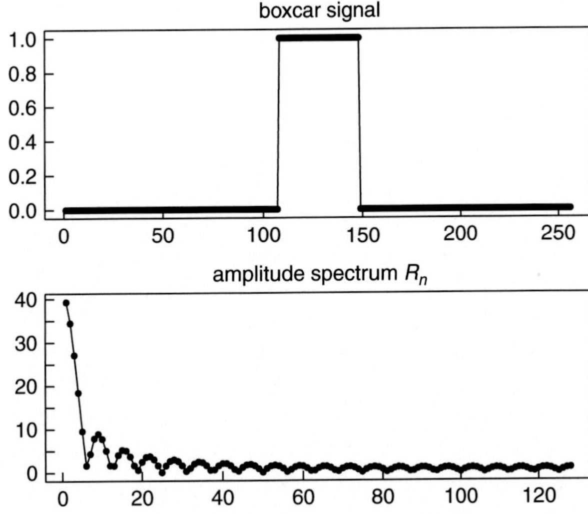

Figure 2.8 (Top) Boxcar signal; (bottom) amplitude spectrum R_n of the discrete Fourier series.

sequence by a boxcar is equivalent to keeping only the segment of the time series within the span of the boxcar and zeroing out the remaining data. The DFT of a boxcar sequence of the form

$$\{a\} = \{0, 0, \ldots, 1_{k_0}, \ldots, 1_{k_1}, 0, \ldots, 0\} \tag{2.9}$$

is

$$A_n = \frac{1}{N}\sum_{k=k_0}^{k_1} \exp\left(-\frac{2\pi i n k}{N}\right). \tag{2.10}$$

The notation 1_k in Equation (2.9) indicates that the k-th element of the time series has unit value. Equation (2.10) can be simplified to a more useful form

$$A_n = \frac{1}{N}\exp\left[\frac{-\pi i n[k_0 + k_1]}{N}\right]\frac{\sin(\pi n[k_1 - k_0 + 1]/N)}{\sin(\pi n/N)}. \tag{2.11}$$

It is easy to see from Equation (2.11) that the amplitude spectrum of the boxcar sequence is

$$R_n = \frac{1}{N}\left|\frac{\sin(\pi n[k_1 - k_0 + 1]/N)}{\sin(\pi n/N)}\right|. \tag{2.12}$$

Notice that the amplitude spectrum R_n is peaked at low frequency but has multiple side lobes extending to higher frequencies (Figure 2.8, bottom). This behavior is due to the zeroes which occur at $n(k_1 - k_0 + 1)/N$ in the numerator of Equation (2.12). The larger the width of the boxcar, i.e. the larger the difference $k_1 - k_0$, the narrower the low-frequency peak in the amplitude spectrum. The limiting case of an infinitely wide boxcar is a uniform sequence whose amplitude spectrum is the impulse signal. Conversely, the smaller the width of the boxcar, the wider the peak in the amplitude spectrum. The limiting case of an infinitely narrow boxcar is an impulse signal, or spike, that generates a uniform amplitude spectrum.

Recall that the formula for the inverse DFT is given by Equation (2.6). In the derivation of this equation (Gubbins, 2004), it is implicitly assumed that the data sequence $\{a\}$ is padded with zeroes outside the interval $n = 0, 1, 2, \ldots, N - 1$. However, if we were to replace $k \rightarrow k + N$ in Equation (2.6), which would be the case if the data sequence $\{a\}$ repeated itself outside the interval $n = 0, 1, 2, \ldots, N - 1$, we find

$$a_{k+N} = \sum_{n=0}^{N-1} A_n \exp\left(+\frac{2\pi i n[k + N]}{N}\right) = a_k \tag{2.13}$$

since $\exp[+2\pi i n] = \cos(2\pi n) + i\sin(2\pi n) = 1$. The result $a_{k+N} = a_k$ implies that the DFT treats the underlying data sequence $\{a\}$ as if it were an N-periodic function. Similarly, we find also that $a_{k-N} = a_k$ and $a_{k+2N} = a_k$, and so forth. This has profound implications for data processing: a DFT is strictly valid only if the data sequence repeats itself periodically with a period of $T = N\Delta t$. If the data sequence is not N-periodic, but truncates to zero at the ends of the interval, then the DFT is strictly incorrect. The geophysicist must recognize that amplitude and phase spectra are *almost always* distorted since the data sequence from a realistic phenomenon is not likely to be N-periodic.

Recall further that the coefficients of the discrete Fourier transform of the data sequence $\{a\}$ of length N are given by Equation (2.5), which shows that the computation of each Fourier coefficient A_n requires a summation of N terms. There are N Fourier coefficients to be computed, which means that the DFT requires N^2 complex operations, each of which involves multiplication and addition. The *fast Fourier transform* (FFT) algorithm (Brigham, 1988) accomplishes the DFT with far less computation, in proportion to $N\log_2 N$. The speedup for lengthy data sequences can be dramatic, for example, if $N = 10^6$ then $N^2 = 10^{12}$ while $N\log_2 N \sim 2.0 \times 10^7$, a speedup of nearly 5 orders of magnitude.

The DFT and its inverse are *information-preserving* in the sense that all information contained in the original data sequence $\{a\}$ is contained in the set of Fourier coefficients generated by the forward transform. Similarly, all information contained in the Fourier coefficients is preserved as the inverse transform is applied. Since no information is lost or gained when transforming between the discrete time and discrete frequency domain, the fact that there are N *real* data points but N *complex* Fourier coefficients suggests that the Fourier coefficients must contain redundant information.

From the N-periodic repetition property of the DFT, described earlier, it follows that $A_{N+n} = A_n$. We find also, by replacing $k \rightarrow N - k$ in Equation (2.6), that $A_{N-n} = A_n^*$ where A_n^* denotes complex conjugation of the Fourier coefficient. The constraint equations imposed by the condition $A_{N-n} = A_n^*$ indicate that the Fourier coefficients for discrete frequencies higher than $\omega_{N/2}$ are determined by the Fourier coefficients of the discrete frequencies less than $\omega_{N/2}$. Thus, the spectrum of Fourier coefficients is $N/2$-periodic in amplitude. There is a constant phase shift of π in the upper half of the spectrum, compared to the lower half, due to the complex conjugation. The Fourier coefficients are generally complex, with two exceptions. Notice that $A_{N/2}$ is real because of the constraint $A_{N/2} = A_{N/2}^*$.

Notice also that A_0 is real since $A_0 = (1/N)\sum_{k=0}^{N-1} a_k$. To summarize, there are two real Fourier cofficients (A_0 and $A_{N/2}$) and $N/2 - 1$ independent complex Fourier coefficients, giving a total of $2(N/2 - 1) + 2 = N$ independent real numbers in the transform, as required.

The highest frequency involved in the DFT is $\omega_{N/2}$, which corresponds to the *Nyquist* frequency f_N given by

$$f_N = \frac{\omega_{N/2}}{2\pi} = \frac{1}{2\Delta t}. \tag{2.14}$$

Signal energy at any frequency $\omega > \omega_{N/2}$ higher than the Nyquist frequency that is embedded in the original data sequence $\{a\}$ is shifted downward into the frequency range $[0, N/2]$ spanned by the Fourier coefficients and becomes mixed with the lower frequency $\omega < \omega_{N/2}$ signal energy. This phenomenon is known as *aliasing*. Notice from Equation (2.14) that the smaller the sampling interval Δt, the higher the Nyquist frequency f_N. In other words, the Fourier transform can identify higher frequency components, without aliasing, as the sampling interval decreases.

It is necessary to remove energy above the Nyquist frequency *prior* to digitizing through the use of an analog *anti-alias filter*. Filters are described in the next section. On the other hand, if there is no energy above the Nyquist frequency present in the physical input signal, then aliasing does not occur.

2.5 Filtering

Filtering is a common step in the processing of geophysical data. The essential goal of signal processing is to improve the signal-to-noise ratio. A *signal* is defined as a sequence of numbers that contains interesting (i.e. wanted and relevant) information about some property of the Earth or a geological process. *Noise*, the residual data left over after the signal has been isolated, contains either no information or else it contains unwanted or irrelevant information. In some seismological investigations, for example, the P- or S-wavefield might be of interest while the Rayleigh wavefield is classed as noise. In other types of seismic investigation, the reverse might be true. In all cases, the data processing steps should be performed with care since invariably some information is lost at each operation.

Typically, it is not possible to predict which part of the physical input acquired by the sensor contains the wanted signal and which part contains the unwanted remainder. Hopefully, filtering a dataset enhances a particular feature within the data that the geophysicist can recognize and *interpret* as a signal originating from some interesting Earth structure or geological process.

There are three main categories of filters. Roughly speaking, a *low-pass* filter reduces the amplitude of all frequency components in a signal that are above a specified cut-off value. A *high-pass* filter reduces the amplitude of all frequency components below a specified cut-off value. A *notch* filter reduces the amplitude of all frequency components inside a specified band. As an example, consider a moving average process. It is essentially a low-pass filter operation. To understand why, recall that an M-point moving average replaces each element of a time series by the average of the element itself and its $(M - 1)/2$ neighboring values on either side. Such an averaging procedure tends to smooth out any rapid, or high-frequency, oscillations in the time series.

Let us think more carefully about the moving-average filter operation in the discrete frequency domain. As explained in Gubbins (2004), a moving-average process is equivalent to a multiplication of the boxcar spectrum with the DFT of the data sequence. Unfortunately, the presence of the side lobes in the boxcar spectrum (see Figure 2.8, bottom) allows high-frequency energy in the data sequence to pass. An excellent low-pass filter would have a sharp cut-off at high frequencies without the side lobes. Thus, the moving average is actually a poor low-pass filter.

The ideal low-pass filter would be a step-off function in the frequency domain, with unit amplitude in the pass band and zero amplitude elsewhere. In the time domain, however, such a filter generates spurious oscillations, known as the *Gibbs phenomenon*. A compromise between the length of the sequence in the time domain and the sharpness of the frequency cut-off is required. The Butterworth filter, defined by the energy spectrum,

$$|F_L(\omega)|^2 = \frac{1}{1 + \left(\dfrac{\omega}{\omega_c}\right)^{2n}} \tag{2.15}$$

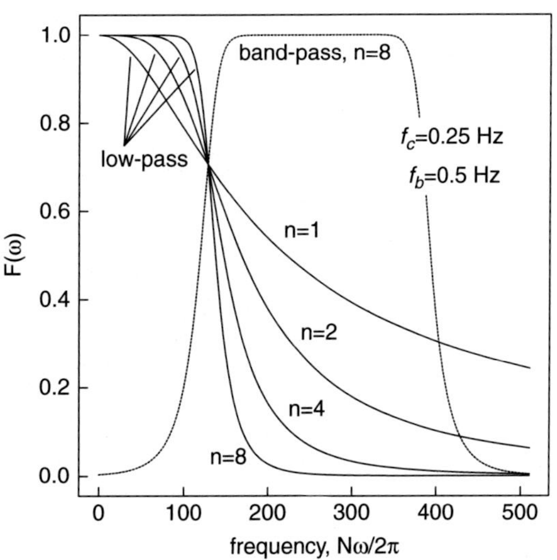

Figure 2.9 Amplitude spectra of Butterworth n-pole low-pass filters and a bandpass filter, where $f_c = \omega_c/2\pi$ and $f_b = \omega_b/2\pi$.

achieves a tapered cut-off with cut-off frequency ω_c. The parameter n controls the sharpness of the cut-off. The roll-off is sharper as n increases.

A high-pass filter has the complementary amplitude spectrum with respect to the low-pass filter, $F_H(\omega) = 1 - F_L(\omega)$. A bandpass filter is obtained by shifting the peak in the energy spectrum to the center frequency ω_b,

$$|F_B(\omega)|^2 = \frac{1}{1 + \left(\dfrac{\omega - \omega_b}{\omega_c}\right)^{2n}}. \tag{2.16}$$

A notch filter has the complementary amplitude spectrum with respect to the bandpass filter, $F_N(\omega) = 1 - F_B(\omega)$. The amplitude spectra of some Butterworth filters are shown in Figure 2.9.

An example of the application of a low-pass filter is shown in Figure 2.10. The original signal $f(t)$ shown as the solid curve is given by

$$f(t) = -20\sin(t/2) + 10\sin t + 3\cos(t - 5) - 2\cos(3t + 1) + \sin(4t - 32) + 5\sin 5t, \tag{2.17}$$

for the sequence of t-values

$$\{t_k\} = \{20\pi k/N\}_{k=0}^{N-1}, \tag{2.18}$$

where $N = 512$ is the number of sample points. The solid dots in Figure 2.10 correspond to a single-pole ($n = 1$) Butterworth low-pass-filtered signal with cut-off frequency $\omega_c = \pi/30$. Lowering the cut-off frequency results in the attenuation of higher frequencies. For example, the long dashed line shows the filtered signal with $\omega_c = \pi/100$, while the light dashed line shows the filtered signal with $\omega_c = \pi/300$. As expected, higher-frequency signal energy is removed as the cut-off frequency is reduced.

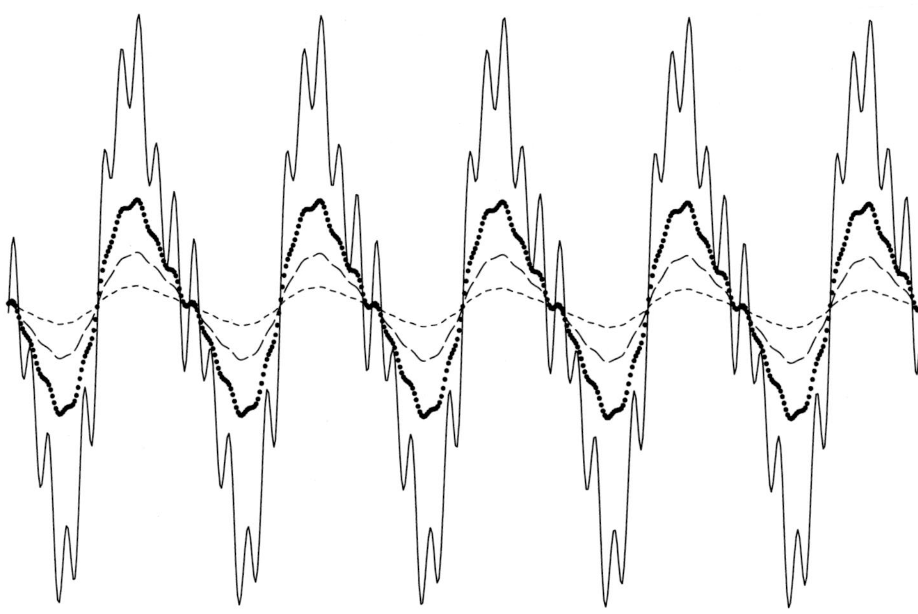

Figure 2.10 Effect of low-pass filtering.

2.6 Convolution

The passage of seismic energy through the Earth can be regarded as a low-pass filtering of the source excitation. In general, the higher-frequency components of the seismic wave are attenuated at short distances from the source whereas the lower-frequency components are able to propagate further. The low-pass filtering effect distorts and spreads out the transmitted source pulse.

In this section we show that the process of filtering is equivalent to the time-domain convolution of a data sequence with a second, usually shorter, sequence. We will recognize that the output from a geophysical instrument is simply the convolution of the physical input signal with the impulse response of the instrument. In this sense, the measurement process itself may be viewed as a filtering operation.

In an idealized case, an explosive source can be modeled as a unit impulse $\delta(t - t_0)$ at time t_0, defined as

$$\delta(t - t_0) = \begin{cases} 1 & \text{if } t = t_0 \\ 0 & \text{otherwise} \end{cases}. \tag{2.19}$$

The unit impulse function $\delta(t - t_0)$ is illustrated in Figure 2.11a. The effect of the propagation of the impulse through the Earth to a receiver is characterized by an *impulse response* $h(t)$. The digitized impulse response can be modeled as the superposition of a set of K_H scaled and delayed impulse functions, given by

$$h(t) = \sum_{k=1}^{K_H} h_k \delta(t - t_k) \tag{2.20}$$

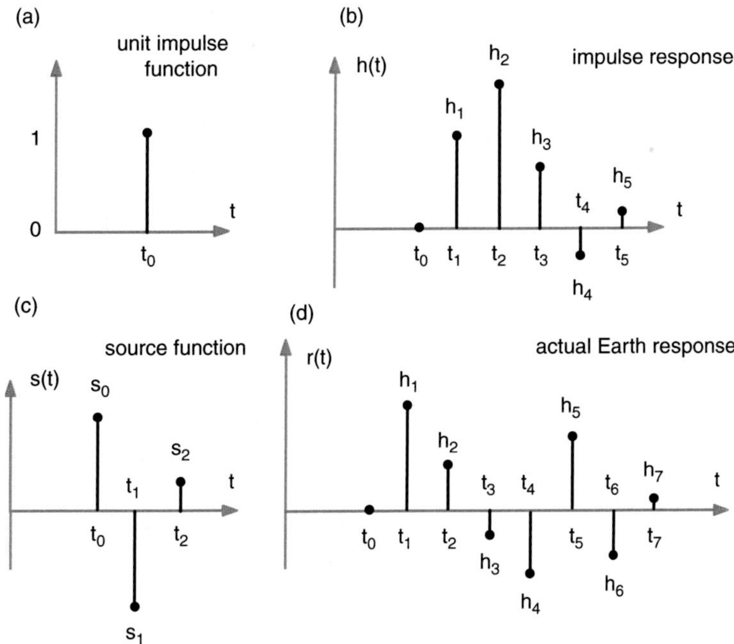

Figure 2.11 (a) Unit impulse function $\delta(t - t_0)$; (b) impulse response $h(t)$; (c) source function $s(t)$; (*d*) actual Earth response $r(t)$.

where h_k are the scale factors and t_k are the time-shift factors. An example of an impulse response is shown in Figure 2.11b for which $K_H = 5$, the sequence $\{h\}=\{h_1, h_2, h_3, h_4, h_5\}=\{0.20, 0.28, 0.17, -0.05, 0.03\}$, and the time shifts are $t_k = k\Delta t$.

A realistic source function $s(t)$ can also be digitized as a series of scaled and time-shifted unit impulse functions,

$$s(t) = \sum_{k=1}^{K_S} s_k \delta(t - t_k). \tag{2.21}$$

Notice that the source function $s(t)$ starts at t_0, while the impulse response $h(t)$ starts at $t = t_1 > t_0$; this is a consequence of causality since effects cannot precede their causes and a finite time is required for signals to propagate from the source to the receiver. An example of a source function for which $K_S = 3$, and the sequence $\{s\}=\{s_0, s_1, s_2\} = \{1.0, -1.0, 0.4\}$, is shown in Figure 2.11c.

The actual response of the Earth $r(t)$, measured at the receiver, to the realistic source pulse can then be viewed as a superposition of scaled and time-shifted impulse responses. The main assumption is that the system is *linear* and *time-invariant*, which means that the impulse response scales proportionately with the amplitude of the impulse function and that the shape of the impulse response does not depend on the time t_0 at which the impulse is launched. The response $r(t)$ is thus the convolution

$$r(t) = h(t) * s(t) = \sum_{k=0}^{K_S} s_k h_{t-k}, \tag{2.22}$$

which is shown in Figure 2.11d.

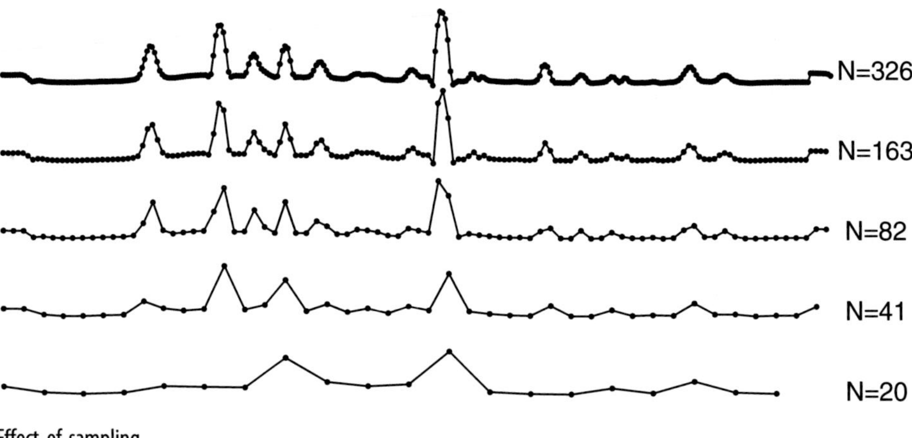

Figure 2.12 Effect of sampling.

2.7 Sampling and aliasing

Physical signals are continuous in time, while signals recorded by geophysicists are *sampled* at fixed intervals of time Δt. Important information is often lost by increasing the sampling interval, as illustrated in Figure 2.12. The upper curve shows an actual spatial series of electromagnetic geophysical measurements ($N = 326$) which we can suppose to consist of target signatures superimposed on background noise. Roughly, if we assume that each isolated peak in the data series is caused by a "target" residing in the Earth, a crude interpretation of the data in the top curve would seem to suggest that at least 8–10 targets are present, some of which are larger than others. The remaining curves in the figure show the progressive deterioration of information for each doubling of the sampling interval. The coarsely sampled data series at the bottom, for example, has lost much of the original target information. In fact, a cursory inspection would suggest that only two, or possibly three, targets are present in the subsurface.

There are some situations, however, in which information is not lost if the sampling interval is increased. In Appendix A we prove the *Shannon sampling theorem* which states that no information is lost by regular sampling provided that the sampling frequency is greater than twice the highest frequency component in the waveform being sampled.

2.8 Data windows and spectral analysis

A basic problem in DFT analysis is the spectral leakage that occurs during analysis of signals of finite-time duration. Realistic signals generally do not vanish at the beginning and the end of the measurement period. Recall that the DFT assumes time sequences to be N-periodic. The periodic extension of the signal introduces "hidden" discontinuities at the

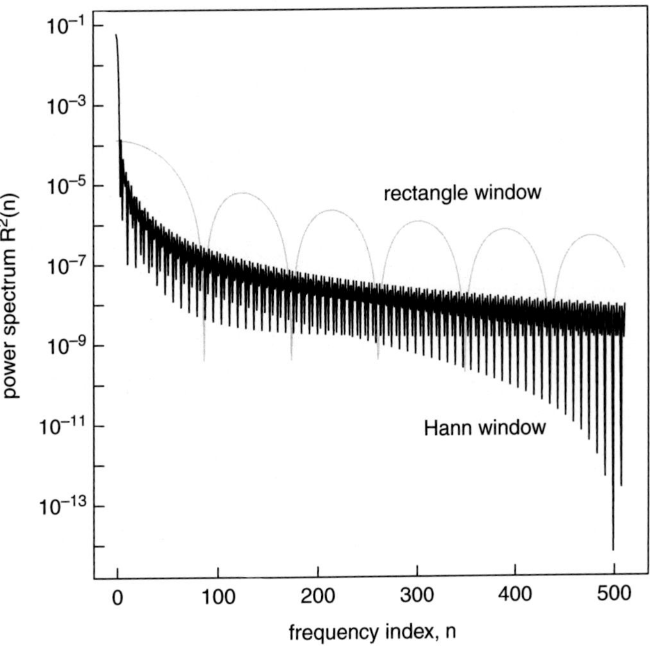

Figure 2.13 Power spectra of rectangle and Hann windows.

beginning and the end of the time series. These hidden discontinuities contribute spurious broadband energy to the power spectrum of the signal. The situation is carefully explained by Harris (1978).

An improved estimate of the power spectrum, with less spectral leakage, can be obtained by applying a window function to the time series. The purpose of the window function is to smoothly reduce, or taper, the time series to zero at both ends, thereby removing the hidden discontinuities. A number of window functions have been developed over the past decades to perform data tapering. A popular one is the raised cosine, or Hann window, given by the formula

$$w_n = \frac{1}{2}\left[1 - \cos\left(\frac{2n\pi}{N}\right)\right], \quad \text{for } n = 0, 1, \ldots, N-1. \tag{2.23}$$

The power spectrum of the Hann window, compared to the power spectrum of a rectangle window (essentially a boxcar sequence), is shown in Figure 2.13. Notice that the side lobes of the Hann-window spectrum are much smaller than those of the rectangle-window spectrum. It is the side lobes that are responsible for spectral leakage.

The effect of applying a Hann window to a geophysical time series is shown in Figure 2.14, left. The time series shown in gray is of length $N = 2048$ and corresponds to hourly fluctuations in the north component of the geomagnetic field measured at an observatory in northern Europe. The time series shown in black at the left is the tapered sequence $\{wa\}$, obtained by multiplication of the original time series $\{a\}$ by the Hann window sequence $\{w\}$, where $\{w\}$ is given by Equation (2.23).

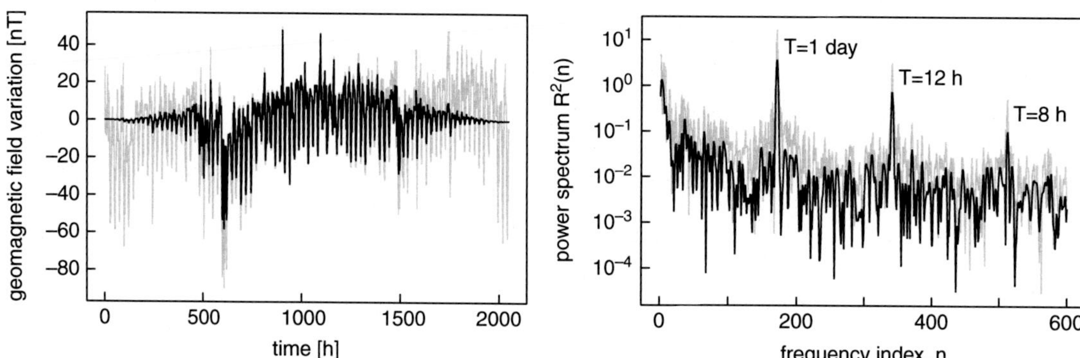

Figure 2.14 Effect of Hann windowing on geomagnetic time series: (left) original and Hann-windowed time series; (right) corresponding power spectra.

The power spectra for the original time series $\{a\}$ and the Hann-windowed sequence $\{wa\}$ are shown in Figure 2.14, right, as the gray and black lines respectively. Notice in both cases that power is concentrated, as expected, in the daily variation ($T = 1$ day) and its overtones. The background continuum, however, is greatly reduced in the windowed power spectrum compared to that of the original time sequence, showing that spectral leakage has indeed been reduced by the windowing procedure.

2.9 De-spiking time series

In most cases, geophysical times series are well-behaved in the sense that large outliers, or spikes, occur relatively infrequently. Sometimes, however, an occasional outlier does appear in a dataset. Suppose it can be established with some confidence that the outlier is not part of the wanted signal. Then it is desired to identify and remove the extraneous spike prior to further processing. Goring and Nikora (2002) have discussed several methods for detecting spikes in time series. In this section we describe one of these methods, the *RC filter technique*, and apply it to the spiky geophysical time series (open circles) shown in Figure 2.15. For our purpose here, it is not necessary to speculate as to the physical origin of the spikes. The dataset was generated from a magnetic survey (see Chapter 3) over a buried steel pipeline on the Texas A&M campus.

Suppose the spiky time series is of length N and denoted by the sequence $\{a\}$ as in Equation (2.1). The idea behind the RC filter method is to check each element a_k (for $k = 0, \ldots, N-1$) of the time series and determine whether it is an outlier based on the amount that it varies from a smoothed version of $\{a\}$. A spike would, *by definition*, have a large variance according to some matching criterion. Specifically, denote the sequence $\{L(a)\}$ as a low-pass-filtered version of the time series $\{a\}$. Then, the *variance* $\{\sigma\}$ is defined as the sequence

$$\{\sigma\} = \{L(a^2)\} - \{L^2(a)\}, \tag{2.24}$$

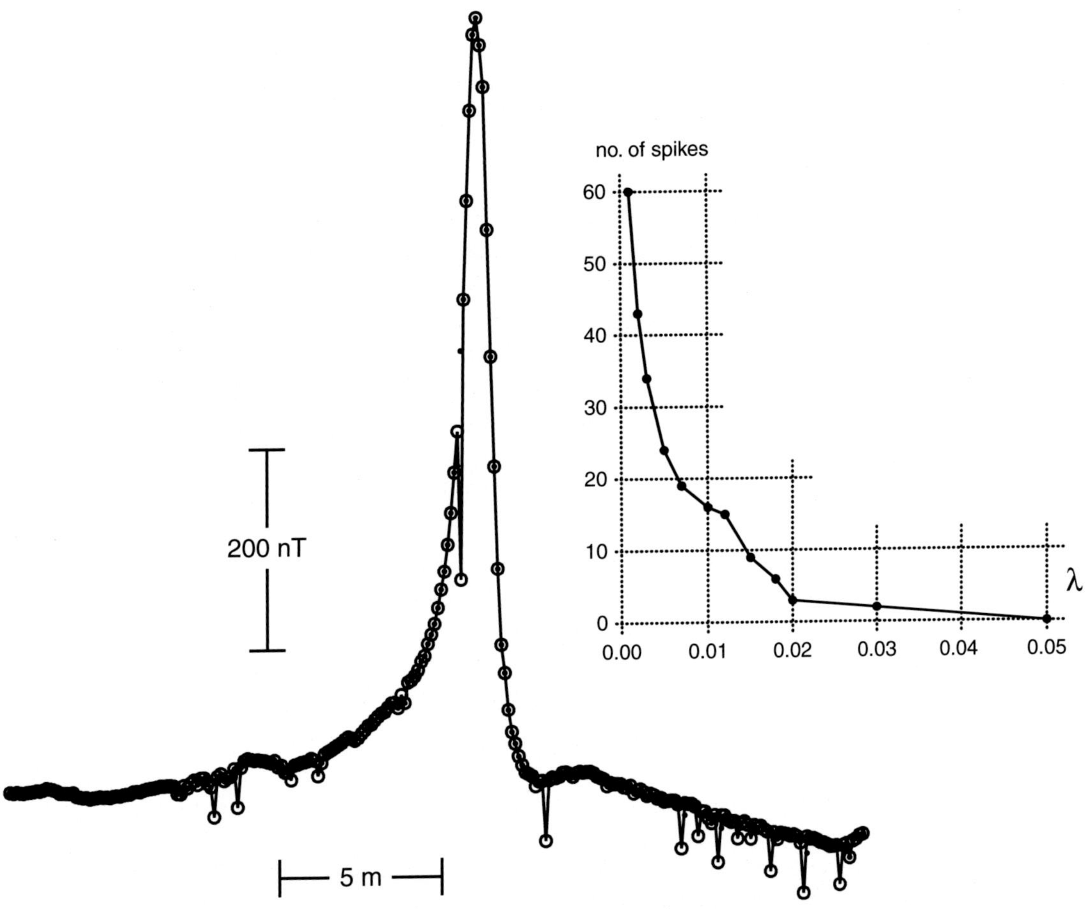

Figure 2.15 De-spiking a magnetics data set ($N = 256$) using $\lambda = 0.005$. The inset shows the number of spikes as a function of parameter λ, see text for details.

where $\{L(a^2)\}$ is the low-pass-filtered version of the sequence $\{a^2\}$ consisting of squared elements of $\{a\}$, while $\{L^2(a)\}$ consists of the squared elements of $\{L(a)\}$. The *sample variance* of the k-th element of the time series is σ_k. The element a_{k+1} is declared *not* a spike if it satisfies the following criterion:

$$L_k(a) - \lambda\sigma_k < a_{k+1} < L_k(a) + \lambda\sigma_k, \tag{2.25}$$

where λ is a pre-defined threshold. In other words, the element a_{k+1} is not a spike if it falls within $\pm\lambda\sigma_k$ of the smoothed sequence $\{L(a)\}$.

In the magnetics example shown in Figure 2.15, a total of 24 spikes were identified using the RC filter method with a threshold value of $\lambda = 0.005$. The small solid dots in the figure show the de-spiked time series in which each spike is replaced by the mean value of its two immediate neighbors. The inset shows the number of spikes identified by the RC-filter technique for different choices of threshold λ. The smaller the threshold, the smaller the interval $\pm\lambda\sigma_k$, and hence the greater the likelihood that the sample point a_{k+1} would fall outside that interval. Hence, the number of detected spikes decreases with increasing λ, as shown. A one-pole Butterworth low-pass filter with 8 Hz cut-off frequency was used in this example.

2.10 Continuous wavelet transform (CWT)

The tools of Fourier analysis perform well on *stationary* geophysical time series. A stationary time series is one whose statistical properties do not depend on the epoch during which the observations are made. In other words, an underlying process whose statistical properties are invariant with respect to time generates a stationary time series. No particular epoch can be distinguished, statistically speaking, from any other.

Many geophysical signals, however, whether they are in the form of a time series $f(t)$ or a spatial series $f(x)$ of measurements, exhibit variations that occur across different temporal or spatial scales. An elementary example is that of a single spike occurring at time $t = t_0$ embedded in random, time-invariant noise. A spike renders the time series non-stationary since observations made during the epoch of the spike have different statistics to those made either before or after the spike occurred. Generally, a non-stationary time series is one whose statistical moments, such as mean, variance, etc., depend on time. Non-stationary data series that contain *time-localized features* such as spikes are not strictly amenable to Fourier analysis.

Recall that the basis functions of Fourier analysis are complex exponentials of the form $\exp(\pm i\omega t)$. Such basis functions are *global* since they are non-zero for all time $-\infty < t < \infty$. A time-localized function such as a spike, however, is non-zero only for the brief time interval, or instant, at which the spike occurs. Thus, it is difficult to synthesize a spike using a sum of complex exponentials since it is required for the sum to be exactly zero outside the brief time interval containing the spike. This requirement implies that complete destructive interference of the superposition of global basis functions must be achieved almost everywhere.

The use of time-localized basis functions, or *wavelets*, can circumvent the difficulty faced by Fourier analysis in decomposing signals containing time-localized events. In fact, certain wavelets are scale-independent and can be used to analyze a signal regardless of the scale of its inherent variability. Wavelets are designed to study signals containing features at different scales. This section introduces wavelet analysis and the continuous wavelet transform (CWT) of a geophysical time series. A good guide to wavelets with selected applications in geophysics is provided by Kumar and Foufoula-Georgiou (1997).

The CWT of a time series $f(t)$ is defined (Kaiser, 1994) by

$$\tilde{f}(s,t) = \int_{-\infty}^{\infty} du \psi_{s,t}(u) f(u), \tag{2.26}$$

where, for a fixed value of s, the transformed time series $\tilde{f}(s,t)$ reveals the details contained in the original signal $f(t)$ at scale s. The function $\psi_{s,t}(u)$ in Equation (2.26) is a *wavelet*

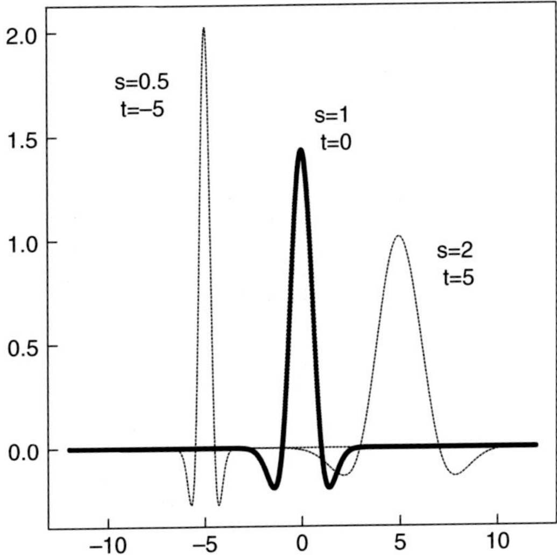

Figure 2.16 Scaled and translated Mexican hat wavelet functions, using $p = 0.5$.

$$\psi_{s,t}(u) = |s|^{-p}\psi\left(\frac{u-t}{s}\right), \tag{2.27}$$

where $\psi(u)$ is the *mother wavelet.* The mother wavelet plays a role in wavelet analysis that is similar to that of the complex exponential in Fourier analysis. The mother wavelet should have zero mean and unit energy, that is

$$\int_{-\infty}^{\infty} \psi(u)du = 0, \quad \text{and} \quad \int_{-\infty}^{\infty} |\psi(u)|^2 du = 1. \tag{2.28}$$

Many different mother wavelets have been used in signal analysis; one particular choice is the Mexican hat wavelet

$$\psi(u) = \frac{2}{\sqrt{3}}\pi^{-1/4}(1-u^2)\exp\left(-\frac{u^2}{2}\right). \tag{2.29}$$

The factors p and s in Equation (2.27) control the shape of the wavelet while the parameter t shifts it along the horizontal axis. As shown in Figure 2.16, increasing the value of s stretches the Mexican hat wavelet and reduces its amplitude. If s is negative (not shown in the figure), the wavelet has the same shape as the corresponding wavelet for positive s except that it is reflected about the horizontal axis. The factor p is always positive and its effect is to reduce (increase) the amplitude of the wavelet as it is stretched (compressed) along the horizontal axis.

A wavelet analysis is carried out on the geophysical time series that was analyzed earlier (Figure 2.12, top). The time series of length 326 is displayed again in Figure 2.17, top. Treating the original time series as the function $f(u)$ in Equation (2.26), the integration to find $\tilde{f}(s,t)$ is carried out numerically for several values of s. The results are shown in

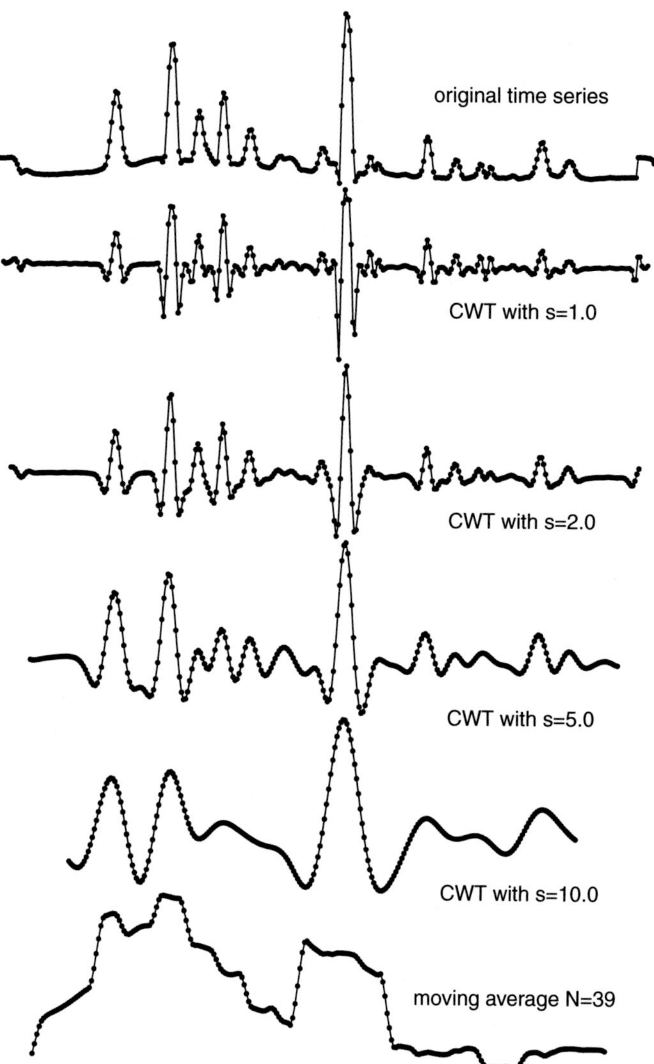

Figure 2.17 Continuous wavelet transforms and 39-point moving-average filter applied to the original geophysical time series shown at top.

Figure 2.17, middle panels. Notice that more and more of the fine details of the original time series are lost with increasing values of s. Each curve should be viewed as a function of t for a fixed value of s. Shown for comparison at Figure 2.17, bottom is a 39-point moving-average low-pass filter of the original time series. The formula for an N-point moving-average filter is

$$\tilde{f}(t) = \frac{1}{N} \sum_{u=-(N-1)/2}^{(N-1)/2} f(t+u). \tag{2.30}$$

The wavelet transform clearly provides a much smoother and more easily interpretable representation of coarse-level details in a time series than does the moving-average filter.

Problems

1. What is the dynamic range, in dB, of a mechanical recording system based on a 0.7 mm pen and 8.5 × 11 in graph paper? How large would the paper have to be in order for the dynamic range of this recording system to equal that of a 24-bit seismometer?
2. Explain why the impulse-signal amplitude response R_n is flat while the phase response Φ_n is linear with a negative slope, as shown in Figure 2.7.
3. Explain how a series of hourly spikes, due to the passing of a nearby electric train, affects the Fourier amplitude spectrum of an idealized geomagnetic signal consisting of diurnal ($T = 1$ day) and semi-diurnal ($T = 12$ h) components that is sampled once per minute over a 10-day duration. What would happen to the spectrum if the 6 pm train was 10 minutes late each evening?
4. Show that the Fourier transform of the comb function $\sum\delta(t - n\Delta t)$ in the time domain is a comb function in frequency $[2\pi/\Delta t]\sum\delta(\omega - n\Delta\omega)$, where $\Delta\omega = 2\pi/\Delta t$. The sums run from $n = -\infty$ to $n = +\infty$.
5. Show that a moving-average filter of a given time series is equivalent to the convolution of a boxcar signal with the time series.
6. **Computer assignment**. Write a computer program in the language of your choice that takes a discrete Fourier transform of a boxcar time series of length N of the form of Equation (2.9). Run the program for length $N = 1024$ and the following three cases: (a) $k_0 = 450$ and $k_1 = 574$; (b) $k_0 = 492$ and $k_1 = 532$; (c) $k_0 = 508$ and $k_1 = 516$. Compare your computed amplitude spectrum R_n against the analytic solution in Equation (2.12). Comment on the size and width of the sidelobes in the amplitude spectrum for the three cases. For each scenario, describe and explain the behavior of the phase spectrum. For part (c), shift the boxcar in time (try the cases $k_0 = 510$, $k_1 = 518$ and $k_0 = 512$, $k_1 = 520$) and comment on the resulting changes in the phase spectrum. Can you explain why the phase spectrum changes like this?

3 Magnetics

The purpose of the magnetic geophysical technique is to explore the spatial distribution of magnetized rocks and buried ferrous metal objects, based on magnetic measurements made at or near the surface, and then to make a geological or anthropogenic interpretation in terms of the objectives of the investigation. The magnetic method is perhaps the oldest of geophysical exploration techniques (Nabighian *et al.*, 2005).

The magnetic method has become widely used in near-surface geophysics for several reasons (Hansen *et al.*, 2005): (a) buried targets or geological structures of interest often have readily detectable magnetic signatures owing to the high sensitivity of modern magnetometers; (b) the measurements are fast, reliable, and non-invasive; (c) magnetic data are often straightforward to interpret using qualitative and quantitative techniques, especially when large amounts of high-resolution data are acquired over a wide area such that a continuous plan view of the site and its surroundings can be obtained.

3.1 Introduction

Traditional exploration-scale application of magnetics includes such tasks as spatial mapping of igneous extrusives and intrusives and determining the depth to crystalline basement in sedimentary basins. Igneous rocks, due to their high iron content and lack of metamorphic alteration, can generate important magnetic anomalies. Sedimentary rocks and unconsolidated sediments generally are weakly magnetized. For near-surface applications, the high sensitivity of present-day instrumentation enables buried faults, contacts, sedimentary structures, building foundations, pipelines, and even areas of anthropogenically disturbed earth, to be successfully imaged. The following case studies illustrate the use of the magnetic technique in near-surface geophysics.

Example. Aeromagnetic survey of geothermal field.

As part of a geothermal resource exploration effort by the United States Geological Survey (USGS) on behalf of the Department of Energy (DOE), Grauch (2002) conducted a high-resolution aeromagnetic survey of a sedimentary basin in Nevada in order to assess the possible role of faulting in the subsurface flow of hot fluids. Most of the faults within the basin are shallow buried and are difficult to detect with seismic techniques. The total-field aeromagnetic data (Figure 3.1a) reveal areas of widespread igneous rocks on the periphery

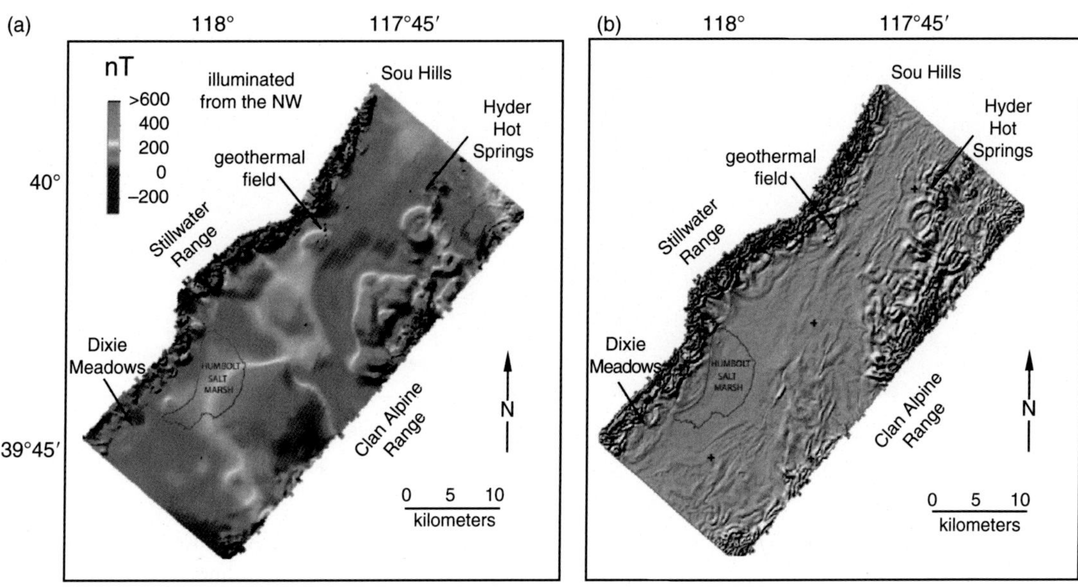

Figure 3.1 (a) Total-field aeromagnetic data after processing in shaded-relief format: Dixie Valley, NV geothermal field; (b) horizontal gradient filter applied to aeromagnetic data to enhance edges and lineaments. After Grauch (2002). See plate section for color version.

of the basin, in addition to subtle, lineated magnetic contrasts that originate from the juxtaposition of different sediment types.

To enhance the visualization of basin-fill sedimentary structures, a horizontal gradient operation is applied to the aeromagnetic data, with the result shown in Figure 3.1b. As discussed later in this chapter, the gradient filter is an important interpretive tool since it highlights localized regions of high magnetic gradient; in this case, shallow buried faults. The faults are indicated as the lineated anomalies in the central part of the basin and are consistent with previously known faulting patterns. Larger, deeper magnetic sources such as igneous basement rocks produce high total-field amplitudes but the corresponding anomalies are wide and gently varying. Thus, the horizontal gradient filter provides an effective procedure for separating shallow and deep magnetic sources while enhancing the visualization of linear features.

Example. Archaeogeophysics.

Magnetic geophysics has also seen increasing utilization in archaeological prospecting (Linford, 2006). A site at which archaeology based on the magnetics technique has been particularly successful is the Roman city of *Viroconium* at Wroxeter, UK (Gaffney *et al.*, 2000). An aerial photograph of the site is shown in Figure 3.2a. Fluxgate gradiometry surveys were conducted in order to investigate the open areas surrounding the central basilica and bounded by defensive ramparts. Processed magnetics data are shown in Figure 3.2b.

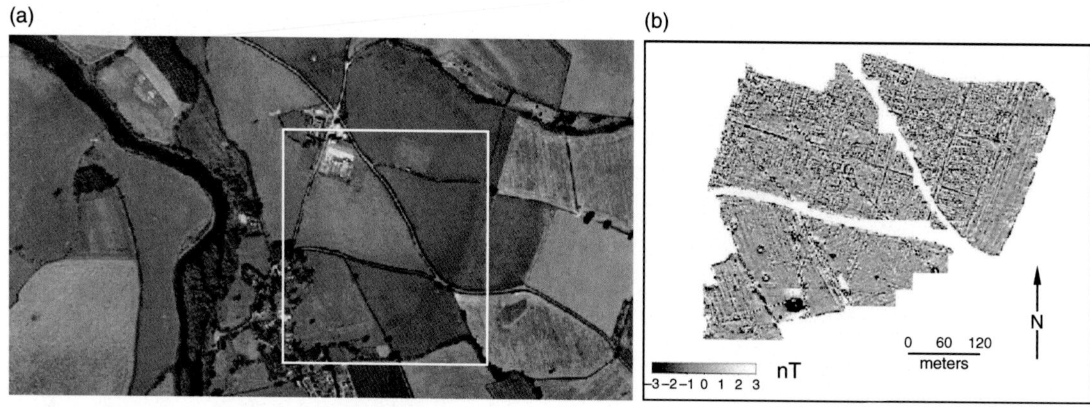

Figure 3.2 (a) Aerial photo of archaeological site at the Roman city of Wroxeter, Shropshire, UK. (b) Magnetic map obtained by combining data from four different magnetometer configurations. From Gaffney *et al.* (2000).

Line spacing is 1.0 m and station spacing is 25 cm. Several image-enhancing data-processing steps were performed but these are not described in detail by the authors.

The success of the magnetic imaging is due to the low magnetic mineral content of the building materials compared with the higher magnetic mineral content of the settlement soils. The interpretation of the magnetics data at Wroxeter indicate that *Viroconium* was not, as previously thought, a garden city containing large open spaces. Instead, the geophysical survey reveals important details on the grid layout and zonation of the ancient city.

3.2 Fundamentals

Bodies containing magnetic minerals, chiefly Fe–Ti oxides (Clark, 1997), may be viewed as vast assemblages of microscopic aligned *magnetic dipoles*, which are the elementary building blocks of magnetic sources. To properly understand magnetics data it is necessary to grasp the concept of an individual magnetic dipole. A magnetic dipole (Cullity and Graham, 2009) is an infinitesimal loop of area A [units of m^2] carrying an electric current I that is measured in amperes [A], see Figure 3.3.

The unit vector $\hat{\mathbf{m}}$ defines the orientation of the dipole and it is directed perpendicular to the plane of the loop, in alignment with the loop axis. The *dipole moment* $\mathbf{m}$ [Am^2] is defined as the vector

$$\mathbf{m} = IA\,\hat{\mathbf{m}}\,, \tag{3.1}$$

where $m = |\mathbf{m}| = IA$ is the strength of the dipole. An important property of the magnetic dipole is that the area A of the loop is presumed to be vanishingly small while the current I

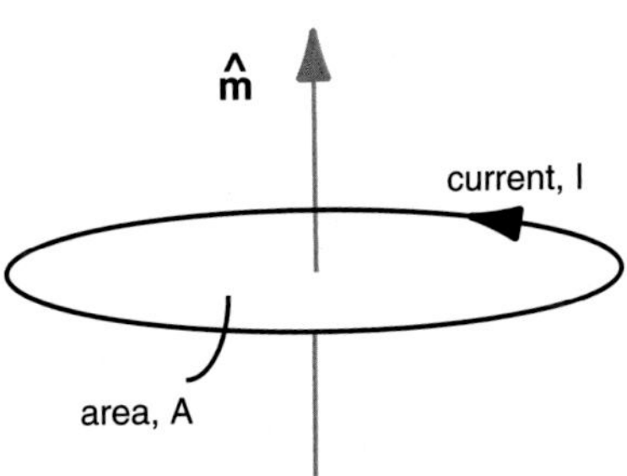

Figure 3.3 A magnetic dipole.

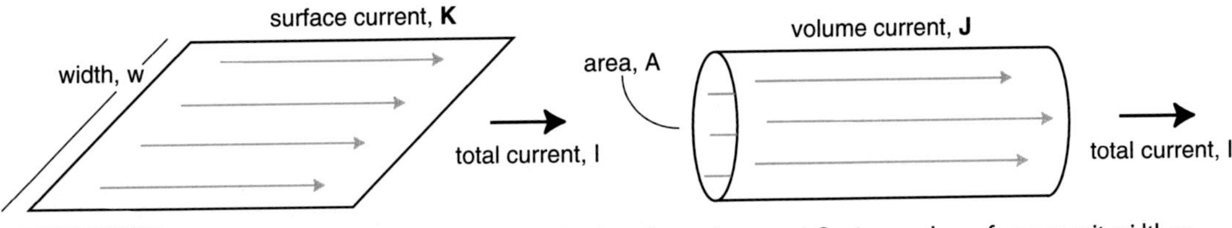

Figure 3.4 The surface current density **K** (left) is defined as the total current I flowing on the surface per unit width w perpendicular to the current flow; the volume current density **J** (right) is defined as the total current I flowing within the volume per unit area A perpendicular to the current flow.

is sufficiently large such that the product $m = IA$ remains finite. A magnetic dipole tends to rotate into alignment with an externally imposed magnetic field. This is a fundamental characteristic of magnetism.

The magnetic signature of a heterogeneous magnetized body is, to first order, identical to that of an equivalent surface electric current of density $\mathbf{K}(\mathbf{r})$ [A/m] flowing in closed loops around the body, plus an equivalent volume current of density $\mathbf{J}(\mathbf{r})$ [A/m^2] flowing in closed loops inside the body. The definitions of surface current density **K** and volume current density **J** are illustrated in Figure 3.4.

The *magnetization* **M** of a magnetized body is the net dipole moment per unit volume and is given by

$$\mathbf{M} = \frac{1}{V}\sum_{i=1}^{N} \mathbf{m}_i(\mathbf{r}_i), \tag{3.2}$$

where V is the total volume of the body, while the moments of the N microscopic magnetic dipoles that constitute the body are $\mathbf{m}_i(\mathbf{r}_i)$, $i = 1, \ldots, N$; and $\mathbf{r}_i$ is the position vector of the i-th dipole. The constituent magnetic dipoles are not perfectly aligned in a single direction. Various factors, including remanent magnetization, corrosion, random thermal motions, and material heterogeneities contribute to the lack of perfect dipole alignment. The

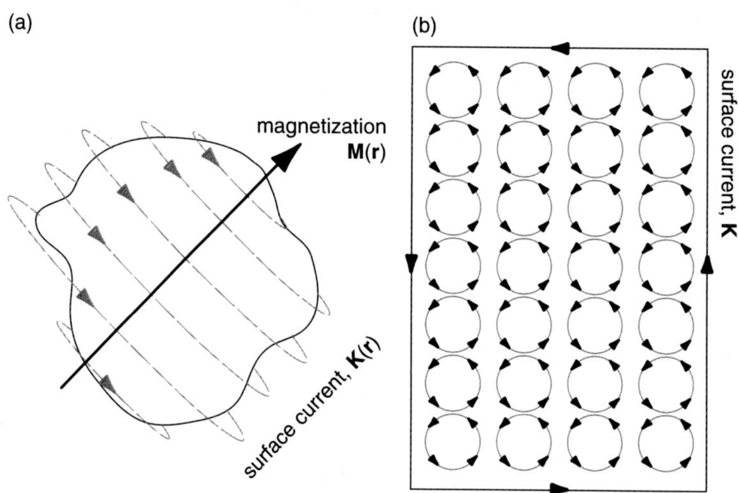

Figure 3.5 (a) Magnetized (iron, titanium-bearing) rock and its equivalent surface electric current **K(r)** and magnetization **M(r)** vectors. For a heterogeneous rock there is also an equivalent volume current **J(r)**. (b) Cancellation of internal electric currents in the case of a homogeneous magnetization **M**.

net alignment of the dipoles, however, is such that the vector sum of the moments is non-zero. On the other hand, the constituent dipoles are randomly oriented inside a non-magnetic body such that **M** may be considered to vanish everywhere, to a good approximation.

Figure 3.5a shows the equivalent surface current $\mathbf{K}(\mathbf{r})$ of a homogeneous magnetized body. In this case, $\mathbf{M} = M\hat{\mathbf{M}}$ with M constant, while the equivalent volume current density vanishes, $\mathbf{J}(\mathbf{r}) = 0$, since the internal current distribution cancels out. The cancellation can be visualized by considering a slice through a homogeneous magnetized body (Figure 3.5b) in a plane perpendicular to the magnetization direction $\hat{\mathbf{M}}$. As indicated, only the surface currents along the boundary of the heterogeneous body do not cancel, so that $\mathbf{K} \neq 0$ in this case.

A common idealization of a magnetized geological body, or a buried steel object, is a macroscopic magnetic dipole of moment **M**. This idealization is valid if the magnetic field **B** is measured at a sufficiently great distance from the body, for example, at least several times its characteristic length scale. Consider a buried steel drum represented by a magnetic dipole located at point $(x', y', z') = (0, 0, h)$ as shown in Figure 3.6. Henceforth, we use unprimed coordinates $\mathbf{r}_P = (x, y, z)$ to describe points at which magnetic observations are made and primed coordinates $\mathbf{r}_Q = (x', y', z')$ to describe points occupied by magnetic sources. The geometry of a general observation point P and a general source point Q is portrayed in Figure 3.7.

In Figure 3.6, for simplicity the magnetic dipole moment is presumed to be aligned with the long axis of the drum. In reality, a steel drum would possess a magnetization vector $\mathbf{M}(\mathbf{r}')$ whose magnitude and direction varies from point to point within the drum. The magnetization vector depends on a wide range of factors including the details of the drum manufacture, its corrosion state, its shape, material heterogeneities, and its magnetic history.

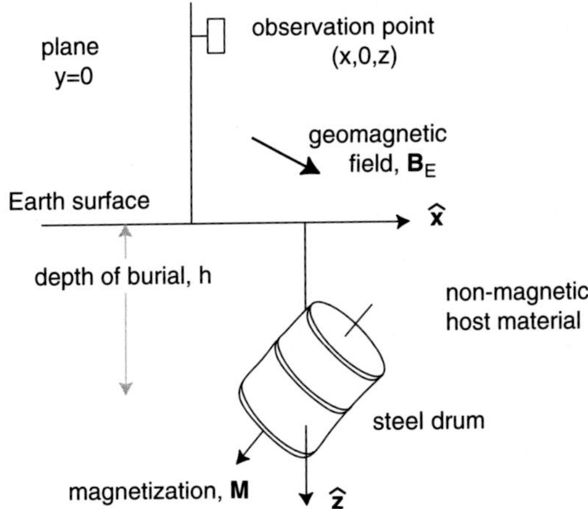

Figure 3.6 A steel drum buried at depth h beneath Earth's surface is represented by a magnetic dipole of moment **M**. A magnetometer is placed at observation point P(x,0,z). The ambient geomagnetic field is $\mathbf{B}_E$.

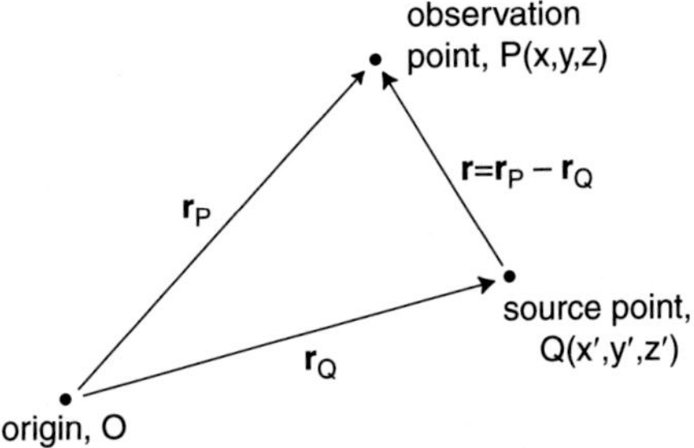

Figure 3.7 Unprimed coordinates describe the location $\mathbf{r}_P$ of observation point P and primed coordinates describe the location $\mathbf{r}_Q$ of source point Q.

A magnetometer senses the magnetic field **B**(P) at observation point P produced by the magnetic dipole. Let us investigate further this magnetic field. In the early nineteenth century, physicists discovered that two current-carrying wires exert a force on one another. In addition to attractive and repulsive forces, it was determined by experiment that loops of current tend to cause each other to rotate. The magnetic field **B**(P) at observation point P due to the presence of a magnetic dipole of moment **M** located at point Q is defined in terms of the *torque* that would be experienced by a small test loop located at P. This magnetic field is given by Blakely (1995; pp.72–75) as

$$\mathbf{B}(P) = \frac{\mu_0 M}{4\pi r^3}\left[3(\hat{\mathbf{M}}\cdot\hat{\mathbf{r}})\hat{\mathbf{r}} - \hat{\mathbf{M}}\right], \tag{3.3}$$

where the position vector **r** is

$$\mathbf{r} = \mathbf{r}_P - \mathbf{r}_Q = (x - x')\,\hat{\mathbf{x}} + (y - y')\,\hat{\mathbf{y}} + (z - z')\,\hat{\mathbf{z}} \tag{3.4}$$

such that $\hat{\mathbf{r}} = \mathbf{r}/r$ with $r = |\mathbf{r}|$.

The quantity μ_0 in Equation (3.3) is the *magnetic permeability of free space* [measured in henrys per meter, or H/m] and is a measure of the strength of the magnetic coupling

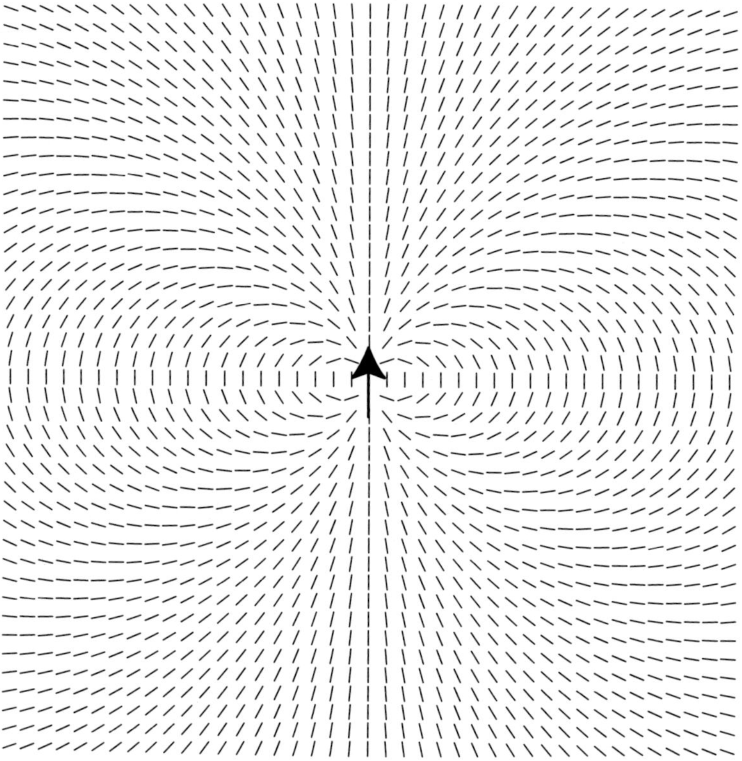

Figure 3.8 Lines of force of a magnetic dipole.

between two current loops located in free space. Its numerical value is $\mu_0 = 4\pi \times 10^{-7}$ H/m. Sometimes it is convenient to introduce a different magnetic field, conventionally denoted by **H**. The relationship between **B** and **H** in linear isotropic magnetic media, which corresponds to most geomaterials, is simply $\mathbf{B} = \mu\mathbf{H}$ where μ is the permeability of the medium. As noted in more detail below, for magnetite-bearing geological formations or man-made steel objects, we typically have $\mu > \mu_0$.

Physically, **B** and **H** can be distinguished as follows. The magnetic field **B** measures the strength of the force on an elementary dipole embedded in a magnetic material of permeability μ. The contributions to **B** include both externally applied currents and the equivalent surface and volume currents that are intrinsic to the magnetic material. The magnetic field **H** considers only the former and neglects the latter; i.e. the elementary dipole is presumed to reside in free space. In magnetic geophysical prospecting, the magnetometer is almost always located in air, a non-magnetic medium, so that the distinction between **B** and **H** as the measured quantity is not significant. Their difference would be important if the magnetometer were embedded within a magnetic material.

Lines of force of a magnetic dipole in a plane containing the dipole axis are illustrated in Figure 3.8. Each line of force is parallel to the direction of the magnetic field **B** at the point at which its midpoint is plotted.

3.3 Instrumentation

The *proton precession magnetometer* is a relatively simple instrument that measures the total intensity $|\mathbf{B}|$ of the magnetic field. Its operation is based on the fact that a proton H^+ in an external magnetic field undergoes a precession about its spin axis, like a spinning top (see Figure 3.9a).

A proton may be regarded as a spinning magnetized sphere (Figure 3.9b) with an intrinsic magnetic moment m_P and an intrinsic spin angular momentum I_P, both of which are atomic constants. The gyromagnetic ratio is defined as their ratio $\gamma_P = m_P/I_P = 0.2675222$ [1/s nT]. Suppose the proton is placed in a magnetic field $\mathbf{H} = \mathbf{B}/\mu_0$. The effect of the magnetic field is to exert a torque on the spinning proton which tends to align the intrinsic magnetic moment into the direction of the external field. The proton starts to process about its spin axis, just as a spinning top does in the presence of the gravitational torque $\tau = mgx$. The angular frequency of the proton precession is $\omega_P = \gamma_P|\mathbf{H}|$, which is known as the *Larmor frequency.*

The proton precession magnetometer measures the precession frequency ω_P, from which the total intensity of the magnetic field is easily found using the formula $|\mathbf{H}| = \omega_P/\gamma_P$. The operating principle may be briefly summarized as follows (e.g. Campbell, 2003). The intrinsic magnetic moment of an ensemble of protons is first brought into rough alignment with a strong external field $\mathbf{H}_0$ of $\sim 10^7$ nT. The external field is abruptly shut off. The proton moments then start to precess about the direction of the much weaker, ambient magnetic field, $\mathbf{H}$. The precession is observed as an oscillating voltage (of order millivolts) in a pickup coil wrapped around the ensemble of protons. The oscillation frequency is ω_P, the Larmor frequency. The same coil may be used for the excitation and the pickup and typically consists of $\sim$ 500 turns wound around a cylinder containing a proton-rich liquid

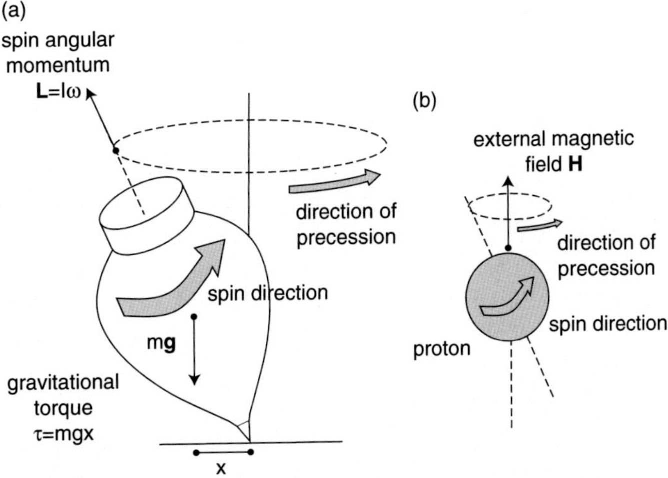

Figure 3.9 (a) Precession of a top of mass m and moment of inertia I spinning with angular frequency ω in a gravitational field $\mathbf{g}$; (b) precession of a spinning proton in a magnetic field $\mathbf{H}$.

Figure 3.10 The Geometrics G-858 cesium vapor magnetometer shown being used in gradiometer mode.

such as water, alcohol, or oil. In a typical instrument, the volume of liquid is ~ 0.5 L, and the frequency of precession is on the order of kilohertz. The sensitivity of the resulting magnetic-field measurement is ~ 0.1 nT.

A more commonly used instrument for magnetic near-surface geophysical surveys, shown in Figure 3.10, is the *cesium vapor magnetometer*. This instrument, like the proton precession magnetometer, registers the total intensity $|\mathbf{B}|$ of the magnetic field. However, the measurement involves the precession of magnetic moments of electrons, rather than protons. The instrument is based on the principles of *optical pumping* (Bloom, 1962), in which cesium vapor in a glass cell is irradiated by a light source. A full description of its operating principle requires a detailed knowledge of nuclear magnetism, but the following simplified discussion from Telford *et al.* (1990) outlines the main ideas.

Initially, the cesium atoms are distributed evenly into two closely spaced energy levels, 1 and 2, according to whether the spin of the valence electron is aligned parallel or anti-parallel to the ambient magnetic field. A circularly polarized irradiating light is then shone on the atoms such that the atoms in state 1 are preferentially elevated to a higher energy level 3. Atoms in state 2 preferentially remain in that lower state. The excited atoms are short-lived, however, and, when one returns to the ground state, it falls with equal probability into state 1 or state 2. If the excited atom falls into state 1, it is likely to be elevated once again into the higher energy state. If the atom falls into state 2, it is likely to remain there. By this mechanism, an accumulation of atoms in state 2 occurs.

The energy required to elevate a cesium atom from ground state 1 into its excited state 3 is absorbed from the irradiating light source. As the population of atoms in state 2 grows, fewer atoms are to be found in state 1, and consequently the absorption of energy drops off. This is manifest as an increasing transmission of light across the sample of cesium atoms, such that eventually in the fully pumped configuration a maximum current across the cesium-containing cell is registered by a photosensitive detector.

Looking at the system of cesium atoms in closer detail, the presence of the ambient magnetic field causes the spin axes of the valence electrons to precess about the direction of the field at the Larmor frequency. At one point in the precession cycle, the spin axes are most nearly aligned parallel to the polarization direction of the irradiating light, while at another point in the cycle the spin axes are aligned most nearly anti-parallel to the polarization direction. When the spin axes are best aligned with the polarization direction, promotion of the atoms to energy level 2 is favored, absorption is high, and the current across the cell reaches its minimum value. When the spin axes are mostly anti-parallel to the polarization direction, the photocurrent is maximized. The net result is that the current registered at the photocell flickers at the Larmor frequency, which is in the radio frequency (RF) range, ~ 100 kHz. Recall from the discussion of the proton precession magnetometer that the relationship between the Larmor frequency and the intensity $|\mathbf{B}|$ of the ambient magnetic field is known and has a simple form. Thus, a measurement of the frequency at which the light intensity across the cell flickers is equivalent to a measurement of the ambient magnetic field intensity.

In order to re-establish the equilibrium of the cesium atoms so that equal numbers of them are in states 1 and 2, the RF signal from the photocell is fed back to a coil wrapped around the glass cell containing the cesium atoms. The RF signal oscillates at the Larmor frequency, which exactly corresponds to the small energy difference between levels 1 and 2. The RF feedback has the effect of randomizing the atomic states, effectively undoing the pumping effect, and preparing the cell for another pumping cycle. In this way, the cesium vapor magnetometer is able to make continuous magnetic-field measurements.

A fast sampling rate of 10–100 Hz and a high sensitivity ~ 0.01 nT can be achieved with optically pumped magnetometers. This makes such an instrument ideal for capturing, at high spatial resolution, anomalies caused by subtle magnetic variations in the soil and upper sedimentary layers. Accordingly, the cesium vapor magnetometer is popular for geological mapping and archaeological, agricultural, forensic, and other applications. A comprehensive overview of optical magnetometers is given by Budker and Romalis (2007).

It is now possible to fabricate millimeter-sized alkali vapor magnetic sensors that are battery operated and disposable (Shwindt *et al.* 2004). The miniaturization by a factor of 10^4 over traditional bulky, high-power-consumption field magnetometers is enabled by taking advantage of micromechanical techniques. Other instruments in routine use for near-surface geophysics include the fluxgate magnetometer, which measures one or more of the components of the vector **B**. The SQUID (superconducting quantum interference device) magnetometer is very precise but requires superconducting materials and cryogenic cooling. A number of advanced solid state magnetic detectors are presently under active development.

3.4 Magnetic gradiometry

A magnetic *gradiometer* is formed by constructing an array of two or more magnetometers. A gradiometer measurement is the difference $\mathbf{B}_i - \mathbf{B}_j$ between the magnetic field $\mathbf{B}_i$ recorded at the *i*-th sensor and the magnetic field $\mathbf{B}_j$ recorded at the *j*-th sensor of the array. Compared to single magnetometer measurements, gradiometer data reveal substantially finer details concerning the spatial variations of the magnetic signature of magnetized subsurface bodies.

3.5 Geomagnetic field

In near-surface geophysics, oftentimes the length scale of measurement is 10–100 m. It is necessary to characterize larger-scale global and regional fields in order to isolate the local magnetic field caused by the magnetized bodies of interest. We define *regional scale* as lengths greater than the geophysical survey dimensions but much less than the *global scale*, which is comparable to a large fraction of Earth's radius. Together, the global and regional fields comprise the geomagnetic field.

In 1838, Carl Friedrich Gauss determined that the vast bulk of the geomagnetic field, termed the *main field*, is of internal origin. Later researchers determined that the major cause of the main field and its multi-decadal changes, or *secular variation*, is related to the slow convection of liquid iron in the outer core. Furthermore, an important cause of spatial variations of the geomagnetic field spanning the 10^2–10^4 km length scales is permanent magnetization of iron-bearing geological formations located within the lithosphere. A small, rapidly time-varying component of the geomagnetic field is external in origin. The external field is due largely to geomagnetic storms and related solar-originating disturbances, including electric currents that flow in the ionosphere at ~ 100 km altitude in response to diurnal solar heating of the upper atmosphere.

The conventional manner of describing the geomagnetic-field vector is in terms of its elements (X, Y, Z) which are, respectively, the (north, east, downward)- components of the geomagnetic field $\mathbf{B}_E$. The horizontal intensity of the geomagnetic field, denoted by H, is $H = \sqrt{X^2 + Y^2}$. The total intensity is $F = \sqrt{X^2 + Y^2 + Z^2} = |\mathbf{B}_E|$. The inclination I, or geomagnetic dip, is the vertical angle between the geomagnetic-field direction and the horizontal plane, $I = \tan^{-1}(Z/H)$. The declination is defined to be the azimuth D of the magnetic meridian, the vertical plane containing the geomagnetic-field vector, and is given by $D = \sin^{-1}(Y/H)$. The geomagnetic-field elements defined in this way (Figure 3.11) are based on an assumption that Earth is perfectly spherical; however, the departure of the actual shape from a sphere is not important in near-surface geophysical prospecting.

Global databases of magnetic-field measurements have been assembled from long land-based, ship-borne, aircraft and satellite surveys. As discussed below, the data can be compactly represented using a family of mathematical functions called spherical

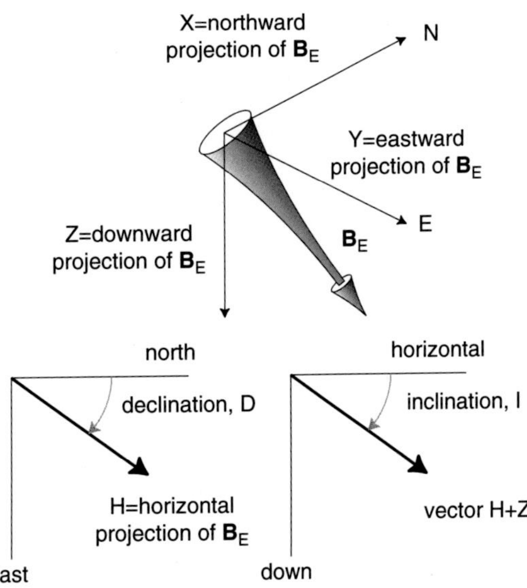

Figure 3.11 Elements (X, Y, Z) of the geomagnetic field $\mathbf{B}_E$. After Blakely, 1995.

harmonics. Such functions are particularly useful for describing magnetic data acquired on or near the surface of a spheroid, such as Earth. A spherical harmonic expansion provides an efficient means of compressing the vast global database into a set of simple basis functions. By international convention, a standard spherical harmonic representation of the global magnetic database has been developed and it is termed the IGRF, or *international geomagnetic reference field.*

It is known from potential theory (Blakely, 1995) that a magnetic field $\mathbf{B}(P)$ can be written as the negative gradient of a scalar potential in any region that does not contain magnetic sources, such as the air just above Earth's surface but beneath the ionosphere.

The magnetic scalar potential V obeys Laplace's equation $\nabla^2 V = 0$. In the region $r > R$, where $R = 6371.2$ km is Earth's radius, we can write $\mathbf{B}(P) = -\nabla V(P)$ with $V(P)$ being called the *geomagnetic potential.*

A general spherical harmonic expansion for the geomagnetic potential in geocentric (r, θ, ϕ) coordinates is (Finlay *et al.*, 2010)

$$V(r,\theta,\phi) = R\sum_{n=1}^{N}\left[\frac{R}{r}\right]^{n+1}\sum_{m=0}^{n}[g_n^m \cos m\phi + h_n^m \sin m\phi]P_n^m(\cos\theta) \tag{3.5}$$

where the $P_n^m(\cos\theta)$ are associated Legendre functions of degree n and order m, while θ is colatitude and ϕ is longitude. The functions $1/r^{n+1}$, $\cos m\phi$, $\sin m\phi$ and $P_n^m(\cos\theta)$ appear multiplied together in Equation (3.5) because their products constitute the general solution to Laplace's equation in spherical coordinate systems (Arfken *et al.*, 2012).

The spherical harmonic functions are denoted by $\Psi_n^m(\theta,\phi) = \cos m\phi P_n^m(\cos\theta)$ and $\Psi_n^m(\theta,\phi) = \sin m\phi P_n^m(\cos\theta)$. For the interested reader, the full development of Equation (3.5) is explained in Appendix B.

The so-called Gauss coefficients $\{g_n^m, h_n^m\}$ in Equation (3.5) are obtained by carefully adjusting their values until the global database of magnetic measurements agrees, at every location as closely as possible, with the negative gradient of Equation (3.5). The index N describes the number of terms to be used in the summation in Equation (3.5). Increasing the value of N provides a more detailed description of spatial variations in the geomagnetic field. Generally, the main field due to fluid motions in the outer core is well described by the Gauss coefficients up to and including degree $n \sim 13$ while the magnetization of the lithosphere is adequately characterized by higher-degree coefficients (Langel and Estes, 1982).

Since the geomagnetic field changes slowly with time, due to the fluid motions in the outer core, the international community adopts a new IGRF model at the end of every five-year *epoch*. Each IGRF description provides time-derivatives of the model, which describe the *secular variation* that varies greatly with location but is typically <100 nT/yr. Global maps of intensity F, inclination I, and declination D for the IGRF-11 model (Finlay *et al.*, 2010) valid for the five-year epoch 2010.0–2015.0 are shown in Figure 3.12. These maps can be dowloaded from the website www.wdc.kugi.kyoto-u.ac.jp operated by the World Data Center for Geomagnetism in Kyoto, Japan. For reference, the first few Gauss coefficients of the IGRF-11 model appear in Table 3.1. An online calculator based on IGRF-11, which estimates current and past values of the geomagnetic-field elements at any user-specified geographical location, is available at www.ngdc.noaa.gov/geomag-web.

Keeping only the $n = 1$ terms in Equation (3.5), it is readily shown that the geomagnetic field is grossly similar to a *geocentric dipole* of moment vector $\mathbf{m}_E$ (see Figure 3.13). A dipole of this moment is described by the formula

$$V_D = \frac{\mu_0}{4\pi r^2}[\mathbf{m}_E \cdot \hat{\mathbf{r}}]. \tag{3.6}$$

The general morphology of the geomagnetic field is thus similar to that of a hypothetical bar magnet placed with a ~ 10° tilt angle near the center of the Earth. The usual convention for labeling the poles of a bar magnet is such that lines of **B** emanate from the north pole and converge on the south pole (Tipler and Mosca (2007)). Thus, the north pole of the hypothetical bar magnet points into the southern hemisphere. The reader should not need to be reminded that the main geomagnetic field is actually produced by convective fluid motions in the outer core, not the bar magnet placed near the center of the Earth shown in the figure. The bar magnet is simply a convenient fiction to better visualize the largest-scale structure of the geomagnetic field.

The *geomagnetic poles* are the locations at which the geocentric dipole axis intersects Earth's surface. Notice that the convention for labeling the geomagnetic poles is opposite to that of the bar magnet, so that the geomagnetic north pole resides in the northern hemisphere close to the geographic north pole.

An improved fit to the global magnetic database is obtained if the dipole is not geo-centered but instead shifted northward by ~ 400 km along the rotation axis from Earth's center. The resulting field is that of the so-called *eccentric dipole*. The intensity of the *non-dipole field*, which is the sum of the terms in Equation (3.5) after the geocentric dipole ($n = 1$) contribution is removed, comprises about 10% of the actual geomagnetic-field intensity.

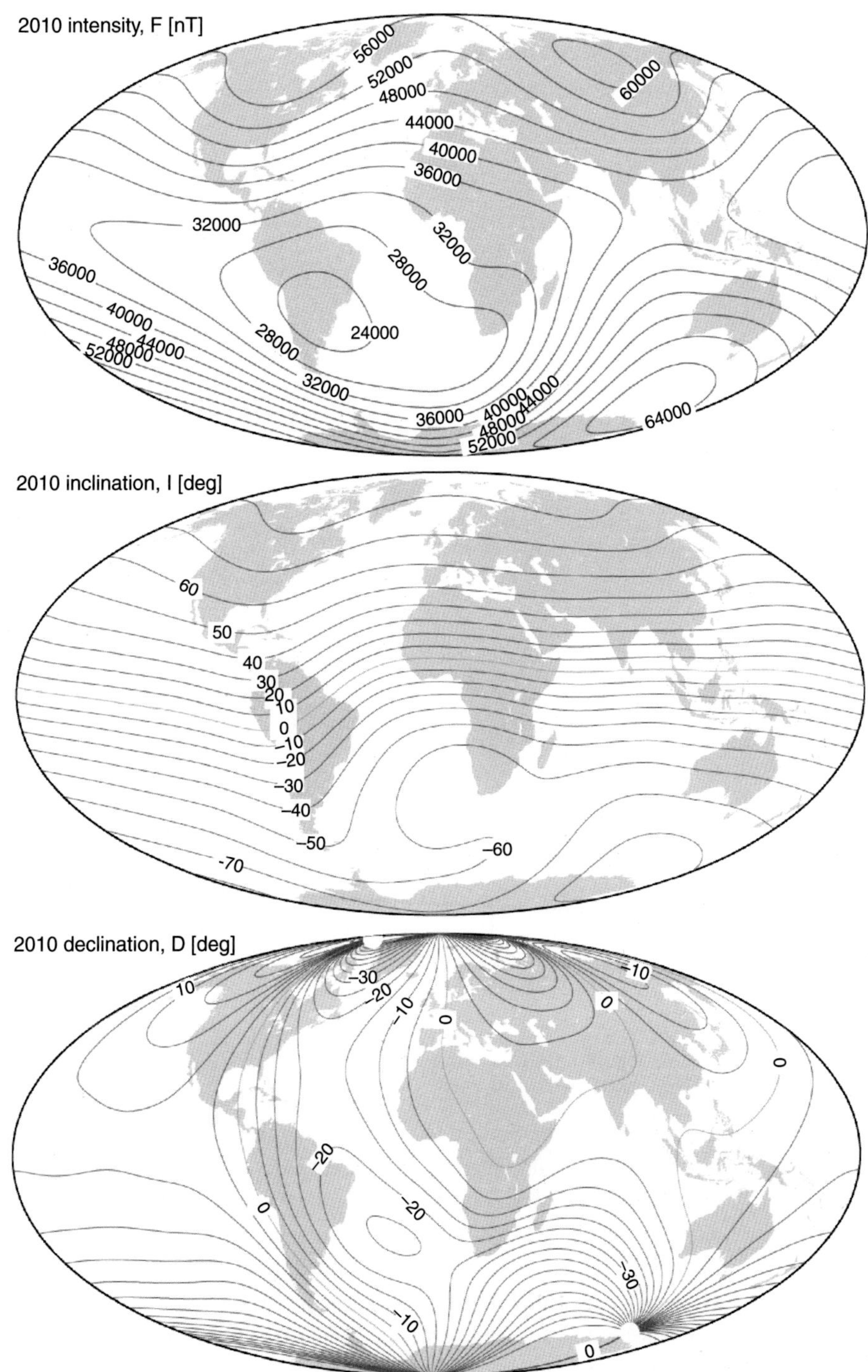

Figure 3.12 IGRF-11. (Top) total intensity; (middle) inclination; (bottom) declination. After Finlay et al. (2010).

Table 3.1 First few Gauss coefficients of the IGRF-11 geomagnetic field model, from Finlay *et al.*, 2010. The associated Legendre functions $P_n^m(\cos\theta)$ are shown in the rightmost column

N	m	g_n^m [nT]	h_n^m [nT]	$P_n^m(\cos\theta)$
1	0	−29496.5	0	$\cos\theta$
1	1	−1585.9	4945.1	$\sin\theta$
2	0	−2396.6	0	$(1/2)(3\cos^2\theta - 1)$
2	1	3026.0	−2707.7	$3\cos\theta\sin\theta$
2	2	1668.6	−575.4	$3\sin^2\theta$
3	0	1339.7	0	$(1/2)(5\cos^3\theta - 3\cos\theta)$
3	1	−2326.3	−160.5	$(3/2)(5\cos^2\theta - 1)\sin\theta$
3	2	1231.7	251.7	$15\cos\theta\sin^2\theta$
3	3	634.2	−536.8	$15\sin^3\theta$

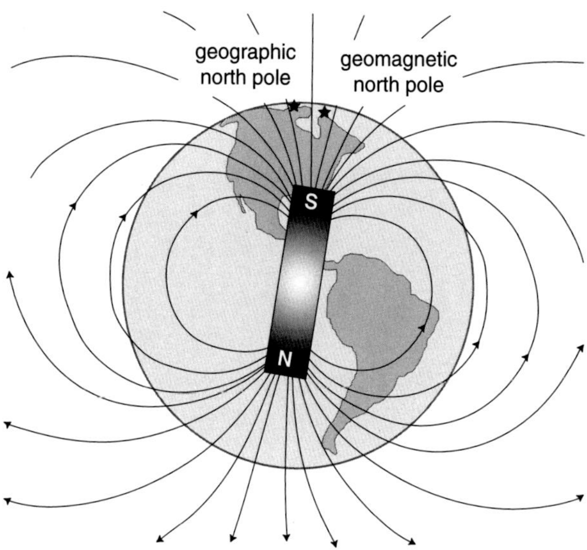

Figure 3.13 The geocentric dipole as a bar magnet.

The foregoing discussion has provided a brief overview of the global-scale geomagnetic field. Regional-scale magnetic-field modeling provides higher spatial resolution that may be valuable in characterizing the background signals for enhanced interpretation of near-surface magnetic survey data. A digital magnetic database that merges a variety of airborne and marine observations has been compiled for much of the North American continent and its offshore regions (http://mrdata.usgs.gov/magnetic). This map displays magnetization anomalies with wavelengths up to ~ 150 km. These anomalies are due mainly to crustal-scale igneous structures such as batholiths, major faults, folds, and dikes. Other regional and worldwide magnetic databases have also been compiled.

Despite impressive recent advances, current geomagnetic databases still contain large uncertainties due to irregular data coverage and problems associated with merging datasets acquired by different groups using different equipment and data-processing techniques. The United States Geological Survey is planning future low-altitude (<500 m above terrain) and high-altitude (~ 15–22 km) data-acquisition campaigns in order to refine, respectively, the short-wavelength- and long-wavelength-anomaly information contained in their digital magnetic maps.

3.6 Total-field anomaly

Let us now return to the steel-drum exploration problem in Figure 3.6. Field measurements indicate that the magnetic field $\mathbf{B}_T$ due to a 55-gallon steel drum buried 1–2 m in a non-magnetic soil is typically of strength ~ 100–1000 nT. However, as shown in Figure 3.12 (top), the strength, or intensity, of the background geomagnetic field $\mathbf{B}_E$ in the central USA is $F \sim 55\,000$ nT. The field due to the drum thus represents only a few parts per thousand of the total field intensity sensed by the magnetometer, assuming no other magnetic targets are present.

A total-field magnetometer measures the strength of the magnetic field $\mathbf{B}_{OBS}$ due to all sources present. In the idealized case where the steel drum is the only appreciable source that affects the magnetometer, then the observed field intensity is simply $|\mathbf{B}_{OBS}| = |\mathbf{B}_E + \mathbf{B}_T|$. To reveal the magnetic anomaly due to the drum, it necessary to remove the background geomagnetic field intensity. The *total-field anomaly* T accordingly is defined as

$$T = |\mathbf{B}_{OBS}| - F. \tag{3.7}$$

We can view Equation (3.7) as the difference between the "meter reading" $|\mathbf{B}_{OBS}|$ taken in the field and the "chart reading" F obtained from the current IGRF geomagnetic field model. Since the magnetic-field intensity due to the drum is much smaller than the geomagnetic field intensity, we have $|\mathbf{B}_T| << F$. Under this condition, it is readily shown that the total-field anomaly can be written as

$$T = \mathbf{B}_T \cdot \hat{\mathbf{B}}_E \tag{3.8}$$

where $\hat{\mathbf{B}}_E$ is the unit vector in the direction of the geomagnetic field, that is, $\mathbf{B}_E = F\hat{\mathbf{B}}_E$. Thus, Equation (3.8) shows that the total-field anomaly T is simply the *projection* of the target magnetic field vector $\mathbf{B}_T$ in the direction of the ambient geomagnetic field. It should be clear to the reader that the total-field anomaly T depends on the geographic location in which the magnetic survey is acquired.

3.7 Interpretation of magnetic anomalies

A common near-surface investigation scenario requires the characterization of elongate magnetized geological structures such as faults or dikes, archaeological or geotechnical targets such as buried walls or foundations, or buried steel objects such as pipelines. If the

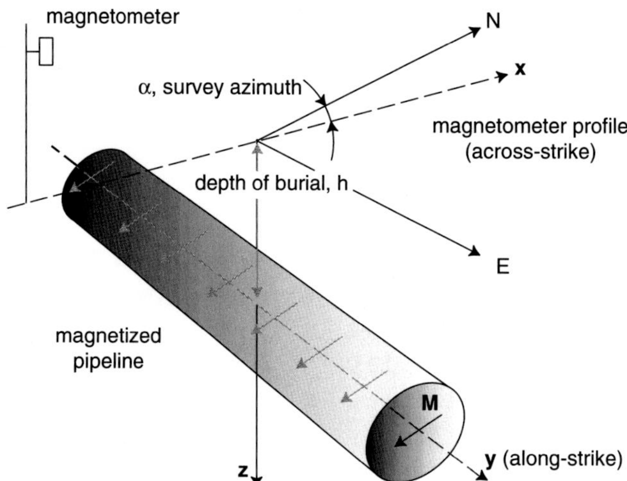

Figure 3.14 A buried pipeline as an example of a 2-D magnetic target.

physical properties of the body under investigation do not vary significantly along the long axis of the body, it can be approximated by a two-dimensional (2-D) structure with the strike axis parallel to the long axis. Typically, the strike axis is horizontal and aligned with the y-axis. In this case the shape of the body is completely determined by its cross-section in a vertical x/z-plane, which for convenience is taken as the $y = 0$ plane.

It is common practice, where possible, to acquire total-field magnetic readings over a 2-D body along a profile across its strike. Consider the acquisition of total-field-intensity readings $|\mathbf{B}_{OBS}|(x)$ across the strike of a long, buried, uniformly magnetized pipeline, as shown in Figure 3.14. The across-strike direction $\hat{\mathbf{x}}$ is oriented at some azimuth α measured with respect to geographic north, as shown in the figure. It is of interest to calculate the expected total-field anomaly $T(x)$ of this important near-surface target.

Recall that we approximated a steel drum as a single magnetic dipole of moment **M**. A uniformly magnetized, long pipeline can be approximated as an infinite *line of dipoles* of magnetization $\mathbf{M} = M'\hat{\mathbf{M}}$ where M' [Am] is the magnetic dipole moment per unit length. Blakely (1995; p.96) shows that the magnetic field $\mathbf{B}_{2D}(P)$ at observation point $P(x, z)$ in the $y = 0$ plane due to a line of dipoles located at $Q(x', z')$ is given by

$$\mathbf{B}_{2D}(P) = \frac{\mu_0 M'}{2\pi r_{2D}^2}[2(\hat{\mathbf{M}}\cdot\hat{\mathbf{r}}_{2D})\hat{\mathbf{r}}_{2D} - \hat{\mathbf{M}}], \tag{3.9}$$

where $\mathbf{r}_{2D} = r_{2D}\hat{\mathbf{r}}_{2D} = (x - x')\hat{\mathbf{x}} + (z - z')\hat{\mathbf{z}}$. The symmetry of the problem, and other fundamental considerations, dictate that the magnetization vector **M** cannot have a $\hat{\mathbf{y}}$ component and without loss of generality we write $\mathbf{M} = M_x\,\hat{\mathbf{x}} + M_z\,\hat{\mathbf{z}}$.

In order to calculate the expected total-field anomaly $T(x)$ along the measurement profile indicated in Figure 3.14, which is oriented at azimuth α with respect to geographic north, we recall Equation (3.8) and recognize that $T(x) = \mathbf{B}_{2D}\cdot\hat{\mathbf{B}}_E$ where $\mathbf{B}_{2D}(P)$ is given by Equation (3.9). Assume for simplicity that the magnetization vector is aligned with the ambient geomagnetic field such that $\hat{\mathbf{M}} = \hat{\mathbf{B}}_E$. This implies that the magnetization of the

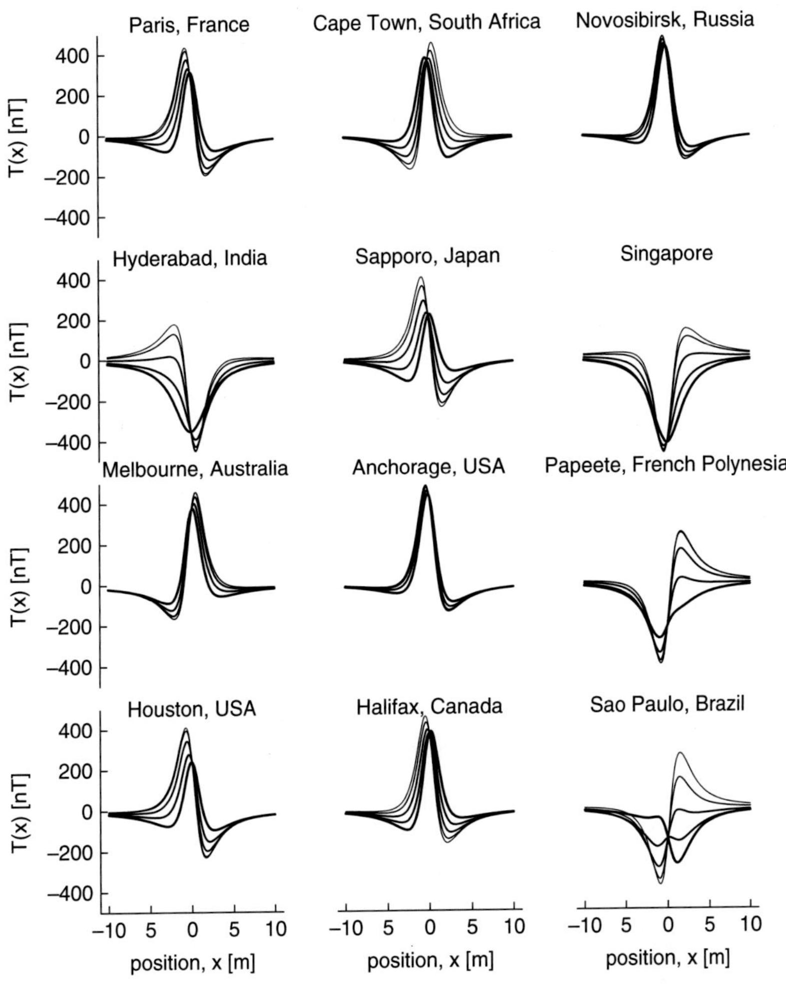

Figure 3.15 Total-field anomaly due to a magnetized pipeline at different geographic locations and, for each location, different survey azimuths.

pipeline is purely induced with no remanence (see below). Under this condition, it is readily calculated from Equations (3.8) and (3.9) that

$$T(x) = \frac{\mu_0 M'}{2\pi r_{2D}^2}[2(\hat{\mathbf{F}}_{2D}\cdot\hat{\mathbf{r}}_{2D})^2 - 1], \tag{3.10}$$

where $\hat{\mathbf{F}}_{2D} = \mathbf{F}_{2D}/F_{2D}$. The 2-D geomagnetic field intensity vector is defined as

$$\mathbf{F}_{2D} = [X\cos\alpha + Y\sin\alpha]\,\hat{\mathbf{x}} + Z\,\hat{\mathbf{z}}, \tag{3.11}$$

where X, Y, and Z are respectively the northward, eastward, and downward vertical geomagnetic elements, as defined earlier. The expected total-field anomaly $T(x)$ for a range of profile azimuths and geographic locations is shown in Figure 3.15. The modeled pipeline is buried at depth 1.0 m and the magnetometer is located at height 1.0 m above

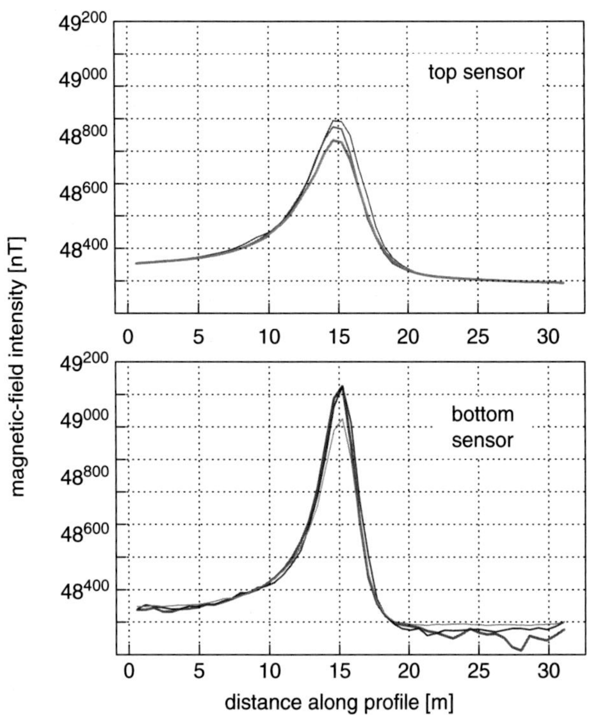

Figure 3.16 Magnetic-intensity measurements made at Riverside campus, Texas A&M University, over a buried pipeline.

Earth's surface. The IGRF-11 model is used to compute the geomagnetic field at each of the geographic locations. The different curves at each location indicate, using progressively heavier lines, different survey azimuths of $\alpha = 0.0$, 22.5, 45.0, 67.5, and 90.0°, respectively. The differences between the curves provide a glimpse of the worldwide range of possible anomaly shapes that can be generated by acquisition of total-intensity measurements along a profile oriented across the strike of a long, buried pipeline that is uniformly magnetized in the same direction as the ambient magnetic field.

Magnetic-intensity readings using a Geometric G-858 cesium vapor magnetometer, in vertical gradiometer mode with sensor heights $h_1 = 1.0$ m and $h_2 = 1.5$ m, were taken by geophysics undergraduate students on Riverside campus at Texas A&M University. The measurement profile crossed, at its midpoint, a buried natural-gas pipeline oriented at azimuth $\alpha \sim 225°$. The profile was repeated three times, with good repeatability, as shown in Figure 3.16. The pipeline generates an anomly of ~ 700 nT on the bottom sensor and ~ 500 nT on the top sensor. The shape of the observed curves is in qualitative agreement with the theoretical curve for Houston, USA shown in Figure 3.15.

3.8 Reduction to the pole

A reduction-to-the-pole (RTP) filter essentially removes the effect of the geomagnetic-field direction at the measurement site. In this section we derive the RTP filter for a total-field-anomaly dataset $T(x)$ that has been collected along an x-directed horizontal profile at a

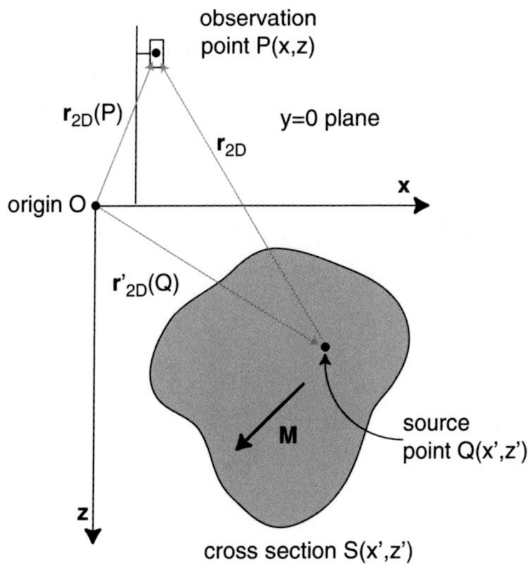

Figure 3.17 A 2-D uniformly magnetized body with *y*-strike direction.

given geographic location. It is assumed that the causative magnetized body is 2-D with $\hat{\mathbf{y}}$ being the along-strike direction so that the profile is acquired across the strike of the body.

As before, the magnetometer profile is oriented at some azimuth α, measured positive clockwise with respect to geomagnetic north. It is further assumed that the data have been Fourier-transformed in the $\hat{\mathbf{x}}$ direction such that $T(k)$ is available, where $k = k_x$ is the across-strike wavenumber.

Let us start with the basic governing equation for the magnetic potential $V(P)$ at point P due to an arbitrary 2-D uniformly magnetized body S of magnetization $\mathbf{M}$ (Blakely, 1995; p. 97),

$$V(P) = \frac{\mu_0}{2\pi} \int_S \frac{\mathbf{M} \cdot \hat{\mathbf{r}}_{2D}}{r_{2D}} dS, \tag{3.12}$$

where $\hat{\mathbf{r}}_{2D}$ is the unit vector directed from an arbitrary source point $Q(x', y')$ lying *inside* the magnetized body S toward an observation point $P(x, z)$ lying *outside* the body (Figure 3.17). It is assumed that the magnetized body is invariant in the strike direction. Without loss of generality it is further assumed that the total-field-anomaly data are acquired in the vertical plane $y = 0$. The magnetized body is represented by the two-dimensional cross-sectional area $S(x', y')$. An elemental area of the surface is $dS = dx'dz'$

The magnetic field at observation point P is the negative gradient of the magnetic potential, $\mathbf{B}(P) = -\nabla V(P)$. The total-field anomaly at point P is defined by Equation (3.8). In terms of the potential,

$$T(P) = -\hat{\mathbf{F}}_{2D} \cdot \nabla V(P), \tag{3.13}$$

where $\hat{\mathbf{F}}_{2D}$ is the direction of the ambient geomagnetic field $\mathbf{F}_{2D}$, as given by Equation (3.11).

The uniform magnetization of the body is represented by the vector $\mathbf{M} = M_x\, \hat{\mathbf{x}} + M_z\, \hat{\mathbf{z}}$, where M_x is the horizontal component of magnetization aligned with the magnetometer profile and M_z is the vertical component. Since the body is infinitely extended in the y direction, and uniformly magnetized, the along-strike magnetization component M_y makes no contribution to the magnetic field outside the body. Thus, without loss of generality we can set $M_y = 0$. The integrand in Equation (3.12) becomes

$$\mathbf{M}\cdot\hat{\mathbf{r}}_{2D} = \frac{1}{r_{2D}}[M_x(x-x') + M_z(z-z')] \tag{3.14}$$

such that the magnetic potential is

$$V(P) = \frac{\mu_0}{2\pi}\int_S \frac{M_x(x-x') + M_z(z-z')}{(x-x')^2 + (z-z')^2}dx'dz'. \tag{3.15}$$

We use the expressions (3.11) and (3.15) to find the total-field anomaly

$$T(P) = -\hat{\mathbf{F}}_{2D}\cdot\nabla V(P) = -\frac{1}{F_{2D}}\left[F_x\frac{\partial V}{\partial x} + Z\frac{\partial V}{\partial z}\right], \tag{3.16}$$

where $F_x = X\cos\alpha + Y\sin\alpha$. The partial derivatives $\partial V/\partial x$ and $\partial V/\partial z$ are readily computed. Notice that the derivative operators $\partial/\partial x$ and $\partial/\partial z$ slide through the integration in Equation (3.15) since the derivatives are taken with respect to the unprimed (observation) coordinates (x, z) while the integration is respect to the primed (source) coordinates (x', z'). After some elementary differentiation of polynomials we find

$$\frac{\partial V}{\partial x} = \frac{\mu_0}{2\pi}\int_S dx'dz'\frac{M_x\{(z-z')^2 - (x-x')^2\} - 2M_z(x-x')(z-z')}{r_{2D}^4}; \tag{3.17}$$

$$\frac{\partial V}{\partial z} = \frac{\mu_0}{2\pi}\int_S dx'dz'\frac{M_z\{(x-x')^2 - (z-z')^2\} - 2M_x(x-x')(z-z')}{r_{2D}^4}. \tag{3.18}$$

Inserting these two expressions into Equation (3.16) results in

$$T(P) = -\frac{\mu_0}{2\pi F_{2D}}\int_S dx'dz'\frac{(ZM_z - F_xM_x)\{(x-x')^2 - (z-z')^2\} - 2(F_xM_z + ZM_x)(x-x')(z-z')}{r_{2D}^4}. \tag{3.19}$$

At this point, the total-field anomaly has been evaluated. Now we specify that the observations are made along the x-profile at the constant elevation $z = 0$, so that

$$T(x) = -\frac{\mu_0}{2\pi F_{2D}}\int_S dx'dz'\frac{(ZM_z - F_xM_x)\{(x-x')^2 - z'2\} + 2(F_xM_z + ZM_x)(x-x')z'}{\rho_{2D}^4} \tag{3.20}$$

where $\rho_{2D} = \sqrt{(x - x')^2 + z'2}$. Taking a Fourier transform of both sides of Equation (3.20) results in

$$T(k) = -\frac{\mu_0}{2\pi F_{2D}} \int_S dx'dz'[(ZM_z - F_x M_x)G_1(k) + 2(F_x M_z + ZM_x)z'G_2(k)], \quad (3.21)$$

where

$$G_1(k) = \mathrm{F}\left\{\frac{(x - x')^2 - z'2}{\rho_{2D}^4}\right\}; \quad (3.22)$$

and

$$G_2(k) = \mathrm{F}\left\{\frac{x - x'}{\rho_{2D}^4}\right\}. \quad (3.23)$$

The notation $G(k) = \mathrm{F}\{g(x)\}$ denotes an instruction to take the Fourier transform of the function $g(x)$ and thereby generate the wavenumber-domain function, or spectrum, $G(k)$. The Fourier transforms in Equations (3.22) and (3.23) are analytic. The expressions are

$$G_1(k) = -\pi k \exp(-ikx')\exp(-|k|z')\mathrm{sgn}(k); \quad (3.24)$$

$$G_2(k) = -\frac{i\pi k}{2z'}\exp(-ikx')\exp(-|k|z'), \quad (3.25)$$

where sgn $(k) = +1$ for $k > 0$ and sgn $(k) = -1$ for $k < 0$. Inserting Equations (3.24) and (3.25) into Equation (3.21) yields

$$T(k) = -\frac{\mu_0}{2\pi F_{2D}} \int_S dx'dz'[-\pi k e^{-ikx' - |k|z'}\mathrm{sgn}(k)][(ZM_z - F_x M_x) + i\mathrm{sgn}(k)(F_x M_z + ZM_x)]. \quad (3.26)$$

In the above expression we have used the identity sgn $(k) = 1/\mathrm{sgn}(k)$.

We are now ready to consider the reduction of the total-field anomaly $T(k)$ to the geomagnetic pole. For simplicity, we require that the magnetization **M** of the body is totally induced by the present-day geomagnetic field **F**, that is, there is no remanent component. The magnetization vector **M** [units A m] and the geomagnetic field vector **F** [units nT] are therefore aligned such that $\mathbf{M} = \beta\mathbf{F}$, where β is the scale factor with units of [A m/nT]. Inserting $M_x = \beta F_x$ and $M_z = \beta Z$ into Equation (3.26) results in

$$T(k) = \frac{-\mu_0\beta}{2\pi F_{2D}} \int_S dx'dz'[-\pi k e^{-ikx' - |k|z'}\mathrm{sgn}(k)][Z^2 - F_x^2 + 2i\mathrm{sgn}(k)F_x Z]. \quad (3.27)$$

Let the vertical component of the geomagnetic field at the north geomagnetic pole be denoted by F_{ZP}. If the total-field-anomaly measurements had, in fact, been acquired at the geomagnetic north pole, where $X = Y = 0$ and $F = Z = F_{ZP}$, then Equation (3.27) reduces to

$$T_R(k) = \frac{-\mu_0 \beta}{2\pi} \int_S dx' dz' F_{ZP}[-\pi k \exp(-ikx') \exp(-|k|z') \mathrm{sgn}(k)] \tag{3.28}$$

where we have labeled the reduced-to-pole anomaly as $T_R(k)$. A comparison of Equations (3.27) and (3.28) immediately shows

$$T_R(k) = T(k) \frac{F_{2D} F_{ZP}}{Z^2 - F_x^2 + 2i\mathrm{sgn}(k) F_x Z}. \tag{3.29}$$

Defining the RTP filter $\Psi_R(k)$ by the formula $T_R(k) = T(k)\Psi_R(k)$, we finally obtain the desired result

$$\Psi_R(k) = \frac{F_{2D} F_{ZP}}{Z^2 - F_x^2 + 2i\mathrm{sgn}(k) F_x Z}. \tag{3.30}$$

Notice that the RTP filter correctly reduces to $\Psi_R(k) = 1$ if the data were actually measured at the geomagnetic north pole, where $F_x = 0$ and $F = Z = F_{ZP}$. The RTP-filtered dataset $T_R(x)$, which essentially has the effects of the local geomagnetic direction removed, is obtained by taking the inverse Fourier transform

$$T_R(x) = \mathrm{F}^{-1}\{T_R(k)\} = \mathrm{F}^{-1}\{T(k)\Psi_R(k)\}. \tag{3.31}$$

Thus, an RTP filtering of the dataset $T(x)$ in the spatial domain is equivalent to a multiplication in the wavenumber domain of the Fourier-transformed data $T(k)$ with the filter $\Psi_R(k)$.

Example. Line of dipoles.

A simple check of the 2-D RTP filter can be made using the 2-D magnetic model consisting of an infinitely long line of dipoles extending along-strike. The total-field anomaly $T(x)$ is given by Equation (3.10). The RTP total-field anomaly is found by setting $X = Y = 0$ and $F = Z = F_{ZP}$

$$T_R(x) = \frac{\mu_0 M'}{2\pi r_{2D}^2} \left[2\left(\frac{z'}{r_{2D}}\right)^2 - 1 \right]. \tag{3.32}$$

A computation of $T(x)$ and $T_R(x)$ over the profile $-12.0 < x < 12.0$ m has been made using the parameter values $\alpha = 18°$; $X = 24091$ nT; $Y = 3149$ nT; $Z = 42742$ nT; $F_{ZP} = 49164$ nT; $(x', z') = (0.0, 1.0)$ m; and $M' = 1.0$ A/m. The results are shown in Figure 3.18. An excellent agreement is found between $T_R(x)$ computed directly using Equation (3.32) and the same quantity computed indirectly via the RTP-filtering procedure summarized in Equations (3.30–3.31). Notice that the original anomaly profile $T(x)$ is skewed whereas the RTP-anomaly profile $T_R(x)$ is symmetric about $x = 0$. This shows that the RTP operation effectively removes the effect of the local geomagnetic direction from magnetics data.

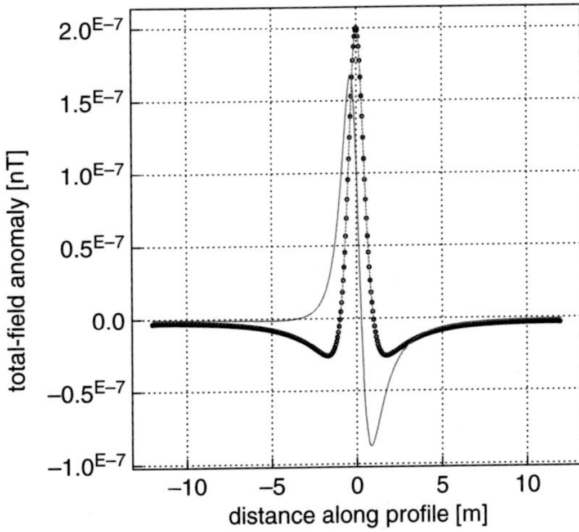

Figure 3.18 Test of the RTP filter. The lighter curve is $T(x)$ computed using Equation (3.10). The heavier curve is $T_R(x)$ computed directly using Equation (3.32). The symbols are the RTP-filtered $T_R(x)$ anomaly computed using Equations (3.30) and (3.31).

3.9 Depth rules

If the shape of the causative magnetic body is known, or can be approximated, then a simple estimate of the depth to the body can be made in terms of the width of the magnetic anomaly along a profile on the surface. A *depth rule* is readily developed for the magnetized pipeline. It is easily shown using Equation (3.32) that the depth to the pipeline is approximated by the FWHM of $T_R(x)$, which is the full width of the total-field anomaly at half its maximum value. This is illustrated in Figure 3.19. Depth rules for other target shapes are given in Sharma (1997).

3.10 Magnetic properties of rocks, soils, and buried steel objects

The magnetics method of applied geophysics is sensitive to spatial variations in the magnetization of rocks and soils. Essential overviews of the origin and properties of magnetic minerals, with relevance to environmental processes and exploration geophysics, respectively, are provided by Dekkers (1997) and Clark (1997).

Suppose a magnetic material is exposed to a weak magnetic field **H**, such as the geomagnetic field $\mathbf{B}_E/\mu_0$. The resulting induced magnetization **M** is proportional to the applied magnetic field, such that $\mathbf{M} = \chi\mathbf{H}$. The constant of proportionality is the magnetic susceptibility χ, a dimensionless quantity usually quoted in units [SI] to signify its value

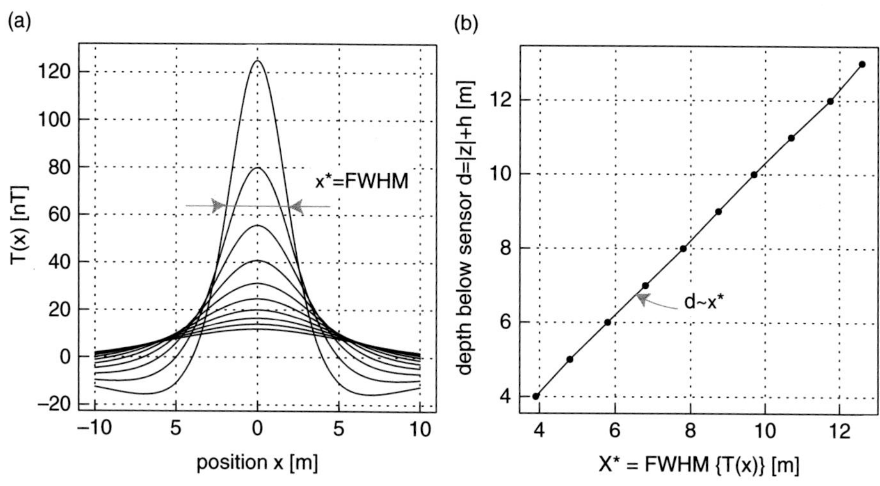

Figure 3.19 Depth rule for long magnetized pipeline, reduced to the pole. (a) Total-field anomaly T(x) for different burial depths $h = 1, 2, \ldots, 10$ m. Sensor height is 3 m above ground surface at $z = 0$. (b) Linear relationship between depth below sensor $d = |z| + h$ and full width at half-maximum (FWHM) x^* of the total-field anomaly.

when both **M** and **H** are specified in the SI units [A/m]. Susceptibility is a bulk physical property of geomaterials. Nominally non-magnetic materials that possess a slightly negative susceptibility are termed *diamagnetic* (e.q. quartz, -1.5×10^{-5} SI) while *paramagnetic* minerals possess a positive susceptibility that is usually $\sim 5 - 100 \times 10^{-5}$ SI (e.g. olivine, pyroxene) but can become as high as $\sim 1000 \times 10^{-5}$ SI (e.g. siderite). *Ferromagnetic* materials are those with much higher susceptibilities whose atomic structure becomes highly ordered under exposure to an applied magnetic field and whose resulting magnetization does not disappear after the external field is removed.

Magnetite Fe_3O_4 and maghemite γ-Fe_2O_3 are two of the major ferromagnetic minerals responsible for the bulk magnetic signature observed by magnetometers. In general, the Fe–Ti oxides play the dominant role in rock magnetism (Figure 3.20). There are other weakly magnetic minerals found in moderate abundance such as lepidocrocite γ-FeOOH, goethite α-FeOOH, and hematite α-Fe_2O_3. Natural pedogenic (soil-forming) processes involving bacteria (Fassbinder *et al.*, 1990) and fires mediate, respectively, organic and inorganic chemical reactions that convert relatively abundant weakly magnetic minerals into stronger magnetic oxides. Anthropogenic activity can also enhance and concentrate magnetic iron oxides within the soil (Tite and Mullins, 1971).

A common near-surface applied geophysics activity involves detection and characterization of anthropogenic buried ferrous metal or metal-bearing objects such as steel drums, storage tanks, unexploded ordnance (UXO), reinforced concrete, and pipelines. Such activity could be part of a larger brownfields site investigation or the buried object itself could be the focus of interest.

Iron-bearing minerals are found generally in most soils and sediments and their spatial distribution has a profound effect on the bulk magnetic properties and consequently their magnetic geophysical signatures. Strongly magnetic soils on the island of Kaho'olawe, Hawaii are derived from a tholeiitic basalt parent rock which has a magnetite content of

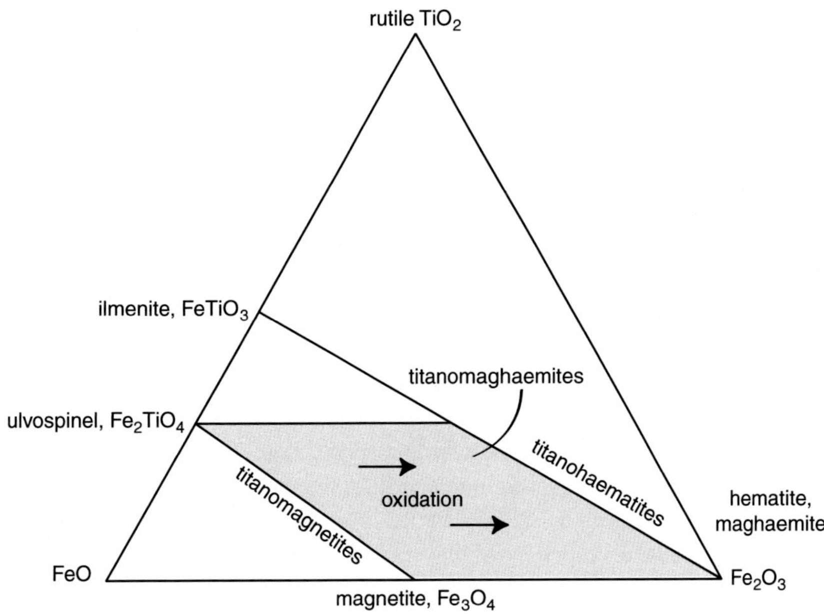

Figure 3.20 Ternary Fe–Ti–O diagram; the shaded region shows the major minerals that contribute to rock magnetism. After Clark, 1997. © Commonwealth of Australia (Geoscience Australia) 1997.

~ 40 g/kg (van Dam *et al.*, 2005). The magnetic soil generates a variable background magnetic signature that can overwhelm the signal from buried ferrous metal targets such as UXO (Butler, 2003).

To achieve the best possible interpretation of magnetics data, several considerations should be kept in mind. For example, it is essential to understand how long-term corrosion can degrade the magnetization of a buried steel object. It is also important to recognize that steel objects tend to become magnetized in several ways: along their long axis; in the direction of the prevailing geomagnetic field; or in the direction of a magnetic field that existed in the past. The magnetic field of a steel object depends more strongly on its shape than its composition or the details of its magnetic history (Eskola *et al.*, 1999).

3.11 Remanent magnetization

Remanent magnetization is the permanent magnetization that remains after an applied field is removed. The total magnetization of an iron-bearing geomaterial is the vector sum of the induced $\mathbf{M}_I$ and remanent $\mathbf{M}_R$ magnetizations. Their relative contributions are described by the Koenigsberger ratio $Q = |\mathbf{M}_R|/|\mathbf{M}_I|$, which reflects the largely unknown magnetic history of the object (Ravat, 1996) and is often in the range ~ 1–1000. Remanence is a complicated phenomenon that can render magnetic anomalies difficult to interpret in terms of the source geometry of subsurface magnetized bodies. The interested reader is referred to Clark (1997) for a complete discussion of remanent magnetization.

3.12 Image-enhancement filters

A magnetic-anomaly map, in general, contains spatially overlapping signatures from a potentially large number of causative bodies of widely varying size, shape, depth of burial, and magnetization intensity and direction. The map could be the superposition, for example, of a steel pipeline buried in a magnetic soil that, in turn, is derived from an underlying igneous parent material. The map will also contain noise typically from a number of sources including extraneous small-scale magnetic heterogeneities (clutter), terrain variations, transient magnetic fields originating from natural or artificial sources in the external environment, drift in the instrument signal, and large-scale regional anomalies caused by deeply buried geological bodies. To better reveal the geometry and magnetic properties of the causative subsurface structures, and to reduce the noise, a number of image-enhancement filtering operations have been developed by geophysicists. A filter acts on the original magnetic anomaly data to produce a second, filtered dataset that hopefully provides the interpreter with a more intuitive grasp of the subsurface bodies of primary interest.

It is well known that differentiating an arbitrary function $f(x)$ is a roughening operation. In other words, the derivative $f'(x)$ is less smooth than the original function. In this sense, examining the graph of the derivative $f'(x)$ can help to reveal fine-scale structure subtly embedded in the function $f(x)$. For this reason, many of the image-enhancement filters for magnetics data contain spatial derivatives. While derivatives are straight forward to construct directly from adjacent samples of the function $f(x)$ using a finite difference technique, in practice they can be computed more efficiently using Fourier-transform methods.

Consider $\mathrm{F}\{f\}$ to be the 2-D Fourier transform of a potential field $f(x,y)$. In the (k_x, k_y) wavenumber domain, the Fourier transforms of the horizontal spatial derivatives are

$$\mathrm{F}\left\{\frac{\partial f}{\partial x}\right\} = ik_x\,\mathrm{F}\{f\};\mathrm{F}\left\{\frac{\partial f}{\partial y}\right\} = ik_y\,\mathrm{F}\{f\}. \tag{3.32}$$

Since the potential field $f(x,y)$ obeys Laplace's equation $\nabla^2 f = 0$, the fourier transform of the vertical spatial derivative on the surface $z = 0$ can be shown to be (Grant and West, 1965)

$$\mathrm{F}\left\{\frac{\partial f}{\partial z}\right\} = \sqrt{k_x^2 + k_y^2}\mathrm{F}\{f\}. \tag{3.33}$$

The procedure for evaluating the spatial derivatives, then, is to take a 2-D Fourier transform of the function $f(x,y)$ to get $\mathrm{F}\{f\}$, and then to use Equations (3.32) and (3.33) to determine the spatial derivatives in the wavenumber domain. These expressions can then be inverse Fourier-transformed to obtain $\partial f/\partial x$, $\partial f/\partial y$, and $\partial f/\partial z$ in the spatial domain.

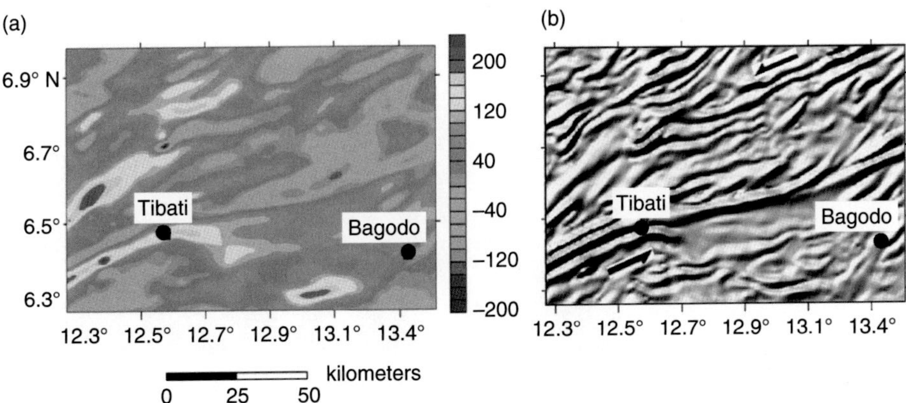

Figure 3.21 (a) Total-field anomaly map in the Adamawa region, Cameroon; (b) shaded-relief map of the total horizontal derivative, *THD*. The area is located in the Foumban shear zone. Black arrows show direction of shearing (from Noutchogwe *et al.* 2010). See plate section for color version.

The *total horizontal derivative* (*THD*) and the *analytic signal* (*AS*) are particular combinations of horizontal and vertical spatial derivatives, namely,

$$THD(x,y) = \sqrt{\left(\frac{\partial f}{\partial x}\right)^2 + \left(\frac{\partial f}{\partial y}\right)^2}; \tag{3.34}$$

$$AS(x,y) = \sqrt{\left(\frac{\partial f}{\partial x}\right)^2 + \left(\frac{\partial f}{\partial y}\right)^2 + \left(\frac{\partial f}{\partial z}\right)^2}, \tag{3.35}$$

where $f(x, y)$ represents gridded magnetic data, such as the total-field anomaly.

Maps of $THD(x, y)$ and $AS(x, y)$ are useful since they enhance the edges of subsurface magnetized bodies relative to a map of the original anomaly $f(x, y)$. An example of the use of *THD* in delineating faults and structural contacts for a hydrogeological investigation in Cameroon is shown in Figure 3.21. Furthermore, for $AS(x, y)$ it can be shown (Roest and Pilkington, 1993) that the map locations of the edges in the analytic signal correspond to the actual locations of the edges of the causative body, regardless of the direction in which it is magnetized. An application of the analytic signal to identification of kimberlite pipes in a diamond exploration context is described in Keating and Sailhac (2004).

Cooper and Cowan (2011) have introduced a *generalized derivative operator* (*GDO*) consisting of a sum of weighted spatial derivatives divided by the analytic signal,

$$GDO(x,y) = \frac{\left(\frac{\partial f}{\partial x}\sin\theta + \frac{\partial f}{\partial y}\cos\theta\right)\cos\phi + \frac{\partial f}{\partial z}\sin\phi}{\sqrt{\left(\frac{\partial f}{\partial x}\right)^2 + \left(\frac{\partial f}{\partial y}\right)^2 + \left(\frac{\partial f}{\partial z}\right)^2}}, \tag{3.36}$$

where (θ, ϕ) are parameters which weight the various spatial derivatives of the magnetic-anomaly data.

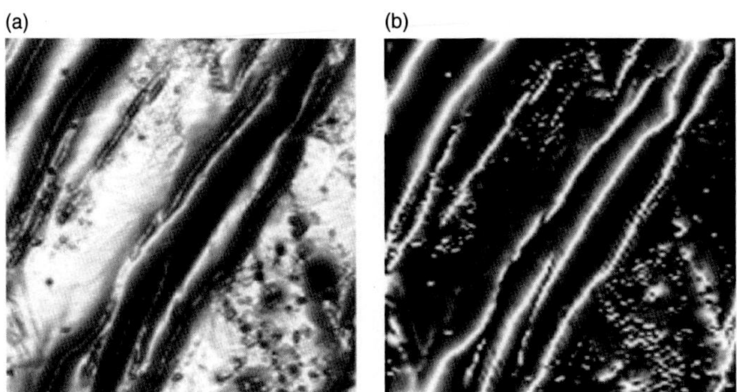

Figure 3.22 (a) Shaded-relief aeromagnetic data from the Bushveld igneous complex in South Africa; (b) the generalized derivative filtered data (from Cooper and Cowan 2011). The size of the surveyed area is 32.5 km^2.

The generalized derivative operator has been applied to the aeromagnetic data that is shown in shaded-relief format in Figure 3.22a. A geological interpretation of the normalized aeromagnetic data, acquired over part of the Bushveld igneous complex in South Africa, indicates the presence of southwest–northeast trending dikes. The *GDO*-filtered data, using $\theta = 225°$ and $\phi = 90°$, are shown in Figure 3.22b. The *GDO* filter provides a sharper image of the linear features evident in the aeromagnetic data.

3.13 Upward continuation

Some properties of filters were introduced in the previous chapter on data analysis. In general, a filter is a multiplicative process applied to either 1-D transect or 2-D gridded data that have been transformed into the Fourier wavenumber domain. The filtered data is simply the inverse Fourier transform of the product of the filter kernel with the Fourier-transformed data.

A filtering operation of particular interest is *upward continuation*, which transforms anomalies that are measured on one horizontal surface into anomalies that would have been measured had the instrument platform been located on some higher horizontal surface that is farther from the sources. The upward-continued anomaly provides additional geometric information about relatively long-wavelength sources at depth since the signatures of local, near-surface bodies tend to be attenuated by the process of upward continuation.

Upward continuation is used to filter out the effects of near-surface heterogeneities which may not be of primary geological interest. It is also used to merge the data from two magnetic surveys, such as a ground-based and an airborne survey, by transferring them to a common altitude datum.

Consider a continuous function $f(\mathbf{r})$ that is harmonic (i.e. it satisfies Laplace's equation) within some region R. From potential theory (Blakely, 1995) we know that the value of the function f at any point P inside R can be determined by the values of f and $\nabla f \bullet \hat{\mathbf{n}} = \partial f / \partial n$ on the surface S, where $\hat{\mathbf{n}}$ is the outward normal. This theorem lays the foundation for upward continuation; it states that the potential field inside a source-free region can be obtained from a knowledge of the potential field and its normal derivative on a surface enclosing the region.

The simplest form of upward continuation is transferring magnetic data from one level surface z_0 to a higher level surface $z_0 - \Delta z$, where $\hat{\mathbf{z}}$ is directed downward. In order to compute $f_U(x, y, z_0 - \Delta z)$, which is the desired result, we need the value of f everywhere on the level surface z_0. The upward continuation integral is then (Blakely, 1995; Grant and West, 1965)

$$f_U(x,y,z_0-\Delta z) = \frac{\Delta z}{2\pi}\int_{-\infty}^{+\infty}\int_{-\infty}^{+\infty}\frac{f(x',y',z_0)}{[(x-x')^2+(y-y')^2+\Delta z^2]^{3/2}}dx'dy', \tag{3.37}$$

where $\Delta z > 0$. Equation (3.37) shows that the potential field at a higher level $z_0 - \Delta z$ can be computed by integration of the potential field measured at the lower level z_0.

In practice, we can never know the potential field $f(x', y', z_0)$ *exactly* at every point (x', y') of an infinite horizontal plane. However, often we know the potential field from measurements on a regular grid of points on the plane. In that case, we can use Fourier analysis to perform the upward continuation. A very simple result is obtained if the upward continuation is carried out in the wavenumber domain.

To see this, first note that Equation (3.37) has the form of a convolution equation

$$f_U(x,y,z_0-\Delta z) = \int_{-\infty}^{+\infty}\int_{-\infty}^{+\infty} f(x',y'z_0)\Psi_U(x-x',y-y',\Delta z)dx'dy' = f(x,y,z_0)*\Psi_U \tag{3.38}$$

where

$$\Psi_U(x,y,\Delta z) = \frac{\Delta z}{2\pi}[(x-x')^2+(y-y')^2+\Delta z^2]^{-3/2} \tag{3.39}$$

can be viewed as the fundamental upward continuation operator. Recall that the Fourier transform of a convolution is equal to the product of the Fourier transforms of the two functions being convolved,

$$\mathrm{F}\{f_U\} = \mathrm{F}\{f\}\mathrm{F}\{\Psi_U\} \tag{3.40}$$

where $\mathrm{F}\{f_U\}$ is the Fourier transform of the upward-continued field. It turns out that there is a simple expression for $\mathrm{F}\{\Psi_U\}$. Equation (3.39) is re-written as

$$\Psi_U(x,y,\Delta z) = -\frac{1}{2\pi}\frac{\partial}{\partial z}\left[\frac{1}{r}\right]_{z=\Delta z} \tag{3.41}$$

so that

$$\mathrm{F}\{\Psi_U\} = -\frac{1}{2\pi}\frac{\partial}{\partial z}\mathrm{F}\left\{\frac{1}{r}\right\}_{z=\Delta z} = -\frac{1}{|k|}\frac{\partial}{\partial z}\exp(-|k|z)|_{z=\Delta z} \tag{3.42}$$

$$\mathrm{F}\{\Psi_U\} = \exp(-|k|\Delta z), \quad \text{with } \Delta z > 0. \tag{3.43}$$

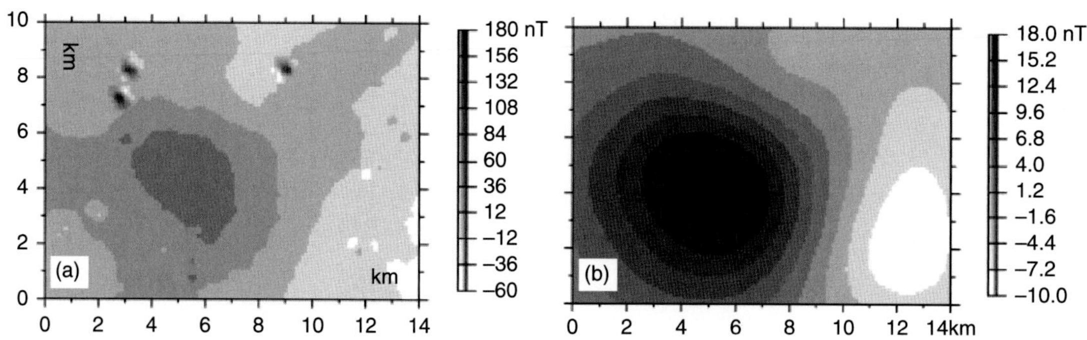

Figure 3.23 (a) Magnetic-intensity data from east China acquired on the ground; (b) upward continuation of the data to altitude 2 km. From Wang (2006).

This equation shows that the upward continuation operator in the frequency domain is just an exponential attenuation. Note that larger-wavenumber contributions (large $|k|$) are damped faster than small-wavenumber contributions (small $|k|$). This is consistent with the idea that upward continuation acts as a smoothing filter to remove effects of near-surface heterogeneities. A potential field that has been upward continued to a large altitude retains predominantly its long-wavelength, or small-wavenumber, components.

The standard procedure for upward continuation in the Fourier domain, therefore, is to take a Fourier transform of the measured potential field on the original-level surface, multiply the result by $\exp(-|k|\Delta z)$, and then take an inverse Fourier transform of the product to get the potential field on the higher-level surface.

Other techniques have been developed for upward continuation besides the Fourier approach described above. An example of upward continuation of total-intensity data using a spline-based method (Wang, 2006) is shown in Figure 3.23. Notice that the ground-based data contain a number of small-scale anomalies which are attenuated in the upward-continued data. The latter retain only the long-wavelength components of the original signal.

3.14 Euler and Werner deconvolution

Earlier in this chapter a simple rule was introduced to estimate the depth to a causative subsurface body based on the width of its total-field anomaly signature. Thompson (1982) presented the *Euler deconvolution* method to estimate depths to magnetic sources along a 1-D measurement profile. To develop the method, we first ascertain that the total-field anomaly $T = \mathbf{B}_T \cdot \hat{\mathbf{B}}_E$ for a magnetic dipole located at (x', y', z') satisfies the homogeneity equation

$$(x - x')\frac{\partial T}{\partial x} + (y - y')\frac{\partial T}{\partial y} + (z - z')\frac{\partial T}{\partial z} + NT = 0, \tag{3.44}$$

with $N = 3$ where it is assumed that the regional magnetic field does not vary over the survey area. The parameter N is termed the *structural index*. The general solution to (3.44) is

Table 3.2 Magnetic structural indices

Causative body	Equivalent source	Structural index, N
Fault	None	1
Semi-infinite sheet	Line of poles	1
Finite sheet	Line of dipoles	2
Semi-infinite thin dike	Pole	2
Finite thin dike	Dipole	3

$$T = Gr^{-N} \tag{3.45}$$

where G does not depend on r. The structural index N measures the rate of fall-off of the magnetic total-field signal with distance r from the source. In turn, the rate of fall-off is a function of the spatial extension of the source. A number of elementary source geometries besides the dipole, along with their structural indices, are given in Table 3.2.

As shown in Figure 3.24, a semi-infinite magnetized vertical sheet can be modeled as a line of poles (see Blakely, 1995 for the use of fictitious magnetic charges, or poles) while a finite vertical sheet can be modeled as a line of dipoles. A semi-infinite thin dike can be modeled as a single pole while a finite thin dike can be modeled as a single dipole. A more detailed calculation is required to show that the magnetic signature of a fault juxtaposing two differently magnetized rock units falls off as $1/r$ and hence has structural index $N = 1$. All the cases are shown in the figure.

Suppose the causative body is 2-D in extent with strike direction $\hat{\mathbf{y}}$, in which case $\partial/\partial y = 0$ in Equation (3.44). In the case that the observation point lies in the plane $z = 0$, we have

$$(x - x')\frac{\partial T}{\partial x} - z'\frac{\partial T}{\partial z} + NT = 0, \tag{3.46}$$

which upon re-arranging gives

$$x'\frac{\partial T}{\partial x} + z'\frac{\partial T}{\partial z} = x\frac{\partial T}{\partial x} + NT. \tag{3.47}$$

The objective of Euler deconvolution is to determine the unknown source location (x', z') and structural index N from profile measurements of the total-field anomaly $T(x)$ and its vertical derivative $\partial T(x)/\partial z$. The latter can be obtained, for example, from gradiometer measurements. The three unknown parameters $\{x', z', N\}$ are determined by least-squares fitting the constraint Equation (3.47) to ~ 7 data points along the profile. This is an example of a stable, over-determined non-linear inverse problem, as shown in Chapter 13. By sliding along the profile a window consisting of seven contiguous data points, a laterally continuous map of $\{x', z', N\}$ can be constructed. This yields important information on geological structure. The Euler deconvolution method has been extended to three dimensions by Reid *et al.* (1990) using gridded total-intensity $T(x, y)$ data.

The *Werner deconvolution* method (Ku and Sharp, 1983) is based on an assumption that the geological target can be represented approximately by a collection of buried, dipping,

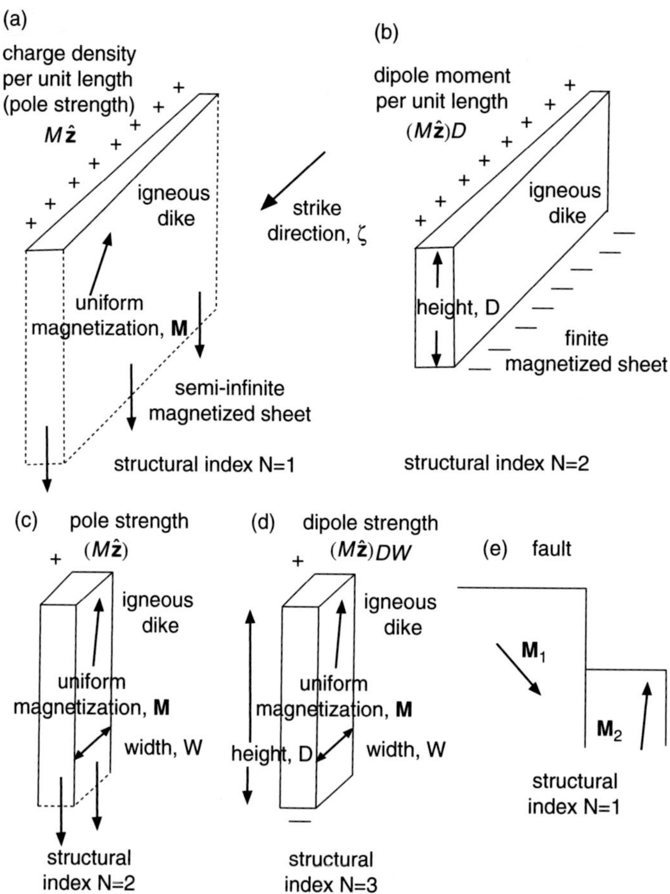

Figure 3.24 Causative bodies, equivalent sources, and structural indices: (a) semi-infinite magnetized sheet; (b) finite magnetized sheet; (c) semi-infinite thin dike; (d) finite thin dike; (e) fault. The semi-infinite bodies extend laterally to $x \rightarrow \pm\infty$ and downward to $z \rightarrow \infty$ as indicated by the dashed lines. The $+/-$ symbols indicate fictitious magnetic charges.

uniformly magnetized sheets of semi-infinite vertical extent. Suppose a profile of total-field-anomaly measurements ΔT_i for $i = 1, \ldots, N$ has been recorded across the strike of such a sheet. According to this technique, the data are divided into groups of four consecutive measurements along the profile, with each group providing an estimate of the source location, depth, dip, and magnetization. This is straightforward since there exists a simple analytic solution for the magnetic field of a uniformly magnetized, dipping sheet. Groups of consecutive measurements are treated as a window sliding across the profile. When the source location and dip inferred from each group of four data points are plotted in cross-section, the individual estimates cluster around the actual location and dip of the sheet-like body. The geology of a more complex source region cannot be modeled as a single sheet. In this case, the Werner-deconvolution procedure provides a representation of the causative structure as a collection of dipping magnetized sheets, ready for further geological interpretation.

Later, in Chapter 5 on the self-potential method, it will be shown that information on the depths and geometry of causative sources can be revealed using a powerful method of analysis of potential fields based on the continuous waveform transform.

3.15 Illustrative case histories

Example. UXO magnetic survey.

The remediation of UXO-contaminated lands is a pressing problem in former conflict zones worldwide. The military must also clear UXO from practice bombing ranges on bases that are scheduled to be transferred to civilian use. Figure 3.25 shows the total-field-anomaly map from a magnetometer survey conducted at a live UXO site (Butler, 2003). The magnetic responses from individual ferrous-metal targets have amplitudes of up to ±150 nT and are clearly resolved. The measurements are made at ultra-high spatial resolution with line spacing ~ 0.25 m and station spacing ~ 0.1 m. Many of the target responses assume the characteristic asymmetry pattern of a non-vertical dipole. The consistent orientation of the asymmetry could reflect the declination and inclination of the ambient geomagnetic field, which is constant over the extent of the surveyed area. The

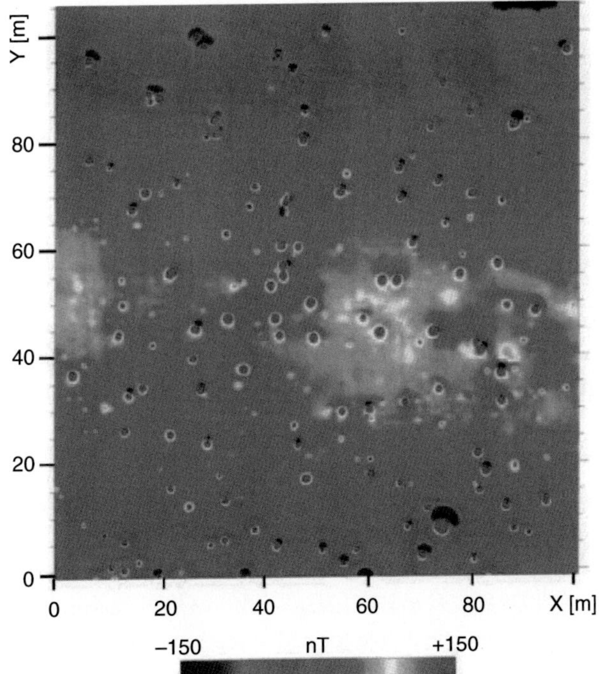

Figure 3.25 Total-field-anomaly map over a live UXO site, Jefferson Proving Ground, Indiana. From Butler (2003). See plate section for color version.

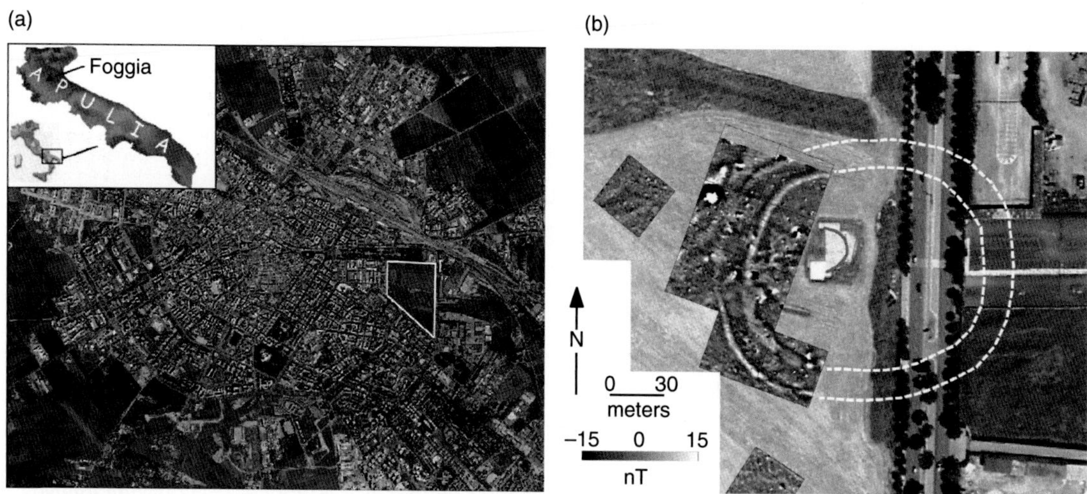

Figure 3.26 (a) Aerial view of Foggia, Italy showing magnetometer survey site. (b) Magnetic signature of a double-ditch structure of archaeological significance. After Ciminale and Gallo (2008).

asymmetric magnetic signatures may also be caused by a preferred alignment of the long axes of ellipsoidal UXO targets.

Example. Urban archaeomagnetics.

The urban environment is a challenging environment in which to conduct geophysical surveys due to the abundance of cultural noise. Ciminale and Gallo (2008) describe a magnetic survey on the grounds of a disused horse racetrack within the city limits of Foggia, Italy (see Figure 3.26a). The purpose of the survey is to detect and describe remnants of the Neolithic human occupation known to have occurred at this site. A characteristic double-ditch structure is clearly evident in the processed magnetograms shown in Figure 3.26b. The ditch has a significant magnetic signature of ±15 nT due to the contrast in magnetization between the host soil and the fill material. The Neolithic ditch structure appears to extend beneath the present-day road, as shown by the dashed lines in the figure.

Problems

1. Show that the magnetic field of a steel drum is given by $\mathbf{B}(P) = \frac{\mu_0 M}{4\pi r^5}\left[3x_P(z_P - h)\,\hat{\mathbf{x}} + \{2(z_P - h)^2 - x_P^2\}\,\hat{\mathbf{z}}\right]$, assuming that the steel drum is represented by a magnetic dipole located at the source point $P(x_P, 0, z_P)$. The dipole is of strength M and is pointing vertically downward in the $+\,\hat{\mathbf{z}}$ direction. The magnetometer is located at observation point $Q(0, 0, h)$.

2. Prove that the x-component of the magnetic field from a z-directed magnetic dipole equals the z-component of an x-directed magnetic dipole.
3. Use elementary mechanics to show that the angular velocity of the precession of the spinning top is $\omega_P = mgr/L$, where $L = I\omega$ is the angular momentum, I is the moment of inertia of the top, and r is the distance from the origin to the center of mass of the top. The top is spinning on its axis with rotation vector $\boldsymbol{\omega}$. Hint: the rate of change of angular momentum of the top is given by the cross-product $d\mathbf{L}/dt = \boldsymbol{\omega}_P \times \mathbf{L}$.
4. Use the table of Gauss coefficients to show that the geocentric dipole ($n = 1$ terms only) points toward latitude 79.1° S and longitude ϕ=108.9° E. Describe the two geographic regions in which the geomagnetic poles reside.
5. Show that the total-field anomaly due to a buried magnetic target is given by $T = \mathbf{B}_T \cdot \hat{\mathbf{B}}_E$, where $\mathbf{B}_T$ is the target field and $\mathbf{B}_E$ is the geomagnetic field. Assume that the condition $|\mathbf{B}_T| << F$ holds. Hint: neglect second-order terms involving $T = \mathbf{B}_T \cdot \hat{\mathbf{B}}_E$ and use the binomial approximation $\sqrt{1+\varepsilon} \approx 1 + \varepsilon/2$ which is valid for small $\varepsilon << 1$.
6. The across-strike total-field-anomaly $T(x)$ of a magnetized pipeline is given by Equation (3.10) where $\mathbf{F}_{2D} = [X\cos\alpha + Y\sin\alpha]\,\hat{\mathbf{x}} + Z\hat{\mathbf{z}}$ is the 2-D geomagnetic field vector, and it is assumed that the pipeline magnetization is purely induced. The parameter α is the azimuth of the x-directed profile with respect to geomagnetic north. Prove analytically that the depth to the target is roughly equal (within ~ 4%) to the FWHM of the $T(x)$ anomaly, after reduction to the pole.

4 Electrical resistivity method

In this chapter the electrical resistivity method, a mainstay of near-surface applied geophysics for many decades (Keller and Frischknecht, 1966; Bhattacharya and Patra, 1968) is described. The technique has enjoyed a resurgence in popularity since the mid 1990s (Loke, 1999; Dahlin, 2001; Zonge *et al.*, 2005) due to rapid and impressive advancements in data acquisition, forward modeling, and inversion capabilities.

The fundamental steps involved in the resistivity method may be outlined as follows. An electric current I [amperes, A] is directly injected into the ground through a pair of electrodes and the resulting voltage V [volts, V] is measured between a second pair of electrodes. The impedance $Z = V/I$ [V/A] of the Earth is formed; it is the ratio of the voltage output V measured at the potential electrodes to the current input I at the current electrodes. The impedance is then transformed into an apparent resistivity ρ_a [ohm-meters, Ωm] which is an intuitively understood indicator of the actual underlying electrical resistivity structure $\rho(\mathbf{r})$ of the Earth, where $\mathbf{r}$ is the position vector. Different arrangements of the electrodes permit the apparent resistivity to be determined at different depths and lateral positions. A map of the apparent resistivity plotted at these locations is termed a *pseudosection* (Loke, 1999). The pseudosection is then inverted to obtain a two- or three-dimensional (2-D or 3-D) resistivity section $\rho(\mathbf{r})$ of the ground. Finally, a geological interpretation of the resistivity section is performed that incorporates, as far as possible, prior knowledge based on outcrops, supporting geophysical or borehole data, and any information gained from laboratory studies of the electrical resistivity of geological materials (see Table 4.1).

Table 4.1 Resistivity of common geological materials

Geomaterial	Resistivity [Ωm]
Clay	1–20
Sand, wet to moist	20–200
Shale	1–500
Porous limestone	$100–10^3$
Dense limestone	$10^3–10^6$
Metamorphic rocks	$50–10^6$
Igneous rocks	$10^2–10^6$

4.1 Introduction

The electrical resistivity method has a long history in applied geophysics, including the pioneering work in 1912 by Conrad Schlumberger of France. A few years earlier than that, Swedish explorationists had experimented with locating conductive bodies by moving around a first pair of potential electrodes while keeping a second pair of current electrodes in a fixed location (Dahlin, 2001).

The two case histories described below introduce the reader to examples of recent usage of the resistivity method. The first example is a study of a hydrogeological problem at a human-impacted site of historical significance. The second example relates to the use of resistivity data for imaging hazardous liquid contaminants at a nuclear-waste site in the USA.

Example. Investigation of an historic WWII site.

The D-Day invasion site at Pointe du Hoc, France (Figure 4.1a) is an important WWII battlefield and remains today a valuable cultural resource but its existence is jeopardized by the risk of potentially devastating cliff collapses. The resistivity method was used there to study the effect of groundwater infiltration on the cliff stability. The great amount of buried steel, concrete, and void spaces at the site renders hydrogeological interpretation of the resistivity data challenging.

A resistivity profile was acquired by laying out a line of electrodes passing within a few meters of a 155-mm gun casemate (Figure 4.1b). The resistivity section shown in

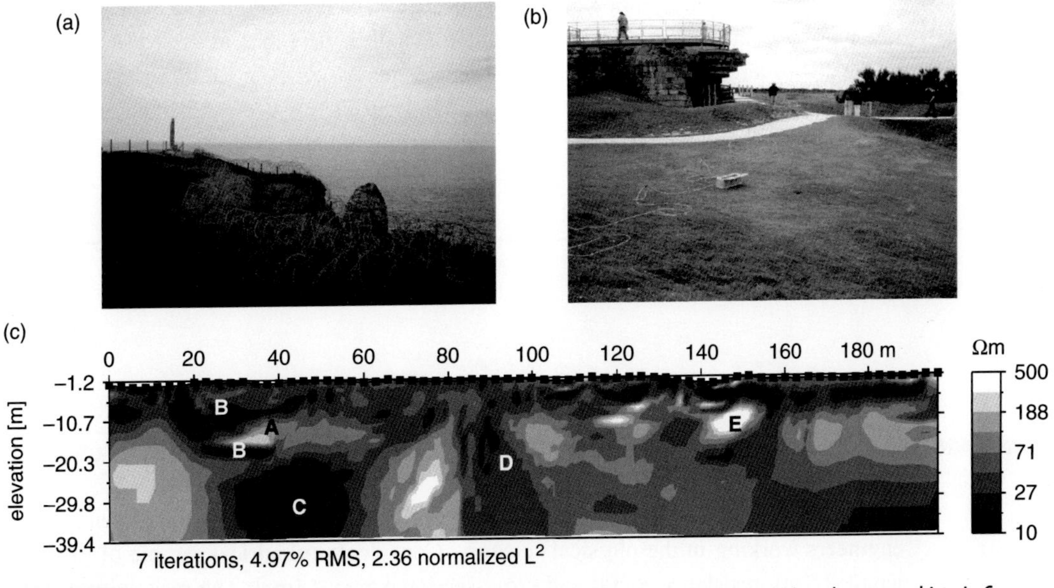

Figure 4.1 (a) WWII battlefield, Pointe du Hoc, France. (b) Resistivity data acquisition passing close to an historic German fortification. (c) Resistivity section showing natural and cultural subsurface features. Labels A–E described in the text. After Everett *et al.* (2006).

Figure 4.1c was constructed and interpreted by Everett *et al.* (2006), as follows. The small zone of high resistivity (A) showing at ~ 15 m depth, with the low-resistivity halo (B) surrounding it, is the geophysical signature of the casemate and its foundations. The larger, deeper low-resistivity zone (C) extending from 25 to 60 m along the profile is likely of geological origin, perhaps a zone of groundwater accumulation. The vertical zone of high conductivity (D) at ~ 90 m is not immediately associated with any known cultural features; it is interpreted as a vertical conduit for groundwater that flows from substantial depths to the surface. The highly resistive zone (E) at distance 145–150 m along the profile is explained by a large slab of buried concrete.

Example. Investigation of the Hanford nuclear site.

Discharge of millions of liters of hazardous liquid electrolytes since the 1940s has occurred at the Hanford nuclear facility in eastern Washington State, USA. Subsurface resistivity imaging of the resulting contaminant plumes in the vadose zone beneath the site remains a challenging task due to the presence of storage tanks, pipelines, metal fences, and other cultural infrastructure. To directly access the deep vadose region beneath the near-surface zone of cultural noise, Rucker *et al.* (2010) took advantage of the large number of existing steel-cased monitoring wells at the site. They utilized the steel casings as long cylindrical electrodes in a novel well-to-well (WTW) pole–pole configuration.

A total of 110 steel casings from wells with lengths up to 90 m were used as electrodes in the WTW survey. The resulting voltage measurements were of reasonable quality, with only ~ 10% of the ~ 12 000 readings being rejected due to high repeat errors. The result of a 3-D inversion of data from 87 centrally located wells is shown in Figure 4.2. Two major low-resistivity anomalies can be identified in this plan-view map at depth 1.4 m. The first, in the lower-left region of the survey, corresponds to the area of a historical non-point source dispersal of nitrate-contaminated (1–2 mol/L) wastewater. The second low-resistivity anomaly occurs in the vicinity of leaking storage tanks T-103 and T-106. The tanks are documented to have discharged into the vadose zone a volume of liquid contaminant exceeding 440 kL.

4.2 Fundamentals

The resistivity technique is founded on basic principles familiar to all scientists and engineers working in the physical sciences. Consider a cylindrical sample of material of length L [m], resistance R [Ω] and cross-sectional area A [m^2]. The resistivity ρ [Ωm] is a material property equal to $\rho = RA/L$, see Figure 4.3. The spatially variable resistivity

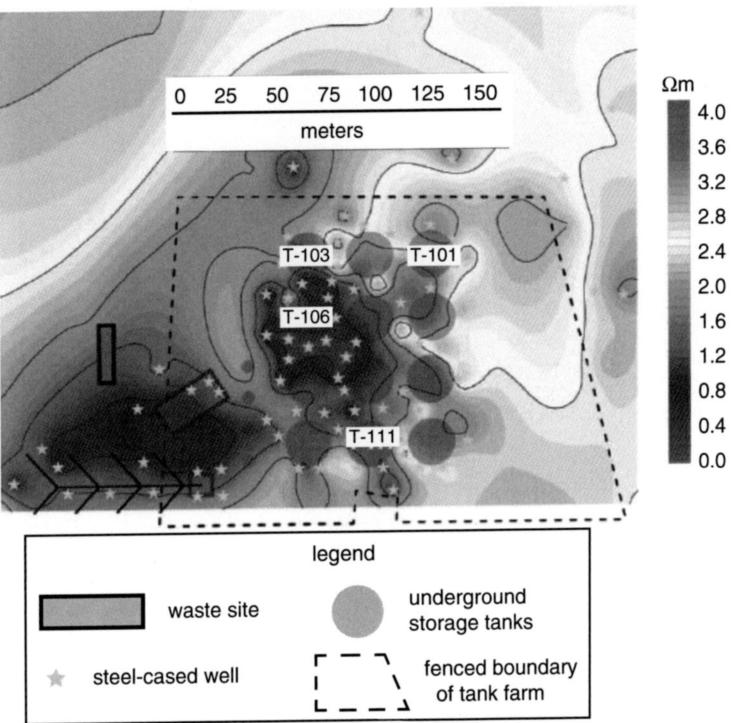

Figure 4.2 WTW resistivity inversion at the Hanford nuclear facility; depth slice at 1.4 m. After *Rucker et al.* (2010). See plate section for color version.

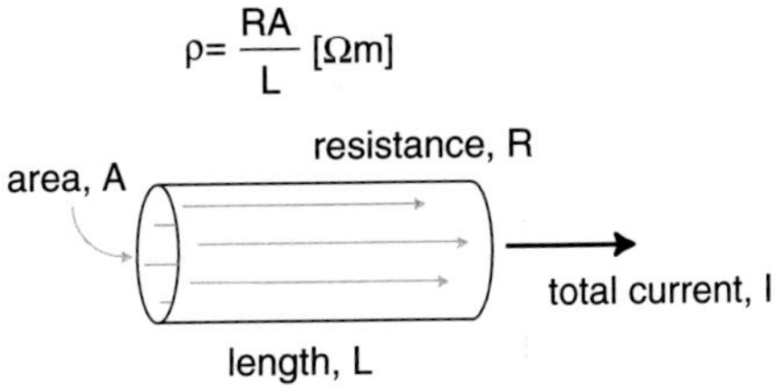

Figure 4.3 Definition of resistivity ρ.

$\rho(\mathbf{r})$ of the subsurface is the physical property that is sensed by the resistivity method. The reciprocal of the resistivity is the electrical conductivity $\sigma = 1/\rho$, which by convention is the preferred quantity used in the electromagnetic and ground-penetrating radar geophysical techniques (see Chapters 8 and 9). Electrical conductivity is a measure of the ability of a material to sustain long-term electric current flow. Thus, electric current can flow readily in low-resistivity zones and is weak or absent in high-resistivity zones.

A general scenario is shown in Figure 4.4, in which a battery is connected to two electrodes that serve as a current source/sink pair. The electric current streamlines (line segments) and equipotentials (contours) are displayed in the figure for a current injection of $I = 1$ A and uniform resistivity $\rho = 1\ \Omega$m.

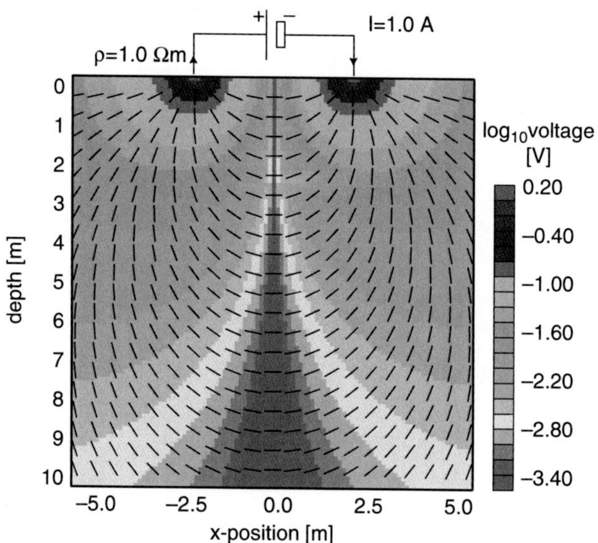

Figure 4.4 Potential and streamlines of electric current for a point source and a point sink of current. See plate section for color version.

To understand how the resistivity method is used to estimate Earth resistivity, first recognize that the subsurface current density **J** is related to the electric field **E** by Ohm's law $\mathbf{J} = \sigma \mathbf{E}$ so that

$$\mathbf{E} = \frac{\mathbf{J}}{\sigma} = \rho \mathbf{J} = \frac{I\rho\, \hat{\mathbf{r}}}{4\pi r^2}. \tag{4.1}$$

Ohm's law, stated in Equation (4.1), is nothing more than the generalization to continuous media of the familiar law as it applies to a simple resistive circuit, $V = IR$, where V is voltage, I is current, and R is resistance.

Next, consider an electric current I injected, at the origin of a spherical coordinate system, into a hypothetical whole-space of uniform resistivity ρ. Suppose the return electrode is placed at infinity. The situation is depicted in Figure 4.5. In the vicinity of the injection point, the current will spread out symmetrically in all three dimensions. At point P at distance r from the injection point, using Equation (4.1) the current density **J** is

$$\mathbf{J} = \frac{I\, \hat{\mathbf{r}}}{4\pi r^2}, \tag{4.2}$$

where $4\pi r^2$ is the area of a spherical surface of radius r. The numerator of Equation (4.2) expresses the magnitude and direction of the current at point P while the denominator expresses the cross-sectional area through which the current uniformly flows.

What is the voltage V measured at observation point P in Figure 4.5? Voltage is *defined* as the work done by the electric field **E** in moving a test charge from infinity to point P. Work is defined by the product of work and distance, or in our case the line integral

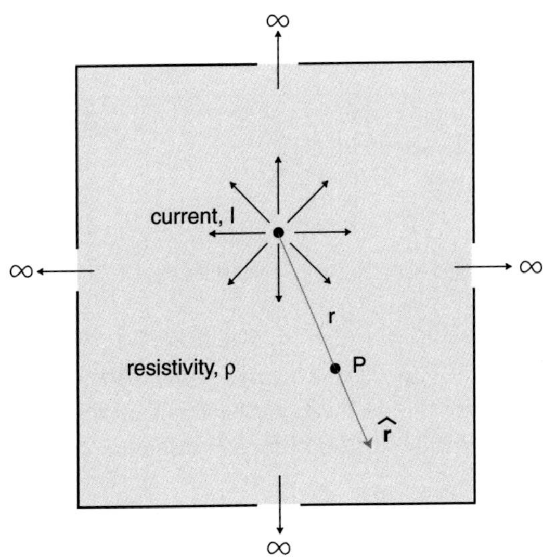

Figure 4.5 Current injection into a wholespace of uniform resistivity ρ.

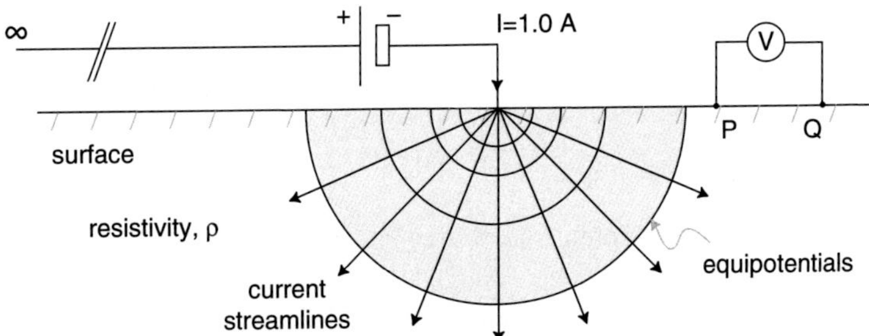

Figure 4.6 Voltage measured between points P and Q for a point source of electric current injected into a halfspace of uniform resistivity ρ.

$$V = \int_C \mathbf{E} \cdot d\mathbf{s}, \tag{4.3}$$

where C is any path from infinity terminating at point P. Hence the voltage at P is

$$V = \int_r^\infty \mathbf{E} \cdot d\mathbf{r} = \int_r^\infty \frac{I\rho}{4\pi r^2} dr = \frac{I\rho}{4\pi r}. \tag{4.4}$$

Now suppose that the injection point is located on the surface of a halfspace representing the Earth, as shown in Figure 4.6.

The electric current, which cannot flow through the non-conducting air, flows radially outward through a *hemisphere* of radius r and surface area $2\pi r^2$. Hence, the current density

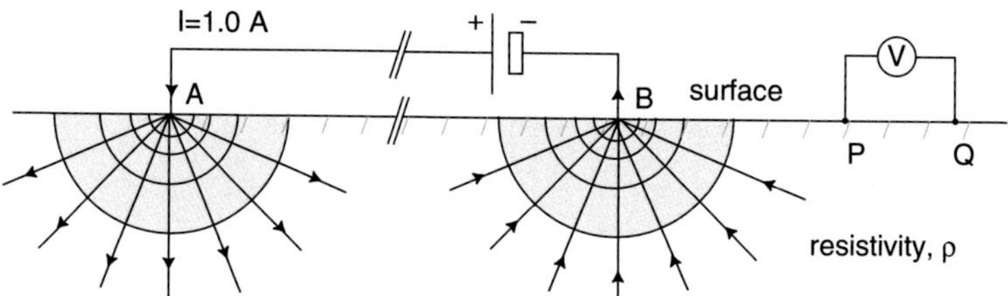

Figure 4.7 Voltage measured between points P and Q for a point source A and point sink B of electric current.

in this case is $\mathbf{J} = I\hat{\mathbf{r}}/2\pi r^2$ so that, using Equations (4.1) and (4.4), the voltage measured at point P is $V = I\rho/2\pi r_P$ where r_P is the distance from the current source to the potential electrode P. This is the basic equation of resistivity. The voltage measured across the terminals P and Q of the voltmeter in Figure 4.6 is the difference

$$V_{PQ} = V_P - V_Q = \frac{I\rho}{2\pi}\left[\frac{1}{r_P} - \frac{1}{r_Q}\right], \tag{4.5}$$

where r_Q is the distance from the current source to the potential electrode Q.

Apparent resistivity. Equation (4.5) is derived under the assumption that the Earth has a uniform resistivity ρ. In reality, the resistivity distribution inside the Earth is heterogeneous. We can re-arrange Equation (4.5) to solve for an *apparent resistivity* ρ_a

$$\rho_a = \frac{2\pi V_{PQ}}{I}\left[\frac{1}{r_P} - \frac{1}{r_Q}\right]^{-1} = \kappa Z, \tag{4.6}$$

which is interpreted to be the resistivity that would have been measured if the Earth were in fact homogeneous. Notice that the apparent resistivity can be written as a product of the measured Earth impedance $Z = V/I$ and a *geometric factor* κ that depends only on the arrangement of the current and potential electrodes. In the configuration shown in Figure 4.6, which is known as a *pole–dipole* arrangement, the geometric factor is simply

$$\kappa = 2\pi\left[\frac{1}{r_P} - \frac{1}{r_Q}\right]^{-1}. \tag{4.7}$$

Although the pole–dipole arrangement has received attention from near-surface geophysicists, a more general exploration scenario is the arbitrary four-electrode configuration shown in Figure 4.7 which includes a point source and the return point sink of current. The geometric factor is $\kappa = 2\pi/[1/r_{AP}–1/r_{AQ} –1/r_{BP} + 1/r_{BQ}]$.

Four-electrode arrays. Historically, a number of four-electrode configurations have proven popular for a wide range of applications of geophysics. As we describe below, computer-controlled configurations of hundreds of electrodes are now in routine use (Loke, 1999). Nevertheless it remains worthwhile to briefly discuss a few of the traditional four-electrode configurations (Figure 4.8) in order to gain insight into the capabilities of the resistivity method and to explore the advantages and disadvantages of the various electrode configurations in terms of depth penetration, lateral resolution, ease of deployment, and

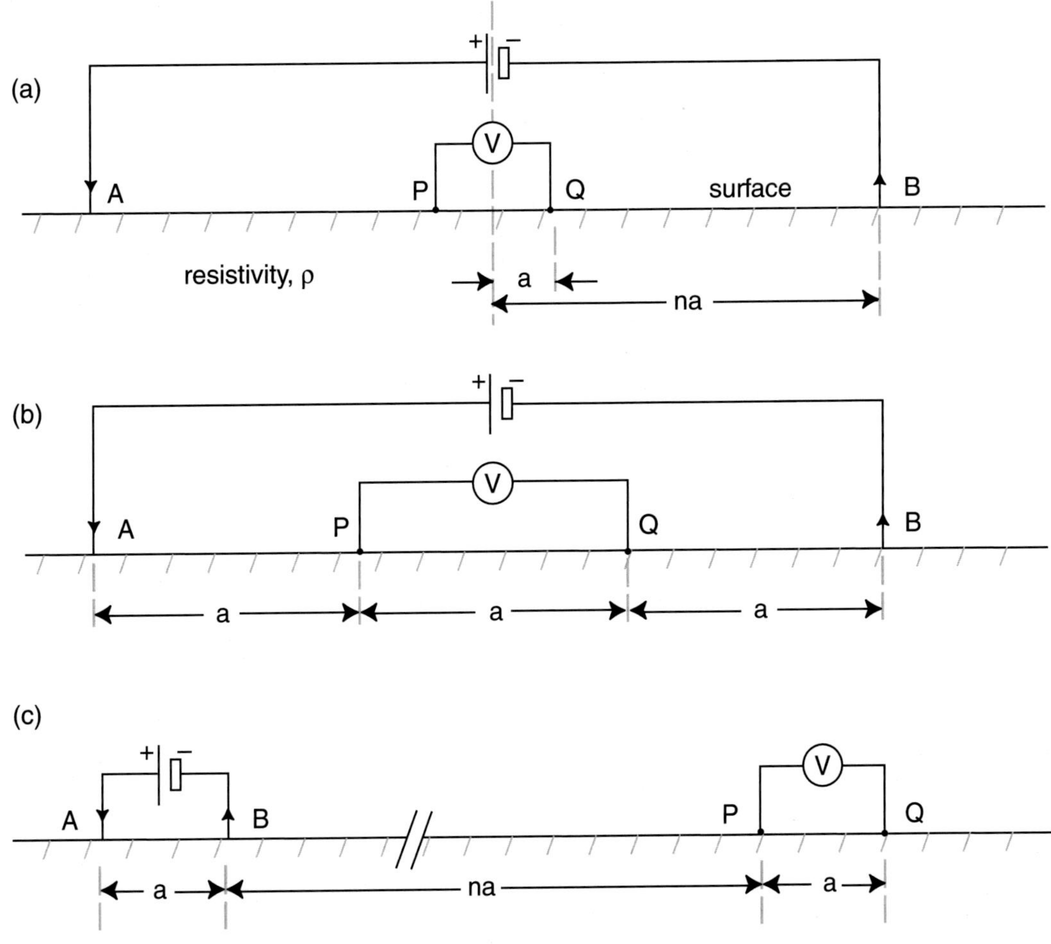

Figure 4.8 Traditional four-electrode configurations: (a) Schlumberger; (b) Wenner; (c) dipole–dipole.

signal-to-noise ratio. A review of these and many other popular electrode configurations is given in Zonge *et al.* (2005).

The Schlumberger array (Figure 4.8a) is designed for *sounding*, that is, determining the Earth resistivity depth profile $\rho(z)$ beneath a single location. The potential electrodes PQ are kept centered at a fixed location with constant separation $2a$. The current electrodes AB are centered at the same location but voltage readings are made as the separation between them is expanded about the common midpoint. In this way, apparent resistivity is determined as a function of the current-electrode separation. It is traditional to display Schlumberger array data as a graph of the form $\rho_a(AB/2)$, where $AB/2$ is one-half of the current-electrode separation. The geometric factor for the Schlumberger array is,

$$\text{Schlumberger}: \kappa = (n-1)(n+1)\pi a/2. \tag{4.8}$$

A Schlumberger sounding can achieve excellent depth penetration with sufficiently large AB separations. The array has limited lateral resolution however, as it is designed for vertical sounding. The Schlumberger array is cumbersome in the field since its traditional deployment requires lengthy wire connections that must be re-positioned for each measurement. The signal-to-noise ratio is moderate to good. The voltage reading is taken in the middle of the array which seems to indicate that a good signal level should be achieved. However, as shown in Figure 4.4, voltages are generally low at the midpoint between the injection and withdrawal electrodes, at least over a uniformly resistive Earth.

The Wenner array (Figure 4.8b) is designed for lateral profiling to determine the Earth resistivity $\rho(x)$ at a roughly constant depth of penetration. There is a fixed separation of a between adjacent electrodes, with the potential electrodes PQ placed inside the current electrodes AB as in the Schlumberger array. Apparent resistivity is determined as the array is moved along a lateral profile. It is easy to see that the geometric factor for the Wenner array is

$$\text{Wenner: } \kappa = 2\pi a. \tag{4.9}$$

The penetration depth of the Wenner array depends on the spacing a; the larger its value, the deeper the penetration. In simple terms, as the spacing between the injection and withdrawal electrodes increases, electric current is driven deeper into the subsurface. The Wenner array is quite effective at mapping lateral contrasts in resistivity within the depth of penetration. The array is moderately easy to deploy as the trailing electrode can be leapfrogged to the front as the configuration is advanced along the profile. This means that only one electrode movement is required per measurement. Signal-to-noise ratio is generally good since the potential electrodes PQ are located in the central part of the array and, unlike those in the Schlumberger array, are relatively widely spaced for a given current-electrode separation AB.

The dipole–dipole configuation is shown in Figure 4.8c. The current electrodes AB and the potential electrodes PQ have the same spacing a but the two pairs are widely separated by a distance na, where $n >> 1$. The geometric factor for the dipole–dipole array is

$$\text{dipole–dipole: } \kappa = \pi n(n+1)(n+2)a. \tag{4.10}$$

The dipole–dipole array offers advantages of both Schlumberger depth sounding and Wenner lateral profiling. However, the signal-to-noise ratio deteriorates at large values of n and the voltage measurements across the electrodes PQ are susceptible to distortion by small-scale, near-surface heterogeneities.

4.3 Sensitivity functions

Measurements made using the resistivity technique are sensitive to spatial averages of the near-surface electrical resistivity distribution. The details of the averaging process depends on the type of array used. Furman *et al.* (2003) have examined the spatial sensitivity of standard four-electrode arrays, plus that of a non-standard, partially overlapping array. A sensitivity function S is defined as the magnitude of the perturbation in the voltage

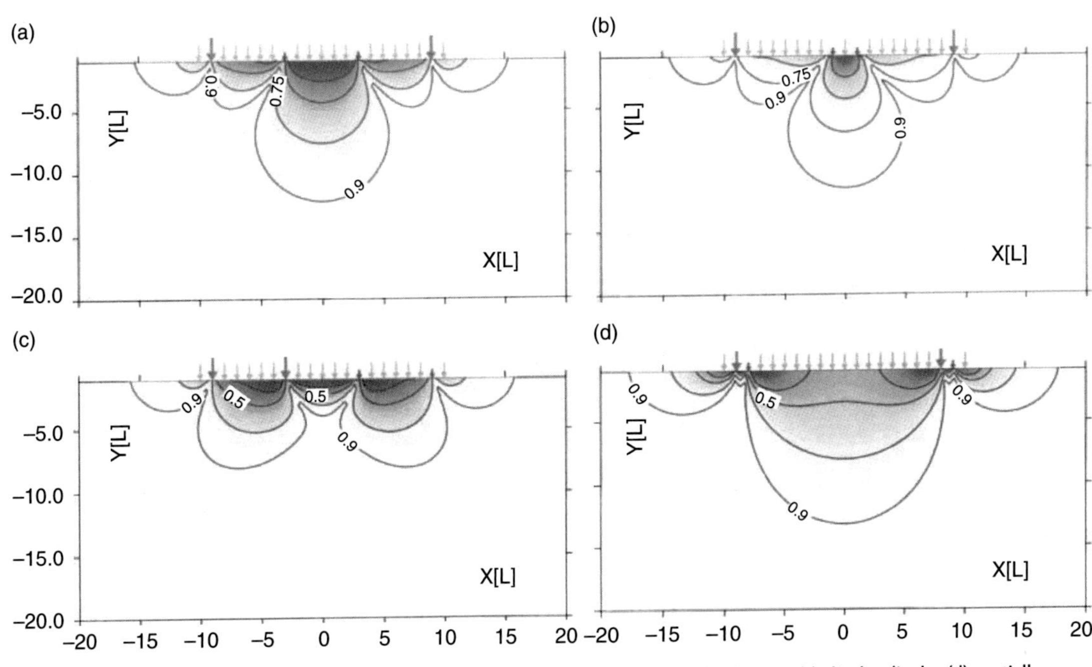

Figure 4.9 Sensitivity functions for four-electrode arrays: (a) Wenner; (b) Schlumberger; (c) dipole–dipole; (d) partially overlapping. Larger, bold arrows show current injection and withdrawal points; smaller, bold arrows show potential measurement points. After Furman *et al.* (2003).

measurement due to a small perturbation in the subsurface resistivity distribution. For this purpose, Furman *et al.* (2003) use a 2-D analytic solution for a small buried cylindrical heterogeneity of radius L and they slightly perturb the resistivity of the cylinder. The cumulative sensitivity function $CS(x, y)$ corresponds to the summation of the sensitivities due to perturbations in voltage measurements caused as the cylindrical heterogeneity is placed at a number of locations in the subsurface.

Plotted in Figure 4.9 is the cumulative sensitivity function $CS(x, y)$ for the four array types analyzed by Furman *et al.* (2003). The plots show contours of $CS(x, y)$ at the 25, 50, 75, and 90% levels. The intepretation of the 90% level, for example, is that 10% of the measurement sensitivity of the array lies outside the 90% contour. In other words, these plots give a good indication of the region of the subsurface to which a given measurement array is sensitive. The Wenner array (Figure 4.9a), for example, is sensitive to the ground mainly beneath the center of the array but has moderate sensitivity across the entire array, which confirms its utility as a lateral profiling configuration. In Figure 4.9b it can be seen that the Schlumberger array, as expected for a vertical sounding configuration, is most sensitive to the near-surface immediately below the potential electrodes and less sensitive elsewhere. As shown in Figure 4.9c, the dipole–dipole array is most sensitive to the regions immediately beneath the current and potential electrodes, with less sensitivity to the intervening region between the two dipoles. The area enclosed by the 90% contour is small, as expected, since it is well known that the dipole–dipole array generally has poor signal-to-noise ratio and high sensitivity to near-surface heterogeneities. As shown by the large size of the 90% contour in Figure 4.9d, the non-standard partially overlapping configuration offers uniformly high sensitivity across a wide and deep region beneath the entire array.

4.4 Multi-layer models

The above discussion outlined the formulas necessary to compute voltages that would be recorded over a homogeneous Earth. In the case of multiple layers of uniform resistivity, the calculations are more involved. Generally, the electric potential $\varphi(\mathbf{r})$ inside a source-free region of the Earth obeys the differential equation

$$\nabla\cdot(\sigma\nabla\varphi) = 0; \tag{4.11}$$

where $\sigma(\mathbf{r}) = 1/\rho(\mathbf{r})$ is the spatially variable electrical conductivity. Within each uniform layer of a layered medium, Equation (4.11) reduces to the Laplace equation $\nabla^2\varphi = 0$. For point-source excitation, we can use 2-D cylindrical coordinates (r, z) with azimuthal symmetry. The general solution to the Laplace equation in that case is (Bhattacharya and Patra, 1968)

$$\varphi(r,z) = \int_0^\infty [A(\lambda)\exp(-\lambda z) + B(\lambda)\exp(+\lambda z)]J_0(\lambda r)d\lambda, \tag{4.12}$$

where $J_0(\lambda r)$ is the zeroth-order Bessel function. A special case occurs for a uniform halfspace of resistivity ρ; the primary potential φ_P due to a single point of current I injected at the origin $(r, z) = (0,0)$ is

$$\varphi_P(r,z) = \frac{I\rho}{2\pi}\int_0^\infty \exp(-\lambda z)J_0(\lambda r)d\lambda. \tag{4.13}$$

Using the Bessel function identity

$$\frac{1}{R} = \int_0^\infty \exp(-\lambda z)J_0(\lambda r)d\lambda, \tag{4.14}$$

the primary potential reduces to the expression that we derived earlier, namely,

$$\varphi_P(r,z) = \frac{I\rho}{2\pi R}, \tag{4.15}$$

with $R = \sqrt{r^2 + z^2}$.

The formula for the apparent resistivity ρ_a over a two-layer Earth is obtained by writing the general solution for $\varphi(r, z)$ in each uniform layer and then matching fundamental boundary conditions at the layer interfaces. The solutions in the first and second layers are, respectively,

$$\varphi_1(r,z) = \frac{I\rho_1}{2\pi} + \int_0^\infty \Big[(1 + A_1)\exp(-\lambda z) + B_1\exp(+\lambda z)\Big]J_0(\lambda r)d\lambda; \tag{4.16}$$

$$\varphi_2(r,z) = \frac{I\rho_1}{2\pi}\int_0^\infty A_2\exp(-\lambda z)J_0(\lambda r)d\lambda; \tag{4.17}$$

which have the form of a sum of the primary $\varphi_P = I\rho_1/2\pi R$ and a secondary potential. Notice that Equation (4.17) does not include a term of the form $\exp(+\lambda z)$ since the potential should vanish at great distance from the source, $z \to +\infty$. The boundary conditions are the continuity of normal electric current at $z = 0$ and the continuity of tangential electric field and normal current at the layer interface, $z = h_1$. These conditions are

$$\frac{\partial \varphi_1}{\partial z}\Big|_{z=0} = \frac{-I\rho_1\partial(R)}{2\pi R}\Big|_{z=0}; \tag{4.18}$$

$$\varphi_1\big|_{z=h_1} = \varphi_2\big|_{z=h_1}; \tag{4.19}$$

$$\frac{1}{\rho_1}\frac{\partial \varphi_1}{\partial z}\Big|_{z=h_1} = \frac{1}{\rho_2}\frac{\partial \varphi_2}{\partial z}\Big|_{z=h_1}. \tag{4.20}$$

The three boundary conditions (4.18–4.20) are sufficient to determine the unknown coefficients (A_1, B_1, A_2). Applying the boundary conditions results in the solution at the surface $z = 0$,

$$\varphi(r) = \frac{I\rho_1}{2\pi}\int_0^\infty k_{12}(\lambda)J_0(\lambda r)d\lambda \tag{4.21}$$

where

$$k_{12}(\lambda) = \frac{1 - u_{12}\exp(-2\lambda h_1)}{1 + u_{12}\exp(-2\lambda h_1)} \tag{4.22}$$

and

$$u_{12} = \frac{\rho_1 - \rho_2}{\rho_1 + \rho_2} \tag{4.23}$$

is interpreted as a reflection coefficient.

Electric-current streamlines for two cases, a resistive layer over a conducting halfspace, and a conducting layer over a resistive halfspace, are shown in Figure 4.10. It is readily observed that the electric current patterns in the underlying medium 2 are similar for both cases. In the overlying medium 1, the horizontal electric current flow is better developed in the conductive case (Figure 4.10b) than in the resistive case (Figure 4.10a). This accords with the definition of electrical conductivity as a material property that measures its capability to sustain long-term, or steady-state, electric-current flow.

For the four-electrode Schlumberger configuration deployed over a two-layer Earth, the apparent resistivity formula is the sum of terms like (4.21), namely,

$$\rho_a = \frac{I\rho_1}{2\pi}\int_0^\infty k_{12}(\lambda)[J_0(\lambda r_1) - J_0(\lambda r_2) - J_0(\lambda r_3) + J_0(\lambda r_4)]d\lambda. \tag{4.24}$$

Generalizing this procedure to the case of three layers, the reader should be able to derive the corresponding formula

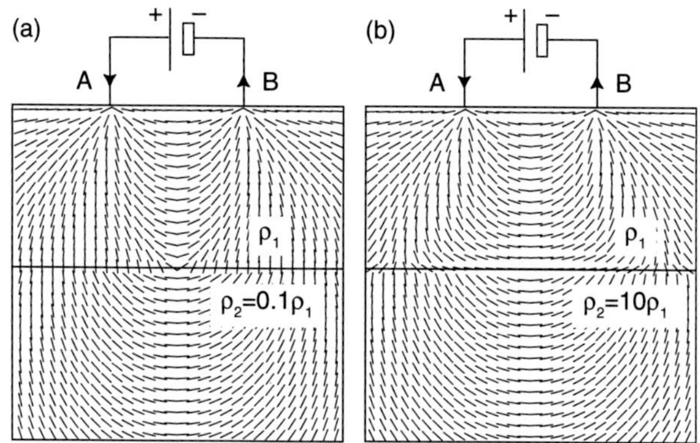

Figure 4.10 Electric-current streamlines for current source and sink over a two-layer Earth: (a) resistor-over-conductor; (b) conductor-over-resistor.

$$\rho_a = \frac{I\rho_1}{2\pi}\int_0^\infty k_{123}(\lambda)[J_0(\lambda r_1) - J_0(\lambda r_2) - J_0(\lambda r_3) + J_0(\lambda r_4)]d\lambda \qquad (4.25)$$

where

$$k_{123}(\lambda) = \frac{1 - u_{123}\exp(-2\lambda h_1)}{1 + u_{123}\exp(-2\lambda h_1)} \qquad (4.26)$$

$$u_{123} = \frac{\rho_1 - \rho_2 k_{23}}{\rho_1 + \rho_2 k_{23}} \qquad (4.27)$$

$$k_{23}(\lambda) = \frac{1 - u_{23}\exp(-2\lambda h_2)}{1 + u_{23}\exp(-2\lambda h_2)} \qquad (4.28)$$

$$u_{23} = \frac{\rho_2 - \rho_3}{\rho_2 + \rho_3}, \qquad (4.29)$$

where h_2 is the thickness of the second layer.

Further generalization of the formula (4.25) to an arbitrary number of layers should be obvious.

Schlumberger sounding curves $\rho_a(AB/2)$ for three-layer Earth models are shown in Figure 4.11, for the different classes of resistivity models shown in Table 4.2. The sounding curves are obtained by solving Equation (4.25), for each type of resistivity model, with AB the distance between the current electrodes. Notice that the sounding curves resemble smoothed versions of the actual resistivity structure, which facilitates their qualitative geological interpretation.

The formulas for apparent resistivity over a layered Earth, such as (4.24) or (4.25), requires the evaluation of Hankel transforms of the form

$$f(r) = \int_0^\infty K(\lambda)J_0(\lambda r)d\lambda. \qquad (4.30)$$

A Hankel transform can be regarded as a type of *digital filter* which takes a known kernel function $K(\lambda)$ and converts it into a function $f(r)$ that is to be determined. Following the

Table 4.2 Classes of three-layer Schlumberger sounding curves, after Bhattacharya and Patra (1968)

Model type	Resistivity values
H (minimum)	$\rho_1 > \rho_2 < \rho_3$
Q (double descending)	$\rho_1 > \rho_2 > \rho_3$
A (double ascending)	$\rho_1 < \rho_2 < \rho_3$
K (maximum)	$\rho_1 < \rho_2 > \rho_3$

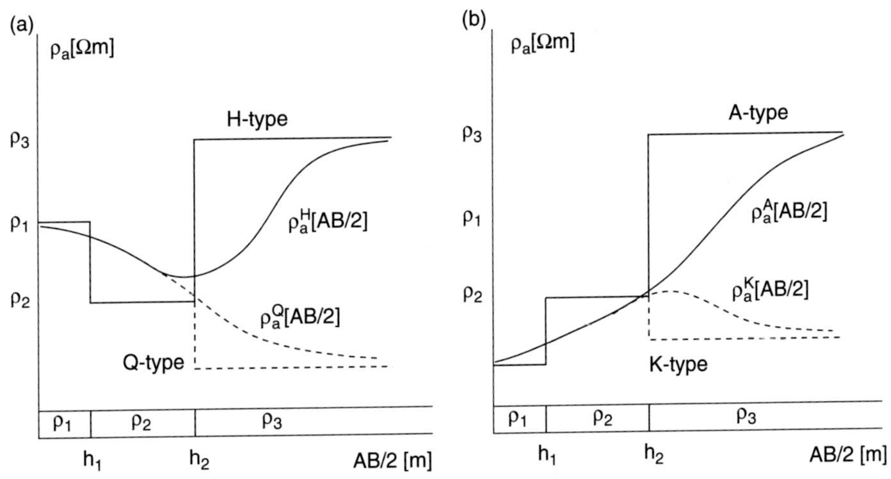

Figure 4.11 Schlumberger three-layer apparent-resistivity sounding curves: (a) H-type and Q-type models; (b) A-type and K-type models (see text for details). After Bhattacharya and Patra (1968).

approach of Guptasarma and Singh (1997), the unknown function $f(r)$ is written as a linear combination of filter weights W_i, $i = 1, \ldots, N$, that is,

$$f(r) = \frac{1}{r} \sum_{i=1}^{N} W_i K(\lambda_i) \tag{4.31}$$

where the kernel is sampled at the logarithmically spaced points

$$\lambda_i = \frac{1}{r} 10^{a+(i-1)s}. \tag{4.32}$$

The Hankel transform in Equation (4.30) can be solved using the weights found by Guptasarma and Singh (1997), for filter lengths of $N = 61$ and $N = 120$. The digital filter can then be checked against several analytic cases, such as

$$\int_0^\infty \exp(-c\lambda) J_0(\lambda r) d\lambda = \frac{1}{c^2 + r^2} \tag{4.33}$$

$$\int_0^\infty \lambda \exp(-c\lambda^2) J_0(\lambda r) d\lambda = \frac{1}{2c} \exp\left(-\frac{r^2}{4c}\right) \tag{4.34}$$

$$\int_0^\infty \lambda \exp(-c\lambda) J_0(\lambda r) d\lambda = \frac{c}{(c^2 + r^2)^{3/2}}. \tag{4.35}$$

The filter weights W_i and the two parameters (a, s) in Equation (4.32) were found by Guptasarma and Singh (1997) based on a least-squares fit to the particular transform given by Equation (4.35). There are a large number of other numerical techniques for evaluating Hankel transforms that can be found in the literature, some of these are listed in Chapter 8.

4.5 Azimuthal resistivity

Many applications of near-surface geophysics require an understanding of the hydrogeological behavior of fractured geological formations. Fluid flow in fractured systems is important for studies of safe waste disposal, contaminant transport, and groundwater discovery and management. Fracturing of a low-permeability rock formation can dramatically enhance the hydraulic conductivity. Bulk properties of a fractured aquifer, such as fracture distribution, aperture distribution, effective porosity, and permeability, are difficult to obtain directly. However, these important parameters may be measured indirectly using the *azimuthal resistivity* method.

A fractured, or jointed, rock formation is inherently *anisotropic* if the fractures are preferentially aligned along one or more certain directions. Consider a vertically jointed rock formation, an idealization of which is shown in Figure 4.12 (top right). The along-strike direction in this case is assumed to be in the y-direction, as indicated by the coordinate axes. Similarly, across-strike is x-directed. Other classes of anisotropy are

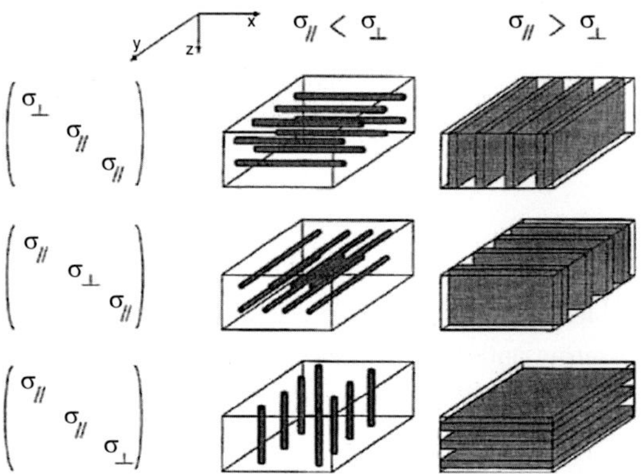

$$\begin{pmatrix} \sigma_\perp & & \\ & \sigma_{//} & \\ & & \sigma_{//} \end{pmatrix} \quad \begin{pmatrix} \sigma_{//} & & \\ & \sigma_\perp & \\ & & \sigma_{//} \end{pmatrix} \quad \begin{pmatrix} \sigma_{//} & & \\ & \sigma_{//} & \\ & & \sigma_\perp \end{pmatrix}$$

Figure 4.12 Representation of an anisotropic medium by conducting rods and sheets. After Everett and Constable (1999).

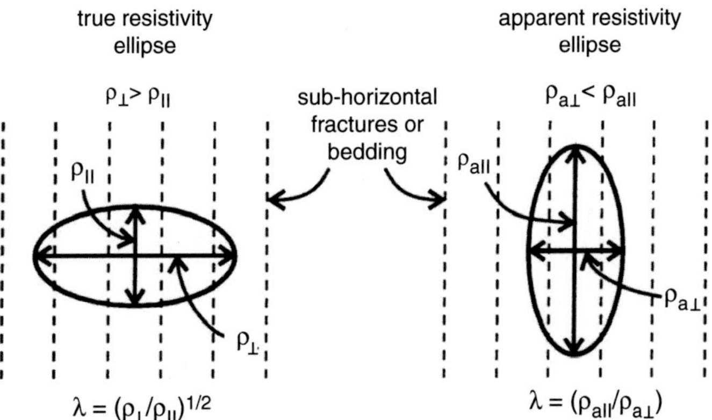

Figure 4.13 True and apparent resistivity ellipses, based on the Wenner linear electrode array, showing the paradox of anisotropy. The coefficient of anisotropy is λ. After Watson and Barker (1999).

shown in the remaining panels of Figure 4.12. The corresponding *uniaxial* conductivity tensors σ are also shown with its principal axes aligned along-strike and across-strike relative to the anisotropic geological structure. Notice that the electrical conductivity of an anisotropic rock formation, by definition, depends on the direction in which it is measured.

The apparent resistivity for the Wenner array aligned at angle θ with respect to the strike of the anisotropy of a vertically fractured medium is given by Bhattacharya and Patra (1968) as

$$\rho_a(\theta) = \sqrt{\frac{\rho_\perp \rho_{||}}{\cos^2\theta + (\rho_\perp/\rho_{||})\sin^2\theta}}. \tag{4.36}$$

Note that for $\theta = 0$, the Wenner array is aligned with the fractures along the strike of the anisotropy, but according to the previous equation we find $\rho_a(0) = \sqrt{\rho_\perp \rho_{||}}$, not $\rho_{||}$ as would be expected. Similarly, when the array is aligned across the strike of the fractures such that $\theta = \pi/2$ we find $\rho_a(\pi/2) = \rho_{||}$, not $\rho\perp$ as expected. These two results are collectively termed the *paradox of anisotropy*; see Figure 4.13 and note the different orientations of the true and apparent resistivity ellipses. The general rule (Wasscher, 1961) is that the resistivity measured in the direction aligned with a four-electrode linear array is the geometric mean of the resistivities in the two directions perpendicular to the electrode array.

A square electrode array, in which the current electrodes *AB* and the potential electrodes *PQ* are arranged to form a square of side *a*, often provides a greater sensitivity to anisotropy than a linear array. Boadu *et al.* (2005) acquired azimuthal resistivity data using a square array to map subsurface fractures in crystalline metamorphic rocks to support a groundwater resources assessment in an agricultural district of Ghana. Their results are shown in Figure 4.14. The polar axes in Figure 4.14a–d show apparent resistivity as the array is rotated about its central point. Note that the minor axis of the apparent resistivity ellipsoid aligns quite well with the fracture direction when a square-array electrode configuration is used.

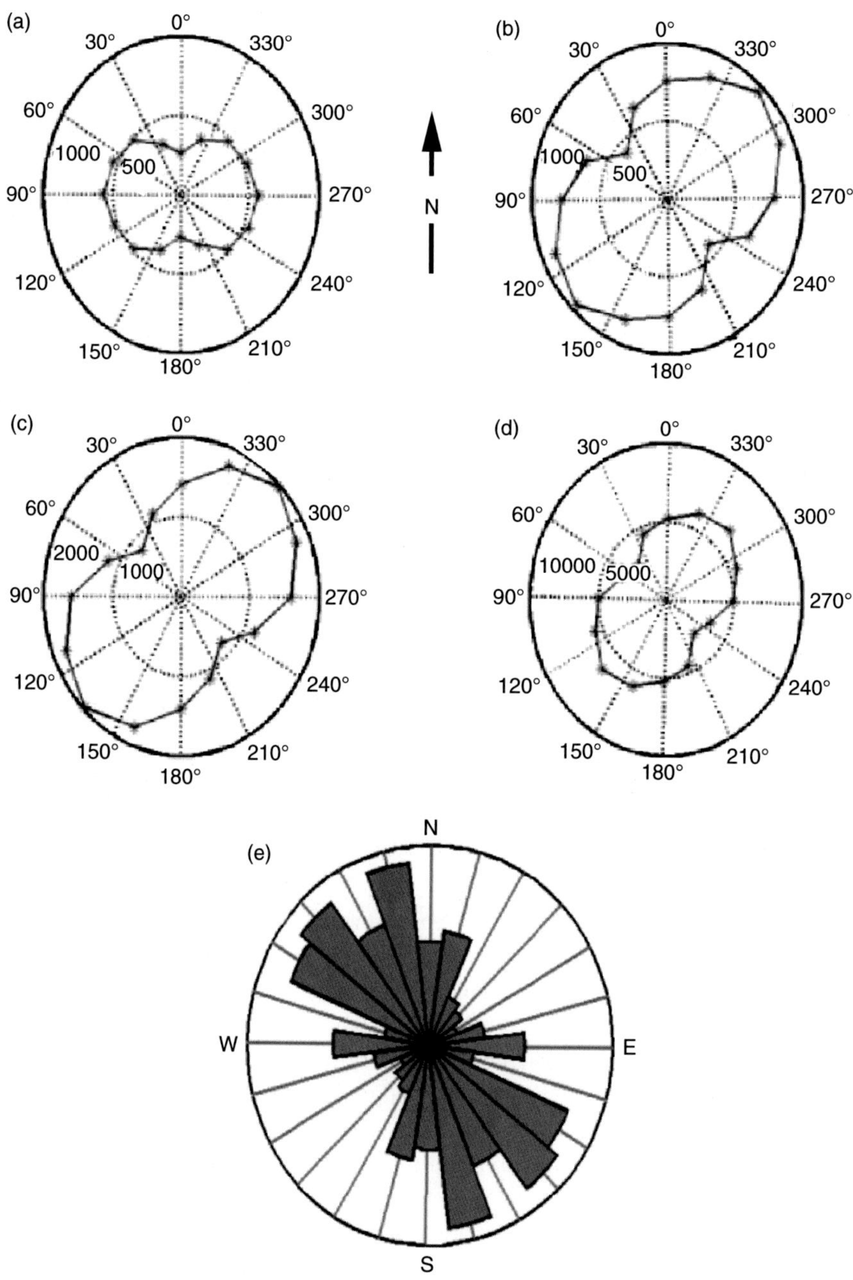

Figure 4.14 The top panels (a–d) show azimuthal resistivity ellipses from a fractured rock formation in Ghana. The bottom panel (e) shows geological fracture orientations mapped in the field. After Boadu *et al.* (2005).

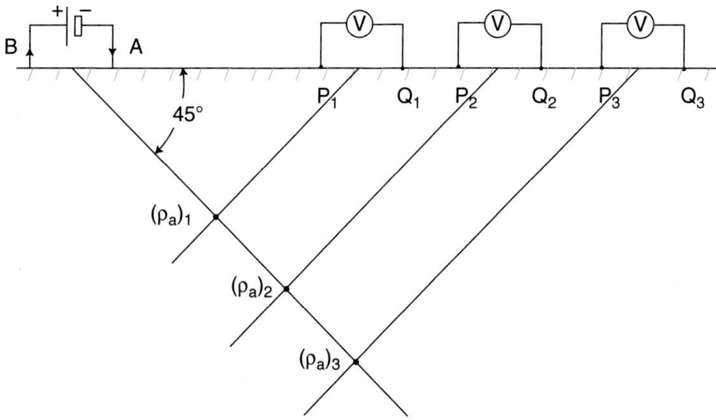

Figure 4.15 Construction of a dipole–dipole resistivity pseudosection.

4.6 Resistivity pseudosections

A convenient way to display multi-electrode resistivity data is by means of a resistivity pseudosection. As shown in Figure 4.15, the measured apparent resistivity ρ_a associated with a current AB and a potential electrode pair PQ is plotted at the intersection of two lines, each making a 45° angle with the ground surface and passing through the center of one of the electrode pairs. This procedure is repeated for each pair of current and potential electrodes. The resulting pseudosection provides a rough indication of the true Earth resistivity since good sensitivity to the ground structure of a given voltage measurement occurs close to the midpoint of the four-electrode configuration at a depth of approximately one-half the separation of the current–potential electrode pairs.

To provide an example of resistivity imaging, a dataset was acquired using the multi-electrode Sting R8/IP system (www.agiusa.com), as illustrated in Figure 4.16. The acquisition protocol involved a computer-controlled sequence of Schlumberger and dipole–dipole electrode configurations, as discussed in more detail below. The observed pseudosection is shown in the top panel of Figure 4.17. The larger solid squares at the top of the pseudosection correspond to electrode locations while the smaller symbols that appear in the interior of the pseudosection mark the locations where the measured apparent resistivity is plotted. The pseudosection is displayed by contouring the apparent resistivity data. The trapezoidal shape of the pseudosection reflects the fact that the measured data are only minimally sensitive to ground structure in the two triangular regions beneath either end of the profile.

4.7 Electrical-resistivity tomography (ERT)

Traditional electrical-resistivity soundings use a conventional electrode configuration such as Schlumberger, Wenner, dipole–dipole, pole–dipole, or pole–pole (see Figure 4.8). A sounding, in which the electrode-separation lengths are varied without moving the array

Figure 4.16 (Left) Multi-electrode resistivity system from AGI Geosciences, Inc.; (right) close-up of an electrode installation.

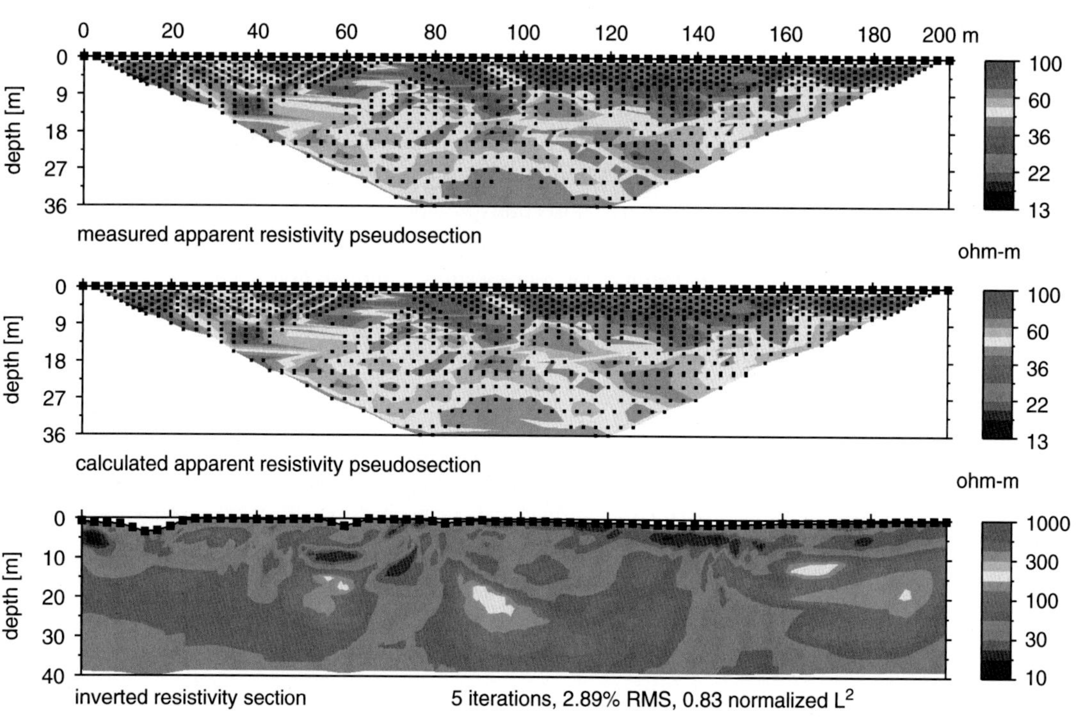

Figure 4.17 Measured apparent resistivity pseudosection (top) for a hybrid Schlumberger-DD electrode configuration, along with the inverted resistivity image (bottom). The calculated pseudosection (middle) is based on solving the forward problem for the resistivity structure shown at the bottom. Note the good match between the measured and calculated pseudosections. See section 4.7 for further details on resistivity inversion. See plate section for color version.

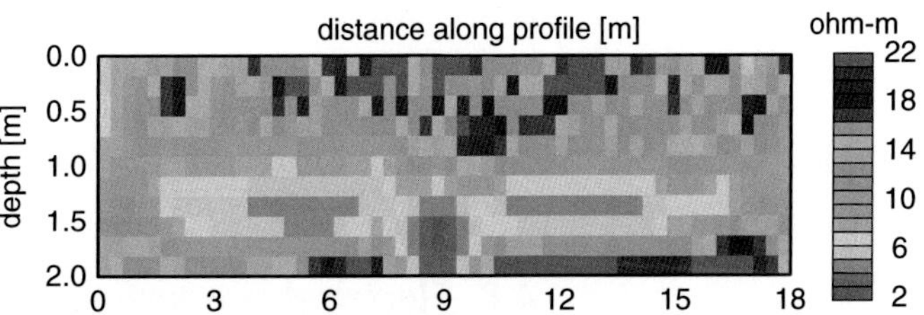

Figure 4.18 Inversion of dipole–dipole resistivity data acquired over a buried pipeline at the Texas A&M Riverside campus. The pipeline is the conductive (*red*) zone at depth 1.5–2.0 m midway along the profile. Electrode spacing is 0.3 m. See plate section for color version.

midpoint, provides a local 1-D electrical-resistivity depth model, $\rho(z)$. Alternatively, lateral profiling of $\rho(x)$ over a narrow depth interval can be achieved by traversing the electrode array along a horizontal profile without changing the electrode separations. Neither the sounding nor the profiling method alone provides an accurate indication of subsurface resistivity distribution in complex geological terrains.

Resistivity imaging of complex subsurface structures has recently advanced with the development of multi-electrode acquisition systems and 2-D and 3-D inversions. The resistivity technique for near-surface applications has surged in popularity due to these advances. Pioneering work on electrical-resistivity tomography (ERT) was published by Daily and Owen (1991) who considered a cross-borehole electrode configuration.

ERT imaging is performed by matching the measured apparent resistivity pseudosection to a computed pseudosection that is obtained by solving, for a given Earth resistivity structure $\rho(\mathbf{r}) = 1/\sigma(\mathbf{r})$, the governing scaled-Laplace equation (4.11). The electric-potential distribution $\varphi(\mathbf{r})$ is evaluated at the locations of the potential electrodes and then transformed into a computed apparent resistivity. The model $\rho(\mathbf{r})$ is then adjusted, and the apparent resistivity re-computed, ideally until it matches the measured apparent resistivity to within a pre-defined acceptable tolerance. The process of imaging is discussed in detail in the inversions chapters of this book. The bottom two panels of Figure 4.17 show a resistivity image and its calculated apparent resistivity response. The spatial structure of the resistivity image, in this example, reflects variations in subsurface moisture content.

A second field example of ERT imaging is shown in Figure 4.18. A dipole–dipole data set was acquired along a profile over a buried pipeline on Riverside campus at Texas A&M University. The inversion algorithm dc2dinv, which is available at www.resistivity.net, was used to construct the image. The profile uses 56 electrodes and is oriented orthogonal to the strike of the target with 0.3-m electrode spacing. The signature of the pipeline is the low-resistivity zone that appears at the midpoint of the profile, at depth ~1.5–2.0 m beneath the surface. The laterally variable structure that is evident in the upper part of the image (depths < 0.5 m) is likely caused by near-surface soil heterogeneity and irregular coupling of the electrodes to the ground.

There is a possibility for interpretation errors to occur if ERT data are aquired on a sparse set of orthogonal 2-D lines or transects but then subjected to fully 3-D modeling and inversion. Gharibi and Bentley (2005) recommend that to avoid artifacts the line spacing should be not more than two–four times the electrode spacing. Measurements should be made, if possible, using a wide range of azimuths of the line joining the midpoints of the current–electrode pair and the potential–electrode pair. Furthermore, the electrode spacing should not be greater than the dimensions of the smallest feature to be imaged.

4.8 Electrical properties of rocks

In the shallow subsurface, the most important geological factor that controls the bulk electrical resistivity is the spatial distribution of pore-fluid electrolytes. The aqueous pore fluids may be contained in pores, fractures, or faults. Electrical geophysical investigations have been used in various hydrogeological, environmental, geotechnical, and civil-engineering applications.

A fundamental assumption of the electrical and electromagnetic methods of geophysics, including the electrical-resistivity technique, electromagnetics (EM), and ground-penetrating radar (GPR), is that the underlying geological medium is electrically neutral, containing vast but equal numbers of positive and negative charge carriers. Some of the charges are free or quasi-free to migrate, or drift, from place to place within the geological medium. Other charges are bound to lattice atoms or other microscopic, localized "charge centers" (Jonscher, 1977), or they are held at material interfaces. Bound charges play no role in the electrical-resistivity or EM techniques but as the reader will see in later chapters they are significant in shaping GPR, induced polarisation (IP), and self-potential signals. Electrical conductivity σ measures the capability of a material to sustain long-term current flow via the charge migration mechanism. By longstanding convention, the electrical resistivity $\rho = 1/\sigma$ is the material parameter used to interpret data acquired using the electrical-resistivity technique but herein we use conductivity and resistivity interchangeably. Two significant types of charge polarization, atomic and molecular, and two types of charge migration, semi-conduction and electrolytic conduction, are illustrated in Figure 4.19.

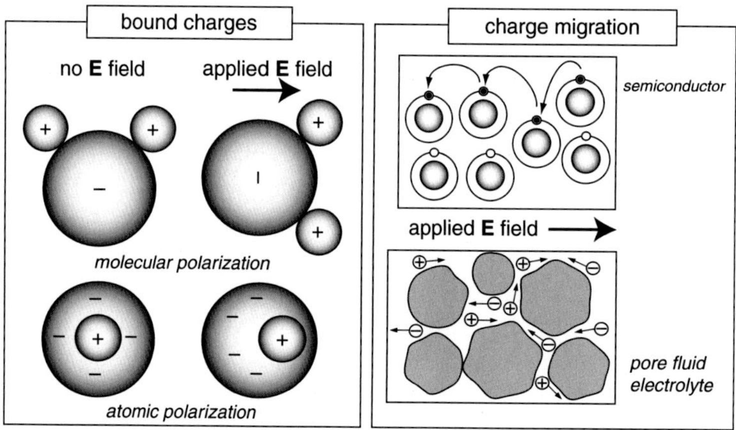

Figure 4.19 Electrical-polarization (left) and -migration (right) mechanisms.

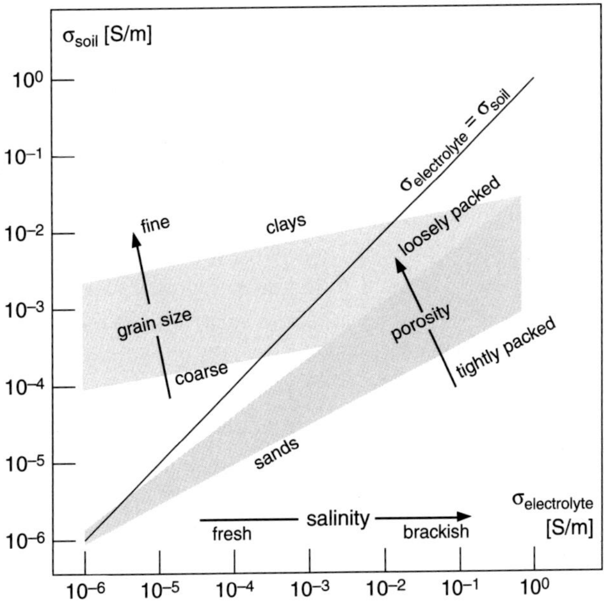

Figure 4.20 Electrical conductivity of saturated soils, after Santamarina *et al.* (2005).

Electrical conduction in most rocks is *electrolytic*, with ions in the pore fluids being the predominant charge carriers. The solid matrix of grains is typically semi-conducting, with notable exceptions being metallic grains and the surface of certain clay minerals, which are conducting. The pore space in a rock is generally much more conductive than the solid grains owing to the presence of dissolved ions in the pore-fluid solution.

Electrolytic conductivity increases, or equivalently, resistivity decreases with increasing salinity, porosity ϕ, and temperature T. A high-salinity pore fluid has a greater concentration of ions available for conduction; a rock with high interconnected porosity often has an abundance of paths for conduction; higher temperature enhances the mobility of ions. The variation of bulk electrical conductivity as a function of salinity, for sand-dominated and clay-dominated saturated soils, is outlined in Figure 4.20.

Purely siliciclastic rock units such as clean, unfractured sandstones typically exhibit a fairly regular intergranular, or primary, porosity that developed as the original sediment compacted and lithified. The pore space in such rocks usually forms an interconnected network with open intergranular spaces that are connected by clear throats. A well-sorted sandstone of this type has a high permeability.

In clean, unfractured sandstones characterized by water saturation S_W and porosity ϕ, the traditional Archie's law (Archie, 1942) gives the bulk electrical conductivity σ [siemens per meter, S/m] as

$$\sigma = a\sigma_W S_W^n \phi^m. \tag{4.37}$$

This empirically based relationship with saturation exponent $n \sim 2$ has long been used by petroleum geoscientists and more recently by hydrogeophysicists to describe the bulk electrical conductivity σ of hydrocarbon reservoirs and aquifers. The parameter m, historically known as the cementation exponent, depends on the grain shape and generally lies within the range $m \sim 1.2$–2.3 for sandstones (Worthington, 1993). The leading coefficient a can vary widely depending on the pore cementation, tortuosity, grain size and shape, fluid wettability, clay content, and numerous other factors.

The quantity $\sigma_W \sim 0.3$–1.0 S/m (Keller and Frischknecht, 1966) in Equation (4.37) is the electrical conductivity of the pore electrolyte, which is controlled by the salinity, or more generally the total dissolved solids in the pore water. Commonly, an intrinsic formation factor $F = \sigma_W/\sigma$ for a fully water-saturated ($S_W = 1$) rock unit is defined. Since pore fluids are electrically conductive relative to the solid rock matrix, we have $\sigma_W > \sigma$ and hence $F > 1$. From Archie's law (4.37), for a fully saturated rock we see that the formation factor is related to porosity by $F \sim \phi^{-m}$.

Clay minerals originating from secondary diagenetic processes can coat the sand grains and clog the pore throats, reducing the porosity and permeability. A general rule of thumb is that high-porosity sandstones are likely to be clean while low-porosity sandstones are likely to be shaly. Clay minerals also have an inherent negative surface charge which contributes an additional electrical conduction pathway that is not found in clean sandstones. Thus, clay-bearing formations generally have a considerably higher bulk electrical conductivity than clean sandstones.

Waxman and Smits (1968), Worthington (1993), and others have demonstrated that important modifications to Archie's formula are necessary in the case of shaliness. These studies suggest that the intrinsic formation factor $F = \sigma_W/\sigma$ should be regarded only as an *apparent* formation factor F_a and that shale effects might be responsible for much of the observed variability in the Archie parameters (a, m). The shale effect on the formation factor is described by the Waxman–Smits equation

$$F_a = F\left[1 + \frac{\tilde{\sigma}}{\sigma}\right]^{-1} = \frac{\sigma_W}{\sigma + \tilde{\sigma}}, \tag{4.38}$$

where $\tilde{\sigma}$ is the "excess conductivity" of the rock that accounts for the surface conduction associated with the clay particles, including the effects of the cation-exchange reactions that occur between the clay and the saturating electrolyte. Numerical values for $\tilde{\sigma}$ can be computed from basic electrochemical tables if the composition and concentration of the saturating electrolyte is known. When shale effects are absent, $\tilde{\sigma} \rightarrow 0$ and $F_a \rightarrow F$ in Equation (4.38), as expected.

Geological interpretation of resistivity data is more problematic in carbonate terrains than in sands. Carbonate rock units commonly exhibit secondary porosity, such as moldic vugs caused by dissolution of fossil remains, that develop after the carbonate rock is formed. The secondary porosity can carry a considerable fraction of the permeability. The bulk permeability depends strongly on whether the vugs are separate or touching each other (Lucia, 1983). It is very difficult to make an assessment from studying outcrops and drilled core samples about the role of vugs in shaping the bulk hydrogeological behavior of a carbonate formation. Furthermore, the range of intergranular textures found in carbonate rocks varies greatly from coarser grain-dominated to finer mud-dominated fabrics.

In carbonates, the relationship between electrical resistivity and porosity is quite complicated. Archie's law has limited predictive value in carbonates as the m value can vary widely and it is difficult to ascertain for a given formation. Cementation exponent values as high as $m \sim 4$–7 have been observed in Middle East oil reservoirs (Focke and Munn, 1987). Asquith (1995) has noted that large values of m are associated with higher separate-vug porosities, while a lower m value is associated with touching-vug porosity, such as fractures.

4.9 Electrical–hydraulic field-scale correlation studies

A long-standing debate amongst near-surface geophysicists has arisen about whether it might be possible to interpret geoelectrical measurements in terms of aquifer bulk properties. Resolution of this issue is one of the main focus points of the rapidly expanding field of hydrogeophysics (Slater, 2007). The capability to convert electrical geophysical data into estimates of bulk aquifer properties such as transmissivity or storativity, for example, would represent a major breakthrough in the field of water-resources research.

Many laboratory studies over the past several decades have claimed to discover links between electrical and hydraulic properties of rock samples. Early investigations by Katz and Thompson (1986), for example, received a great deal of attention. They proposed a relationship of the form

$$k \sim \frac{l_c^2}{226}\left(\frac{\sigma}{\sigma_W}\right), \tag{4.39}$$

where k [darcies, D $= \mu m^2$] is the fluid permeability, l_c [μm] is a characteristic length of the pore space, and σ is the electrical conductivity of a rock sample saturated with a brine solution of electrical conductivity σ_W. Katz and Thompson (1986) explain how l_c is determined in the laboratory from mercury-injection experiments.

Equation (4.39) suggests that a positive correlation exists between the fluid-transport property k and electrical conductivity. Huntley (1986) has shown, however, that the electrical–hydraulic (σ, k) relation in rock samples can also be negatively correlated. The latter case is readily understood in clay-bearing samples where the presence of clay increases the bulk electrical conductivity, due to enhanced surface conduction effects, but at the same time decreases the bulk fluid permeability since clay particles generally clog the pores. A good discussion of the complexities involved in establishing petrophysical relationships between electrical and hydraulic properties, at the sample scale, is provided by Lesmes and Friedman (2005).

While laboratory experiments can provide a fundamental basis for understanding electric–hydraulic correlations at the sample length scale (~ 0.1–10 cm), of greater interest here are linkages at much larger field scales (~ 10 m–1 km) between geoelectrical measurements and aquifer properties.

A simple interpretive rule, which strictly applies only under ideal geological conditions, has been suggested by MacDonald *et al.* (1999). Consider a clean, sandy aquifer. Standard hydrological theory predicts, in this case, a direct relation between porosity ϕ and hydraulic conductivity K. Furthermore, based on Archie's law (4.37), an inverse relation is expected between porosity ϕ and resistivity ρ, where ρ is the reciprocal of conductivity σ. Thus, in idealized clay-free aquifers the product $K\sigma$ is approximately constant. On the other hand, it is well known that in clay-rich aquifers the hydraulic conductivity and electrical resistivity are both controlled by the clay content rather than the porosity. Since K and σ both increase in direct relation to the clay content, to first order, we have that the quotient K/σ should remain roughly constant in ideal clay-rich aquifers.

MacDonald *et al.* (1999) rely on the foregoing analysis to relate a bulk aquifer property to the bulk electrical resistivity inferred from surface-based electrical geophysical measurements. Consider the aquifer transmissivity $T = Kh$, where h is the aquifer thickness. Suppose also that a 1-D-layered resistivity sounding curve has been acquired over the aquifer. Such a curve could be constructed, for example, from an inversion of Schlumberger or dipole–dipole apparent-resistivity measurements. If the resistivity of the layer that corresponds to the aquifer is comparatively high, the aquifer is likely to have a low clay content and it can be shown that the best-resolved geoelectrical parameter is the transverse resistance $R_t = h\sigma$ of the layer. Similarly, the longitudinal conductance $L_c = h/\sigma$ is the best-resolved parameter in the case of a clay-rich, comparatively conductive layer.

For the clay-free aquifer, we can combine the finding that $K\sigma$ is constant with the definitions $R_t = h\sigma$ and $T = Kh$ to deduce that

$$T = C_1 R_t. \tag{4.40}$$

In other words, the aquifer transmissivity is in direct proportion to the geophysically determined transverse resistance of the aquifer. Similarly, for the clay-rich aquifer it follows that

$$T = C_2 L_c, \tag{4.41}$$

or the transmissivity is in direct proportion to the longitudinal conductance. The appropriate constant C_1 or C_2 can be determined by comparing the resistivity sounding curve to hydraulic data at locations where the transmissivity has been determined by pump tests. Then, using geophysics and either Equation (4.40) or (4.41) depending on the aquifer type, an estimate of the aquifer transmissivity can be extrapolated to areas outside the region of influence of the available pump tests. This is a valuable exercise since pump tests are much more expensive to carry out than surface geophysical measurements. MacDonald *et al.* (1999) have used this method to estimate the spatial distribution of transmissivity of the gravel aquifer at Desborough Island on the Thames River near London. The result is shown in Figure 4.21.

A different approach for interpreting geoelectrical measurements in terms of aquifer properties was suggested by Soupios *et al.* (2007) who combined information from wells with 1-D inversions of Schlumberger vertical electric sounding data acquired at a number

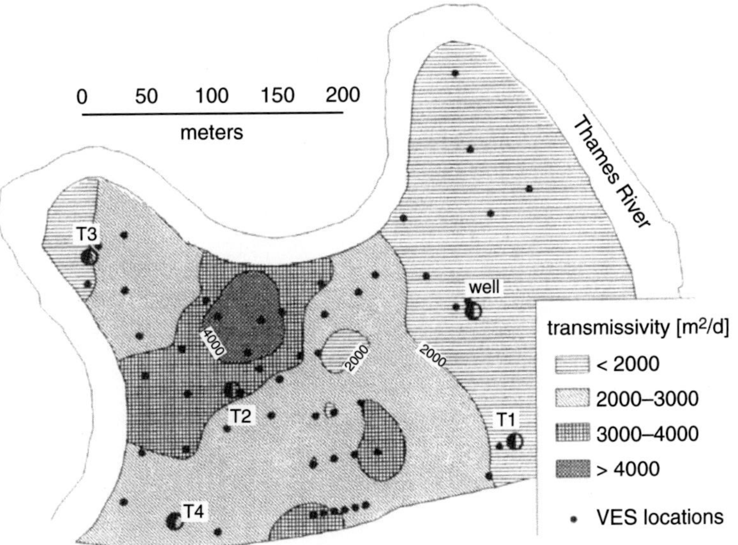

Figure 4.21 Transmissivity of the subsurface gravel aquifer at Desborough Island, UK. Point estimates are obtained from pump tests at locations T1–T4. Spatially distributed estimates are obtained from ~ 50 electrical geophysical soundings (VES), marked by the small black dots. After MacDonald *et al.* (1999).

of sites on the island of Crete. The conversion from resistivity to hydraulic conductivity was performed using the geophysically inferred value of the porosity ϕ, as follows. In the clay-bearing aquifer at Crete, Archie's law (4.37) no longer applies and consequently the Waxman–Smits equation was used. Re-arranging Equation (4.38) gives

$$\frac{1}{F_a} = \frac{1}{F} + \tilde{\sigma}\rho_W, \tag{4.42}$$

such that a plot of the inverse of the apparent formation factor F_a against the pore-fluid resistivity ρ_W yields an intercept of $1/F$. The apparent formation factor F_a at each sounding location on Crete was evaluated using bulk resistivity ρ from inversion of the Schlumberger data along with the fluid resistivity ρ_W measured from a nearby well. The intrinsic formation factor F was obtained from a linear regression based on Equation (4.42), and then converted into porosity ϕ using Archie's law. Then the hydraulic conductivity k [m/day] of the aquifer was computed using the standard hydrology equation (Domenico and Schwartz, 1998)

$$k = \frac{\delta_W g d^2 \phi^3}{180\mu(1-\phi)^2}, \tag{4.43}$$

where d [m] is the grain size, $g = 9.80$ m/s^2 is gravity, $\delta_W = 1000$ kg/m^3 is the fluid density, and $\mu = 0.0014$ kg/m s is the fluid dynamic viscosity. Finally, the aquifer transmissivity $T = kh$ [m^2/d] was evaluated using Equation (4.43) along with the known thickness of the aquifer obtained from the nearby wells. The resulting spatial distribution of aquifer transmissivity is shown in Figure 4.22.

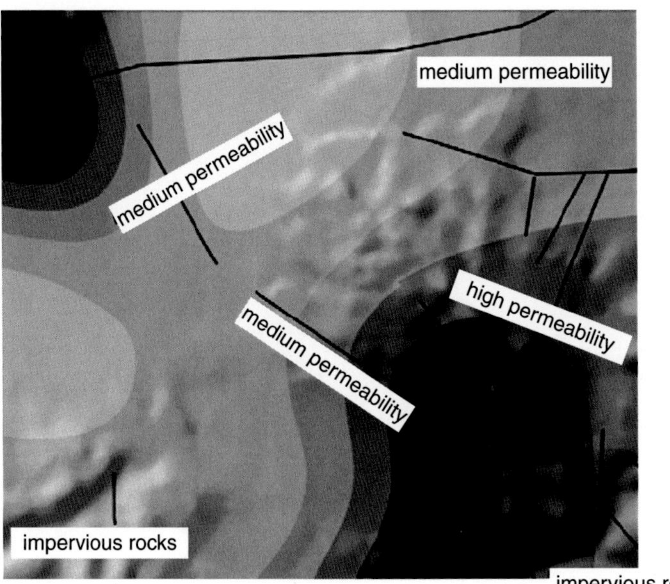

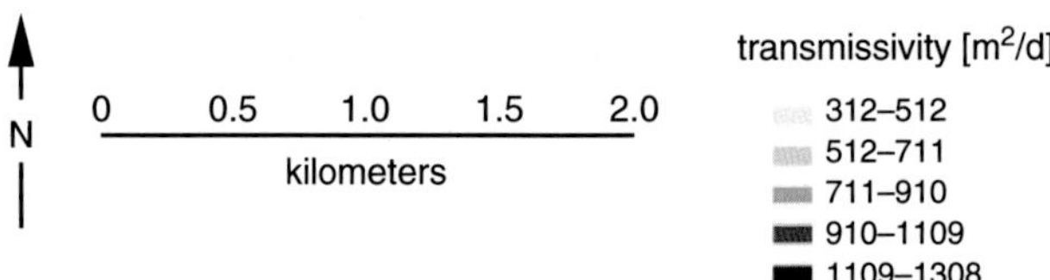

Figure 4.22 Transmissivity of the Keritis basin aquifer in Crete based on Schlumberger electrical soundings. After Soupios *et al.* (2007).

4.10 Optimal electrode placement

Extensive field tests and sensitivity studies have indicated that each of the traditional electrode configurations has strengths and weaknesses. With the advent of computer-controlled multi-electrode acquisition capabilities (e.g. Loke, 1999), data can be recorded using a variety of standard and non-standard configurations. The combined dataset can then be inverted to obtain a 2-D or 3-D subsurface model.

Consider a linear array of d equally spaced electrodes. There are a total of

$$D = d(d-1)(d-2)(d-3)/8 \tag{4.44}$$

possible four-point (one current source, one current sink, and two potential) electrode configurations. For example, if $d = 50$ electrodes are laid out, this leads to $D = 690\,900$ possible four-point measurements, each of which takes ~ 1 s to acquire. Commercial systems routinely use $d > 100$ electrodes. Thus, acquisition of a *comprehensive dataset* of D responses is impractical.

It is of interest, therefore, to determine whether limited combinations of electrode configurations can supply subsurface resistivity images that are comparable in quality to images that would be generated from a comprehensive dataset. An experimental design procedure has been developed by Stummer *et al.* (2004) to achieve this goal. A non-linear objective function is defined and maximized using a global optimization technique.

The performance of traditional electrode configurations is evaluated using a synthetic model (Figure 4.23a, left panel) consisting of a thin, moderately conductive ($\rho = 100\ \Omega$m, typical of silt) near-surface layer overlying a resistive ($\rho = 1000\ \Omega$m, typical of dry gravel) basement. The highly conductive ($\rho = 10\ \Omega$m) block C and the highly resistive ($\rho = 10\,000\ \Omega$m) block D are embedded in the model to provide lateral heterogeneity, as shown.

A simulated resistivity configuration using $d = 30$ (such that $D = 82\,215$) electrodes with spacing $\Delta x = 5$ m is deployed along the top of the model. Omitting certain combinations that yield intrinsic poor data quality, such as crossed-dipole and very-long-offset dipole–dipole configurations, a simulated dataset of size $D = 51\,373$ is generated using a standard finite-difference modeling code. The synthetic data $\mathbf{d}_{OBS}$, consisting of logarithmic voltage readings, are then inverted (see Chapters 11–13) using an iterative scheme

$$\mathbf{m}_{i+1} = \mathbf{m}_i + G^{-g}(\mathbf{d}_{OBS} - \mathbf{d}[\mathbf{m}_i]) \tag{4.45}$$

where $\mathbf{m}_i$ is a vector of parameters which describe the resistivity model, and $\mathbf{d}[\mathbf{m}_i]$ is the computed data based on model $\mathbf{m}_i$.

The quantity $G^{-g} = (G^T G + C_M^{-1})^{-1} G^T$ is the generalized stable inverse of the Jacobian matrix G defined by

$$G_{ij} = \frac{\partial \ln V_i}{\partial \ln \rho_j}, \tag{4.46}$$

and C_M is the model covariance matrix, which includes damping or smoothing constraints on the model parameters. In Equation (4.46), V_i is the voltage reading for the i-th electrode configuration and ρ_j is the electrical resistivity of the j-th model cell of the total of P cells. Thus, the dimension of matrix G is $D \times P$. Inversion results for the synthetic test model are shown in Figure 4.23b–e, left panel. Note that the best model recovery is associated with the comprehensive dataset. The conductive body at the left is not seen using the Wenner and/or dipole–dipole datasets.

In the experimental design procedure of Stummer *et al.* (2004), a homogeneous half-space of uniform resistivity is used as a starting model. Assume an initial dataset of size $D_0 << D$, based on a traditional array such as dipole–dipole. The key design step is to determine which additional electrode configurations would provide the most new information about the subsurface. A quantitative measure of information is provided by the Jacobian sensitivity matrix.

Suppose that the 2-D resistivity model consists of $P = pq$ cells. The Jacobian of the comprehensive dataset is denoted by G^C. The Jacobian of the initial dataset is G^0. The two Jacobians have dimensions $dimG^C = D \times pq$ and $dimG^0 = D_0 \times pq$, respectively. We can define a new $(D - D_0) \times pq$ matrix G^1 according to $G^C = [G^0|G^1]$. The matrix G^1

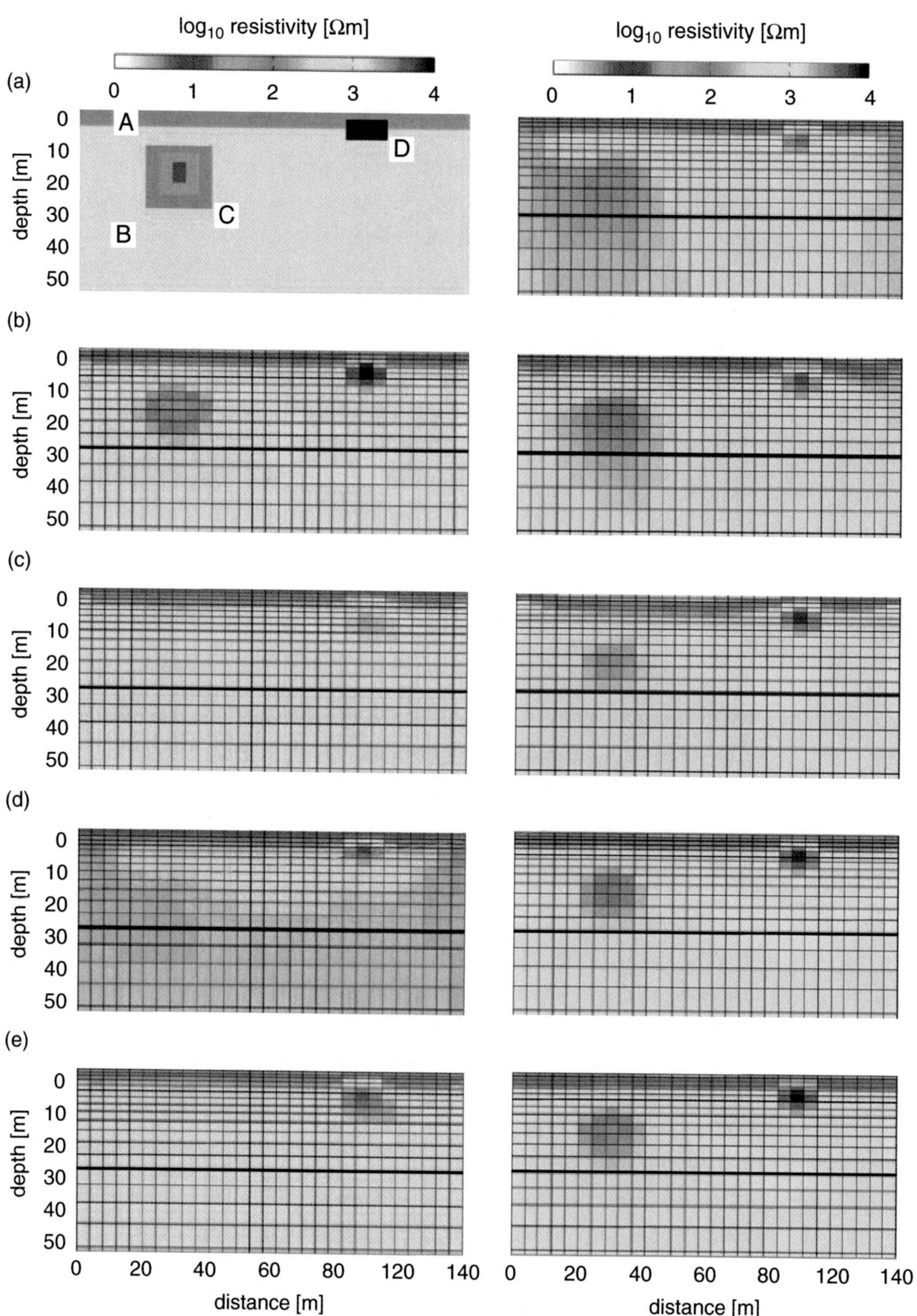

Figure 4.23 (Left panel) Resistivity inversion test, after Stummer *et al.* (2004), (a) starting model. Inversion results using: (b) comprehensive dataset, $D = 51\ 373$; (c) Wenner $D = 135$; (d) dipole–dipole $D = 147$; (e) Wenner/dipole–dipole $D = 282$ electrode configurations. (Right panel) Resistivity inversion based on optimal experimental design, after Stummer *et al.* (2004), (a) $D = 282$; (b) $D = 670$; (c) $D = 1050$; (d) $D = 5740$; (e) $D = 10\ 310$ electrode configurations. See plate section for color version.

corresponds to the remaining, unacquired dataset. We can also define the resolution matrices $R^C = (G^C)^{-g} G^C$ and $R^0 = (G^0)^{-g} G^0$. The i-th electrode configuration in the unacquired dataset is considered to provide important information if it generates a large value of the *goodness function* $\Psi(i)$, where

$$\Psi(i) = \sum_{j=1}^{pq} \frac{|G_{ij}^1|}{G_j} \left[1 - \frac{R_{jj}^0}{R_{jj}^C} \right]. \tag{4.47}$$

The normalization factor

$$G_j = \frac{1}{D} \sum_{i=1}^{D} |G_{ij}^C| \tag{4.48}$$

compensates for the natural tendency of near-surface cells to dominate the sensitivity calculation. Note that $\Psi(i)$ is large when the elements of the corresponding row of matrix G^1 are large. Since a diagonal element $R_{jj} = 1$ indicates a perfect resolution of the j-th model parameter, the term $1 - R_{jj}^0 / R_{jj}^C$ appearing in Equation (4.47) forces the goodness function $\Psi(i)$ to favor electrode configurations that can constrain unresolved model parameters.

The goodness function $\Psi(i)$ in Equation (4.47) is only one of many that could be usefully defined in an optimal experimental design. We can rank the electrode configurations in the unacquired dataset according to their values of the goodness function. The ones that generate the highest goodness function are then used to generate the next dataset. An electrode configuration is rejected if it is linearly independent (as determined by their respective rows in the comprehensive Jacobian matrix) to a configuration used in the initial dataset. An example of optimized inversion using the synthetic example is shown in Figure 4.23, right panel.

4.11 Underwater resistivity techniques

A review of developments in marine electrical and electromagnetic geophysical techniques, including resistivity, IP, and EM methods, is provided by Butler (2009). A summary of the literature shows that there has developed in recent years two common modes of operation for underwater resistivity surveys. In a mode of operation (Day-Lewis *et al.*, 2006; Passaro, 2010) that is particularly suited for a number of shallow-water applications including mapping coastal freshwater discharge and nautical archaeology, an array of floating electrodes is towed on the water surface behind a vessel (Figure 4.24). Such systems can achieve continuous resistivity profiling of the subbottom resistivity structure and have detected seabed anomalies caused by shipwrecks. Other studies have attempted to use bottom-towed electrode arrays but these can be easily damaged as they are dragged across the potentially rugged seafloor.

In another mode of operation, the electrode array is stationary. In some cases an array of grounded electrodes is employed that makes direct electrical contact with the subbottom, as in conventional land surveys. In deeper water, electrodes may be suspended on vertical cables extending from buoys or a vessel at the sea surface into the water column. This approach was used by Baumgartner and Christensen (1998). The choice of which marine

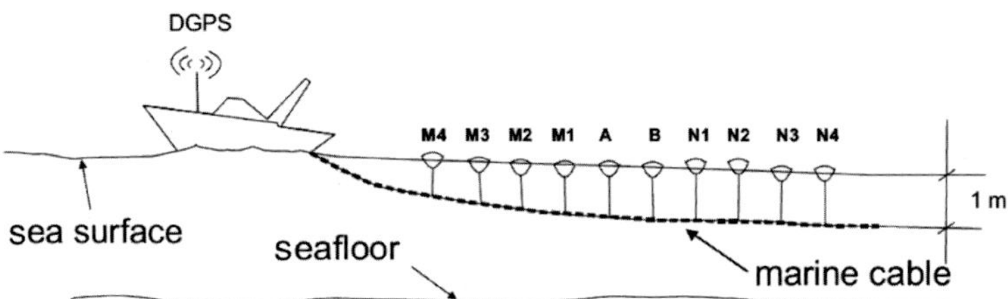

Figure 4.24 A surface-towed electrode array for marine resistivity surveys. DGPS = differential GPS navigation system. After Passaro (2010). A, B = current electrodes; M, N = measurement electrode pairs.

resistivity survey geometry to adopt for a particular project should be dictated by forward modeling of the sensitivity of the apparent resistivity pseudosection to perturbations in the expected geological scenarios, as well as logistical and budget constraints.

4.12 Illustrative case histories

Example. Tunnel construction in the Alps.

Geophysical investigations using the resistivity method were carried out in the Col di Tenda region of the Alps near the Italy–France border where the construction of a new highway tunnel is planned. Better information on the subsurface geology is required in order to determine accurate geotechnical rock-mass quality parameters so that the safety of the excavation and long-term integrity of the finished structure is ensured. A resistivity section oriented perpendicular to the planned tunnel route is shown in Figure 4.25a. Electrode spacing is 12.0 m and depth of penetration is ~ 200–300 m.

The interpretation of the resistivity section (Figure 4.25b) is based on geological outcrop and stratigraphy from boreholes. High-angle normal faults, pervasive in this region, cause the sharp lateral resistivity contrasts seen in the geoelectric section. The low-resistivity zone at the base of the fault-bounded central graben is caused by fluid circulation within an intensely fractured Jurassic limestone layer which underlies higher-resistivity Eocene calcarenites. These resistivity and complementary seismic data were able to provide detailed structual and geological information to assist the planning of this major civil-engineering project.

Example. Brownfield redevelopment.

Many urban areas contain abandoned sites that are legacies from past industrial activities. These sites often contain hazardous materials, such as polycyclic aromatic hydrocarbons and heavy metals, in addition to rubble, metal scraps, old building foundations, and other construction debris. Often it is desired to rehabilitate these sites for re-use such as commercial or residential redevelopment. Boudreault *et al.* (2010) have described geophysical work performed at such a site in downtown Montreal, Canada (Figure 4.26a)

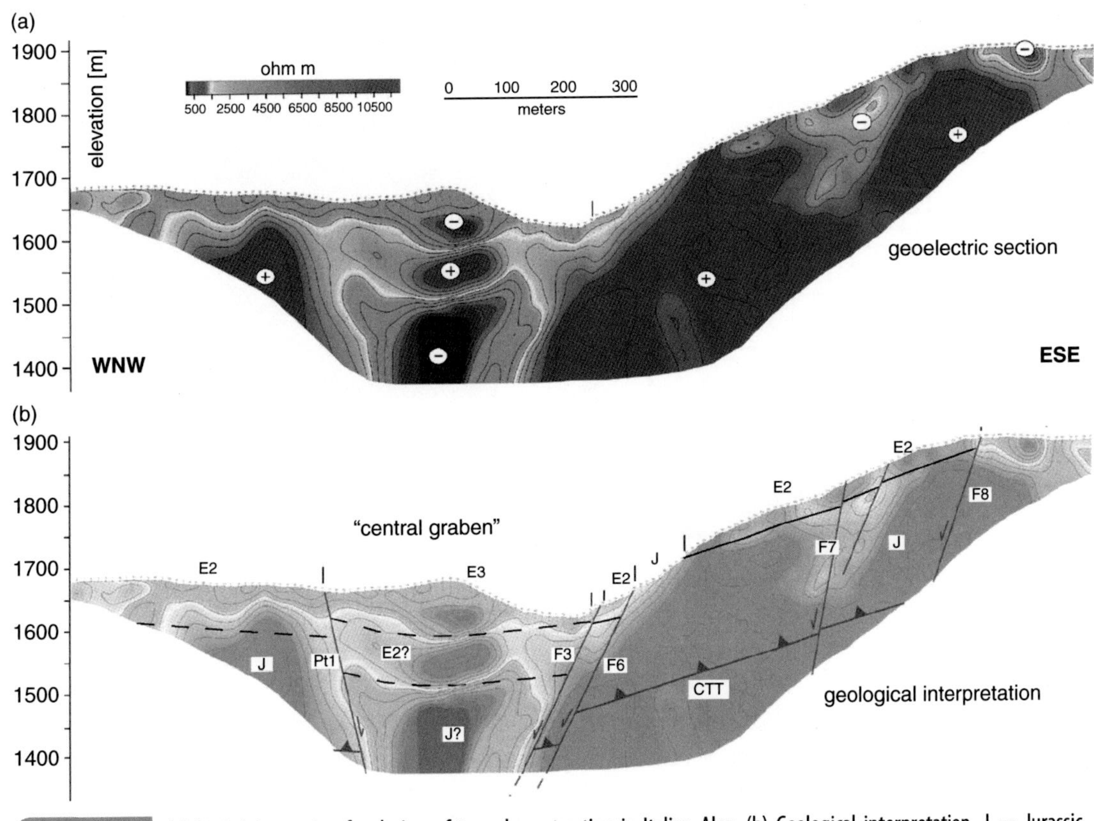

Figure 4.25 (a) Resistivity section for design of tunnel construction in Italian Alps. (b) Geological interpretation. J = Jurassic formation; E,F = Eocene formations. After Cavinato *et al.* (2006).

where new commercial development is reclaiming a long-disused parcel of urban real estate. The site is characterized by abundant heterogeneous urban fill (Figure 4.26b) that trench excavations have shown to occupy the upper ~ 2 m beneath the surface. A total of six ERT profiles were acquired, two of which are indicated in Figures 4.26c, d. The ERT images show that the upper fill layer is of higher resistivity than the underlying low-resistivity layer of natural soil. The upper layer is also strongly heterogeneous reflecting the unorganized spatial distribution of the constituent concrete and brick debris. The upper layer is more resistive than the underlying natural soil since construction materials such as brick and soil are inherently resistive (up to 1000 Ωm) compared to the natural soil and, furthermore, the upper layer is less compacted than the underlying soil and therefore has a much smaller water retention capacity, raising its resistivity. This case study shows that electrical geophysics can play a significant role in the detailed subsurface characterization of brownfield sites. The information obtained using geophysics can be used for a number of purposes, for example, it enables better assessments of potential pollutant distributions and it can help to guide the safe excavation of the site as it undergoes redevelopment.

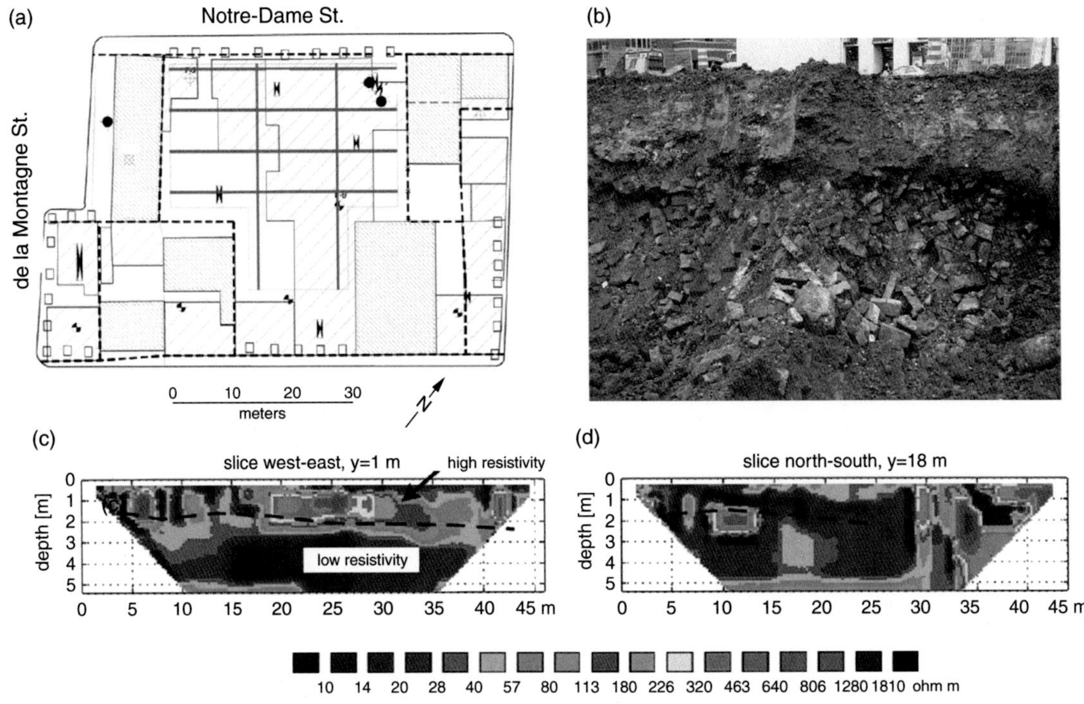

Figure 4.26 (a) Map view of urban redevelopment site, downtown Montreal, Canada. The red lines show ERT profiles. (b) Heterogeneous urban fill containing bricks, concrete, and metal debris. (c) an east–west ERT profile. (d) a north–south ERT profile. The dashed line shows the boundary between the heterogeneous fill and the natural soil, as determined by trench excavations. After Boudreault *et al.* (2010). See plate section for color version.

Problems

1. Show that the geometric factor κ for the arbitrary four-electrode arrangement is given by $\kappa = 2\pi\,[1/r_{AP} - 1/r_{AQ} - 1/r_{BP} + 1/r_{BQ}]^{-1}$.
2. Derive the geometric factors κ for the traditional four-electrode Schlumberger, Wenner, and dipole–dipole configurations.
3. Derive the geometric factor κ for a square array, in which the two current electrodes AB and the two potential electrodes PQ form a square of side a. Consider both cases: (i) the current electrodes are adjacent to each other; and (ii) they are diagonally opposite to each other (the *cross-square array*). Is the result of case (ii) surprising?
4. Consider a pole–pole resistivity experiment over a set of vertical fractures which can be considered as a uniform anisotropic halfspace. The potential at distance r from a point source of current I is given by

$$V(r) = \frac{I\rho_m}{2\pi r}\left[1 + (\lambda^2 - 1)\sin^2\varphi\right]^{-1/2}$$

where the coefficient of anisotropy is $\lambda = \sqrt{\rho_\perp/\rho_{||}}$ and the rms resistivity is $\rho_m = \sqrt{\rho_\perp \rho_{||}}$. The angle that the line between the current and potential electrodes makes with respect to the strike (x-direction) of the fractures is φ. Show that $V(r)$ satisfies the scaled Laplace equation

$$\frac{1}{\rho_{||}}\frac{\partial^2 V}{\partial x^2} + \frac{1}{\rho_\perp}\frac{\partial^2 V}{\partial y^2} + \frac{1}{\rho_{||}}\frac{\partial^2 V}{\partial z^2} = 0.$$

5. The apparent resistivity for the pole–pole experiment over a vertically fractured medium is measured in the along-strike and across-strike directions. (i) Prove that, in each case, the measured apparent resistivity is equal to the geometric mean of the actual resistivities in the other two directions. (ii) On a polar plot, graph the measured apparent resistivity as a function of the angle φ with respect to strike, for the case of soil-filled fractures in a weathered sandstone ($\rho_{||}$ =0.1 Ωm and $\rho_\perp$ =10 Ωm). Comment on the orientation of the resulting ellipse in terms of the *paradox of anisotropy.*
6. Consider a current electrode A placed at one corner of a square of side a. Potential electrodes P, Q, and R are placed at the other three corners, with R diagonally opposite from A. The medium is vertically anisotropic with resistivities $\rho_{||}$ and $\rho_\perp$. The strike of the fractures is aligned with a side of the square. Show that the measured voltages satisfy

$$V_P^2 + V_Q^2 = \left[\frac{\rho_\perp + \rho_{||}}{\rho_m}\right]^2 V_R^2.$$

5 Induced polarization and self-potential

The induced-polarization (IP) and and self-potential (SP) methods of geophysical exploration are based on measurements, normally made at the surface of the Earth, of electric potentials that are associated with subsurface charge distributions. In the IP method, the charge distributions are established by an application of external electrical energy. In the SP method, subsurface charge distributions are maintained by persistent, natural electrochemical processes.

Consider the hypothetical situation shown in Figure 5.1 in which electrical charges are distributed unevenly within the subsurface. Charge accumulations are portrayed schematically in the figure as positive and negative "charge centers." The charges may be volumetrically distributed or they may reside on mineral surfaces and other interfaces. In either case, the regions where charge is concentrated can be viewed as the spatially extended terminals of a kind of natural battery, or geobattery. The sketch shown in the figure greatly simplifies the realistic charge distributions that occur within actual geological formations but it is instructive for the present purpose. Electrical energy supplied from an external source, or energy that naturally arises from a persistent electrochemical process, is required to maintain the "out-of-equilibrium" charge distributions shown in Figure 5.1. Without an energy input, they would rapidly neutralize in the presence of the conductive host medium, and the geobattery would soon discharge.

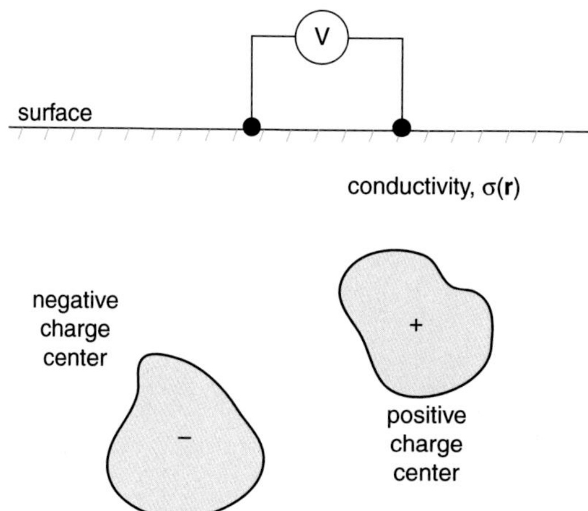

Figure 5.1 Generic subsurface charge distribution.

5.1 Induced polarization (IP): introduction

Suppose a standard resistivity experiment (see Chapter 4) is performed but then the current I is suddenly switched off at time $t = 0$. The voltage across any pair of potential electrodes normally drops instantaneously to zero. However, if the ground is *polarizable*, the voltage drops rapidly from its initial value V_0 to a non-zero value V_1 and thereafter decays slowly, often with a characteristic *stretched-exponential* shape of the form ~$t\exp(-at^{\beta})$ with $0 < \beta < 1$, as shown in Figure 5.2. This transient behavior is known as the time-domain *IP effect*. Notice that the IP decay curve cannot be simply explained by an effective capacitance C in series with an effective resistance R in an equivalent RC circuit of the Earth. The discharge of such a capacitance after current switch off generates a transient voltage that decays purely exponentially as ~$\exp(-t/RC)$ (Tipler and Mosca (2007)), not according to a stretched exponential law.

There is not widespread agreement amongst geophysicists as to the physical explanation for the stretched-exponential shape of the transient IP decay. Significantly however, a stretched-exponential function is nothing more than a linear superposition of ordinary exponential decays. Stretched-exponential functions have been shown by many authors (e.g. Frisch and Sornette, 1997) to generally describe the macroscopic relaxation of a hierarchical or disordered system, each component of which relaxes exponentially at its own characteristic time scale. This is sometimes called the Kohlrausch–Williams–Watts relaxation law.

There are many ways to measure the time-domain IP effect in field studies and a number of different expressions have appeared in the literature. Let $V(t_0)$ be the voltage measured at some fixed time t_0 after switch-off. The *polarizability* η [mV/V] of the ground can be defined as

$$\eta = \frac{V(t_0)}{V_0}. \tag{5.1}$$

The *partial chargeability* M_{12} [ms] of the ground is its polarizability averaged over a pre-defined time window [t_1, t_2] during the stretched exponential decay,

$$M_{12} = \frac{1}{V_0}\int_{t_1}^{t_2} V(t)dt. \tag{5.2}$$

There is a considerable amount of inconsistency throughout the geophysical literature with respect to precise definitions of η and M_{12}. Equations (5.1) and (5.2) are commonly used, but other definitions and nomenclature have been adopted by different authors.

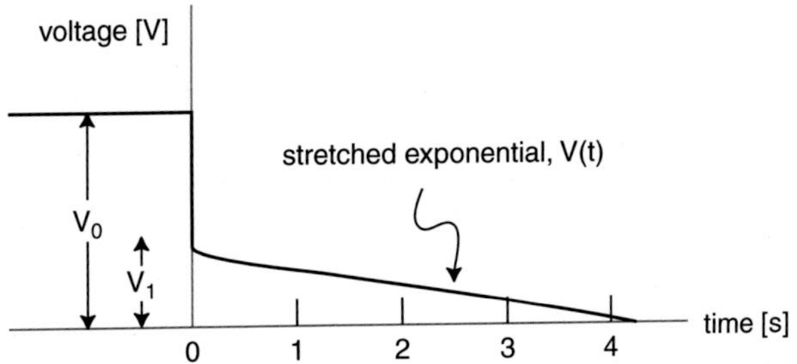

Figure 5.2 A typical IP decay curve.

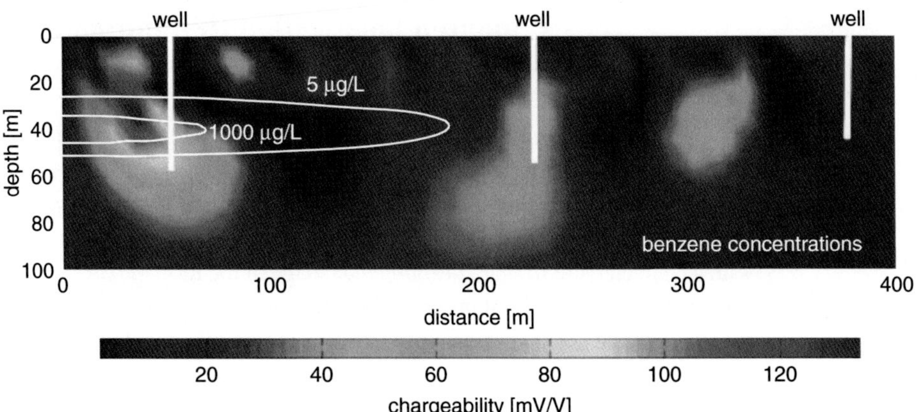

Figure 5.3 IP response of a benzene contaminant plume, Cape Cod, Massachusetts, along with contours of benzene concentration. After Sogade *et al.* (2006). See plate section for color version.

Example. IP mapping of a benzene contaminant plume.

The Massachusetts Military Reservation on Cape Cod was used as a US Air Force facility from 1948 to1973. Following an uncontrolled release of jet fuel, a subsurface plume of volume 265 m^3 and length 1.5 km consisting of benzene and other organic contaminants was detected and initially characterized during the 1980s. Groundwater wells in 1996 showed benzene concentrations as high as 2500 μg/L. The geology is glacial outwash sands and gravels underlain at ~ 60 m depth by clayey lacustrine sediments. A number of laboratory-scale studies on contaminated rock and soil samples (e.g. Vanhala *et al.*, 1992) have found that significant IP anomalies are associated with organic compounds such as benzene, toluene, and trichloroethene. Accordingly, time-domain IP field-scale data were acquired on-site (Sogade *et al.*, 2006) in order to map the spatial extent of the plume. The two-dimensional (2-D) depth section of chargeability shown in Figure 5.3 was constructed from an IP measurement profile in the dipole–dipole configuration of total length 560 m with 24-m electrode spacing. Chargeability is defined in this study as the ratio V_1/V_0 shown in Figure 5.2, which by the definition (5.1) is essentially the polarizability η at time $t_0 = 0$. Notice that the zone of highest chargeability (~ 130 mV/V) shows a very good spatial correlation with the region of highest benzene concentration as determined by sampling of groundwater wells.

In the frequency-domain IP method, an alternating current I of frequency f, generally in the range 1 mHz–1 kHz, is injected into the ground. An IP effect is frequently observed in the frequency domain; it appears as an out-of-phase component of electrical conduction at low frequencies (Vinegar and Waxman, 1984). Using any electrode configuration, a measure of the IP effect in the frequency domain is the *percentage frequency effect PFE* defined by

$$PFE = 100\,\frac{\rho_a(f_1) - \rho_a(f_2)}{\rho_a(f_2)}. \tag{5.3}$$

In Equation (5.3), $\rho_a(f_1)$ and $\rho_a(f_2)$ are the apparent resistivities measured at two frequencies f_1 and f_2 such that $f_1 < f_2$. Recall from the previous chapter that the apparent resistivity at frequency f is given by the formula

$$\rho_a = K\frac{V}{I} = KZ, \tag{5.4}$$

where K is the geometric factor of the electrode configuration and $Z = V/I$ is the impedance. In principle, any convenient electrode-array configuration can be selected for IP measurements. However, the suitability of a given array depends on a number of factors including the signal-to-noise ratio, site geology, electromagnetic-induction coupling effects at frequencies higher than ~ 100 Hz (Revil *et al.*, 2012), and, as described later, electrode polarization effects. These factors are best explored by detailed forward modeling.

A second measure of the IP effect in the frequency domain is the *phase angle* ϕ [mrad] defined, at frequency f, to be the small difference in the phase of the measured voltage with respect to that of the injected current. Note that the phase of the voltage lags behind that of the causative current. In terms of the phase angle, the impedance $Z = V/I$ in Equation (5.4) has the form $Z = |Z|\exp(-i\phi)$. Inserting this expression into Equation (5.4), multiplication by the known array geometric factor K, and finally taking the reciprocal results in

$$\sigma^* = \frac{\exp(i\phi)}{K|Z|} = |\sigma|\exp(i\phi) = \sigma' + i\sigma''; \tag{5.5}$$

where σ^* is conventionally termed the *complex conductivity.* Note that the real and imaginary parts of the complex conductivity are related by

$$\sigma'' = \sigma'\tan\phi. \tag{5.6}$$

In field surveys, the complex conductivity (or equivalently, the amplitude $|\sigma|$ and phase angle φ) is measured at a single frequency. The quantity $\rho^* = 1/\sigma^*$ termed the *complex resistivity* is also frequently mentioned in the literature. The *spectral IP* (SIP) method is essentially a measurement of complex resistivity ρ^* over a range of frequencies. The SIP method is the analogous field-scale geophysical technique to AC impedance spectroscopy or dielectric spectroscopy measurements made in the laboratory on rock samples generally at higher frequencies.

5.2 Phenomenological resistivity dispersion models

A large number of empirical models have been developed to describe the frequency dependence of the electrical resistivity of polarizable geomaterials. Perhaps the simplest is the Debye model that describes a highly idealized medium relaxing according to a single exponential time decay, i.e. $V(t) = V_0 m\exp(-t/\tau)$. In the frequency domain, the Debye complex resistivity $\rho^*(\omega)$ is

$$\rho^*(\omega) = \rho_0\left[1 - m\left(1 - \frac{1}{1 + i\omega\tau}\right)\right], \tag{5.7}$$

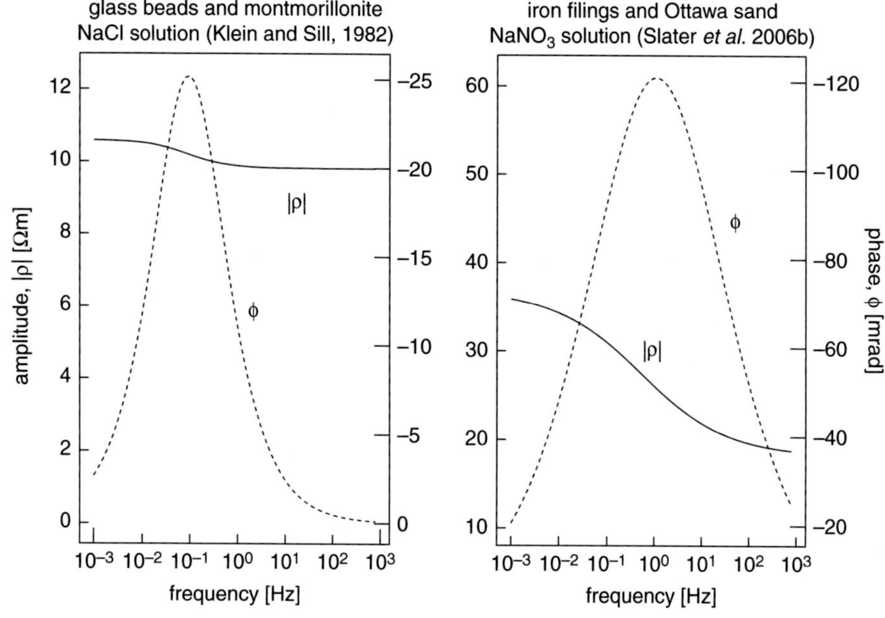

Figure 5.4 Cole–Cole complex electrical resistivity spectra.

where ρ_0 is the DC resistivity at zero frequency. The parameters m and τ [s] are termed the chargeability and relaxation time constant, respectively.

More widely used and applicable to heterogeneous geomaterials is the Cole–Cole phenomenological model (Cole and Cole, 1941; Pelton *et al.*, 1978) written in the form

$$\rho^*(\omega) = \rho_0 \left[1 - m\left(1 - \frac{1}{1 + (i\omega\tau_0)^c}\right)\right], \tag{5.8}$$

with frequency exponent c. The Cole–Cole model describes a material that relaxes according to a heavy-tailed distribution $g(\tau)$ of relaxation times. The distribution function $g(\tau)$ is bell-shaped and symmetric about τ_0 (Revil *et al.*, 2012).

The values of the parameters m, τ_0, and c in Equation (5.8) for a given geomaterial are typically determined by a fit to experimental measurements of electrical properties made in the laboratory. For example, Klein and Sill (1982) report $m = 0.075$, $\tau_0 = 1.8$ s, $c = 0.72$, and $\rho_0 = 10.6$ Ωm for the best fit to electrical resistivity measurements over the frequency range 1 mHz–1 kHz on a mix of glass beads and montmorillonite saturated with a 0.01 mol NaCl solution. Slater *et al.* (2006b) report $m = 0.51$, $\tau_0 = 0.33$ s, $c = 0.424$, and $\rho_0 = 36.9$ Ωm for electrical resistivity measurements between 0.1 Hz and 1 kHz on 10 wt.% iron filings mixed with Ottawa sand and saturated with a 0.01 mol $NaNO_3$ solution. The Cole–Cole electrical spectra using these parameters are shown in Figure 5.4. The Cole–Cole model (5.8) is phenomenological in the sense that its parameters m, τ_0, and c are not intended to be derived from basic physical arguments.

An equivalent electrical circuit (Dias, 2000) for the Cole–Cole dispersion model provides an intuitive conceptualization of the frequency-domain IP effect. The equivalent circuit is shown in Figure 5.5, where

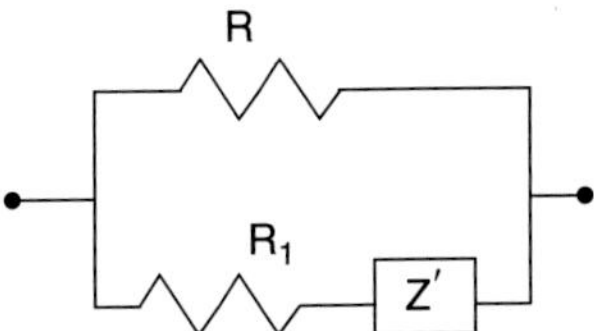

Figure 5.5 Equivalent electrical circuit for the Cole–Cole model.

$$m = \frac{R}{R+R_1}; \quad \tau = \left(\frac{R+R_1}{a}\right)^{\frac{1}{c}}; \quad Z' = \frac{a}{(i\omega)^c}; \tag{5.9}$$

where a is an arbitrary constant, and the frequency exponent obeys $0 \le c \le 1$. Note that the circuit contains a parallel combination of a purely resistive path R, corresponding to quasi-free ionic charge migration, and a second path of resistance R_1 and impedance Z', the latter corresponding to charge accumulation (polarization) at subsurface interfaces. A large number of other phenomenological dispersion models, including various generalizations of the Cole–Cole function (e.g. Kruschwitz *et al.*, 2010), have been proposed over the past several decades to explain IP observations. Equivalent electrical circuits for many of these models, which can become quite complex, are tabulated by Dias (2000).

5.3 Electrode, membrane, and interfacial polarization

The IP effect, measured in either the time or frequency domain, is considered by many geophysicists to be understood as follows. Suppose an electric current is caused to flow within the ground as a result of a voltage applied across a pair of current electrodes. Consider a subsurface interface, such as one between the pore-fluid electro-yte and a charged mineral surface. Such a junction is associated with contrasts in conduction mechanisms and charge-carrier mobilities, and therefore it acts as an electrochemical impedance to current flow. Charges accumulate at or near the mineral–electrolyte interface. In effect, the ground becomes polarized. In this manner, the continuous application of current from the external source forces the system into a non-equilibrium steady state in which charge distributions are built up at mineral–electrolyte interfaces. The charge distributions dissipate once the driving current is switched off.

At IP frequencies, several important mechanisms of subsurface charge accumulation, namely, *electrode polarization*, *membrane polarization*, and *interfacial polarization* have been discussed in the literature (e.g. Olhoeft, 1985). Electrode polarization is found in rocks containing mineral grains of high electrical conductivity. Metallic grains, for example, can partially block or otherwise impede the movement of ions within the pore-fluid electrolyte. The primary conduction mechanism is electronic on the metal side and ionic on the electrolyte side of a metal–electrolyte junction. Electrons are transferred across

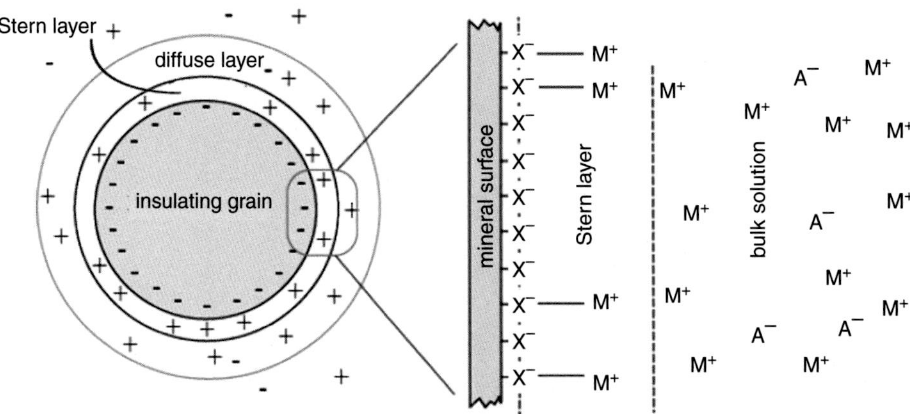

Figure 5.6 The electric double layer at the clay-mineral–electrolyte interface. After Revil *et al.* (2012).

the interface via oxidation–reduction chemical reactions. However, an energy barrier must be overcome in order for the electron transfer to proceed. It is this barrier that causes charges to accumulate at the junction, and thus an electrochemical impedance appears at metal–electrolyte interfaces.

In membrane polarization, the mobility of ions in relatively constricted pores is impeded by less-mobile ions. This can occur when electric current is passed through sediments in which clay is dispersed among larger mineral grains. Consider first the equilibrium case prior to excitation of the system by an external current. The surface of a clay mineral is negatively charged. The negative surface charges attract cations from the pore-fluid electrolyte. As shown in Figure 5.6, some of these cations (M^+) attach weakly to the mineral surface, contributing to the *fixed*, or *Stern layer*. Other cations form a second *diffuse*, or *Guoy–Chapman layer* within the bulk solution. The fixed and diffuse layers comprise the *electric double layer* (e.g. Parsons, 1990; Delgado *et al.*, 2005) that is basic to much of electrochemistry.

Now suppose an external current is applied. The relatively poor mobility of anions and cations in the electric double layer, as shown in the figure, impedes the transport of the externally mobilized, ionic charge carriers. The magnitude of the impedance depends on the size and polarity of the charge carriers. In this manner, clay particles act as ion-selective membranes, passing some charge carriers relatively easily while preferentially impeding others. Leroy *et al.* (2008) have further suggested that the Stern layer becomes polarized when an external current is applied, without an accompanying charge buildup in the diffuse layer.

A third type of polarization, interfacial polarization or the *Maxwell–Wagner effect*, can become important at relatively high IP frequencies, greater than 1 kHz (Hizem *et al.*, 2008; p.5). Interfacial polarization is a purely physical effect caused by electric charges that accumulate at conductivity interfaces within a heterogeneous medium as it is subjected to an applied electric field. We shall see in Chapter 9 that a closely related phenomenon, interfacial dielectric polarization in which charges accumulate at permittivity interfaces, can be important at the low end of ground-penetrating radar (GPR) frequencies.

5.4 IP response and subsurface geological processes

Since the IP effect originates at fluid–grain interfaces, a body that contains widely disseminated metallic or clay minerals should generate a larger IP response than a massive body containing the same total volume of metal or clay. Vinegar and Waxman (1984) report that the IP response is proportional to shaliness, a measure of the clay content, of sandstones. Slater *et al.* (2006b) have investigated in the laboratory both metal–sand and clay–sand mixtures, exhibiting electrode and membrane polarization respectively. They find that a critical parameter controlling the size of the IP effect in both cases is, indeed, an interfacial geometric factor S_p [1/μm]. As described by Pape *et al.* (1987), the factor S_p is the surface area of the mineral grains per unit volume of the saturating fluid; in other words, it is a measure of the mineral surface area that is in contact with pore electrolyte. Slater *et al.* (2006b) show that the IP response increases with the parameter S_p. The relationship $\sigma'' \sim S_p^{0.74}$ provides a good fit to data compiled from the literature by Kruschwitz *et al.* (2010), as shown by the dotted line in Figure 5.7.

A number of authors (e.g. Lesmes and Morgan, 2001; Titov *et al.*, 2002) have attempted to develop first-principle theoretical models that can be used to interpret complex resistivity spectra in terms of pore microstructure and grain-size distributions, with possible implications for inferring hydraulic properties such as permeability. These models have been partially successful but this remains an area of active research. Kruschwitz *et al.* (2010), taking a phenomenological approach, suggest that small pores should lead to faster relaxation of charge polarization and hence small values of the time constant τ_0 in the Cole–Cole equation (5.8). Larger pores would lead to slower relaxation and larger values of τ_0. A number of other attempts have been made to ascribe physical meaning to the Cole–Cole parameters (e.g. Revil *et al.*, 2012).

Following early studies by Olhoeft (1986) and others, there has been fast-growing interest in the application of IP field geophysical methods to characterize subsurface

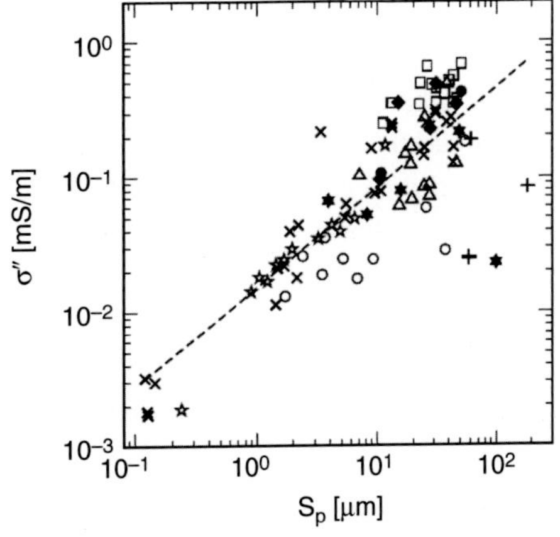

Figure 5.7 Imaginary conductivity as a function of interfacial geometric factor S_p for different geomaterials. After Kruschwitz *et al.* (2010).

organic contaminant plumes and other complex biogeochemical environments. The work of Sogade *et al.* (2006) was described earlier in this chapter. More recently, the spectral IP method was used successfully (Williams *et al.*, 2009) to monitor the biostimulated subsurface activity of microorganisms that reduce metallic iron and sulfate minerals in contaminated groundwater. A primary action of the microorganisms is to convert metallic minerals into a non-metallic sulfide precipitate and hence decrease the magnitude of subsurface electrode polarization. (Doherty *et al.*, 2010) used the IP method in the Wenner-array configuration (electrode spacing $a = 2.0$ m) in conjunction with self-potential, resistivity, and electromagnetic-induction geophysical data to help characterize an organic contaminant plume beneath an abandoned gasworks site in Northern Ireland. The observed chargeability anomalies were explained in terms of an electrochemical model in which ions and electrons are transferred across a thin clay layer separating perched wastewater from the underlying, biodegrading contaminant plume.

5.5 Non-polarizing electrodes

The difference in electric charge transport mechanism between the soil pore-fluid electrolyte and the metal electrodes staked into the ground results in an electrochemical impedance mismatch at the metal–electrolyte interface. If electric current passes in one direction between the soil and the electrode for an extended period of time, charge accumulates at the interface manifesting itself as an electrode polarization. The result is a spurious DC voltage. It is easy to misinterpret the spurious DC voltage as an apparent chargeability of the subsurface. The polarizability of a metal electrode depends on the type of metal employed, as shown in the study by LaBrecque and Daily (2008).

Non-polarizing electrodes reduce the electrochemical mismatch, and hence mitigate the spurious contribution to electrode polarization. Essentially, a non-polarizing electrode is one whose potential does not change upon the passage of an electric current (Bard and Faulkner, 2001). Typical materials used in the fabrication of non-polarizing electrodes are $Pb/PbCl_2$ and $Cu/CuSO_4$. The metal electrode is enclosed within a porous medium or a hard gel that is infused with a solution containing a salt of the same metal. The salt solution buffers the transition from electronic conduction in the metal electrode to ionic conduction in the soil electrolyte. The electric potential of carefully made $Pb/PbCl_2$ non-polarizing electrodes, such as those described by Petiau (2000), remains stable over time frames lasting months to years.

5.6 IP illustrated case history

Example. Archaeological investigation of paleometallurgical activities.

The mine of Castel-Minier in France was an important source of iron, lead, and silver during the Middle Ages. Originating as a by-product of smelting, relic slag heaps contain a

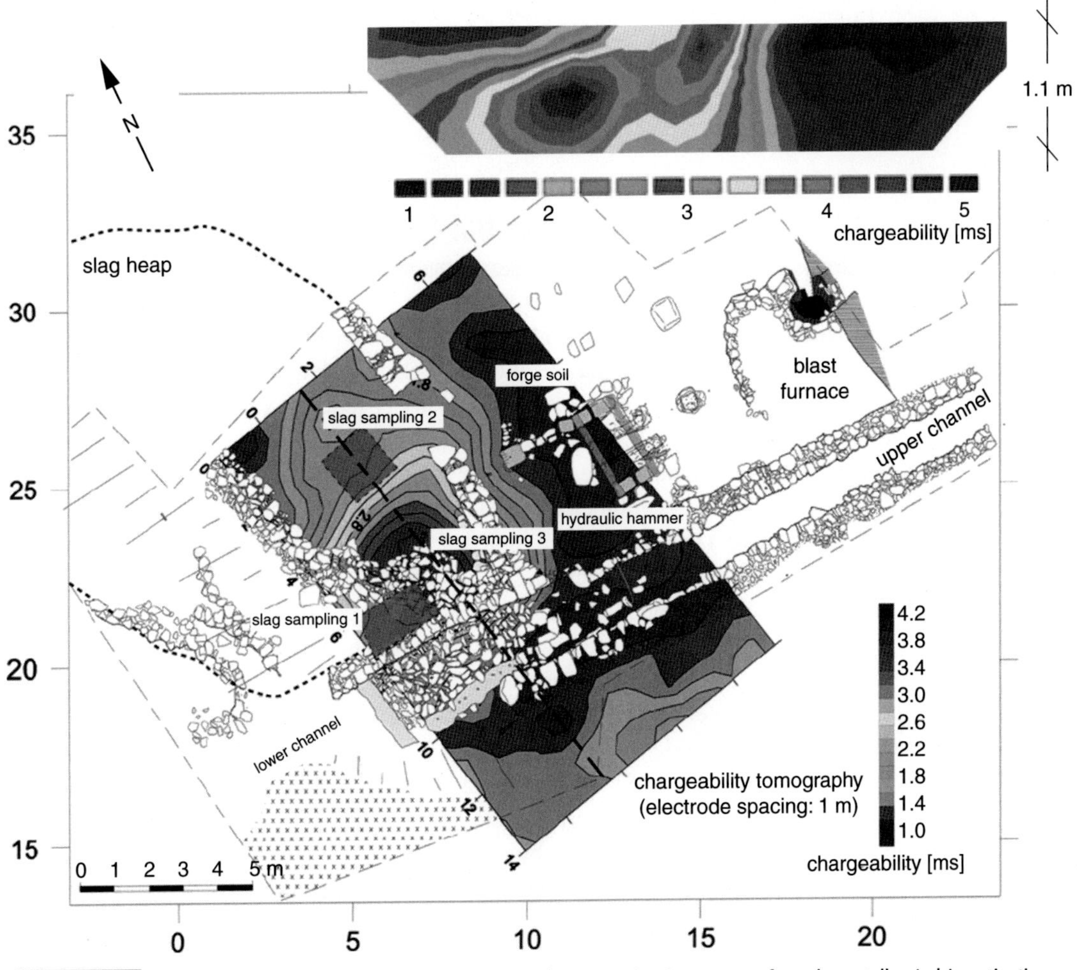

Figure 5.8 Apparent intrinsic chargeability map and vertical cross-section in support of a paleometallurgical investigation. The location of the cross-section corresponds to the N–S dashed line crossing the map. After Florsch *et al.* (2011). See plate section for color version.

wealth of information of interest to metallurgical archaeologists. Of primary interest is to make a determination of the total volume of slag so that the productivity of the former mining operations can be assessed. Slag heaps contain magnetite and other metallic minerals and consequently make an excellent IP target on the basis of electrode polarization. An apparent *intrinsic chargeability* map, including a vertical cross-section, based on the IP survey conducted by Florsch *et al.* (2011) using a Wenner array (electrode spacing $a = 1.0$ m) is shown in Figure 5.8. The intrinsic chargeability m is defined in this study as the partial chargeability M_{12} between $t_1 = 10$ ms and $t_2 = 30$ ms, multiplied by $\Delta t = t_2 - t_1 = 20$ ms, so that $m = M_{12}\Delta t$. The buried slag is easily identified as the zone of high intrinsic chargeability with $m > 3$ ms. The map also shows the spatial relation of the slag heap in the larger context of the site archaeological excavation. Based on a calibration of the IP response from slag sampled from the three shaded regions shown in the figure, Florsch *et al.* (2011) have demonstrated that the slag content increases in proportion to its apparent chargeability. A detailed analysis shows that the sixteenth-century mine produced ~ 7.6 tons of slag, corresponding to ~ 6.9 tons of commercial iron.

5.7 Self-potential (SP): introduction

Under natural conditions, without any introduction of an artifical current into the ground as in the resistivity/IP methods, various modes of electric charge imbalance or accumulation are found to spontaneously develop in the subsurface (Jouniaux *et al.*, 2009). The resulting self-potentials, which can be measured using pairs of electrodes deployed at the surface or less commonly in boreholes, are very stable in time and are sometimes found in the range of ± 100 mV or higher. A voltmeter is used to measure the SP signals; it should have a high sensitivity (0.1 mV) and a high input impedance (typically >100 MΩ) compared to that of the ground between the two electrodes (Revil *et al.*, 2012).

In near-surface applications, commonly, one of the electrodes is held at a fixed location and used as a base or reference station while the other electrode is moved from place to place across an area under investigation. The potential difference between the roving electrode and the reference electrode is the quantity that is measured. In more sophisticated setups, a multi-electrode array of electrodes can also be deployed for monitoring the spatiotemporal changes in SP signals across the survey area. In such cases, the potentials at the various electrodes comprising the array, relative to the potential of the fixed electrode, are recorded periodically by a multiplexed voltmeter.

Self-potential data should always be corrected prior to interpretation for temporal drift and topographic effects (Zhou *et al.*, 1999). A topographic correction is required since SP signals tend to increase in the downhill direction, along the local hydraulic gradient. It has also been commonly observed that SP readings are noisier in heavily vegetated areas, relative to sparsely vegetated areas, due to bioelectric activity. To ensure good electrical contact with the ground, electrodes can be inserted into shallow augured holes filled with a mud slurry.

Example. SP mapping of the groundwater system at Stromboli volcano.

An SP survey conducted on the persistently active Stromboli volcano in the Tyrrhenian Sea, offshore Italy, is reported by Finizola *et al.* (2006) with the goal to identify subsurface groundwater flow patterns. Detection of aquifers on volcanic islands is required for improved water-resource management and also better hazard assessment since water can react violently with magma and cause dangerous phreatic eruptions. The SP data with 20-m station spacing are shown in Figure 5.9 along with a co-located 2-D inversion of resistivity data using the Wenner electrode configuration (see Chapter 4). The hydrothermal system beneath Stromboli summit is easily recognized by the strong SP variations within the region of lowest resistivity, i.e. the rightmost shaded region marked C on the figure.

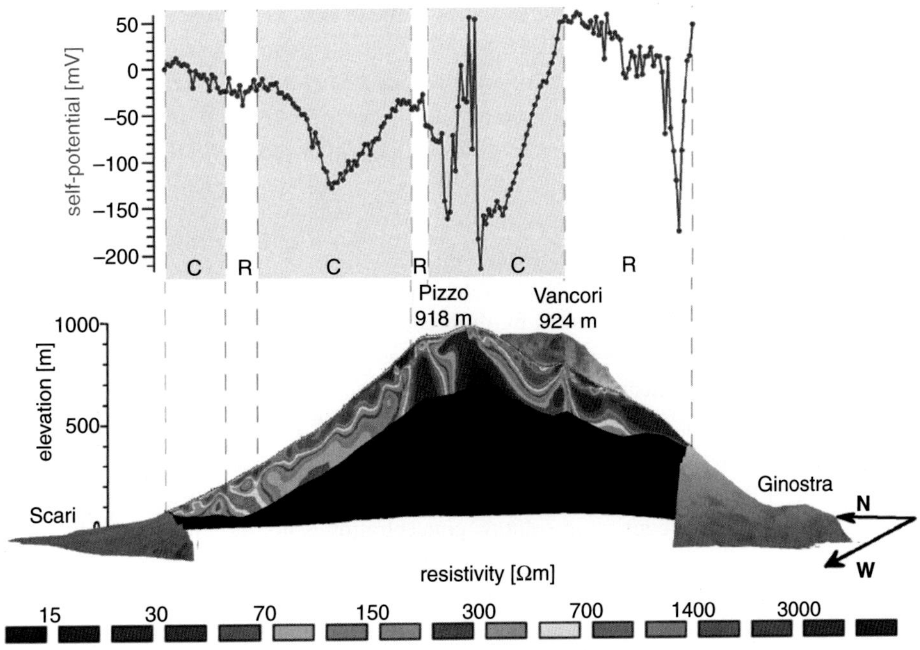

Figure 5.9 SP signals and 2-D electrical resistivity structure at Stromboli volcano, Italy. C = conductive zone; R = resistive zone. After Finizola *et al.* (2006). See plate section for color version.

5.8 Physical mechanisms

Subsurface charge accumulations may be sustained by a number of physical and electrochemical processes. Large SP anomalies, up to hundreds of mV, for example, have long been measured in association with zones of metal sulfide and oxide mineralization. An enduring theory purporting to explain the so-called *mineralization potential* of sulfide bodies is due to Sato and Mooney (1960). Stable SP anomalies of comparable magnitude may also be registered in the presence of metallic cultural noise such as pipelines, steel well casings, metallic fences, and utility boxes (Revil *et al.*, 2012). This type of SP signal is believed to be caused by the movement of electrons in response to spatial variations in the redox potential at heterogeneous metal–electrolyte interfaces.

Another common mechanism that generates a spontaneous potential involves the movement of electrolytic groundwater through a porous rock. A *streaming potential*, for example, develops when pore-fluid electrolyte flows in response to a pressure and/or thermal gradient. Due to the presence of the electric double layer at the mineral surface (Figure 5.6), the fluid flow carries an excess of the mobile counter ions resident in the bulk electrolyte, relative to a deficit of ions of the opposite sign that are fixed in the immobile layer. This differential movement of ions and counter ions is equivalent to an electric current. An effective resistance to the current is provided by the electrolyte viscosity η, its resistivity ρ, and the *zeta potential* ζ (the potential difference between the charged mineral surface and an arbitrary point within the neutral bulk electrolyte). The streaming potential is nothing more than the resulting electrical potential drop between two points along the path of the fluid flow (e.g. Rice and Whitehead, 1965; Levine *et al.*, 1975). A streaming potential of this type often arises as pore electrolytes flow in response to an external mechanical disurbance; this constitutes the fundamental basis for the *seismoelectric effect* to be discussed in Chapter 10.

A different type of SP effect occurs at the liquid–liquid interface between two pore fluids that are characterized by different concentrations of ions in solution. There will be a net transfer of ions across the junction as the more-concentrated solution diffuses into the less-concentrated solution. Since cations and anions generally have different mobilities, the rates of diffusion of the different charged species across the interface are unequal. The differential movement of ions leads to the development of a net charge distribution on the liquid–liquid interface, thereby generating a liquid junction or *diffusion potential.* A sharp contrast in solution concentration is not required, as a diffusion potential can also occur in the presence of a salinity gradient.

It has long been recognized in geophysical well-logging that a spontaneous potential develops at junctions between permeable sand and less-permeable shale beds. Suppose the pore fluid is an ideal electrolyte consisting of NaCl only. Due to its layered clay structure and charged mineral surfaces, the shale is somewhat permeable to the Na^+ cations but it is relatively impervious to the Cl^- anions. Suppose the shale bed separates two sand beds which contain different NaCl concentrations. There will be a net migration of Na^+ cations across the shale bed from the more-concentrated toward the less-concentrated solution. Since shales preferentially pass the cations and restrict the anions, they act much like an ion-selective membranc (Bard and Faulkner, 2001). The differential movement of ions in shale beds generates the *membrane potential* that is measured by a downhole voltmeter.

The *thermoelectric effect* can also be a significant source of SP signals in geothermal systems, volcanoes, near-underground coal fires, and in other geological settings that are characterized by persistent spatial variations in subsurface temperature. The thermoelectric effect is the appearance of a voltage across a sample of geomaterial that is not of a uniform temperature. A temperature gradient causes mobile charge carriers within the geomaterial to migrate from regions of elevated temperature to regions of cooler temperature. The differential movement of charge carriers of different species establishes a non-uniform subsurface charge distribution. The thermoelectric coupling coefficient [mV/°C] is defined as the ratio $\Delta V/\Delta T$ of the voltage to temperature difference. Zlotnicki and Nishida (2003) report values of the order of ~ 0.1–1.0 mV/°C.

Finally, Minsley *et al.* (2007) have shown that subsurface spatial variations in electrical conductivity can also have an effect on potential measurements made at the surface. However, it is often difficult to estimate the magnitude of this effect in practical situations since the underground distribution of electrical conductivity is rarely known.

5.9 Interpretation of SP measurements

Since SP signals are generated by equal and opposite charge distributions, as indicated schematically in Figure 5.1, an SP source may be modeled as an assemblage of electric dipoles. Recall that an electric dipole is the simplest possible distribution of equal and opposite charge, namely a positive point charge q and a negative point charge $-q$ separated by a distance d. The electric dipole moment is given by $m = qd$. A polarized sphere, in which positive charge is distributed over one hemisphere and negative charge is distributed over the other hemisphere, may be represented at large distances relative to its radius by an equivalent electric dipole of moment M.

Polarized sphere. It is straightforward to calculate the electric potential due to a polarized sphere. As shown in Figure 5.10a, suppose the sphere is polarized in the vertical plane $y = 0$ in a direction that makes angle α with respect to the x-axis so that

$$\mathbf{M} = M \cos\alpha\, \hat{\mathbf{x}} + M \sin\alpha\, \hat{\mathbf{z}}. \tag{5.10}$$

The electric potential V of an equivalent electric dipole of moment $\mathbf{M}$ is given by the standard formula (e.g. Blakely, 1995)

$$V = -\frac{\mathbf{M}\cdot\nabla}{4\pi\varepsilon_0}\frac{1}{R} \tag{5.11}$$

where $\varepsilon_0 = 8.85 \times 10^{-12}$ F/m is the permittivity of free space and

$$R = |\mathbf{r}_P - \mathbf{r}_Q| = \sqrt{(x - x_0)^2 + z_0^2} \tag{5.12}$$

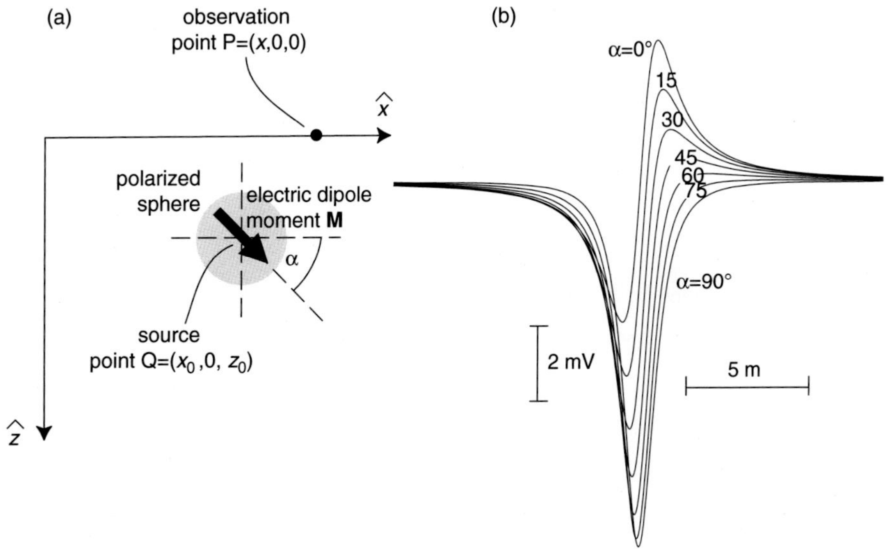

Figure 5.10 (a) A polarized sphere modeled as an electric dipole. (b) The SP response of a buried polarized sphere for different angles of polarization.

is the distance between the dipole at source location $Q = (x_0, 0, z_0)$ and the observation point at Earth's surface, $P = (x, 0, 0)$. Inserting Equations (5.10) and (5.12) into Equation (5.11) results in

$$V(x) = \frac{M}{4\pi\varepsilon_0}\frac{(x - x_0)\cos\alpha - z_0\sin\alpha}{[(x - x_0)^2 + z_0^2]^{3/2}}. \tag{5.13}$$

The function $V(x)$ is plotted in Figure 5.10b for different values of the angle of polarization from $= 0°$, aligned with the $+x$-axis, to $\alpha = 90°$, aligned with the $+z$-axis. The sphere is buried at depth $z_0 = 1.0$ m and located at $x_0 = 0$. The dipole moment is $M = 10^6$ Cm. Finding a similar analytic solution for the SP response of a long horizontal cylinder is left as an exercise for the reader.

Polarized dipping sheet. Another simple geological scenario for which the SP response is amenable to an analytic treatment is fluid flow within a dipping planar fault structure. The fault plane can be modeled as an electrically polarized thin sheet, or ribbon, of half-width a embedded in a host medium of uniform resistivity ρ (Paul, 1965), as shown in Figure 5.11a. One edge of the sheet is represented as a line of negative charges of density $-\Sigma$ [C/m] while the other is a line of positive charges of density $+\Sigma$ [C/m]; both lines extending from $-\infty$ to $+\infty$ in the along-strike ($\pm y$) direction, i.e. in and out of the page.

The electric potential V due to a line of charges has the well-known logarithmic form (e.g. Blakely, 1995)

$$V_\pm = \pm\frac{\Sigma\rho}{\pi}\ln\left(\frac{1}{r_\pm}\right), \tag{5.14}$$

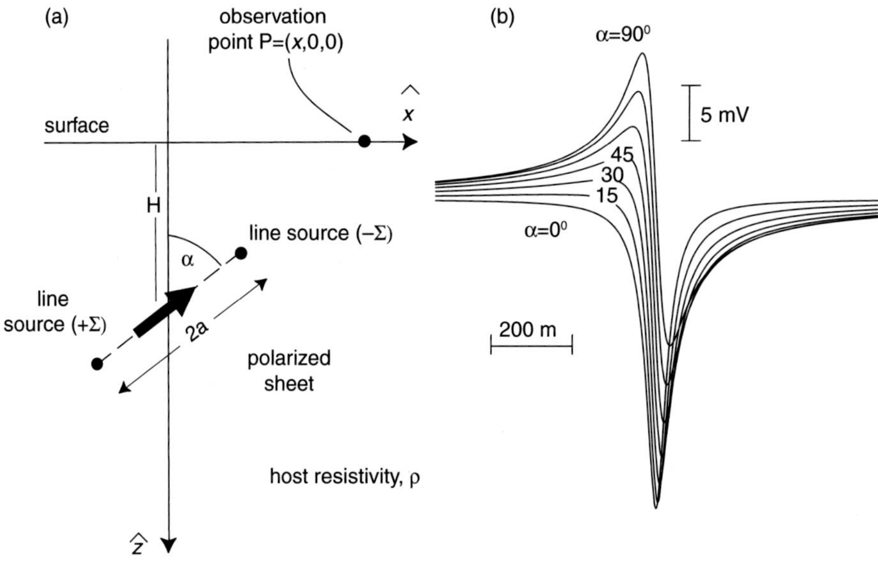

Figure 5.11 (a) A polarized, inclined sheet modeled as electric line charges. (b) The SP response of a buried polarized sheet for different angles of inclination.

where $r_\pm$ is the distance from the observation point to a positive or negative line of charges, respectively. A glance at the geometry in the figure shows that

$$r_- = \sqrt{(x - a\cos\alpha)^2 + (H - a\sin\alpha)^2}; \quad (5.15)$$

$$r_+ = \sqrt{(x + a\cos\alpha)^2 + (H + a\sin\alpha)^2}; \quad (5.16)$$

from which it readily follows

$$\ln\left(\frac{1}{r_-}\right) = -\frac{1}{2}\ln[(x - a\cos\alpha)^2 + (H - a\sin\alpha)^2]; \quad (5.17)$$

$$\ln\left(\frac{1}{r_+}\right) = -\frac{1}{2}\ln[(x + a\cos\alpha)^2 + (H + a\sin\alpha)^2]. \quad (5.18)$$

The total SP signal due to the two lines of charges is the sum of the potentials due to each sheet, namely, $V = V_+ + V_-$. Adding the two contributions using the above equations results in

$$V = \frac{\Sigma\rho}{2\pi}\left\{\ln[(x - a\cos\alpha)^2 + (H - a\sin\alpha)^2] - \ln[(x + a\cos\alpha)^2 + (H + a\sin\alpha)^2]\right\}, \quad (5.19)$$

which is plotted at Figure 5.11b for different values of the angle α. The following parameters are used: $\Sigma = \pm 0.01$ C/m, $\rho = 10.0\ \Omega$m, fault-zone half-width $a = 10$ m, and depth of burial $H = 30$ m.

A number of rules can be inferred from the analytic solutions which relate the width of an SP anomaly to the depth of an idealized source. For example, it is easy to show that the depth to a point charge is given by $d = x_{0.5}/\sqrt{12}$, where $x_{0.5}$ is the FWHM of the corresponding SP anomaly, i.e. its width at one-half its maximum value. Similarly, the depth rule for a vertically polarized sphere is $d \sim 0.65x_{0.5}$.

Self-potential surveys often provide useful hydrological information about fractures and karst conduits since streaming-potential anomalies have long been associated with divergent groundwater flow (Lange and Barner, 1995). As illustrated schematically in Figure 5.12, residual SP anomalies, after drift and topographic corrections have been made, are generally positive over groundwater discharges at springs and negative over groundwater infiltration at sinkholes.

5.10 Continuous wavelet transform analysis

The continuous wavelet transform (CWT) was introduced in Chapter 2 as a means for analyzing non-stationary data series. In this section we will see that the CWT can also be used for interpreting potential field measurements in terms of the geometry of the causative source and its depth of burial. The underlying theory is developed in Moreau *et al.* (1997, 1999) and has since been applied to many types of geophysical data including magnetics by Vallee *et al.* (2004). Here we outline the application of CWT for analysis of SP data

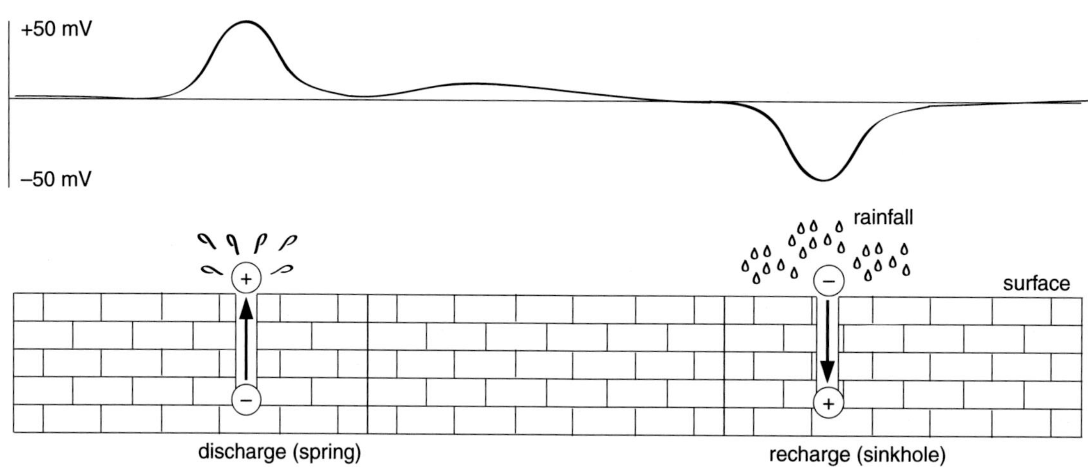

Figure 5.12 (Top) Schematic illustration of streaming-potential residual anomalies associated with groundwater recharge and discharge in karst terrain. (Bottom) The arrows indicate groundwater-flow direction, while the positive and negative symbols indicate equivalent charge polarization.

acquired along a single profile, although the general theory applies to data collected in two spatial dimensions.

To begin, recall that the CWT of an arbitrary function $f(x)$ can be written as

$$\tilde{f}(s,t) = \int_{-\infty}^{\infty} dx \frac{1}{s} \psi\left(\frac{x-t}{s}\right) f(x). \tag{5.20}$$

Suppose the function $f(x)$ is a potential field, such as gravity, magnetics, or SP data. Moreau *et al.* (1997; 1999) have shown that a powerful analysis of potential field data is enabled by using a special wavelet $\psi(x)$ that is based on spatial derivatives of the familiar upward continuation operator.

Recall from Equation (3.38) that a potential field $f(x, y, z_0 - \Delta z)$ on the level surface $z_0 - \Delta z$ can be derived from an upward continuation of measurements of the field $f(x, y, z_0)$ made at the lower level z_0. The formula is

$$f(x,y,z_0 - \Delta z) = \int_{-\infty}^{\infty} \int_{-\infty}^{\infty} f(x',y',z_0) \Psi_U(x-x',y-y',\Delta z) dx' dy' \tag{5.21}$$

where the upward continuation operator is

$$\Psi_U(x,y,\Delta z) = \frac{\Delta z}{2\pi}[(x-x')^2 + (y-y')^2 + \Delta z^2]^{-3/2}. \tag{5.22}$$

Suppose that the potential field is a function of x and z only, i.e. $f = f(x, z)$, which would be the case if the causative body is infinitely extended and uniform in a horizontal strike

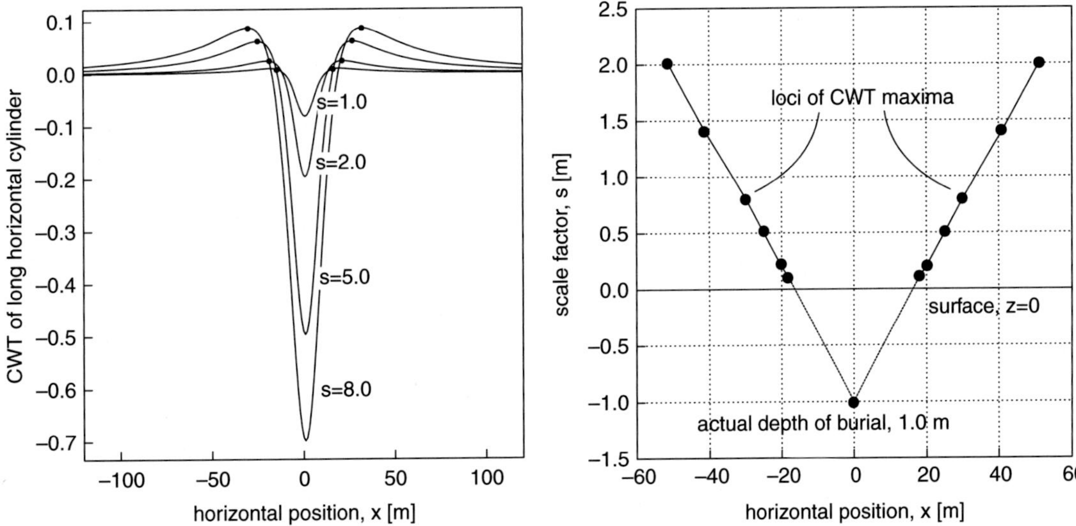

Figure 5.13 (Left) The continuous wavelet transform of a buried long cylinder for various values of *s*, with the symbols marking the maxima. (Right) Straight lines joining the CWT maxima point toward the actual depth of burial of the cylinder.

direction, y. In that case, the integration along the strike direction in Equation (5.21) can be carried out analytically with the result

$$\lim_{L\to\infty}\int_{-L}^{L}\frac{dy'}{[(x-x')^2+(y-y')^2+\Delta z^2]^{3/2}}\sim\frac{2}{(x-x')^2+\Delta z^2}. \tag{5.23}$$

The upward continuation operator therefore simplifies to

$$\Psi_U(x,\Delta z)=\frac{\Delta z}{\pi[(x-x')^2+\Delta z^2]}. \tag{5.24}$$

The analyzing wavelet for use in Equation (5.20) is the horizontal derivative of the upward continuation operator evaluated at $x'=0$, namely $\psi(x)=\partial\Psi_U/\partial x|_{x'=0}$ or,

$$\psi(x)=\frac{2x\Delta z}{\pi[x^2+\Delta z^2]^2}. \tag{5.25}$$

A wide range of other analyzing wavelets based on complex linear combinations of horizontal, vertical, and higher-order spatial derivatives of the upward continuation operator can also be selected; the choice Equation (5.25) is highlighted here due to its simplicity.

Using this choice of analyzing wavelet in Equation (5.20), the continuous wavelet transform $\tilde{f}(s,t)$ of the SP field due to a long, horizontal cylinder (see Problem 1, below, for the analytic formula) buried at depth $a = 1.0$ m beneath horizontal position $x_0 = 0$ is shown in Figure 5.13, left. The maxima of the functions $\tilde{f}(s,t)$ for different values of s are shown in Figure 5.13, right. If straight lines are formed by joining

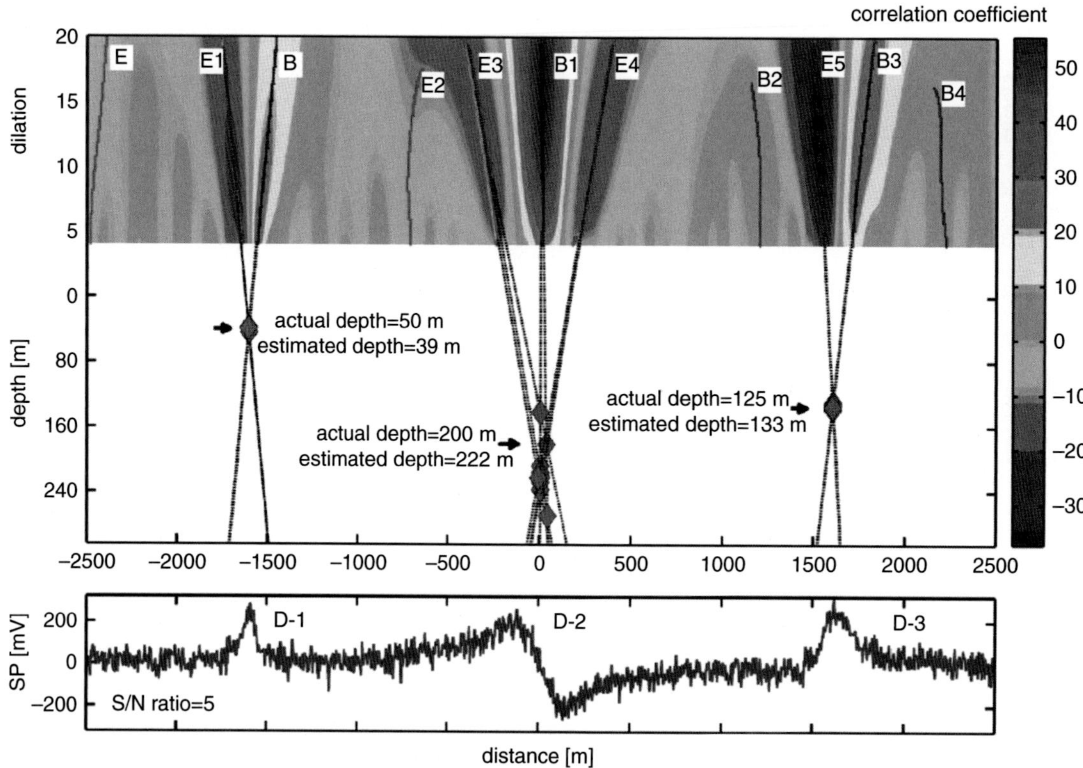

Figure 5.14 Depth determination to SP sources using CWT analysis. After Mauri *et al.* (2010). See plate section for color version.

together the maxima, they point toward the scale value $s = -a$, which is the burial depth of the cylinder. This example is indicative of a general rule (Moreau *et al.*, 1997; 1999) that the loci of maxima of the continuous wavelet transforms of potential fields converge towards their causative sources. The general rule is valid in $n = 1$, 2, and 3 spatial dimensions and applies to both infinitely extended and compact sources of different shapes.

A more detailed example of source-depth determination using wavelet analysis is shown in Figure 5.14. The bottom plot shows theoretical SP signals [mV] for three electric dipoles, each signal being contaminated by Gaussian-distributed noise with a signal-to-noise (S/N) ratio of 5. The source D-1 is a horizontal dipole buried at 50 m depth; D-2 is a vertical dipole buried at 200 m; D-3 is inclined at 45° and buried at 125 m. The CWT of the SP signals is shown in the top panel of Figure 5.14. In this example, the second vertical derivative of the upward continuation operator was selected as the analyzing wavelet. The extremal lines (labeled with B and E in the figure) converge to depths marked by the red diamonds in the figure. The scatter in the positions of the red diamonds is caused largely by the presence of the noise in the synthetic SP signals; however, the depths estimated by the CWT analysis are in good agreement with the model depths.

5.11 SP illustrated case history

Example. SP mapping of an active landslide.

Geoelectrical measurements have become widely used to characterize and monitor active landslides. The SP technique is often used in conjuction with electrical resistivity tomography (ERT) to identify potential slip planes, bedrock surfaces, and zones of fluid discharge and infiltration at sites that are vulnerable to mass movement. Chambers *et al.* (2011) have combined SP, ERT, and conventional geological mapping techniques in their study of an escarpment within the Jurassic mudrocks of northeast England that is triggering earthslides and earthflows. The observed SP signals at this site are caused by streaming potentials associated with downslope groundwater movement.

The SP data were acquired along several transects coincident with 2-D ERT profiles. Non-polarizing Pb/PbCl electrodes were used for both the reference and roving SP electrodes, with 5-m station spacing. Two SP profiles and their co-located resistivity images are shown in Figure 5.15. As a check on repeatability and to estimate temporal drift, SP data were acquired and are shown in the figure for measurements taken in both the upslope (green symbols) and downslope (dark-blue symbols) directions, along with the average readings (solid orange line). The data show good repeatability with a drift of not more than ~ 3 mV. Along both profiles, it is readily seen that the SP signals become more positive with increasing distance downslope. This trend is consistent with a streaming potential caused by groundwater infiltrating at the head of the slope and flowing downslope.

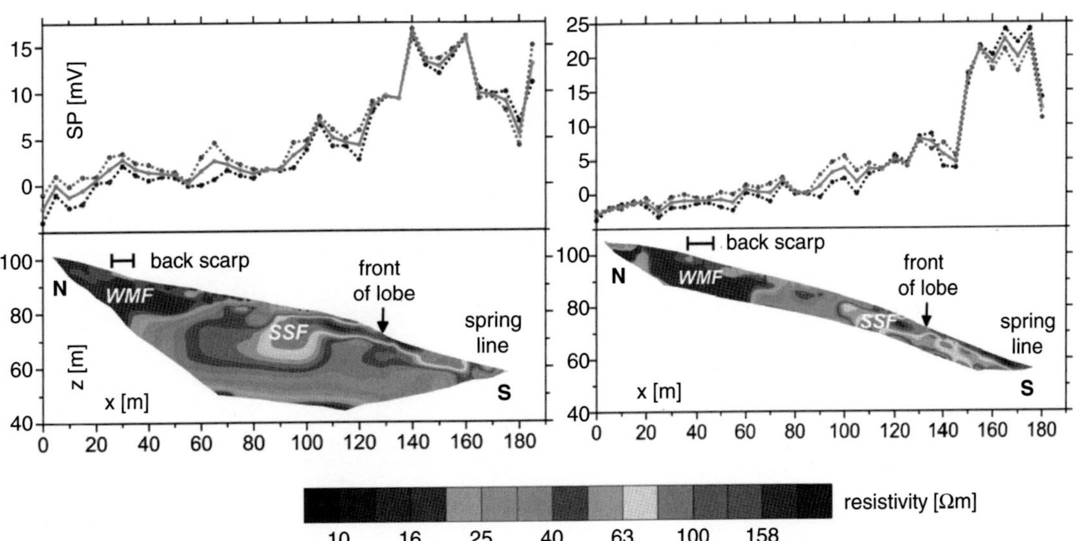

Figure 5.15 SP profiles and co-located 2-D ERT images at an active landslide in northeast England; WMF = Whitby mudstone formation; SSF = Staithes sandstone and Cleveland ironstone formation. After Chambers *et al.* (2011). See plate section for color version.

The sudden increase in SP values in both profiles is associated with fluid discharge though seeps located at the front of the lobe. The SP data are also consistent with the electrical resistivity images, which predominantly reflect lithological variations. For example, the WMF shows low resistivity due to its high clay content. The thin, near-surface zones of high resistivity within the SSF are areas characterized by heterogeneous, well-drained silts and sands. Overall, the combined SP and ERT dataset provides complementary perspectives on, respectively, fluid-flow patterns and the underlying subsurface lithological variations within an active landslide region.

Problem

1. Show that the potential $V(x)$ due to a long horizontal cylinder aligned in the y-direction and buried at location (x_0, z_0) is

$$V(x) = \frac{M'}{2\pi\varepsilon_0}\frac{(x - x_0)\cos\alpha - z_0\sin\alpha}{(x - x_0)^2 + z_0^2},$$

where M' is the electric dipole moment per unit length and the cylinder is polarized at angle α with respect to the x-axis. Hint: consider the cylinder to be a line of dipoles along the y-axis and integrate the standard dipole formula (5.11) accordingly.

6 Seismic reflection and refraction

The seismic-reflection and -refraction methods in near-surface geophysical investigations are based on the introduction of mechanical energy into the subsurface using an active source and the recording, typically using surface geophones, of the resulting mechanical response. Passive-source seismic methods also provide important information; these will be described in Chapter 7. The propagation of mechanical energy into the subsurface consists, to a large part, of *elastic waves*. The essential property of an elastic body is that it returns instantaneously to its original pre-deformed state with the removal of a mechanical force that changed its size and/or shape. A delayed return to the original state is termed *viscoelasticity*. Any permanent deformation, such as ductile deformation or brittle failure, is a measure of the *inelasticity* of the body. Significant permanent deformation of the ground surface can occur in the vicinity of large seismic disturbances such as earthquakes (e.g. Lee and Shih, 2011) but inelasticity can be safely neglected in most near-surface active-source or passive-source studies. An important characteristic of elasticity is the relationship between the strain, or deformation, of a body and the stress, or mechanical force, that produces the deformation of the body. An excellent review of the elementary physics of wave motion is found in French (1971).

6.1 Introduction

There are several possible types of elastic wave motion following the introduction of a seismic disturbance. The particle motion associated with compressional, or P-waves, is aligned with the direction of wave propagation (Figure 6.1a). The particle motion associated with shear, or S-waves, is aligned in a direction perpendicular to the direction of wave propagation (Figure 6.1b). Both vertically polarized (SV, as shown in the figure) and horizontally polarized (SH) motions are possible. The P- and S-waves are known as *body waves* since they are transmitted through the interior of the Earth. As shown by the shaded cells in the figure, P-waves are associated with a change in size and aspect ratio of an elementary material volume while S-waves are associated with a change in shape. With surface Rayleigh, or R-waves, discussed more fully in the next chapter, the particle motion near the surface is retrograde elliptical (Figure 6.1c, top) and only those particles in the region close to the surface of the Earth, at depths comparable to the elastic wavelength, are set into motion. A second type of surface wave motion (Figure 6.1*c*, bottom) is characterized by horizontal particle motion that oscillates transverse to the direction of wave propagation. Such waves are termed Love waves.

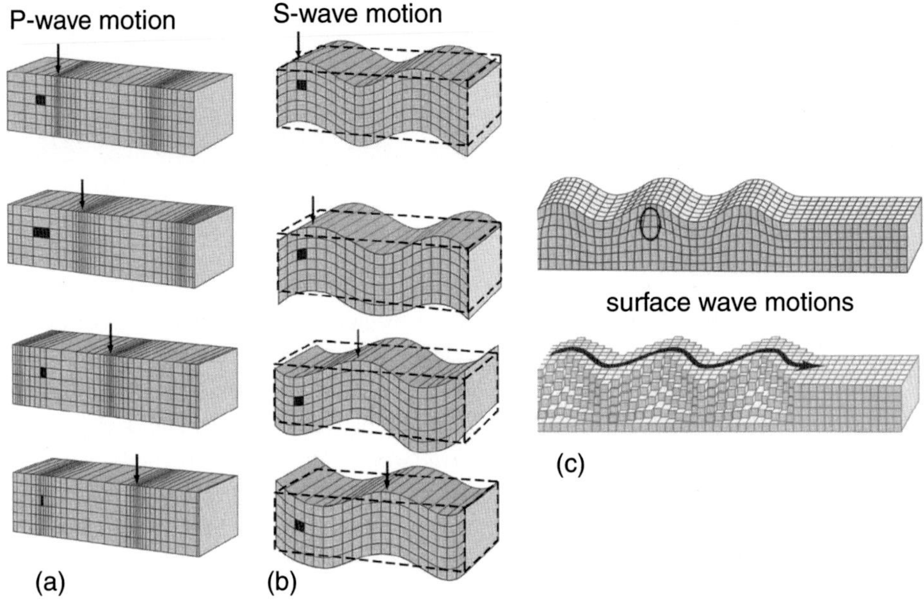

Figure 6.1 Particle motions associated with the propagation of seismic waves: (a) compressional; (b) shear (vertically polarized, SV); (c) two types of surface waves. Direction of propagation in all cases is to the right. After Grotzinger and Jordan (2010).

The following two case histories are presented to introduce the reader to recent applications of active-source seismology for imaging near-surface geological structures.

Example. High-resolution investigation for seismic hazard evaluation.

Reflection seismology offers a useful technique for seismic hazard assessment based on the study of buried fault structures. For example, in the New Madrid zone of contemporary seismicity within the northern Mississippi embayment of the central USA, neotectonic features are oftentimes obscured by Quaternary sediments. An integrated seismic-reflection analysis of P- and horizontally polarized SH-wave profiles was undertaken by Bexfield *et al.* (2006) to identify potentially reactivated Paleozoic bedrock faults that underlie the Quaternary cover. It is important to evaluate the extent of faulting in this area since critical facilities in this region such as dams, power plants, and bridges along the Ohio River are exposed to the significant intraplate earthquake hazard within the New Madrid seismic zone.

The P- and SH-wave profiles provide complementary perspectives on the subsurface at this location. The P-wave velocity responds strongly to the presence of water and the abundant methane gas since these fluids both affect the bulk modulus k. The SH-wave velocity, with its greater sensitivity to shear modulus μ, images primarily the solid rock or sediment component. The bulk and shear moduli, and their effect on seismic-wave propagation, are discussed later in this chapter. The SH-waves, despite a lower signal-to-noise ratio, yield higher spatial resolution in the uppermost 100 m since their wavelength is 0.3–0.5 times those of P-waves.

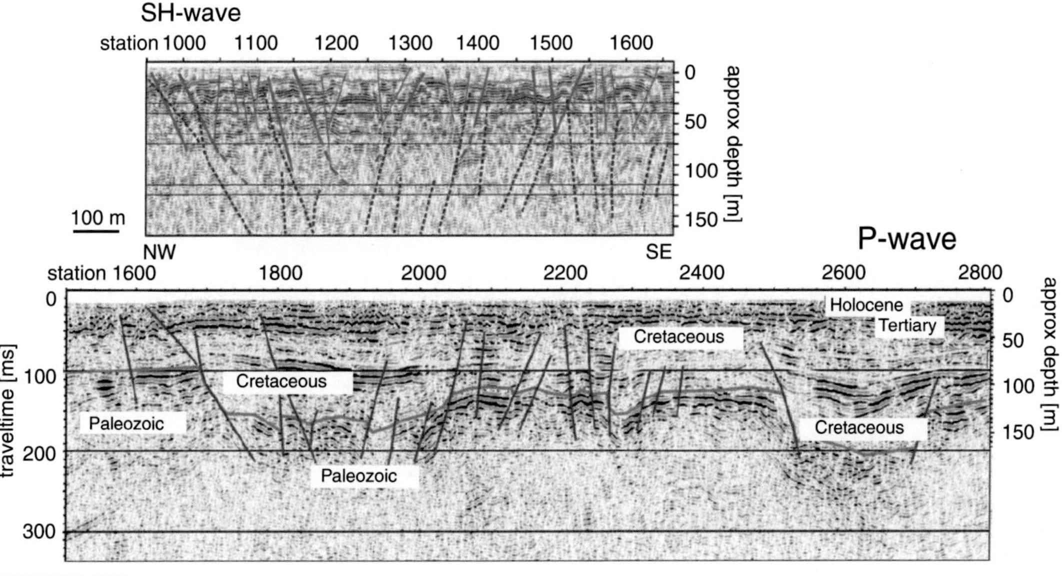

Figure 6.2 SH-wave (top) and P-wave (bottom) seismic profiles in the New Madrid zone of contemporary seismicity revealing reactivated Paleozoic bedrock normal faults. After Bexfield *et al.* (2006). See plate section for color version.

The seismic profiles shown in Figure 6.2 were acquired in wetlands adjacent to the Ohio River. Drilling in the region has indicated that depth to the Paleozoic bedrock is ~ 80–90 m beneath fluvial sediments and the underlying Paleocene to Cretaceous clay layers. The seismic P-wave section (Figure 6.2, bottom) reveals detailed images of bedrock faulting. Prominent in the profile, for example, are breaks in the bedrock reflector that are interpreted as grabens filled with Cretaceous sediments. The SH-wave section (Figure 6.2, top) does not clearly resolve to bedrock depths but does provide high-resolution images of near-surface deformation, including fine-scale faulting in the overlying Quaternary–Tertiary sediments. Many of the faults recognized in bedrock in the P-wave profile appear to propagate upward into the Quaternary sediments and are observed in the SH-wave profile.

Example. Shear-wave imaging of sinkholes in an urban environment.

Sinkholes in urban areas overlying karst or salt-dominated geology constitute another type of natural hazard. Human activities such as construction or groundwater utilization typically increase the potential for new sinkholes to develop or existing ones to become re-activated. Krawczyk *et al.* (2011) describe the development of a shear-wave seismic imaging system that they have used to characterize active sinkholes in a built-up area of Hamburg, Germany. Although the cause of the sinkholes is not yet fully understood, they may be associated with dissolution of shallow salt and caprock structures. A major reason for concern is that microearthquake activity recently observed nearby may be a precursor to the imminent collapse of the sinkholes.

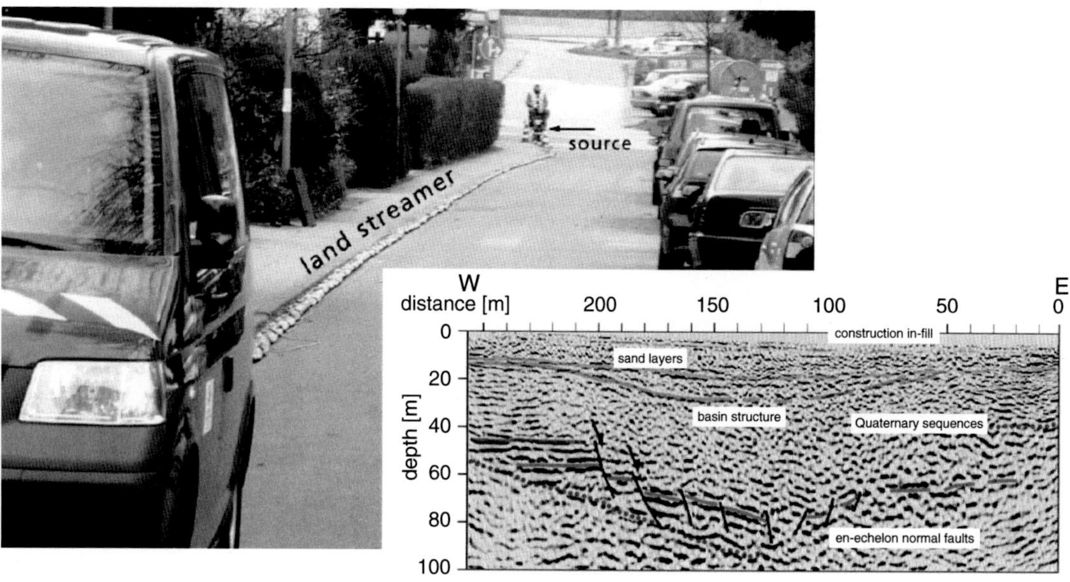

Figure 6.3 Shear-wave imaging of sinkholes in the city of Hamburg. After Krawczyk *et al.* (2011).

A 95-kg vibrator source swept through frequencies between 30 and 120 Hz was used to generate the shear waves. Signals were recorded using a land streamer of 120 geophones towed along a city street behind a vehicle (Figure 6.3, photo). The geophones were spaced 1 m apart and are preferentially sensitive to the detection of SH-polarized waves. The source spacing of 2.0 m resulted in ~ 40–50-fold data coverage (see the discussion on common midpoint profiling later in this chapter). The P-wave signals were largely suppressed by taking differences of traces acquired with opposite shear-wave polarity. After data processing, the migrated seismic section shown in Figure 6.3 was produced. The upper ~ 10 m of the section is interpreted as construction in-fill material and sandy layers. The basin-shaped seismic-reflecting horizon at 20–30 m depth observed within the Quaternary sedimentary sequence may be indicative of a subsidence feature. A pattern of en-echelon normal faults seen at ~ 55–80 m depths correlates with the known depth to the top of the salt/caprock structure. This case study demonstrates how seismic shear-wave analysis can be used for structural mapping in support of natural-hazard assessment in a noisy urban environment.

6.2 Stress and strain

An overview of the fundamental physics of elasticity is now presented. More complete, yet still elementary, treatments are given in Telford *et al.* (1990) and Gudmundsson (2011) while a more advanced approach may be found in Chapman (2004). The essential

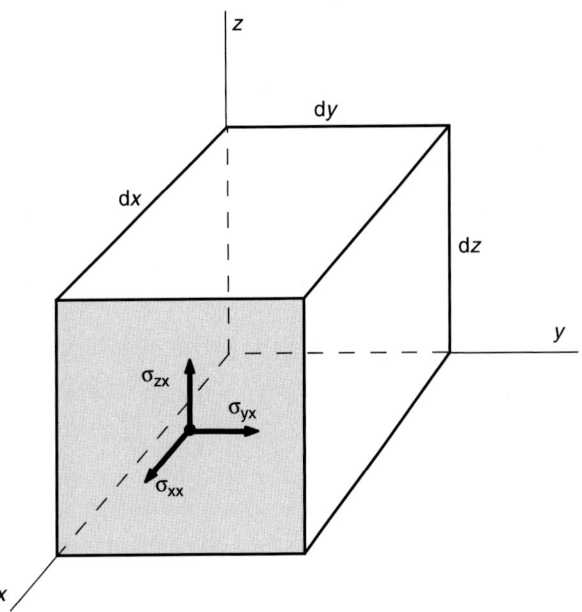

Figure 6.4 The three components of stress in a yz-plane perpendicular to the x-axis.

quantities in elasticity theory are stress and strain. Suppose a mechanical force is applied to an elastic body. Stress σ [Pa, or N/m^2] is defined as the ratio of the applied force [N] to the area [m^2] over which the force acts. A normal stress, or *pressure*, occurs when the applied force is directed perpendicular to the area. The pressure is defined herein to be positive if the normal stress is tensile and negative if it is compressive. A *shear* stress, on the other hand, is directed tangential, or parallel, to the area over which it is applied. A arbitrary stress can be resolved into its normal and shear components. The three stress components (σ_{xx}, σ_{yx}, σ_{zx}) in the highlighted vertical plane perpendicular to the x-axis are shown in Figure 6.4. Note that σ_{xx} is a normal stress (shown here as tensile) while σ_{yx} and σ_{zx} are shear stresses.

Suppose the elastic body is in static equilibrium, so that it is not undergoing active deformation. In this case, the stresses must balance so that there is no net shear or pressure on the body. Accordingly, the three stress components (σ_{xx}, σ_{yx}, σ_{zx}) on the opposing yz-plane to the one highlighted in Figure 6.4 must be equal and opposite to the components shown in the figure. It must also be noted that the tangential stress components, such as σ_{yx} shown in the figure, exert a torque on the elastic body. An inspection of the figure indicates that, in order for the body to be in static equilibrium, the torque due to σ_{yx} must be counterbalanced by a stress component σ_{xy} of equal magnitude acting parallel to the x-axis in the xz-plane so that $\sigma_{yx}= \sigma_{xy}$. In general, a condition for static equilibrium is a symmetric stress *tensor*, $\sigma_{ij} = \sigma_{ji}$, for $i, j = x, y, z$.

The small changes in size and shape that occur in response to stress are called strains. A simple rigid-body rotation through some angle θ or a rigid-body translation from one location to another, without a change in size and shape, does not constitute a strain. Strain ε is a dimensionless quantity defined as the fractional change in the size and shape of a body subject to loading. Suppose P and Q are two different points inside or on the surface of an elastic body. Let the vector $\mathbf{u}_P$ be the displacement of point P and the vector $\mathbf{u}_Q$ be the displacement of point Q in response to an applied stress. The strain is non-zero if $\mathbf{u}_P \neq \mathbf{u}_Q$ for any pair (P, Q). Strain can also be decomposed into normal and shear components.

Let the three Cartesian components of the displacement vector be denoted as $\mathbf{u} = (u, v, w)$. The diagonal element ε_{ii} of the strain tensor is the relative increase in length along the i-axis, where $i = x, y,$ or z, such that

$$\varepsilon_{xx} = \frac{\partial u}{\partial x}; \ \varepsilon_{yy} = \frac{\partial v}{\partial y}; \ \varepsilon_{zz} = \frac{\partial w}{\partial z}. \tag{6.1}$$

The *dilatation* Δ of a body is its fractional change in volume, given by

$$\Delta = \varepsilon_{xx} + \varepsilon_{yy} + \varepsilon_{zz} = \frac{\partial u}{\partial x} + \frac{\partial v}{\partial y} + \frac{\partial w}{\partial z}. \tag{6.2}$$

The diagonal elements of the strain tensor are normal strains and they describe the change in size of the body. The shear strains are the off-diagonal elements of the strain tensor that describe a change in shape of the body. The shear strains are defined as

$$\varepsilon_{xy} = \varepsilon_{yx} = \frac{\partial v}{\partial x} + \frac{\partial u}{\partial y}; \tag{6.3a}$$

$$\varepsilon_{yz} = \varepsilon_{zy} = \frac{\partial w}{\partial y} + \frac{\partial v}{\partial z}; \tag{6.3b}$$

$$\varepsilon_{zx} = \varepsilon_{xz} = \frac{\partial u}{\partial z} + \frac{\partial w}{\partial x}. \tag{6.3c}$$

For the very small strains that are relevant to near-surface seismology, a useful idealized description of the relationship between stress and strain is provided by *Hooke's law*. The law states that a given strain is directly proportional to the stress producing it and, moreover, the strain occurs simultaneously with application of the stress. A principle of superposition also applies: when several stresses are present, each stress produces a strain independently of the others. In an isotropic medium, in which elastic properties do not depend on direction, Hooke's law for normal and shear stresses is written as

$$\sigma_{ii} = \lambda\Delta + 2\mu\varepsilon_{ii}, \text{for } i = x, y, z; \tag{6.4a}$$

$$\sigma_{ij} = \mu\varepsilon_{ij}, \text{for } i \neq j; \tag{6.4b}$$

with $\lambda > 0$. The quantities (λ, μ) are known as the Lamé parameters. The parameter μ, as shown by equation (6.4b), determines the amount of shear strain that develops in response to a given applied shear stress. The parameter μ accordingly is termed the *shear modulus*. For liquids, which offer no resistance to shearing forces, the strain is unbounded and $\mu = 0$.

The Lamé parameter λ is not often used in applied geophysics. Of more practical importance is the *bulk modulus k*

$$k = \frac{3\lambda + 2\mu}{3}, \tag{6.5}$$

which provides a measure of the resistance of a material to a uniform compressive stress.

Table 6.1 Bulk and shear moduli of common geomaterials: (1 GPa=10^9 N/m^2)

	Bulk modulus, k [GPa]	Shear modulus, μ [GPa]
Limestone	~ 50	~ 25
Granite	~ 30	~ 20
Sandstone	~ 1	~ 0.5

Table 6.1 gives a rough indication of shear and bulk moduli for common geomaterials. More extensive tables of elastic moduli appear throughout the geophysical literature.

6.3 Wave motion

An elastic body will not remain in static equilibrium if it is acted upon by unbalanced stresses. Suppose for example that the stress on the back face of the cube in Figure 6.4 is σ_{xx} while the stress on the front face is slightly different, $\sigma_{xx}+(\partial\sigma_{xx}/\partial x)dx$. Suppose similar expressions hold for stresses in other directions. The mass of the cube is $dm = \rho dxdydz$, where ρ [kg/m^3] is the density. To determine the motion caused by the unbalanced stresses, an infinitesimal version of Newton's familiar law $\mathbf{F} = m\mathbf{a}$ applies; for example, the x-component of the force law is (following Telford *et al.*, 1990)

$$\rho\frac{\partial^2 u}{\partial t^2} = \frac{\partial\sigma_{xx}}{\partial x} + \frac{\partial\sigma_{xy}}{\partial y} + \frac{\partial\sigma_{xz}}{\partial z}. \tag{6.6}$$

We can now use Hooke's law and the definition of the strain tensor to re-write Equation (6.6) purely in terms of the displacement u as

$$\rho\frac{\partial^2 u}{\partial t^2} = (\lambda+\mu)\frac{\partial\Delta}{\partial x} + \mu\Delta^2 u, \tag{6.7a}$$

where $\Delta^2 = \partial^2/\partial x^2 + \partial^2/\partial y^2 + \partial^2/\partial z^2$ is the Laplacian operator and Δ is the dilatation defined in Equation (6.2). Similar equations are satisfied by v and w, i.e. the displacements in the y and z directions, respectively. These are

$$\rho\frac{\partial^2 v}{\partial t^2} = (\lambda+\mu)\frac{\partial\Delta}{\partial y} + \mu\Delta^2 v, \tag{6.7b}$$

$$\rho\frac{\partial^2 w}{\partial t^2} = (\lambda+\mu)\frac{\partial\Delta}{\partial z} + \mu\Delta^2 w. \tag{6.7c}$$

Now, differentiate the above three expressions with respect to x, y, and z respectively and add the results together. This procedure gives

$$\rho\frac{\partial^2\Delta}{\partial t^2} = (\lambda+2\mu)\nabla^2\Delta, \tag{6.8}$$

which we immediately recognize as the familiar wave equation

$$\frac{1}{V_P^2}\frac{\partial^2 \Delta}{\partial t^2} = \nabla^2 \Delta \tag{6.9}$$

where $V_P = \sqrt{(\lambda + 2\mu)/\rho}$ is the wave velocity. The associated waves are called dilatational, compressional, or *P*-waves, and V_P is the *P*-wave velocity.

A second set of wave equations may be derived as follows. Subtracting the *z*-derivative of Equation (6.7b) from the *y*-derivative of Equation (6.7c) yields

$$\rho \frac{\partial^2}{\partial t^2}\left(\frac{\partial w}{\partial y} - \frac{\partial v}{\partial z}\right) = \mu \nabla^2 \left(\frac{\partial w}{\partial y} - \frac{\partial v}{\partial z}\right). \tag{6.10}$$

If we define the rotational parameter $\theta_x = \partial w/\partial y - \partial v/\partial z$ then Equation (6.10) simplifies to the wave equation

$$\frac{1}{V_S^2}\frac{\partial^2 \theta_x}{\partial t^2} = \nabla^2 \theta_x \tag{6.11}$$

with the wave velocity $V_S = \sqrt{\mu/\rho}$. There are two other wave equations for θ_y and θ_z, respectively. The waves are called rotational, shear, or *S*-waves, and V_S is the *S*-wave velocity. Notice that the *P*-wave velocity always exceeds the *S*-wave velocity, $V_P > V_S$. The *S*-wave velocity is normally about 40–60% but never exceeds about 70% of the *P*-wave velocity. Physically, V_P is larger than V_S since solid materials generally offer greater resistance to the imposition of compressive forces as opposed to shearing forces.

Near-surface geophysicists routinely use P-waves. As we have already seen, some studies use shear waves but special sources, receivers, and acquisition and processing procedures are required to cancel the P-waves. In reflection or refraction studies, surface Rayleigh R-waves are generally considered as a source of noise, known as *ground roll*, but as shown in the next chapter important information can often be extracted from their analysis and, less frequently, from the analysis of Love waves. A fourth type of seismic wave is a guided wave, or *Lamb wave*. This wave is confined to subsurface thin layers and is sometimes useful for probing underground features such as coal seams, fault zones, and subsurface voids.

6.4 Seismic waves and elastic moduli

It is worthwhile to look briefly at relationships between the seismic wave velocities V_P and V_S and the elastic moduli. These relationships are of great interest to geotechnical engineers and others who require a knowledge of spatially distributed soil mechanical properties. Young's modulus E [N/m^2] is a measure of the longitudinal stress to the longitudinal strain (see Figure 6.5a); roughly speaking, high values of E indicate a stiff material while smaller values indicate a compliant, or soft material. Poisson's ratio σ is a dimensionless measure of the transverse strain to longitudinal strain (Figure 6.5b). The formulas are

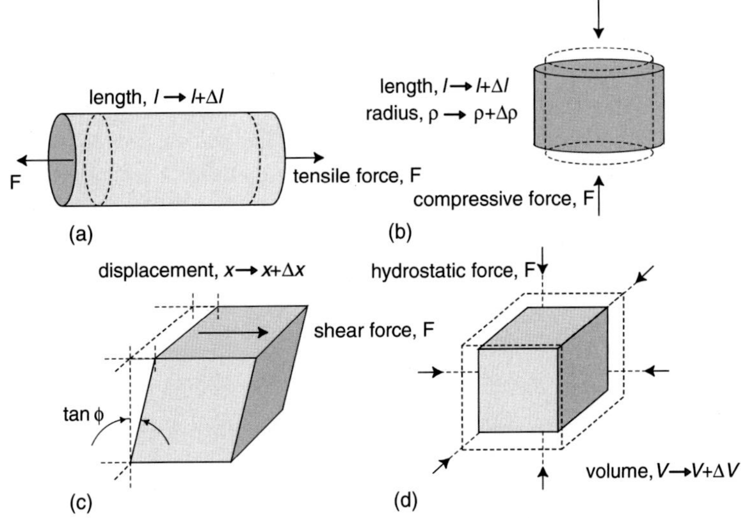

Figure 6.5 Definitions of the elastic moduli: (a) Young's modulus E; (b) Poisson's ratio σ; (c) shear modulus μ; and (d) bulk modulus k.

$$E = \frac{F/A}{\Delta l/l} = \frac{\mu(3\lambda + 2\mu)}{\lambda + \mu}; \tag{6.12}$$

$$\sigma = \frac{\Delta\rho/\rho}{\Delta l/l} = \frac{\lambda}{2(\lambda + \mu)}. \tag{6.13}$$

As mentioned earlier, the shear modulus μ [N/m^2] is a measure of the tangential stress to tangential strain, or shear stiffness, while the bulk modulus k [N/m^2] is a measure of the volume change in response to a change in hydrostatic pressure, or compressibility; see Figure 6.5c, d. The formulas are

$$\mu = \frac{F/A}{\tan\varphi}; \tag{6.14}$$

$$k = \frac{F/A}{\Delta V/V} = \frac{3\lambda + 2\mu}{3}. \tag{6.15}$$

Relationships amongst the seismic velocities and the elastic moduli are

$$V_P = \sqrt{\frac{k + 4\mu/3}{\rho}} = \sqrt{\frac{(1-\sigma)E}{(1+\sigma)(1-2\sigma)\rho}}; \tag{6.16}$$

$$V_S = \sqrt{\frac{\mu}{\rho}} = \sqrt{\frac{E}{2\rho(1+\sigma)}}; \tag{6.17}$$

where ρ is the density of the medium. It is important to note that both seismic velocities V_P and V_S are observed to increase with increasing density, even though it appears from Equations (6.16) and (6.17) that inverse relationships of the form $V \sim 1/\sqrt{\rho}$ exist.

The explanation is that the numerators $k + 4\mu/3$ and μ increase with increasing ρ faster than $1/\rho$ decreases.

Certain tasks in geotechnical engineering, including the design of bridge and building foundations, require a knowledge of the shear strength of the subsurface soil and rock. Determining an accurate value for the Poisson's ratio σ is fundamentally important in such studies. Poisson's ratio ranges from $\sigma \sim 0.1$–0.3 in competent igneous and sedimentary rocks to $\sigma \sim 0.45$ in unconsolidated sediments. From Equations (6.16) and (6.17) we find

$$\frac{V_P}{V_S} = \sqrt{\frac{2(1-\sigma)}{1-2\sigma}}; \tag{6.18}$$

thus it is evident that Poisson's ratio can be estimated *in situ* from a seismic determination of the velocities V_P and V_S. Near-surface applied geophysics is shown by this example to be directly relevant and useful to geotechnical engineers.

6.5 Seismic velocity of geomaterials

As shown in the previous section, the theory of elasticity indicates that the seismic velocities V_P and V_S of a homogeneous medium depend on density ρ and the elastic moduli, k and μ, according to Equations (6.16) and (6.17). However, in many practical situations, it is of interest to determine the bulk seismic velocities of mixtures of geomaterials such as fluid-bearing or clay-bearing sediments and consolidated rocks.

The *Nafe–Drake* curve is an empirical relationship between the P-wave velocity and density of water-saturated marine sediments that has been widely used for many years in exploration and crustal-scale geophysics (see e.g. Fowler, 2005, p.103). Wyllie's mixing law (Wyllie *et al.*, 1958),

$$\frac{1}{V_{bulk}} = \frac{\phi}{V_{fluid}} + \frac{1-\phi}{V_{solid}}, \tag{6.19}$$

is used extensively in well log analysis, where ϕ is porosity. It expresses the seismic P-wave traveltime ($\sim 1/V_{bulk}$) in a fluid-saturated medium as the porosity-weighted average of the P-wave traveltimes in the fluid and solid constituents, $\sim 1/V_{fluid}$ and $\sim 1/V_{solid}$, respectively. Other mixing laws have been proposed in the literature to explain velocity–porosity relations, with velocity generally falling as porosity increases. A good review of the literature on the bulk seismic velocities of various mixtures of geomaterials is given by Knight and Endres (2005).

Marion *et al.* (1992) have measured the seismic velocity of water-saturated, unconsolidated sand–clay mixtures. They find that bulk V_P peaks at a critical value of the clay content, roughly 40%. The following explanation for this behavior is offered. In shaly sands (clay content less than critical value), sand is the load-bearing element and the clay particles are dispersed in the pore space between the sand grains. Increasing the

Table 6.2 Seismic compressional wave velocities

	Velocity [m/s]		Velocity [m/s]
Air	330	Sandstone	1500–4500
Sand (dry)	200–800	Ice	3000–4000
Clay	1100–2500	Limestone	2500–6500
Sand (saturated)	800–1900	Granite	3600–7000
Water	1450	Basalt	5000–8400

clay content of shaly sands fills in the pore space, thereby decreasing the bulk porosity while stiffening the pore-filling material. Both these effects tend to increase the bulk velocity. In sandy shales (clay content higher than critical value), the sand grains are suspended in a clay matrix. In this regime, increasing further the clay content increases the bulk porosity of the mixture since the porosity of pure clay is higher than that of pure sand. This increase in bulk porosity causes the velocity in sandy shales to decrease. Right at the critical value, there is just enough clay to completely fill the space between sand grains. This configuration is the one of minimum bulk porosity, and hence the observed peak in seismic velocity. The work of Marion *et al.* (1992) applies to unconsolidated sand–clay mixtures. Gal *et al.* (1999) point out that, in consolidated sandstones, clay can act as a load-bearing element by coating the sand grains and cementing them together. In such cases, V_P increases rapidly and monotonically with increasing clay content.

A rough guide to the P-wave velocities of selected geomaterials is shown in Table 6.2. The wide ranges in velocities are due in part to variations in lithology, but more important in near-surface geophysics are the general rules that unsaturated, unconsolidated, weathered, fractured, unfrozen, and heterogeneous geomaterials have lower seismic velocities than their saturated, consolidated, unweathered, intact, frozen, and homogeneous counterparts. In the near-surface zone of aeration, bulk seismic velocity is often less than that of water and can become less than that of air. In anisotropic media such as finely interbedded sediments or fractured rocks, the velocity parallel to the strike direction is typically greater by 10–15% than the velocity across the strike.

The effect on seismic velocities of dense non-aqueous-phase liquid (DNAPL) contamination has been examined in the laboratory by Ajo-Franklin *et al.* (2007). They find that bulk V_P of both natural sandy aquifer and synthetic glass-bead samples is reduced by up to ~ 15% as trichloroethylene (TCE) saturation increases to ~ 60%. Such high TCE concentrations have been found at actual contaminated sites but the spatial extent of the accumulations are typically far below the spatial resolution of seismic experiments.

For S-waves, Santamarina *et al.* (2005) quote a range of V_S from less than 50 m/s up to 400 m/s for near-surface saturated soils, rising to 250–700 m/s for lightly cemented soils.

6.6 Reflection and refraction at an interface

The behavior of a seismic wave incident upon an interface between two elastic media is shown in Figure 6.6. One part of the wave energy is *reflected* back into medium 1 while another part is *refracted* into medium 2. Elastic waves are distinguished from acoustic waves in the sense that an incident compressional elastic wave is split into both compressional and shear reflected and refracted components, a process termed P-to-S conversion.

The laws of reflection and refraction may be derived using Huygens' principle. This principle is helpful to understanding the time evolution of seismic *wavefronts*. A wavefront is a surface on which all particles are in the same phase of motion. Huygens' principle states that every point along a wavefront can be regarded as a new source of waves. The future location of a wavefront can therefore be determined by propagating a spherical wavelet from each point on the current wavefront. As shown in Figure 6.7 and following Telford *et al.* (1990), let AB be the seismic wavefront at time t_0. After some time interval Δt, the wave will have advanced a distance $V\Delta t$, as shown. Centered on each sampled point on the current waveform we draw arcs of radius $V\Delta t$. The new wavefront $A'B'$ is simply the envelope of these arcs, as indicated in the figure. The accuracy of the wavefront construction increases as we draw arcs from a finer sampling of points on the current waveform.

Consider now a planar wavefront AB incident upon a plane interface between two materials with P-wave velocities V_1 and V_2, respectively, as shown in Figure 6.8. When the point A arrives at the interface, the wavefront is labeled as $A'B'$. Point B', at this time, is at distance $V_1\Delta t$ from the interface. During the time interval Δt that it takes for B' to reach point R on the interface, part of the energy at A' would have traveled the same distance

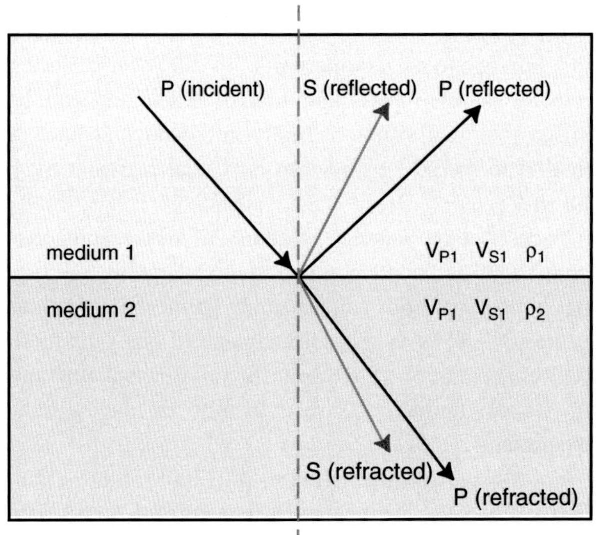

Figure 6.6 P-wave reflection and refraction at an interface, including P-to-S conversion.

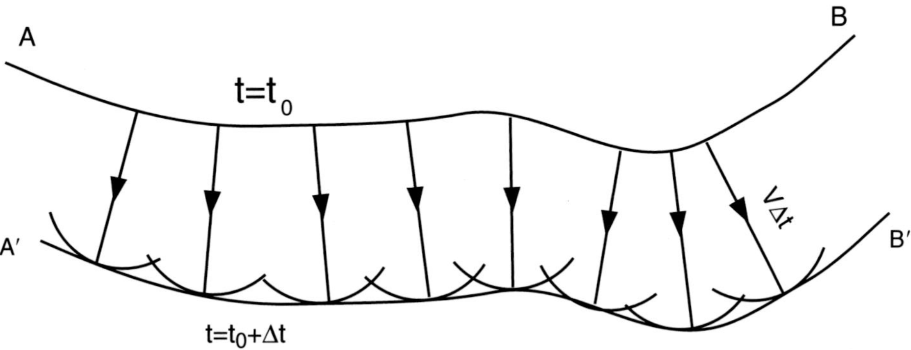

Figure 6.7 Wavefront construction based on Huygens' principle. After Telford *et al.* (1990).

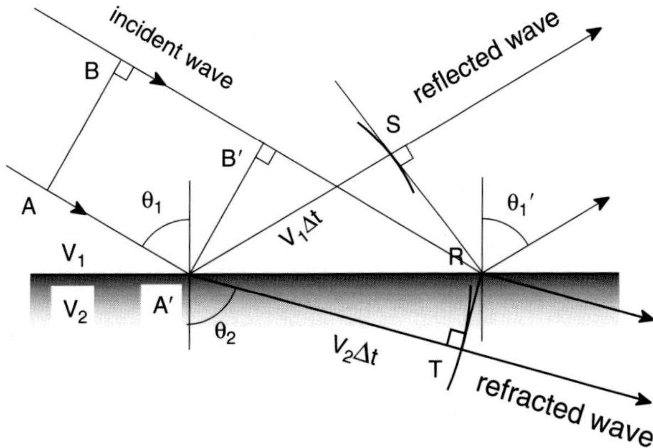

Figure 6.8 Reflection and refraction of a plane wave. After Telford *et al.* (1990).

$V_1\Delta t$ upward back into original medium, while the remainder would have refracted a distance $V_2\Delta t$ into the underlying medium.

To determine the angles of reflection and refraction, we can use Huygens' principle. Again following Telford *et al.* (1990), arcs are drawn with center A' and radii $V_1\Delta t$ and $V_2\Delta t$. The new wavefronts, denoted RS and RT are constructed by drawing tangents to these arcs that intersect point R. A glance at the geometry of the figure shows that the angle of incidence θ_1 is equal to the angle of reflection $\theta_1{}'$; this is the *law of reflection*. Another glance at the figure shows that $V_1\Delta t = A'R \sin \theta_1$ and $V_2\Delta t = A'R \sin \theta_2$. Solving both equations for the quantity $A'R/\Delta t$ gives

$$\frac{\sin \theta_1}{V_1} = \frac{\sin \theta_2}{V_2} = p; \tag{6.20}$$

which is the familiar *Snell's law of refraction*. The quantity p in Equation (6.20) is termed the *ray parameter*. If the medium consists of a number of parallel layers, the ray parameter p does not change from its initial value in the first layer as the wave refracts through the

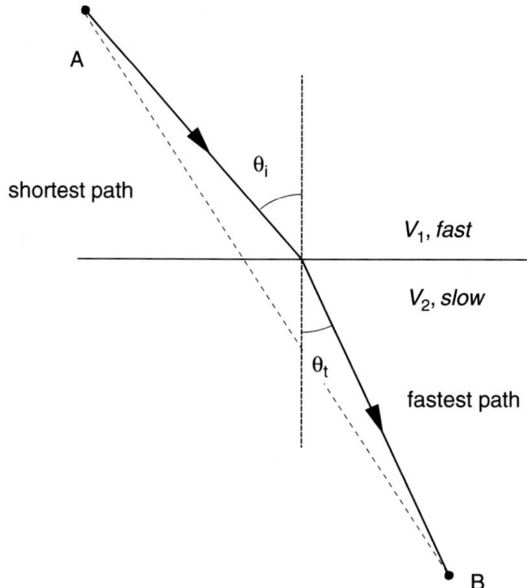

Figure 6.9 Fermat's principle of least time, with incident angle θ_i and transmitted angle θ_t.

stack of layers. An equivalent derivation of the laws of reflection and refraction using Fermat's *principle of least time* (see Figure 6.9), rather than Huygens' principle, is left as an exercise at the end of the chapter.

If a P-wave is incident on an interface, it should be noted that the angles of refraction for both the P-wave and the S-wave can be found by Snell's law using the appropriate P-wave and S-wave velocities in Equation (6.20). Since $V_{P2} > V_{S2}$, it follows that $\sin\theta_{P2} > \sin\theta_{S2}$. Accordingly, as shown in Figure 6.6, the direction of the refracted S-wave is closer to the vertical than that of the refracted P-wave. Similarly, the angle of reflection θ'_{S1} of the S-wave for an incident P-wave may be found from a generalized law of reflection, $\sin\theta'_{P1}/V_{P1} = \sin\theta'_{S1}/V_{S1}$.

For P-waves incident on a low-velocity layer in which $V_2 < V_1$, Snell's law predicts that $\theta_2 < \theta_1$ and thus the wave refracts downward, toward the normal to the interface. Suppose, instead, that the wave is incident on a relatively fast layer in which $V_2 > V_1$. The wave refracts toward the horizontal interface. It is possible to observe an angle of refraction $\theta_2 = 90°$ for the case that the incident angle happens to be $\theta_1 = \sin^{-1}(V_1/V_2)$. The refracted wave in this case travels along the interface between the two media. This is the critically refracted wave and $\theta_C = \sin^{-1}(V_1/V_2)$ is called the *critical angle*. For angles of incidence that are greater than the critical angle, $\theta_1 > \theta_C$, no refraction occurs and there is *total internal reflection* of all the wave energy back into the original medium.

The law of reflection and Snell's law of refraction provide the directions of propagation of the reflected and the refracted waves, but they do not allow us to calculate the relative amplitudes of these waves. The partitioning of energy into the reflected and refracted body waves is described by the Zoeppritz equations (Shuey, 1985). A complete derivation of these equations, which is not undertaken here, requires to solve the elastic wave equation subject to conditions that the normal and tangential components of stress and displacement are continuous across the interface. A detailed calculation reveals that the reflection and refraction coefficients (R, T) depend in a somewhat complicated fashion on the angle of incidence, but for P-waves at normal incidence they reduce to

$$R = \frac{\rho_2 V_2 - \rho_1 V_1}{\rho_2 V_2 + \rho_1 V_1}; \tag{6.21}$$

$$T = \frac{2\rho_1 V_1}{\rho_2 V_2 + \rho_1 V_1}; \tag{6.22}$$

where (ρ_1, ρ_2) are the densities and (V_1, V_2) are the wave velocities of the two media. The reflection coefficient generally obeys $R < 0.2$ for stratification within unconsolidated near-surface geomaterials. The energy reflected at such interfaces is proportional to the square of the wave amplitude and hence $R^2 < 4\%$. Thus, very little of the energy transmitted into the ground by the seismic source is reflected back to the surface where it may be recorded by geophones. On the other hand, the top of competent bedrock (say, $V_P \sim 4500$ km/s) lying below unconsolidated sediments (say, $V_P \sim 800$ km/s) reflects almost 60% of the normally incident seismic wave energy. The Zoeppritz equations further reveal that most of the incident energy is partitioned into reflected and refracted P-waves and only a minor amount is partitioned into reflected and refracted S-waves.

The product ρV that appears in Equations (6.21) and (6.22) is termed the *acoustic impedance*. It is often stated that the seismic-reflection method provides images of subsurface discontinuities in acoustic impedance.

6.7 Diffraction

Seismic energy is *diffracted* if a wave encounters a subsurface feature whose radius of curvature is smaller or not significantly larger than the seismic wavelength, $\lambda = V/f$. Seismic wavelengths are large (e.g. $\lambda = 5.0$ m for $V = 1.5$ km/s and $f = 300$ Hz; typical values encountered in practice) so that understanding diffraction effects is very important in near-surface seismic interpretation. While a rigorous theoretical description of diffraction is beyond the scope of this book, to first order a diffracted wavefront can be constructed using Huygens' principle.

Consider a vertically propagating seismic plane wave that is incident on a sedimentary bed that pinches out, forming a type of wedge, as shown in Figure 6.10. The energy contained in the part of the wavefront that strikes the tip of the wedge scatters in all directions (Keller, 1957). In essence, the wedge tip acts as a point scatterer, or *diffractor*. As shown in the figure, diffracted seismic energy also propagates beneath the wedge.

6.8 Analysis of idealized reflection seismograms

A simplified analysis of seismic-wave propagation uses *rays*, which are semi-infinite lines oriented perpendicular to wavefronts and pointing in the direction of wave propagation. A ray changes direction if the propagating wavefront encounters a change in acoustic

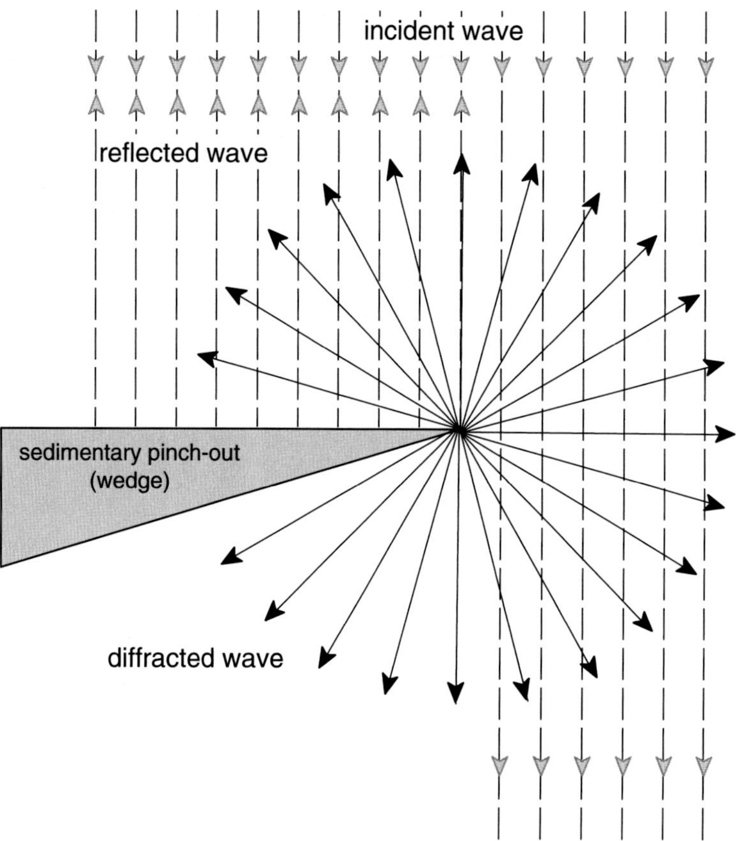

Figure 6.10 Diffraction by a wedge.

impedance. *Ray paths* are used to indicate the various routes of wave propagation between a seismic source (transmitter, TX) and a geophone (receiver, RX). There are three important ray paths in the simple one-layer scenario shown in Figure 6.11. These paths trace the fastest routes of direct, reflected, and critically refracted seismic energy from TX to RX and hence they satisfy Fermat's principle of least time. It proves insightful to analyze the traveltime of the *P*-waves along these paths. Let x be the TX–RX separation distance. The direct *P*-wave traveltime along path 1 is very simply given by the distance divided by the velocity, $\tau_1(x) = x/V_1$.

The traveltime $\tau_2(x)$ for the primary *P*-wave reflection, shown as path 2 in Figure 6.11, is given by

$$\tau_2(x) = \frac{2}{V_1}\sqrt{h^2 + \left(\frac{x}{2}\right)^2}, \tag{6.23}$$

where the $h^2 + (x/2)^2$ term corresponds to the slant distance shown in Figure 6.12.

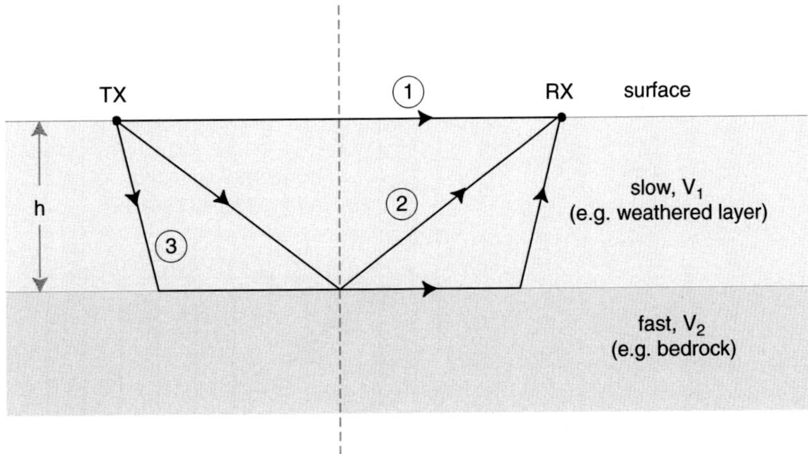

Figure 6.11 Important ray paths between seismic source (TX) and geophone (RX): (1) direct *P*-wave; (2) reflected *P*-wave; (3) critically refracted *P*-wave.

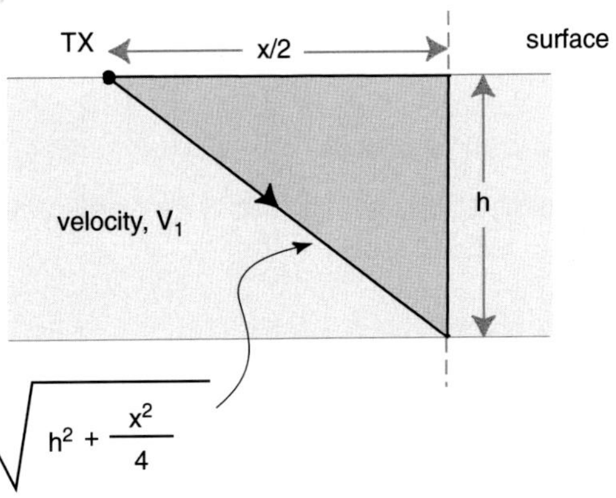

Figure 6.12 Geometry associated with the primary *P*-wave reflection.

The square of the traveltime Equation (6.23), given by

$$\tau_2^2(x) = \tau_0^2(x) + \frac{x^2}{V_1^2}, \tag{6.24}$$

describes a hyperbola with intercept (or zero-offset traveltime) $\tau_0 = 2h/V_1$. The hyperbolic increase in traveltime τ_2 with increasing TX–RX offset x is called the *normal moveout*. An idealized seismogram showing a P-wave reflection is shown in Figure 6.13. The hyperbolic curve in the figure has been computed using the values $h = 1.0$ m and $V_1 = 1500$ m/s. Each vertical trace in the seismogram corresponds to a single geophone response. The collection of traces, displayed in this manner, is termed a *shot gather*.

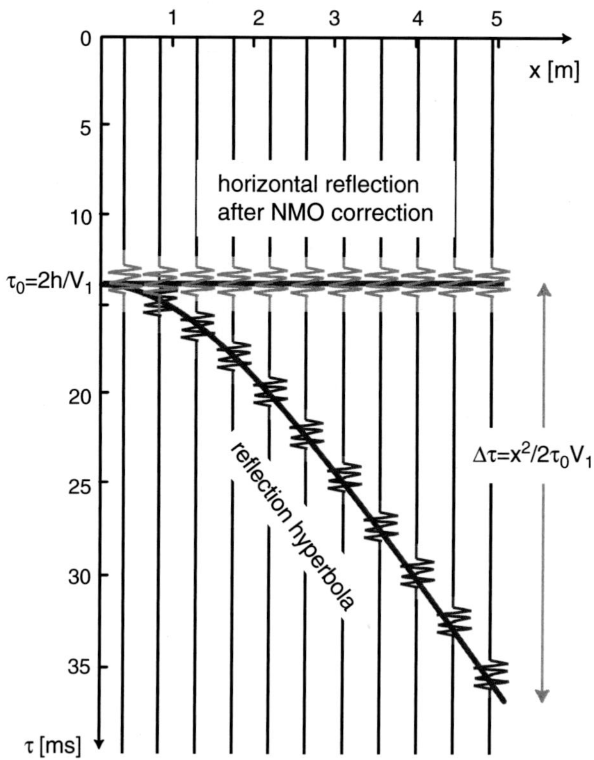

Figure 6.13 Idealized seismogram (shot gather) of primary *P*-wave reflection showing hyperbolic moveout curve and NMO-corrected horizontal reflector.

A plot of τ_2^2 vs x^2 would give a slope of $1/V_1^2$, according to Equation (6.24), and an intercept of $\tau_0^2 = 4h^2/V_1^2$. Thus, an analysis of the slope and intercept of the τ_2^2 vs x^2 plot enables a determination of the layer thickness h and the layer velocity V_1.

Notice that the idealized seismogram, with its hyperbolic *P*-wave reflection event, does not display an accurate image of the subhorizontal velocity contrast between the slow weathered layer and the fast bedrock shown in Figure 6.11. The apparent curvature of the subsurface reflector is due to the variable distance x between the TX and each RX in the geophone array. The subhorizontal velocity contrast would have been accurately imaged if, however, each seismic trace was due to a separate TX located in the same position as each geophone. This hypothetical TX–RX configuration is termed the zero-offset data-acquisition geometry.

A normal moveout (NMO) correction can be applied to each seismic trace. This procedure converts the seismogram into one that would have been measured for zero-offset acquisition across the geophone array. The NMO correction has the effect of straightening out the hyperbolic moveout curves and thereby providing an accurate image of the subhorizontal reflector (see Figure 6.13). Notice that the reflector at the left-most seismic trace (close to zero TX–RX offset, $x = 0$) is already very nearly in its correct position. Multiples, and any other events that have reflected from more than one interface within the subsurface, (not shown in Figure 6.13) do not exhibit a normal moveout.

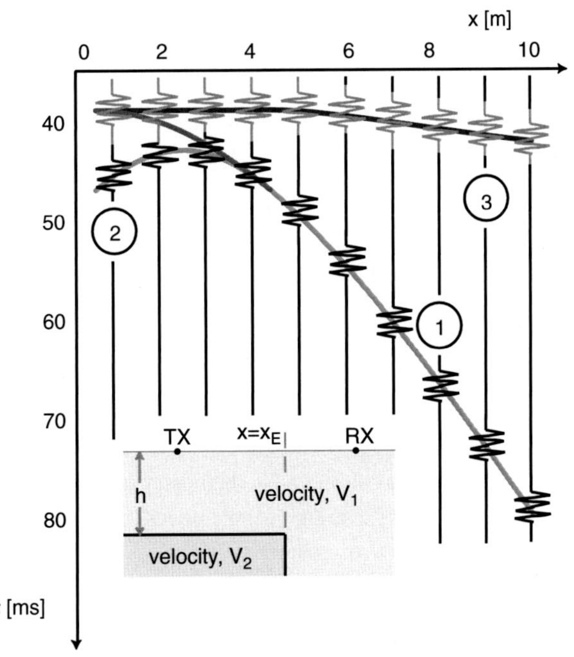

Figure 6.14 Idealized seismogram showing: (1) primary P-wave reflection; (2) diffracted wave from the edge shown in the insert; and (3) the standard NMO-corrected arrivals. For these calculations, $h = 3.0$ m; $V_1 = 1500$ m/s, and location of the edge, $x_E = 2.0$ m. Note that energy appearing in (1) at $x > 2x_E = 4.0$ m has diffracted from the edge.

It is clear from an inspection of the idealized seismogram that the NMO correction $\Delta\tau$ to each seismic trace should be

$$\Delta\tau(x) = \tau_2(x) - \tau_0 \tag{6.25}$$

since the reflection hyperbola is $\tau_2(x)$ and the zero-offset two-way traveltime is τ_0.

Applying an NMO correction is a good diagnostic to distinguish horizontal reflectors in the subsurface from diffraction hyperbolas, multiple reflections, and dipping interfaces. Only the primary reflections from subhorizontal interfaces will align horizontally after the NMO correction. Diffracted arrivals from a lateral discontinuity are illustrated in Figure 6.14. Notice they do not exhibit the normal moveout behavior.

The traveltime equations must be modified if a dipping reflector is present. Consider the path of the primary reflected ray associated with a dipping interface, as shown in Figure 6.15. The location of the reflection point P on the interface is determined by the requirement that the angle of ray incidence, measured with respect to the normal to the interface, equals the angle of reflection. Notice that the geophone (RX) in Figure 6.15 is placed down-dip from the seismic source (TX). Alternatively, a geophone could be placed up-dip from the source.

The down-dip traveltime τ_D from TX to RX is readily computed as the length of the seismic ray path divided by the velocity V of the medium which it traverses. To facilitate the computation and to gain further insight into seismic imaging of a dipping interface, it is

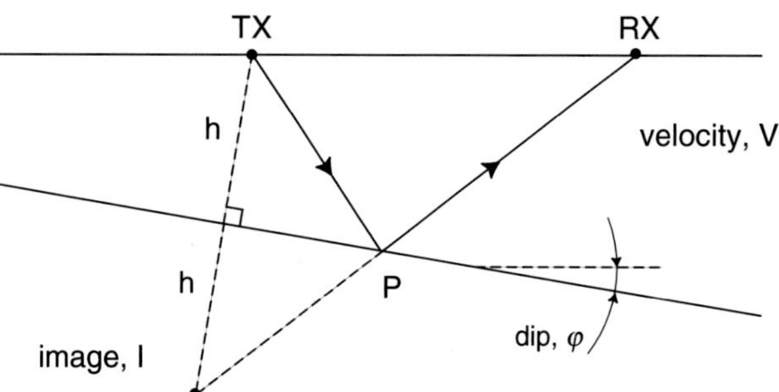

Figure 6.15 Primary reflection from a dipping interface.

useful to introduce the hypothetical image source I shown in the figure. The image source I is placed at the same perpendicular distance h from the dipping interface as the actual source TX. The traveltime τ_D is then calculated using the distance from the image I to the geophone RX as if the image were embedded in a homogeneous medium of velocity V. The result, using the law of cosines, is

$$\begin{aligned}\tau_D = \frac{\text{dist}\,(I - RX)}{V} &= \frac{1}{V}\sqrt{x^2 + (2h)^2 - 2x\,(2h)\,\cos\left(\frac{\pi}{2} + \varphi\right)} \\ &= \frac{2h}{V}\sqrt{1 + \left(\frac{x^2 + 4hx\,\sin\,\varphi}{4h^2}\right)};\end{aligned} \tag{6.26}$$

where x is the distance from TX to RX and φ is the dip angle. Notice that the angle at TX subtended by I and RX is $\pi/2 + \varphi$ [rad]. Similarly, if the source location is kept the same but the geophone RX is now moved to the other side of the TX, such that the seismic ray path is up-dip, the traveltime becomes

$$\tau_U = \frac{2h}{V}\sqrt{1 + \left(\frac{x^2 - 4hx\,\sin\,\varphi}{4h^2}\right)}. \tag{6.27}$$

The dip angle φ can be estimated in practice by measuring both seismic traveltimes τ_U and τ_D; this is accomplished in practice by first placing the TX up-dip from the RX array, and then locating the TX down-dip.

6.9 Vertical and horizontal resolution

According to the Rayleigh criterion (Zeng, 2009), the *vertical resolution* of seismic waves is ~ $\lambda/4$, where $\lambda = V/f$ is the seismic wavelength. The frequency f is the dominant, or central, frequency carried by the seismic wave packet. For a typical near-surface geophysical application, with seismic waves propagating in a material of velocity $V = 1.5$ km/s at $f = 300$ Hz,

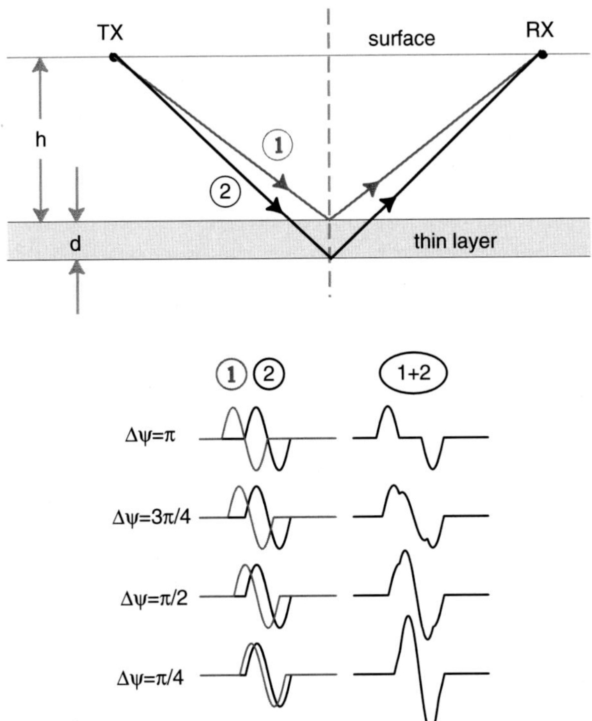

Figure 6.16 Resolution of a thin layer.

the seismic wavelength is 5.0 m. Hence, the vertical resolution is approximately 1.25 m. This means that two interfaces separated by less than 125 cm cannot be individually resolved by the returned seismic signal, as measured by a geophone. This result can be contrasted with exploration-scale geophysics in which dominant source frequencies are commonly 30 Hz or lower. In that case, the seismic wavelength in the same material is greater than 50 m, such that interfaces separated by less than 12.5 m cannot be individually resolved.

While the thickness of a fracture in a rock may not be resolved if the aperture is less than ~ 125 cm, the fracture certainly could be detected as an acoustic impedance discontinuity if the seismic properties of the fracture fill materials contrast sufficiently to those of the host material. In this sense, the ability to detect a fracture must be distinguished from the ability to resolve its aperture.

The vertical resolution of a thin layer is illustrated in Figure 6.16 using an idealized seismic wave packet consisting of a single sinusoidal pulse. Of course, no practical seismic source could generate such an idealized packet; it is employed here simply to illustrate the concept of vertical resolution. The primary ray path 1 in Figure 6.16 indicates a reflection from the top of a thin layer of thickness d whereas the primary ray path 2 shows a reflection from the bottom of the thin layer. The slight difference in the lengths of paths 1 and 2 is manifest as a difference in phase ψ of the two seismic wave packets recorded by the geophone RX. The greater the difference in path length, the greater the difference in phase. If the phase difference is $\Delta\psi < \pi$, the two arrivals are somewhat merged together in the resulting geophone response (which is the sum of the two waves 1 and 2; shown as "1 + 2" in the bottom part of the figure). The geophone response resolves the thin layer as two separate arrivals only as the phase difference increases beyond $\Delta\psi > \pi$. The critical phase difference $\Delta\psi = \pi$ corresponds to a path-length difference of one-half the seismic wavelength, $\lambda/2$. A full wavelength λ would change the phase of the seismic wave packet by 2π. Thus, keeping in mind that the geophone response records the two-way traveltime of seismic wave arrivals, the layer thickness must satisfy $d > \lambda/4$ in order for it to be seen as two distinct events on a geophone. This result is strictly valid for small TX–RX offsets x such that $x << h$, where h is the depth to the top of the thin layer.

The *horizontal resolution* of the seismic-reflection method intuitively cannot be better than ~ $\Delta x/2$, where Δx is the geophone spacing. However, a straightforward analysis indicates that the horizontal resolution of an interface at depth h beneath the surface can also not be better than the Fresnel-zone radius

$$D_0 \sim \sqrt{2\lambda h}, \tag{6.28}$$

which is often larger than $\Delta x/2$. A simple justification of Equation (6.28) can be made as follows. Seismic energy emanates from a source TX into the subsurface in all directions. Three particular ray paths are shown in Figure 6.17a. The path labeled 1 is the primary reflection, and is the fastest purely reflected path to the geophone. A certain amount of energy, however, is carried in the two paths labeled 2 since, in reality, the reflection is not from an idealized point (Lindsey, 1989) but is generated by integration over a circular area which might be termed the seismic footprint. This accords with an extended Huygens' principle (Sein, 1982) in which the incident planar wavefront, where it strikes the interface at a given location, acts as a source of downward-propagating refracted spherical wavelets and upward-propagating reflected spherical wavelets.

As shown in the figure, the two rays labeled 2 strike the reflecting interface at a distance $\pm D/2$ from the primary reflection point at the TX–RX midpoint. Energy carried along these two paths arrives somewhat later than energy carried along path 1. If the phase difference between either of the wave packets propagating along ray path 2 and the wave packet propagating along ray path 1 satisfies $\Delta\psi < \pi$, then the geophone response will not see these packets as distinct arrivals but rather they will appear merged together.

The horizontal resolution at a given depth h is therefore defined as the horizontal distance D along a reflecting horizon within which all reflected energy recorded at a given geophone arrives with phases that are within π of each other. The horizontal resolution is usually referenced to the zero-offset configuration in which the TX–RX separation distance is $x = 0$, but it can also be defined for a given non-zero offset x, as we show in the following.

The length of ray path 1 in Figure 6.17a is given by

$$l_1(x) = 2\sqrt{h^2 + \left(\frac{x}{2}\right)^2} \tag{6.29}$$

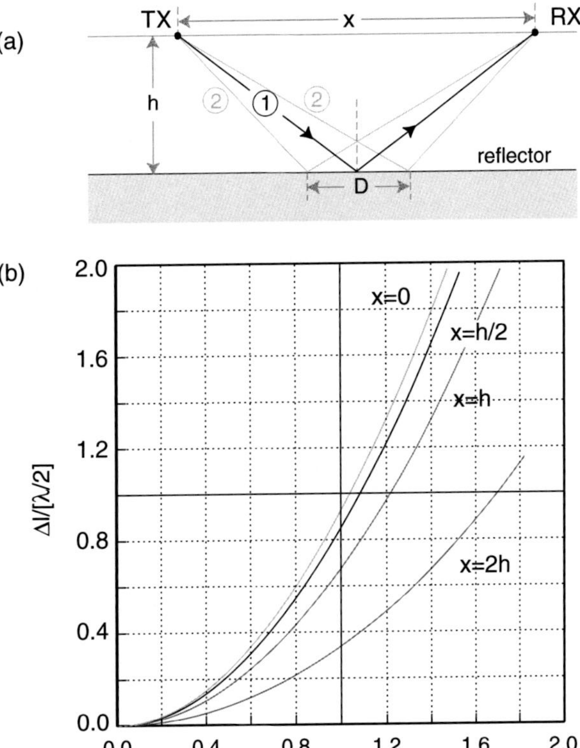

Figure 6.17 Horizontal resolution of the seismic-reflection method. (a) Fresnel-zone concept; (b) path length difference Δl as a function of horizontal distance D.

while the length of either of the ray paths labeled 2 is given by

$$l_2(x) = \sqrt{h^2 + \frac{(x-D)^2}{4}} + \sqrt{h^2 + \frac{(x+D)^2}{4}}. \tag{6.30}$$

The difference $\Delta l = l_2 - l_1$ in the two path lengths, normalized by one-half the seismic wavelength $\lambda/2$, is plotted in Figure 6.17b versus the horizontal distance D/D_0 along the reflecting horizon. The parameters $h = 10$ m and $\lambda = 1.6$ m were chosen such that $D_0 \sim \sqrt{32} \sim 5.5$ m, using Equation (6.28). Different curves are plotted for different TX–RX separation distances x.

Notice that the curve for the zero-offset configuration $x = 0$ passes very close to the critical point $\Delta l = \lambda/2$ when $D = D_0$. This indicates that the horizontal resolution is given, to a very good approximation, by Equation (6.28) in this case. The graph in Figure 6.17b also shows that the horizontal resolution worsens as the TX–RX separation distance x increases. For example, consider the curve for the case $x = 2h$. The curve crosses the critical point $\Delta l = \lambda/2$ at $D \sim 1.7D_0 \sim 9.35$ m. This indicates that the horizontal resolution is

70% worse in the case $x \sim 2h$ compared to the ideal zero-offset case. In other words, the lateral resolution of the seismic-reflection method degrades with increasing distance between the shotpoint and the receiver. It makes intuitive sense that the seismic footprint should grow larger as the incident angle becomes shallower.

6.10 Common midpoint profiling

The accuracy of a reflector image can be improved if the seismic shotpoints and receivers are arranged such that each location along the reflecting horizon is illuminated from a number of different perspectives. This is readily accomplished using the common midpoint (CMP) profiling method. CMP data acquisition involves moving the shotpoint and receiver array forward in regular increments and shooting at each successive move. A subset of the resulting shot records can then be selected to simulate an acquisition that consists of a symmetric configuration of n seismic TX–RX pairs about a common midpoint P, as shown in Figure 6.18. In other words, individual traces that share a common midpoint are collected from the various shot records. This is termed a *CMP gather*. The benefit of this procedure is that a single reflection point on a subsurface interface is sampled n times. A CMP profile constructed in this manner is said to have n-fold data coverage.

After NMO corrections of the form (6.25) are applied, according to the TX–RX offset, each seismic trace in the CMP gather becomes effectively a zero-offset trace. The n NMO-corrected traces are then ready to be averaged, or *stacked*, to enhance the signal-to-noise (S/N) ratio. Generally, the improvement in S/N ratio due to stacking a number of traces acquired with the same TX–RX acquisition geometry can be understood as follows. Suppose the amplitude of a seismic-reflection signal is S, while the average noise amplitude is N. Assuming the noise to be random, the S/N ratio of the stacked trace is related to the S/N ratio of an individual trace by the fundamental formula

$$\frac{S}{N}(n) = \sqrt{n}\frac{S}{N}(1). \tag{6.31}$$

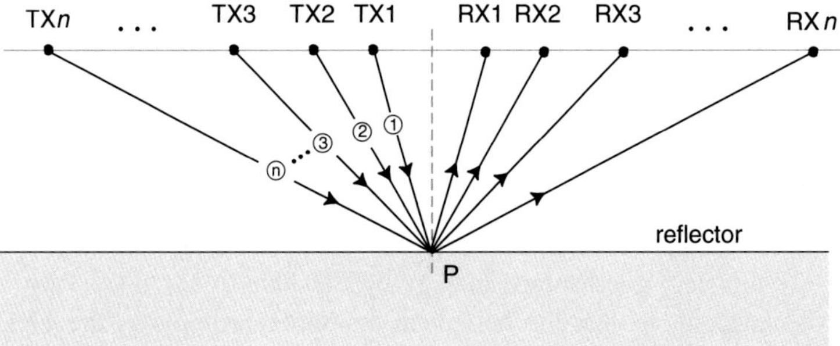

Figure 6.18 Common midpoint profiling, CMP.

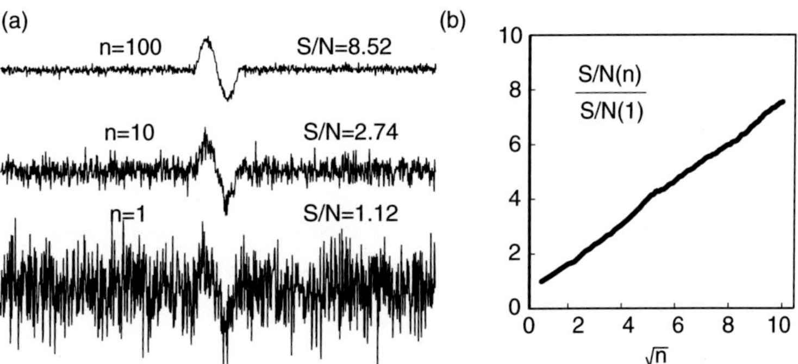

Figure 6.19 (a) Improvement of S/N ratio with stacking number n; (b) Plot of S/N-ratio improvement with $\sqrt{n}$.

There is a diminishing reward for increasing the stack number n, as shown by Equation (6.31). For example, doubling the S/N ratio requires stacking four shot records. Increasing the S/N by an order of magnitude requires stacking 100 shot records.

The effect of stacking on an idealized seismic trace that consists of a single sinusoidal pulse signal embedded in random noise is shown in Figure 6.19a. The signal occupies the middle 10% of the trace, as can be seen in the figure. The background noise is generated by a Gaussian random-number generator. The S/N ratio is *defined* here to be the rms (root mean squared) amplitude of the middle 10% of the trace divided by the rms amplitude of the remainder of the trace. It is evident from the figure that the S/N ratio increases with increasing stack number n, as expected. The sinusoidal pulse is easy to discern in the case $n = 100$, but difficult to detect visually in the case $n = 1$. The behavior of the S/N improvement, defined as the S/N ratio for n traces, normalized by the S/N ratio for a single trace, is shown in Figure 6.19b. Note that the "root-n" improvement in S/N ratio, as prescribed by Equation (6.31), is approximately satisfied.

CMP profiles with n-fold data coverage can offer a tremendous improvement in resolving subsurface reflectors relative to data of single-fold coverage. However, traditional CMP profiling quickly becomes laborious since it is necessary to uproot the entire array of n geophones and shift it forward by one station increment in order to advance the common midpoint by one station increment. CMP data acquisition is greatly simplified, however, with the use of a *roll-along switchbox*.

The effect of the switchbox is to automatically shift a geophone array by one station increment along a survey profile. Instead of manually picking up and moving the geophone array, the equivalent task can be accomplished simply by advancing the knob on the switchbox by one unit. The concept is illustrated in Figure 6.20. To keep the illustration simple, imagine an array of eight geophones connected via the switchbox to a four-channel seismograph (in practice, these numbers are more likely to be in the range of hundreds of geophones and tens or hundreds of channels). Suppose the four active geophones are RX1–RX4 when the switchbox is at position 1, as shown in the top part of the figure. Then, after the shot is recorded at this setting, the switchbox is advanced to position 2, thereby activating RX2–RX5. At the same time, the shotpoint (TX) is moved forward one station increment. This process continues until all switchbox settings have been used. The result

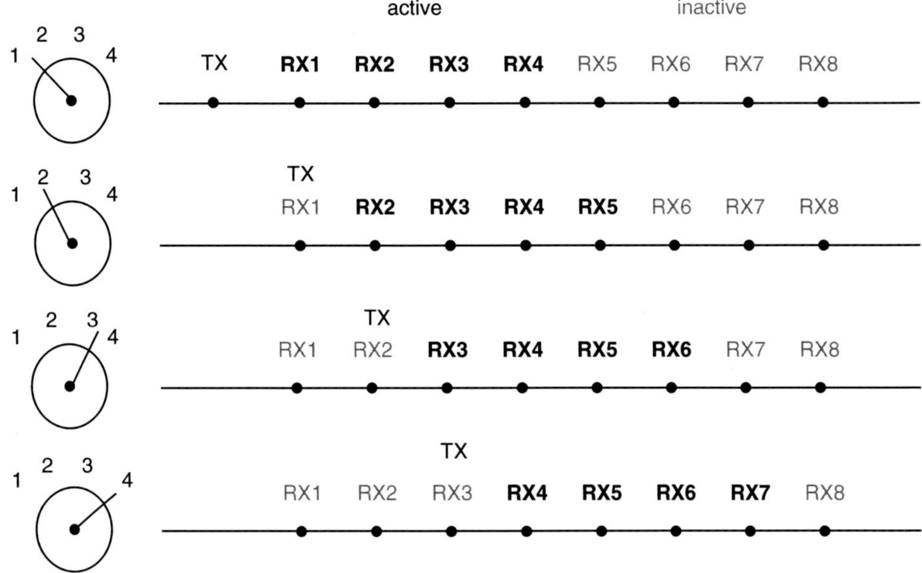

Figure 6.20 Roll-along data acquisition.

(a four-fold CMP profile) is the same as if a smaller array of four geophones had been manually shifted after each shot by one station increment.

6.11 Dip moveout

A major complication to CMP analysis occurs in the presence of *dipping reflectors*. In the presence of dip, the subsurface reflection point unfortunately is not the same for all common-midpoint TX–RX pairs (Figure 6.21). Notice that reflection points P_1 and P_2 are not coincident, and neither of the reflection points lie directly beneath the common midpoint. The NMO correction could be modified to accommodate a single dipping interface but more problematic is the case of multiple dipping reflectors that have different dip angles. Such cases are often encountered in aeolian or fluvial cross-bedded sedimentary systems or salt domes, for example, where gently dipping beds make contact with a steeply dipping structure. The CMP stacking process breaks down and the data quality actually deteriorates with increasing n. The imaging of multiple dipping reflectors can be greatly improved, however, by an application of *dip-moveout* (DMO) processing prior to performing the CMP stack. Here we outline just the essential concepts of DMO since the details of the algorithm are beyond the scope of the book. Good introductions to DMO are provided in Deregowski (1986) and Liner (1999).

The aim of NMO/CMP-stack processing is to identify, adjust the timing, and then gather traces that have the same reflection point on a subsurface reflector. If a dipping reflector of unknown dip angle is present, the location of the reflection point is not known, and hence a

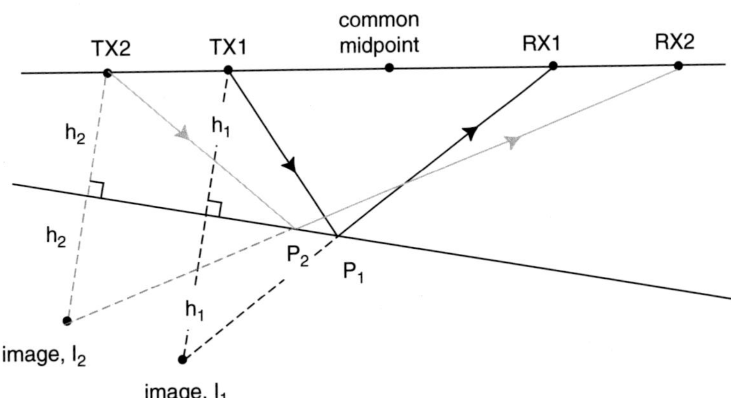

Figure 6.21 On a dipping reflector there is not a single point of reflection for all common-midpoint trace pairs, such as TX1–RX1 and TX2–RX2.

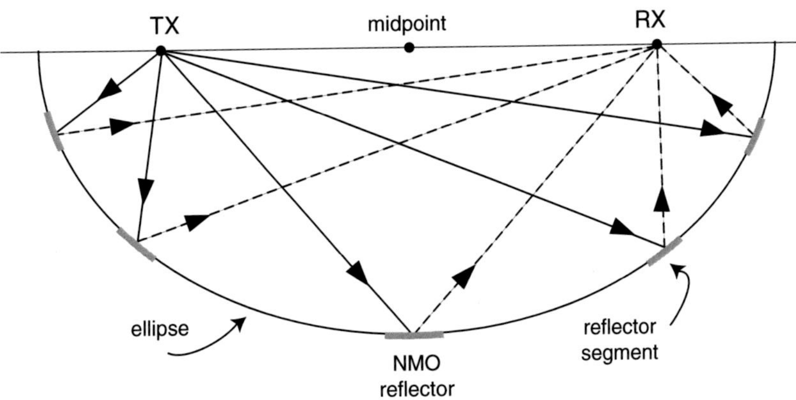

Figure 6.22 A reflection event seen at the RX can come from any point on the ellipse. After Liner (1999).

CMP gather cannot be made. However, only certain locations for the reflection point are possible. Simple geometry shows that the reflection point must be located on the ellipse in Figure 6.22, where the TX and RX are at each focus. A standard NMO correction assumes that the reflection originated from the horizontal segment located immediately below the midpoint. But each of the reflector segments shown in Figure 6.22 is an equivalent possibility, in the sense that they could equally explain the observed traveltime between the TX and RX.

The equivalent reflector segments are shown again in Figure 6.23a. The zero-offset locations (labeled 1–5) associated with each segment are also shown. In Figure 6.23b are shown the traces that would have been recorded had the TX and RX been co-located at each of the locations 1–5. These are the zero-offset traces, and the traveltimes of the first

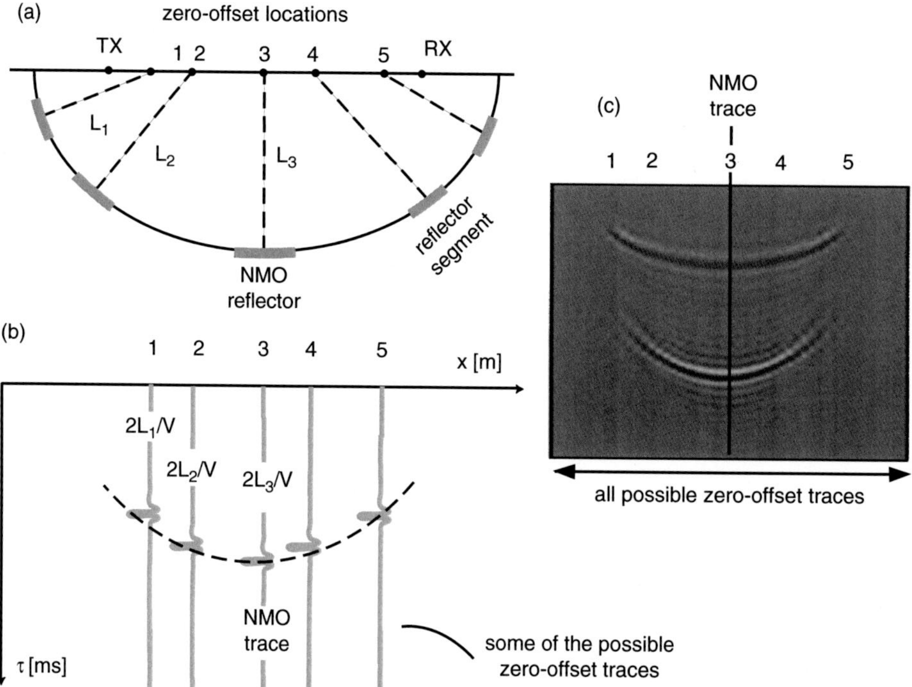

Figure 6.23 (a) Some of the possible zero-offset locations between the original TX and RX in the presence of a dipping reflector; (b) the corresponding zero-offset traces; (c) all possible zero-offset traces, for two dipping reflectors, forming two DMO smiles. After Liner (1999).

three of them are also indicated. The familiar NMO traveltime, for example, is $\tau_3 = 2L_3/V$. Since we don't know the actual dip of the reflector, we still don't know which one of these zero-offset traces to choose as the DMO-adjusted trace, but we do know that one of them must be correct.

Furthermore, the five zero-offset locations and traces that have been so far considered are not the only possibilities. There are many other locations between the original TX and RX. In Figure 6.23c are shown all the possible zero-offset traces (actually, for two dipping reflectors). The portion of this plot that contains the highest amplitudes is whimsically known as a *DMO smile*, for an obvious reason. Again, this plot shows the zero-offset traces for every conceivable dip angle of a reflector.

The foregoing analysis shows that a single reflection event recorded on a single trace can be expanded into a multiplicity of zero-offset traces and then plotted to reveal a DMO smile. The NMO-corrected trace runs through the center of the DMO smile, as shown. By performing a similar DMO analysis on the next trace from the next TX–RX pair along the seismic survey profile, we can construct another DMO smile that partially overlaps the first one. In fact, we can construct an entire sequence of partially overlapping DMO smiles, one for every TX–RX pair along the seismic survey profile. Once we have all these DMO smiles, we simply add them together. This process produces the correct zero-offset image of the dipping reflector horizon in the subsurface. The stacking process works because the reflected energy from the actual dipping horizon is common to all of the DMO smiles such that constructive interference occurs when they are added together. Destructive interference occurs elsewhere. Hence, the correct trace from each of the DMO smiles is naturally selected by the stacking process. Interestingly, multiple dipping horizons with different dip angles are all correctly imaged and appear at their proper locations and orientations within the subsurface. Additional details, including a discussion of the close relationship between DMO and migration, is found in the tutorial article by Liner (1999), to which the interested reader is referred.

6.12 Attenuation

The theory of elasticity governed by Hooke's law predicts reversible stresses and strains in which no energy is lost during the transmission of a seismic wave. An ideal elastic wave nevertheless diminishes in amplitude as it propagates due to its *geometric spreading*. Moreover, its energy is *partitioned* as it is scattered by heterogeneities, and undergoes reflection, refraction, and P-to-S-wave conversion at acoustic impedance boundaries. Real seismic waves, however, also continuously lose energy via the *absorption* of energy by the medium.

Consider a seismic wave propagating in a homogeneous medium and suppose that it has an amplitude A_0 at some distance r_0 from its source. Neglecting the scattering contributions, the amplitude of the wave at some greater distance $r > r_0$ from the source is given by

$$A(r) = A_0\left(\frac{r_0}{r}\right)\exp[-\alpha(r - r_0)], \tag{6.32}$$

where the geometric spreading is described by the $1/r$ falloff and the exponential decay is due to energy absorption with attenuation coefficient α.

The $1/r$ form of the geometric spreading term in Equation (6.32) is due to the fact that the seismic energy associated with a spherical wave at some distance r from a source is distributed over the surface of a sphere of radius r. The surface area of a sphere is $4\pi r^2$. Hence, the seismic energy should fall off as $1/r^2$. The seismic energy is proportional to the square of the wave amplitude, hence the latter decreases as $1/r$. This energy-loss mechanism is termed geometric spreading since it is independent of the elastic properties of the medium.

The seismic attenuation coefficient α, on the other hand, does depend on the elastic properties of the medium and also on frequency. The physical causes of intrinsic seismic attenuation are not fully understood. Friction has been suggested as an absorption mechanism. Recent developments in the theory of *poroelasticity* propose that a large portion of the intrinsic loss in a saturated porous medium is caused by viscous fluid motions that arises in response to mechanical wave excitation (e.g. Pride *et al.*, 2004). This mechanism is termed wave-induced fluid flow. In the scenario described by Muller

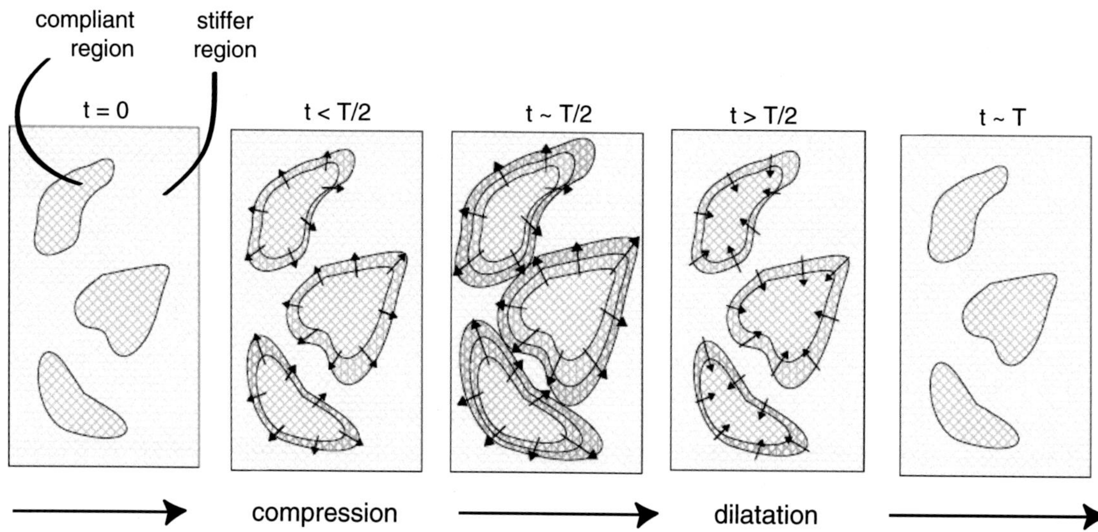

Figure 6.24 Seismic attenuation caused by wave-induced fluid flow in association with mesoscopic heterogeneities, after Muller *et al.* (2010).

et al. (2010) and depicted in Figure 6.24, fluid flows from compliant regions toward stiffer regions (as shown by the arrows in the second and third panels) during the compressional portion of an elastic wave cycle of period T, and vice versa during the expansive, or dilatational portion of the cycle (shown in the fourth panel). Wave-induced fluid flow as an attenuation mechanism is thought to be particularly effective in geomaterials for which the length-scale of the heterogeneities is *mesoscopic*, that is, the heterogeneities are much larger than typical pore sizes but smaller than the seismic wavelength. Decisive field-scale studies which permit the elaboration of the physical mechanisms responsible for attenuation of seismic wavefields have not yet been carried out in near-surface geophysics.

The attenuation of higher-frequency waves is much larger than that of lower-frequency waves. Typical values of the attenuation coefficient vary from $\alpha \sim 0.25$–0.75 decibels/wavelength [dB/m]. Thus, at large ranges compared to the longest seismic wavelength excited by the source, seismic pulses tend to become smoothed out and are of apparently long duration. In this sense, the Earth acts as a *low-pass filter*.

In a *viscoelastic* medium subject to a shear stress, the material undergoes some permanent shear deformation. Other types of irreversible seismic responses are possible in the presence of large strains or strain rates, including brittle failure, buckling or bending of engineered structures, various types of plastic deformation, and soil or sand liquefaction. The inelastic deformation of near-surface geomaterials in response to conventional seismic imaging sources such as hammers, shotguns, and vibrators is normally insignificant and is neglected in most active-source near-surface geophysical surveys.

6.13 Seismic refraction

The seismic-refraction method is similar to the reflection technique, the primary difference being that arrival times of critically refracted waves instead of near-vertical reflected waves are analyzed. The TX–RX offsets in refraction studies are generally larger than those of reflection studies, with the result that the slow-moving surface waves, or *ground roll*, is much less of a concern. The seismic-refraction method is a simple and popular technique used by geophysicists, geotechnical engineers, and others to gather basic site geological information such as depth to bedrock beneath an unconsolidated overburden.

Consider again seismic excitation of the two-layer model shown in Figure 6.11; in particular, we are now interested in the ray path labeled 3. The ray incident at some critical angle i_C travels along the interface as a P-wave with velocity V_2. This is shown as the horizontal portion of ray path 3 in Figure 6.11. From Snell's law, written in the form

$$\frac{\sin i_C}{V_1} = \frac{\sin r_C}{V_2}, \tag{6.33}$$

with the angle of refraction $r_C = 90°$ such that $\sin r_C = 1$, the critical incident angle is given by

$$i_C = \sin^{-1}\frac{V_1}{V_2}. \tag{6.34}$$

As the critically refracted wave propagates along the subsurface interface, energy is transmitted continuously into the upper and lower layers in accordance with Huygens' principle. At some RX location on the surface, sufficiently far from the TX, the first arrival of seismic energy is that which has emerged from the interface at the same critical angle i_C. This is shown as the upward portion of ray path 3 in Figure 6.11.

For small TX–RX separations, the first arrival is the direct *P*-wave, shown as ray path 1 in Figure 6.11. However, at some separation distance, $x = x_C$, known as the *cross-over distance*, the direct wave is overtaken by the critically refracted wave. The latter is sometimes known as a *head wave* since it is the first to arrive at RX locations in the region $x > x_C$.

From an inspection of the geometry of ray path 3 in Figure 6.11, it is straightforward to show that the traveltime $T(x)$ of the head wave is

$$T(x) = \frac{2h}{V_1 \cos i_C} + \frac{x - 2h \tan i_C}{V_2}. \tag{6.35}$$

The first term in the above equation originates from the upward and downward portions of the ray path while the second term corresponds to the horizontally propagating part. Using Snell's law for critical refractions in which

$$\sin i_C = \frac{V_1}{V_2}; \tag{6.36a}$$

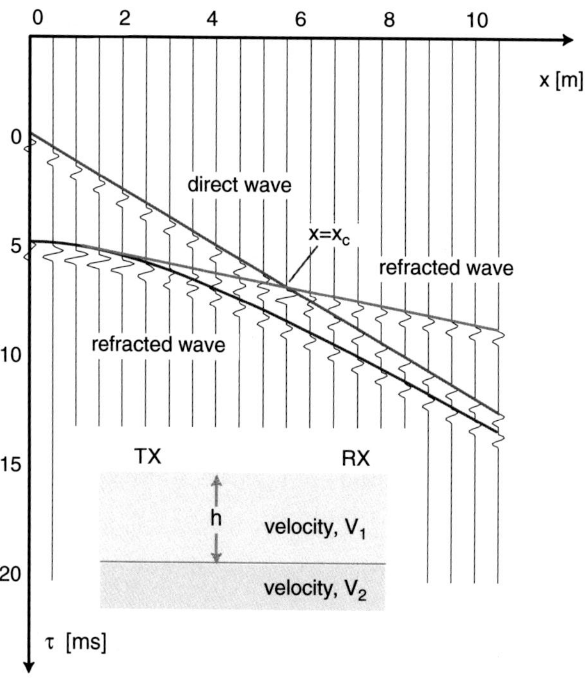

Figure 6.25 Idealized refraction seismogram.

$$\cos i_C = \frac{\sqrt{V_2^2 - V_1^2}}{V_2}; \tag{6.36b}$$

$$\tan i_C = \frac{V_1}{\sqrt{V_2^2 - V_1^2}}; \tag{6.36c}$$

Equation (6.35) reduces to

$$T(x) = \frac{x}{V_2} + \frac{2h\sqrt{V_2^2 - V_1^2}}{V_1 V_2}. \tag{6.37}$$

The traveltime Equation (6.37) predicts that refracted waves, unlike the hyperbolic moveout of reflected waves, move out linearly with TX–RX separation x. The slope of the $T(x)$ curve is $1/V_2$. Thus, one can obtain V_1 from the slope of the direct-wave arrival (which moves out according to $T(x) = x/V_1$; path 1 in Figure 6.11) and V_2 from the slope of the head-wave arrivals. The layer thickness h can be determined by extrapolating the head-wave arrival back to the origin since

$$T_0 = T(0) = \frac{2h\sqrt{V_2^2 - V_1^2}}{V_1 V_2}. \tag{6.38}$$

The moveout curves of the direct wave, critically refracted wave, and reflected wave are illustrated in the idealized seismogram (shot gather) shown in Figure 6.25. The parameters

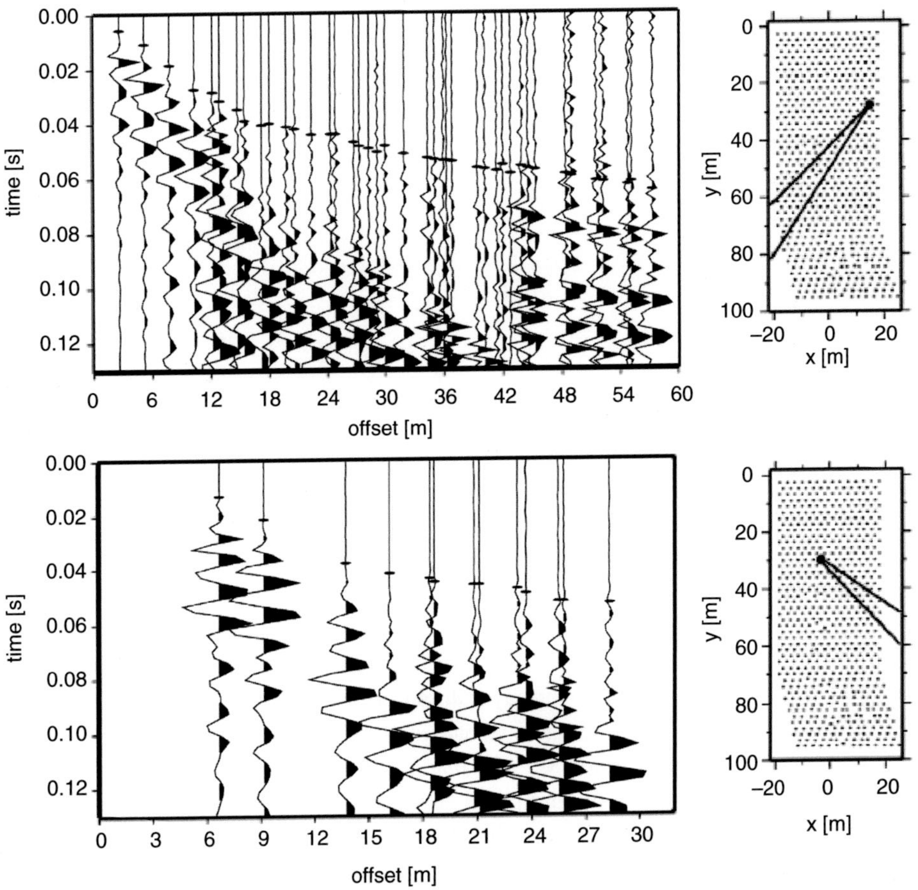

Figure 6.26 Shot gathers showing manually picked first arrivals (left panel), denoted by horizontal marks on the traces, from a seismic-refraction study at a contamination site; (right panel) shotpoints are marked by dark symbols; receivers are selected within a narrow TX–RX azimuth defined by the dark lines. After Zelt *et al.* (2006).

$V_1 = 800$ m/s; $V_2 = 2500$ m/s and $h = 2$ m were chosen for the calculation. Notice that the refracted wave is the first arrival beyond the cross-over distance x_C at which the direct and refracted arrivals are simultaneous. The primary reflected arrival is never the first arrival. Notice also that the refracted arrival does not appear on the traces at short TX–RX offsets, $x < 2h \tan i_C$. Instead, it merges with the primary reflection arrival, as shown in the figure. This behavior is easily understood by an inspection of ray path 3 in Figure 6.11.

In seismic-refraction studies, first arrivals are typically picked manually from observed shot gathers. This can become a time-consuming process for large two- and three-dimensional (2-D and 3-D) surveys. In addition, it is not always possible to precisely identify the first-arriving energy as, by definition, it is a small signal and moreover it often occurs in the presence of noise. Examples of first-arrival picks are provided in Figure 6.26 (left panels) from a 3-D refraction experiment at a contaminated site by Zelt *et al.* (2006). The picked first arrivals are shown by the horizontal markers on the individual traces of the shot gathers. Figure 6.26 (right panels) shows the location of the two shotpoints (large

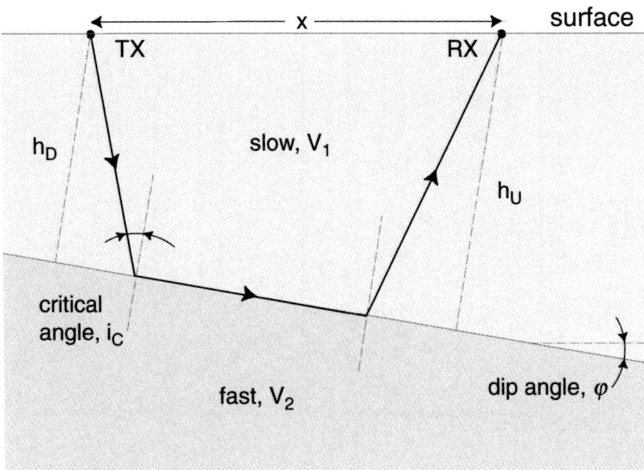

Figure 6.27 Seismic-refraction path for a dipping interface.

dots) relative to the locations of the receivers (small dots). The traces selected for display at the left are from those receivers located within narrow azimuthal cones of the shotpoints. The cones are marked by the solid lines in the right panels.

A dipping subsurface interface can also be analyzed using the seismic-refraction method. Consider the ray path shown in Figure 6.27. The RX is located down-dip from the TX. The dip angle φ measures the inclination of the subsurface bed. The lower medium has a faster seismic velocity than the upper medium. The dip angle φ can be estimated from measurements of up-dip and down-dip traveltimes. The head-wave traveltime curve $T_D(x)$ for a small dip angle φ such that $\cos^2\varphi \sim 1$ is given by

$$T_D(x) = \frac{2h_D \cos i_C}{V_1} + \frac{x \sin(i_C + \varphi)}{V_1}, \tag{6.39}$$

which shows that the down-dip refracted wave moves out with apparent velocity $V_D = V_1/\sin(i_C + \varphi)$. Similarly, the apparent up-dip velocity is $V_U = V_1/\sin(i_C - \varphi)$. The dip angle is then given by

$$\varphi = \frac{1}{2}\left[\sin^{-1}\left(\frac{V_1}{V_D}\right) - \sin^{-1}\left(\frac{V_1}{V_U}\right)\right]. \tag{6.40}$$

The derivation of Equation (6.40) is left as an exercise for the reader.

It is a simple matter to compute traveltime curves for a medium consisting of multiple refracting horizontal strata. The two-layer case is illustrated in Figure 6.28 (bottom). Recall that the traveltime for a single refracted arrival, along the path labeled 1, can be written as

$$t_1(x) = \frac{x}{V_2} + \frac{2h \cos\theta_{C1}}{V_1}, \tag{6.41}$$

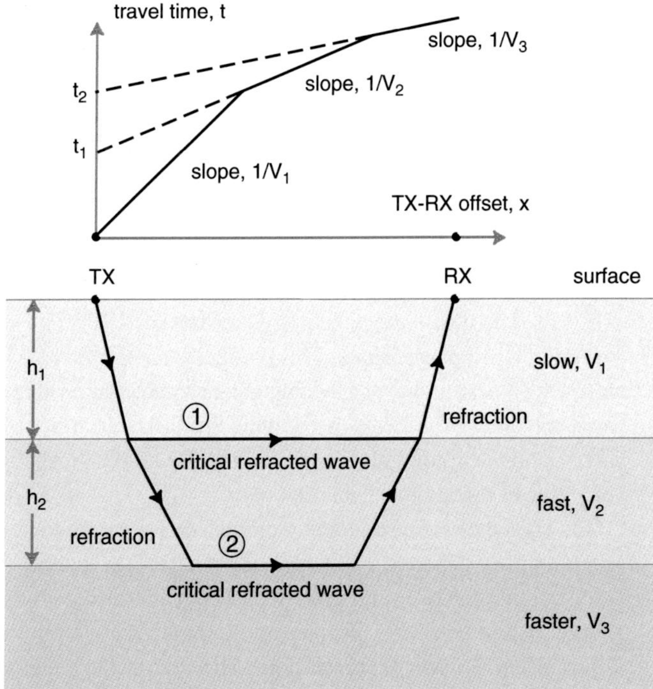

Figure 6.28 Traveltime curves (top) and ray paths (bottom) for the two-layer refraction geometry.

where θ_{C1} is the critical angle of refraction for the interface between media 1 and 2. The curve $t_1(x)$ of Equation (6.41) is shown as the middle straight-line segment in Figure 6.28 (top), with slope $1/V_2$.

Following Telford *et al.* (1990), suppose that a layer of faster velocity $V_3 > V_2$ underlines the two-layer medium. At sufficiently large TX–RX offsets, the deeper refraction will eventually overtake the shallower refraction. The traveltime curve $t_2(x)$ for the refraction along the path labeled 2 is readily computed by applying Snell's law at each of the interfaces to determine the path geometry,

$$\frac{\sin\theta_{C1}}{V_1} = \frac{\sin\theta_{C2}}{V_2} = \frac{1}{V_3}. \tag{6.42}$$

where θ_{C2} is the critical angle of refraction for the interface between media 2 and 3. It is straightforward to generalize the traveltime curve to the case of an n-layer medium. The result is

$$t_n(x) = \frac{x}{V_n} + \sum_{i=1}^{n-1} \frac{2h_i \cos\theta_{Ci}}{V_i}. \tag{6.43}$$

Note that Equation (6.43) provides an accurate description of the observed traveltime curve only in the case in which each bed layer is fast enough and thick enough to contribute first-arriving energy over a significant portion of the overall time–distance curve to permit its slope to be analyzed correctly. It is possible to have a "hidden layer" in a standard refraction analysis. In such cases, the hidden bed is either too thin or its velocity is not sufficiently greater than those of the overlying and underlying beds to contribute a first arrival at any TX–RX offset distance x.

6.14 Practical considerations

The seismic-reflection and -refraction methods work well in fine-grained, saturated sediments where attenuation is low and excellent mechanical coupling to the ground can be achieved by the source and the geophones. The methods fare relatively poorly in loose, dry, coarse-grained, or disturbed sediments.

A sample seismic record acquired at the Texas A&M University main campus in College Station is shown in Figure 6.29. A sledgehammer source was used with geophone spacing of 3.0 m. The near-surface geology consists of ~ 10 m of heavy floodplain clay deposits overlying a soft plastic shale. This particular record shows a high-frequency (>100 Hz) air wave with linear moveout traveling at its characteristic value of ~ 330 m/s; a refracted head wave traveling at ~ 1735 m/s within the fast subsurface shale layer; possible reflected waves; and the ubiquitous low-frequency (20–30 Hz) dispersive ground roll traveling with velocities in the range ~ 160–313 m/s.

The choice of seismic source requires careful consideration. The frequency content of a seismic source depends on the near-surface geology and the source coupling to the ground. Good coupling may be attained with a sledgehammer source using a heavy strike plate with a large surface area. If explosives or a shotgun is used, the charge should be detonated in a tightly packed, water-saturated hole. The role of the water is to fill void spaces in which elastic energy would otherwise be dissipated. It is highly advised to test the performance of a number of different sources prior to conducting data acquisition over the full survey area. For imaging reflections from the uppermost 3–10 m of the subsurface, a source should produce significant energy at frequencies above ~ 250–300 Hz. Baker *et al.* (2000) compared the performance of various impulsive sources, including a 4.5 kg sledgehammer and two different rifles (30.06 and .22 caliber). At a test site in Kansas, the most coherent reflection images were obtained using the .22-caliber rifle with subsonic ammunition since this combination generated the largest amount of high-frequency energy. Important characteristics of some commonly used seismic sources are listed in Table 6.3.

An electromagnetic geophone (see Figure 2.3, left), which is the type of ground-movement sensor most commonly used in near-surface geophysics, detects the relative motion between a magnet and a coil. The magnet, being rigidly coupled to the plastic case of the geophone, is directly coupled to the ground. The coil is wrapped around the magnet and loosely coupled to it by means of a leaf spring. The movement of the magnet relative to that of the coil introduces, by Faraday's law of electromagnetic induction, an electromotive force (emf) in the coil which is recorded as an output voltage.

Good ground coupling of the geophone should be ensured using a long spike. The geophone should be firmly planted into solid or fully saturated ground beneath any organic litter or other poorly consolidated surface materials. Dry sand is a poor environment for geophones because of the high absorption of energy. A simple mechanical model of a geophone consisting of a series arrangement of parallel springs and dashpots is shown in Figure 6.30a. The theoretical response of the model mechanical system to a ground forcing of the form $F_0\exp(i\omega t)$ is easily calculated (Krohn, 1984), as shown below.

Table 6.3 Seismic source characteristics

Source	Repeatability	Frequency [Hz]	Cost
Hammer	Fair–good	50–200	$
Weight drop	Good	50–200	$$
Explosives	Fair	50–200	$$
Shotgun/rifle	Very good	100–300	$$
Vibrator	Poor	80–120	$$$

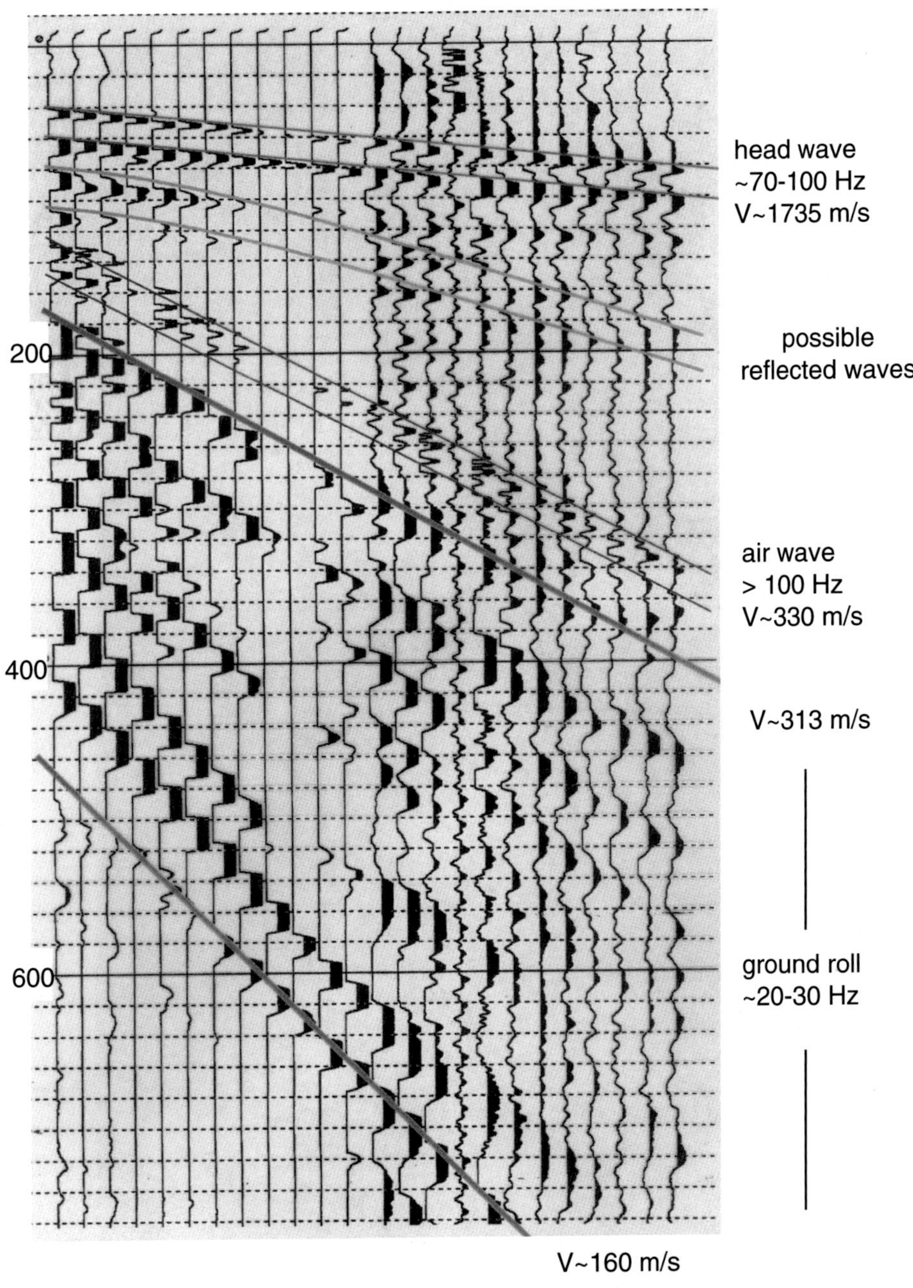

Figure 6.29 Seismic-reflection records at Texas A&M main campus. Adjacent traces are separated by 3.0 m. The vertical axis is two-way traveltime [ms].

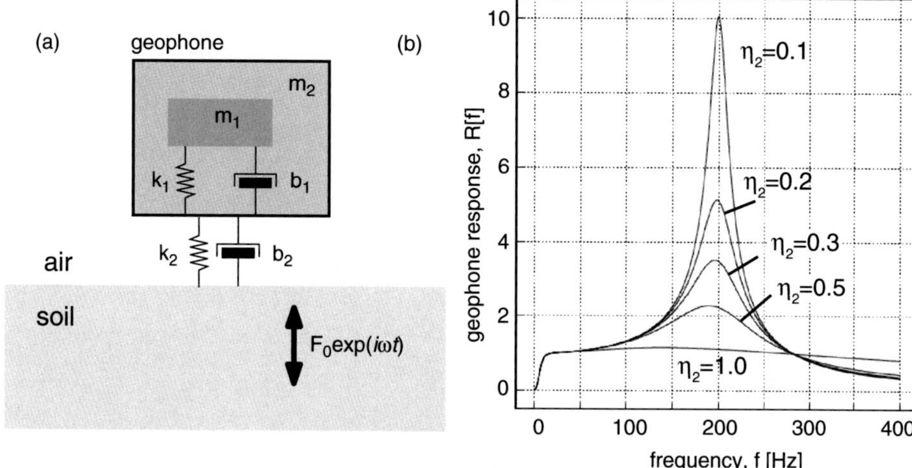

Figure 6.30 (a) A simple mechanical model of a geophone coupled to the ground; (b) calculated geophone response function amplitude $|R(f)|$ for different values of the ground damping factor, η_2.

Mechanical analysis of the geophone. As a first step, neglect the internal mechanisms inside the geophone and suppose the system to consist of a single undifferentiated mass m_2 undergoing damped oscillations in the presence of a driving force $F_0\exp(i\omega t)$ that represents harmonic ground motion at frequency ω. The geophone is then approximated by a simple spring and a viscous resistance in parallel arrangement, see Figure 6.30a. The associated spring force is given by Hooke's law $F_S = -k_2\xi$ with spring constant given by k_2. The restoring force F_S is proportional to the displacement ξ of the geophone; as the spring is stretched, the force becomes stronger and acts in opposition to further extension.

Friction against the motion of the geophone due to its imperfect coupling to the ground results in damping of the oscillations. The friction force F_R is proportional to the velocity of the geophone (French, 1971) such that

$$F_R = -b_2\frac{d\xi}{dt}, \tag{6.44}$$

where b_2 is the *mechanical resistance.* The negative sign indicates that the friction force opposes the motion of the geophone. The equation of motion for the damped oscillation is obtained by the force balance

$$m_2\frac{\partial^2\xi}{\partial t^2} + b_2\frac{\partial\xi}{\partial t} + k_2\xi = F_0\exp(i\omega t), \tag{6.45}$$

which in the frequency domain has the solution

$$\xi(\omega) = \frac{F_0/m_2}{(\omega_2^2 - \omega^2) + i\omega\gamma_2}, \tag{6.46}$$

where $\gamma_2 = b_2/m_2$. In Equation (6.46), the parameter $\omega_2 = \sqrt{k_2/m_2}$ is termed the *resonant frequency.* The maximum amplitude of $\xi(\omega)$ will occur when the in-phase term in the denominator vanishes, that is, at the resonant frequency $\omega = \omega_2$. Note that the system is said to be *critically damped* if the special condition $\omega_2 = \gamma_2/2$ is met. The amplitude of the oscillations of a critically damped system is zero. Equation (6.46) is the complete solution for the displacement of a rigid mass m_2 undergoing damped oscillations driven by harmonic ground motion of the form $F_0\exp(i\omega t)$.

A complete mechanical analysis should take into account the internal magnet and coil of the geophone. The internal spring constant is k_1 and the internal mechanical resistance is b_1. The magnet–coil system can be assumed to be driven by the motion of the external case, described by Equation (6.46), in which case a new balance of forces yields the equation of motion of mass m_1 in Figure 6.30a,

$$m_1\frac{\partial^2 z}{\partial t^2} + b_1\frac{\partial z}{\partial t} + k_1 z = m_2\frac{\partial^2 \xi}{\partial t^2} = -b_2\frac{\partial \xi}{\partial t} - k_2\xi, \tag{6.47}$$

where z is the displacement of mass m_1. Since all terms in Equation (6.47) are harmonic with frequency ω, it is a simple exercise in algebra to solve this equation in the frequency domain. This results in the *geophone response function* $R(\omega)$ given by

$$R(\omega) = \frac{-\left(\frac{\omega}{\omega_1}\right)^2\left[1 + i\left(\frac{\omega}{\omega_2}\right)\eta_2\right]}{\left[1 - \left(\frac{\omega}{\omega_1}\right)^2 + i\left(\frac{\omega}{\omega_1}\right)\eta_1\right]\left[1 - \left(\frac{\omega}{\omega_2}\right)^2 + i\left(\frac{\omega}{\omega_2}\right)\eta_2\right]}, \tag{6.48}$$

where $\eta_1 = \gamma_1/\omega$ is the *damping factor* of the magnet–coil system and $\eta_2 = \gamma_2/\omega$ is that of the ground coupling. The damping factor describes how close a system is to critical damping; the value $\eta = 0$ corresponds to free oscillations (no damping), while $\eta = 2$ corresponds to critical damping. Both the parameter $\gamma_1 = b_1/m_1$ and the resonant frequency $\omega_1 = \sqrt{k_1/m_1}$ of the geophone magnet–coil system are under control of the geophone designer. The physical significance of the response function $R(\omega)$ is that it is proportional to the acceleration of the mass m_1, that is, $R(\omega) \sim \omega^2 z(\omega)$.

The response function amplitude $|R(f)|$, where $f = \omega/2\pi$ is the frequency [Hz], is plotted in Figure 6.30b for different values of the ground-coupling damping factor η_2. A standard geophone is assumed in this calculation, with resonant frequency $\omega_1 = 8$ Hz and damping factor $\eta_1 = 1.4$, or 70% of the critical value. These are typical geophone design values. It is also assumed that the resonant frequency of the ground coupling is $\omega_2 = 200$ Hz. The figure illustrates the effect of imperfect ground coupling on the frequency response of a geophone. As the ground-coupling damping factor η_2 gets larger, the effect is an overall reduction in the amplitude of the geophone response and a flattening of the amplitude spectrum. A peak in the response function amplitude $|R(f)|$ indicates frequencies of ground motion to which the geophone is most sensitive. Thus, the effect of loose coupling of the geophone to the soil (i.e. a large value of η_2) is a great reduction in the sensitivity to the ground motion at its resonant frequency, in this case 200 Hz. The geophone therefore should be firmly planted in the soil, or buried, to keep the damping factor η_2 as low as possible.

Once the geophone output voltage is measured, the signal is amplified, filtered, and stored in a digitized form on a *seismograph*. The main characteristics of a seismograph are dynamic range and the number of channels. The dynamic range is the ratio of the largest measureable signal to the smallest measureable signal, as mentioned earlier in Chapter 2. Near-surface geophysical applications typically utilize 16-bit (96 dB) dynamic range. The number of channels (usually 36–48, or more) is the number of geophones whose response can be simultaneously recorded.

6.15 Seismic data processing

A number of data processing steps must be performed in order to convert seismic-reflection shot gathers into migrated depth sections that are ready for geological interpretation. The discussion here will be brief. The classic reference for exploration-scale seismic data processing is Yilmaz (2001). The course notes of Baker (1999) provide a good overview of data-processing procedures that are relevant to near-surface geophysics. Chapters 2, 9, and 11 of this book also contain information on basic data processing.

Gain control. It is often required to amplify the small geophone signals recorded in a near-surface geophysical survey. Since seismic waves attenuate exponentially with distance, and are subject to spherical wavefront spreading, the return amplitudes from reflectors at depth are generally much lower than the amplitudes returned from shallower reflectors. Similarly, returns on the far-offset geophones are much smaller than those on geophones that are placed close to the source. Amplifier gains on each channel should be set in order to roughly equalize the returned signal amplitudes, thus permitting a better visualization of deeper reflectors especially at the far-offset geophones. This procedure is termed *gain control*.

Bandpass filtering. Filters may be used to suppress unwanted events and highlight events of interest on seismic-reflection records. The essentials of filtering were discussed earlier in Chapter 2. A number of filtering operations are specialized to seismic-reflection data processing. For example, *mutes* are routinely used to blank out refraction first-breaks, airwave and/or ground-roll energy from shot gathers. A *bandpass filter* reduces the amplitude of the frequency components of a signal that reside outside a specified band. In many cases, surface waves can be effectively removed by bandpass filtering. Surface waves are unwanted low-frequency events that can obscure or interfere with higher-frequency seismic reflections. An example of bandpass filtering of a shot gather acquired with a 1-kg hammer source at 0.25-m receiver spacing on the Matanuska glacier is Alaska is shown in Figure 6.31. The predominant frequencies are above 800 Hz for the wanted reflections from within the ice layer and from debris-rich ice at the base of the glacier, while the frequency content of the unwanted surface waves was considerably lower. Accordingly, a bandpass filter with pass band 700–1200 Hz proved effective in attenuating the surface waves while preserving the reflections, as shown on the right side of the figure. In this example, the surface waves were relatively non-dispersive with a group velocity of ~ 1700 m/s while the body waves traveled much faster, at 3600 m/s.

Refraction-statics correction. Refraction statics (Gardner, 1967) are adjustments that are made to the timing of individual seismic traces to take into account factors such as

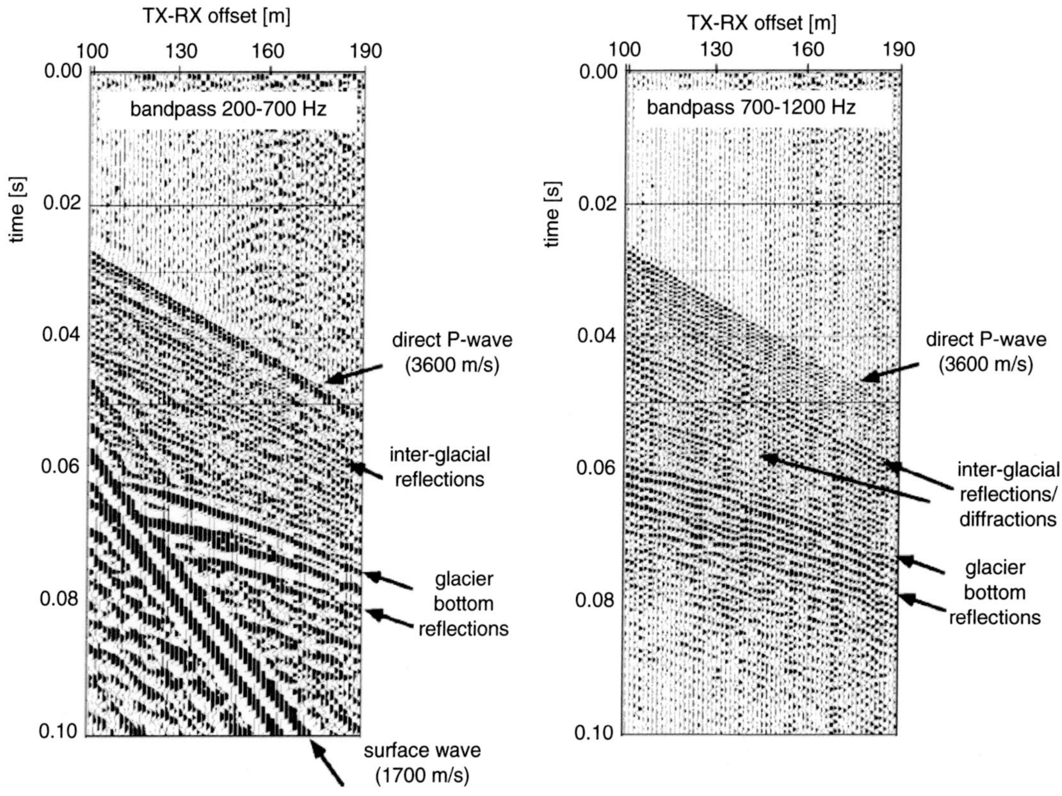

Figure 6.31 Surface-wave attenuation using bandpass filtering. After Baker *et al.* (2003).

near-surface lateral velocity variations, undulations of a shallow refracting horizon, and irregular terrain. The results of a refraction-statics correction yield better estimates of the traveltimes to deeper reflectors. The new traveltimes are the ones that would have been recorded if the near-surface layer were homogeneous and/or the terrain and the shallow refracting horizon were level. With static-corrected data, deeper reflection events from horizontal layers generally exhibit an improved normal moveout, such that NMO stacks are more coherent than they would have been in the absence of the refraction-statics correction. Essentially, a refraction-statics correction enables better control on deeper reflector imaging by removing the deleterious effects of near-surface lateral velocity variations on traveltimes.

Docherty and Kappius (1993) have cast a refraction-statics correction into a linear inverse problem (see Chapter 11). In two dimensions, the situation they treat is illustrated in Figure 6.32. The observed refraction first-arrival traveltime t_{ij} between the TX–RX pair (i, j) is inverted for the down-going and up-going delay times (s_i and r_j, respectively) and the slownesses σ_k along the undulating refractor horizon. The latter is discretized into cells of width d_k. The delay time s_i is the time taken for the signal to propagate from the source down to the refractor horizon while the delay time r_j is the time taken for the signal to propagate upward from the refractor horizon to the receiver at the surface. The time-delay equations are

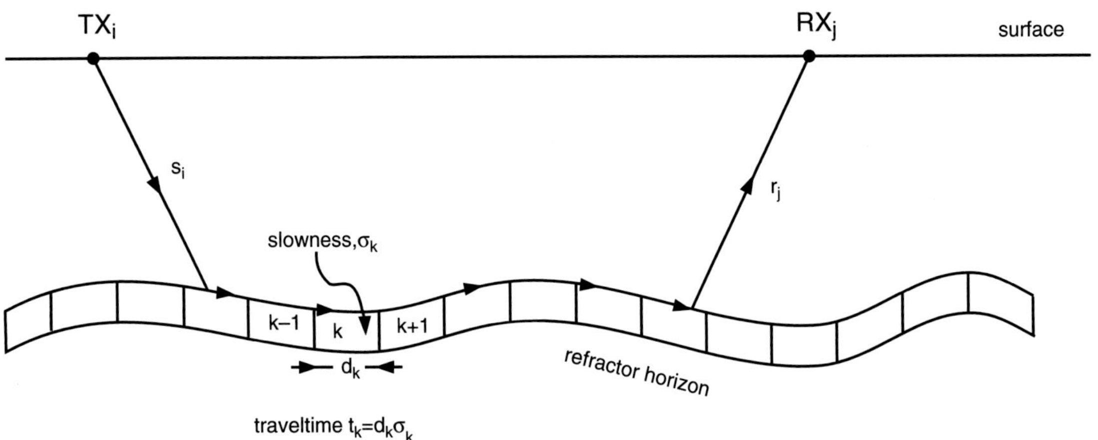

Figure 6.32 Refraction ray path associated with an undulating near-surface horizon for the (i, j)-th TX–RX pair.

$$t_{ij} = s_i + r_j + \sum_{k=1}^{P} d_k \sigma_k, \tag{6.49}$$

for each of the TX–RX pairs and P is the number of cells along the probed refractor horizon. The equations in (6.49) constitute a set of linear constraints that connect the measured delay times t_{ij} to the unknown model parameters (s_i, r_j, σ_k). The system of equations can be solved using one of the linear-inversion techniques discussed later in Chapter 11.

Velocity analysis. Another critical step in the basic seismic data-processing sequence is to determine the *stacking velocity* V that should be used in the NMO correction (Equation (6.25)). Velocity analysis is a process that is performed on CMP gathers. Recall that a CMP gather is a collection of traces that contains reflections from the same point on a subsurface horizon; for horizontal strata, the reflection point lies directly beneath the common TX–RX midpoint. Velocity analysis can also be performed on *CMP supergathers*. A CMP supergather is a collection of adjacent CMP gathers plotted together. If lateral velocity variations are sufficiently small, all reflection points in a CMP supergather are presumed to reside within the same Fresnel zone on the subsurface horizon. Velocity analysis can also be performed in the presence of dipping layers, in which case the stacking velocity V is used in the DMO process.

In a simple form of velocity analysis known as *constant-velocity stacking* (CVS), the stacking velocity is presumed to be uniform throughout the subsurface. A sequence of regularly spaced values of stacking velocity is tried in the NMO correction procedure, with the optimal value being the one that best appears to flatten out the hyperbolas and provide good lateral continuity of reflectors on the CMP (super)gather. The CVS technique is normally applied manually, with the interpreter looking simultaneously and qualitatively comparing different CMP gather or supergather displays, each one having been NMO-corrected using a different stacking velocity.

A less subjective, automated method of velocity analysis involves the construction and analysis of a *semblance* plot. Semblance is a robust (noise-tolerant) measure of the

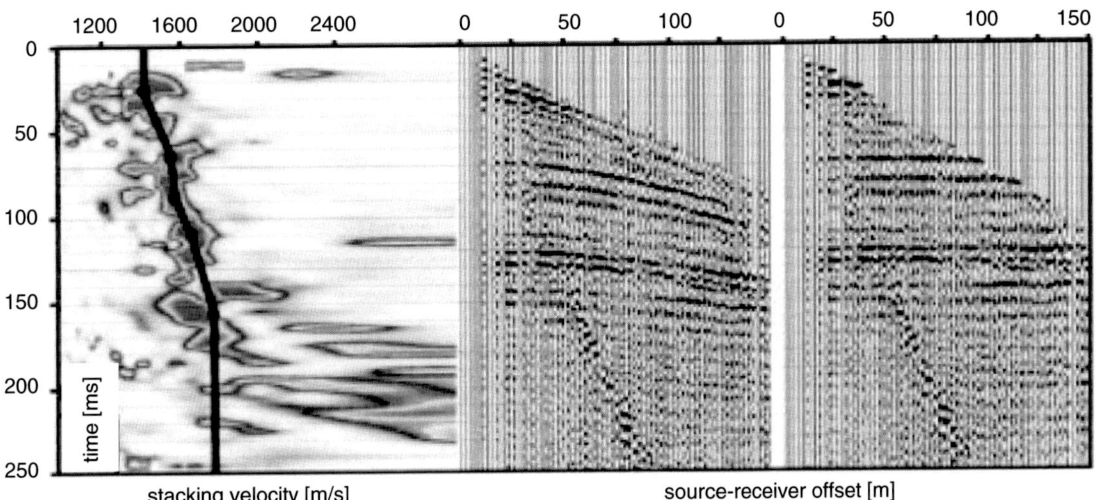

Figure 6.33 (Left) Determination of stacking velocity by passing a continuous curve through regions of high (red) values of semblance; (middle) an uncorrected CMP supergather; (right) the NMO-corrected CMP supergather using the stacking-velocity function in the left panel. After Spitzer *et al.* (2003). See plate section for color version.

similarity between a large number of trial hyperbolas and the actual hyperbolas present on a CMP (super)gather. A semblance contour plot is constructed with the trial stacking velocity on the horizontal axis and the zero-offset time, or equivalently an apparent depth, on the vertical axis. The best stacking velocity, for a given apparent depth, corresponds to the highest value of the semblance. An example of a semblance-based velocity analysis of a CMP supergather appears in Figure 6.33. The stacking velocity curve (solid line, left panel) is chosen as one that joins regions of high semblance values. Using this velocity function in the NMO process tends to flatten out the hyperbolic reflection events seen in the CMP supergather (middle and right panels).

Linear τ–p filtering. Another means of separating primary reflection events from other types of source-generated energy such as surface waves, refractions, direct waves, and guided waves is via a linear *τ–p transformation*. The essential mathematics of the transformation (Diebold and Stoffa, 1981) is summarized in Appendix C for a simple three-layer case without dip. As shown in the appendix, transforming a shot gather from the familiar t–x domain into the linear τ–p domain converts events that have linear moveouts, such as direct, guided, and surface waves, into points. Hyperbolic reflecting events in the t–x domain are mapped into elliptical-shaped curves in the linear τ–p domain. This property facilitates the separation of wanted reflected signals from the remaining source-generated energy, which in reflection imaging is regarded as noise. After filtering, an inverse τ–p transformation is performed to reconstruct the original shot gather in the t–x domain, but without the source-generated noise.

A linear τ–p transformation of a synthetic shot gather is illustrated in Figure 6.34a, from a paper by Spitzer *et al.* (2001). The synthetic data are generated from a four-layer velocity model by a finite-difference simulation and are plotted for convenience using the reduced

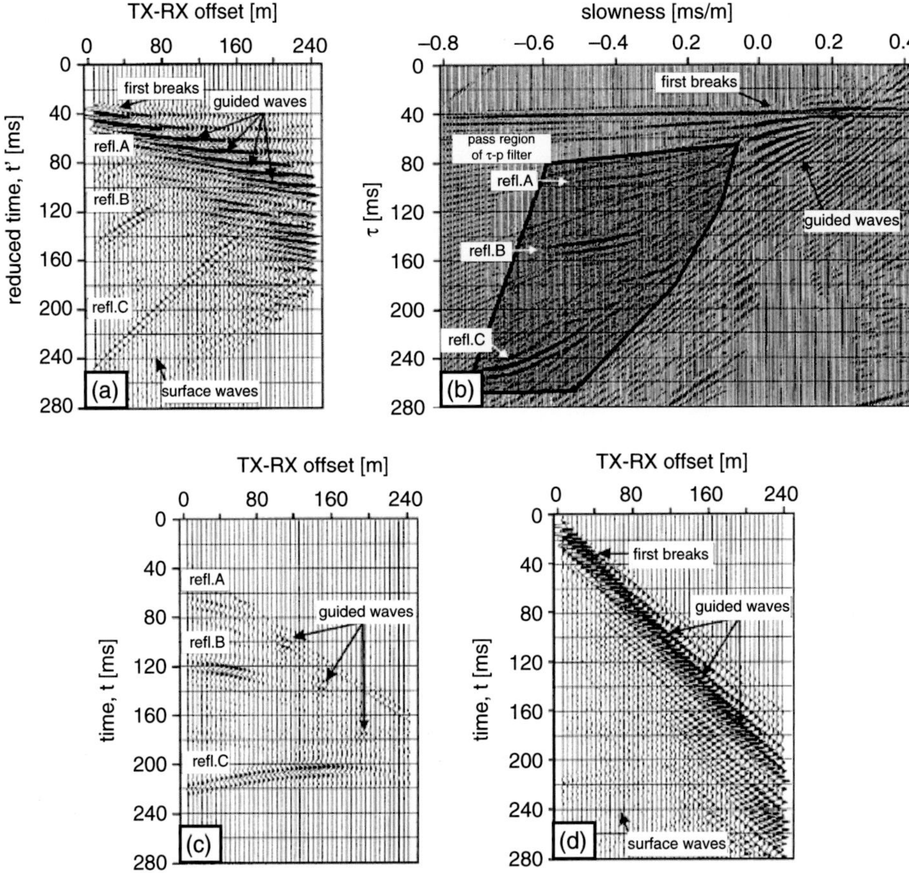

Figure 6.34 The results of linear τ–p processing on synthetic data: (a) finite-difference simulated shot gather plotted using reduced traveltime; (b) result of linear τ–p transformation; the black line outlines the pass region; (c) result of the inverse linear τ–p transformation after filtering; (d) the source-generated noise removed by the linear τ–p processing. After Spitzer *et al.* (2001).

traveltime $t' = t + 30 - x/1700$ [ms]. The velocity model contains a dipping reflector. Moreover, near-surface waveguiding effects and surface waves are included in the simulation. The results of the linear τ–p transformation are shown in Figure 6.34b. The black line outlines the pass region of the τ–p filter. The pass region is selected as one which contains mainly the elliptic-shaped curves characteristic of reflection events (see Appendix C). The results of the inverse linear τ–p transformation are shown in Figure 6.34c. Note that the energy from model reflectors A, B, and C is enhanced. The source-generated noise that was removed using this procedure is shown in Figure 6.34d.

Migration. As described already, CMP stacking of NMO-corrected traces assume that all energy originates from a single point on a subhorizontal reflecting horizon located directly beneath the TX–RX midpoint. This assumption is not correct in the presence of dipping reflectors or in the case of diffractions from edges. The purpose of *migration* is to image

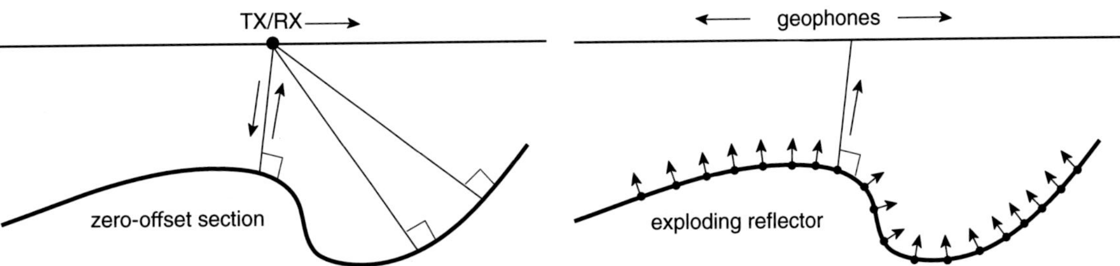

Figure 6.35 The exploding-reflector concept. Adapted from material on the Stanford Exploration Project website, sepwww.stanford.edu.

subsurface dipping reflectors and diffracting points at their correct positions within the subsurface, essentially undoing the effects of wave propagation. Over the past decade, sophisticated algorithms have emerged for migrating seismic data on exploration length scales (Etgen *et al.*, 2009). An excellent survey of the development of migration from its historical roots to its modern application in the oil and gas sector is given in Gray *et al.* (2001).

Migration can be applied either before or after the CMP stack. A small number of near-surface geophysicists have applied DMO/pre-stack migration and some have investigated post-stack migration. It has been found, however, that data quality must be very high for migration to be successful. Especially in the uppermost several tens of meters, strong velocity contrasts and heterogeneities are present, leading to severe static effects and dominant source-generated noise. Migration in this case can introduce significant artifacts and can actually degrade a geological interpretation; thus migration is not always used for very shallow applications. For seismic acquisition layouts that probe to greater depths, in the range of ~ 200–500 m, data quality is typically better such that pre-stack migration is often effective (G. Baker, personal communication).

There are two types of migration: reverse-time and depth migration. Reverse-time migration does not attempt to develop a geologically reasonable subsurface velocity model; rather it uses an ad hoc velocity function that produces a pleasing image containing coherent reflections. Depth migration is more involved as it first tries to estimate and then utilize an accurate model of the subsurface velocity distribution. In areas of structural complexity, however, the velocity distribution can be very difficult to determine. There are many different migration procedures used in geophysics. A detailed exposition of the various possibilities is beyond the scope of the book but I can refer the reader to Etgen *et al.* (2009). Herein, only the elementary *reverse-time Kirchhoff* method is discussed. The discussion is drawn largely from a tutorial on the Stanford Exploration Project website, sepwww.stanford.edu.

Consider the seismic experiment portrayed in Figure 6.35, left, in which seismic energy propagates downward and outward from a source (TX), reflects from a subsurface horizon, and then propagates upward to a co-located receiver (RX). Notice in this scenario that three primary reflection events will be recorded; these are shown on the illustration. The multiplicity of reflected arrivals is a consequence of the undulations in the reflecting horizon. Now suppose the co-located TX–RX pair is moved along the acquistion surface,

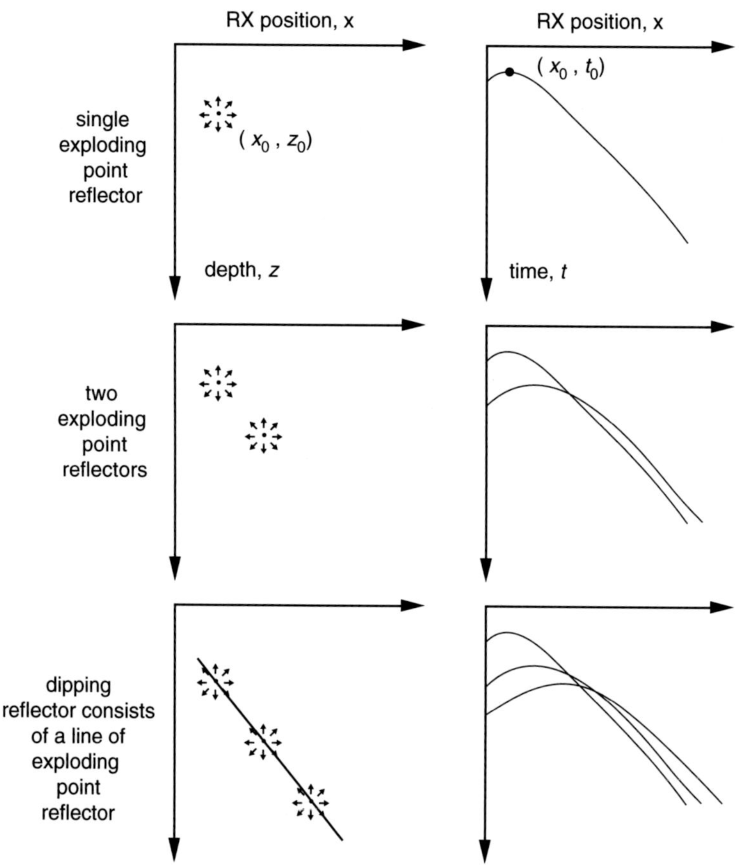

Figure 6.36 Forward modeling using the exploding-reflector concept.

as shown by the arrow pointing to the right. A zero-offset time section can be constructed. However, the undulating horizon would not be correctly imaged in the section since, as we have already seen, the single horizon produces multiple reflection events.

Figure 6.35, right, is a schematic illustration of the *exploding-reflector* concept attributed to Loewenthal *et al.* (1976). The main idea is that we would get exactly the same zero-offset time section if, instead of moving the single TX–RX pair along the profile, each point on the subsurface horizon were somehow to simultaneously explode and we could record the resulting signals with a geophone array spread across the surface. The observed wavefields $u(x, t)$ in the two experiments, the actual one at left and the hypothetical one at right, would be identical (with the exception that the traveltimes in the hypothetical experiment are just one-half those of the actual experiment). The exploding-reflector concept is simple yet powerful. In fact, the reader should be able to judge the validity of the exploding-reflector concept by thinking about Huygens principle, which states that each point on a reflecting wavefront acts as a spherically spreading point source.

Figure 6.36 illustrates how the exploding-reflector concept can be used to predict the wavefield due to reflections from an undulating horizon. At top left is shown a single

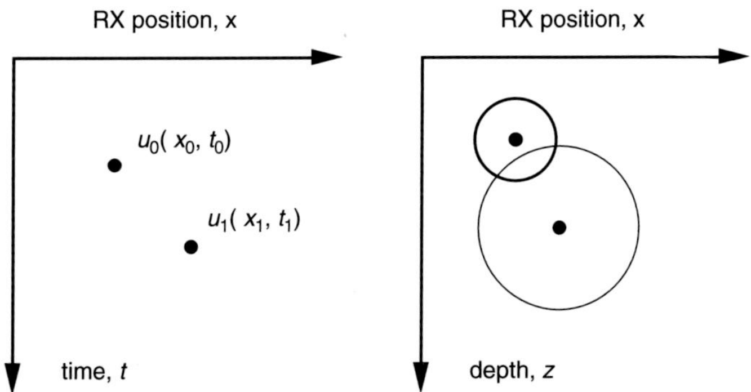

Figure 6.37 Kirchhoff imaging principle.

exploding point reflector at some location (x_0, z_0) in the subsurface. This explosion acts as a point source of spherically expanding seismic energy. An analysis of the normal moveout of this energy reveals that it should appear to a surface geophone array as a diffraction hyperbola with its apex at (x_0, t_0), as shown at top right. The transformation from depth to time is given by the familiar hyperbolic equation

$$V^2t^2 = (x - x_0)^2 + z_0^2. \tag{6.50}$$

The middle panels show the corresponding situation for two exploding point reflectors, in which case two diffraction hyperbolas should be observed by the geophone array. It should now be clear that a dipping reflecting horizon can be modeled as a continuous line of exploding point reflectors. Three of these explosions are shown for simplicity at bottom left. The dipping reflecting horizon appears on the geophone array as the superposition of many diffraction hyperbolas; again, just three of these are shown at bottom right. This example shows how the exploding-reflector concept can be used to predict the wavefield generated by seismic excitation of a physical Earth structure. The next task is to consider the reverse process, namely, how to construct an image of the physical Earth structure based on an observed seismic wavefield.

Figure 6.37 describes the key aspect of the Kirchhoff imaging principle. We work in two dimensions for simplicity although the generalization to three dimensions is straightforward. At left are shown two samples of an observed seismic wavefield $u(x, t)$. At right, these two data points are transformed from time t into depth z using the transformation (6.50). The seismic energy u_0 that is observed at location (x_0, z_0) could have arrived there from any point or points on a circle of radius Vt_0, where V is the velocity of the medium; similar remarks apply for the energy u_1 at location (x_1, z_1). We therefore construct two circular mirrors, as shown: a smaller one of radius Vt_0 centered on (x_0, z_0) that has strength, or reflectivity, u_0; and a larger one of radius Vt_1 centered on (x_1, z_1) that has strength, or reflectivity, u_1. We construct a similar circular mirror for each data sample in the wavefield. Then, we simply add all the circles together to produce the Kirchhoff image.

To understand how Kirchhoff imaging works, consider the following. Suppose the seismic energy observed at $u_0(x_0, t_0)$ and $u_1(x_1, t_1)$ originated from just a single exploding reflector. That reflector must be located at one of the two intersection points of the circles.

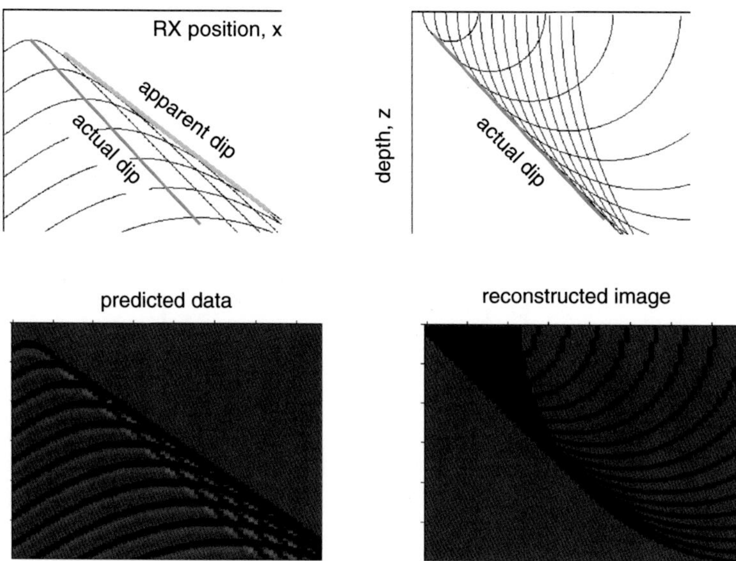

Figure 6.38 Kirchhoff reverse-time imaging of a dipping reflector. Adapted from material on the Stanford Exploration Project website, sepwww.stanford.edu.

Adding the two circles together produces the largest amplitudes at the intersection points. In other words, the two possible locations of the reflector contribute most to the summation. A more detailed imaging example is shown in Figure 6.38. The left panels show, using two different display formats, the modeled wavefield due to a dipping reflector (compare to Figure 6.36, bottom right). This wavefield has again been computed based on the exploding-reflector concept. Notice that the dip of the reflector is not correct. The apparent dip is shallower than the actual dip. At right are shown the Kirchhoff circular mirrors. The reconstructed image is the superposition of the circular mirrors. Notice that the greatest amplitudes in the image are found, as desired, along the actual dip of the reflector.

3-D example. An example of a complete suite of 3-D seismic-reflection data-processing steps is given by Kaiser *et al.* (2011). This study of an active, oblique-slip segment of the Alpine fault zone in New Zealand is one of the first published reports of a 3-D near-surface seismic survey. While a 3-D dataset is clearly more time-consuming to acquire and process than its 2-D counterpart, a 3-D survey permits better imaging of complex fault geometries, including out-of-plane reflectors and diffractors. The acquisition layout in the New Zealand survey included 24 parallel source lines and 27 parallel receiver lines, each of ~ 500 m length, running perpendicular to the fault strike. The source spacing was 8 m and the receiver spacing was 4 m, which resulted in an average fold of ~ 20. The line spacing was ~ 10 m. Due to cost and time constraints, most of the TX–RX pairs were nearly in-line, resulting in a limited azimuthal coverage. Ideally, a 3-D seismic survey would consist of full, densely sampled TX–RX azimuthal coverage.

Pre-stack processing steps included deconvolution, mutes, static corrections, τ–p filtering, velocity analysis on CMP supergathers, and NMO and DMO corrections. The

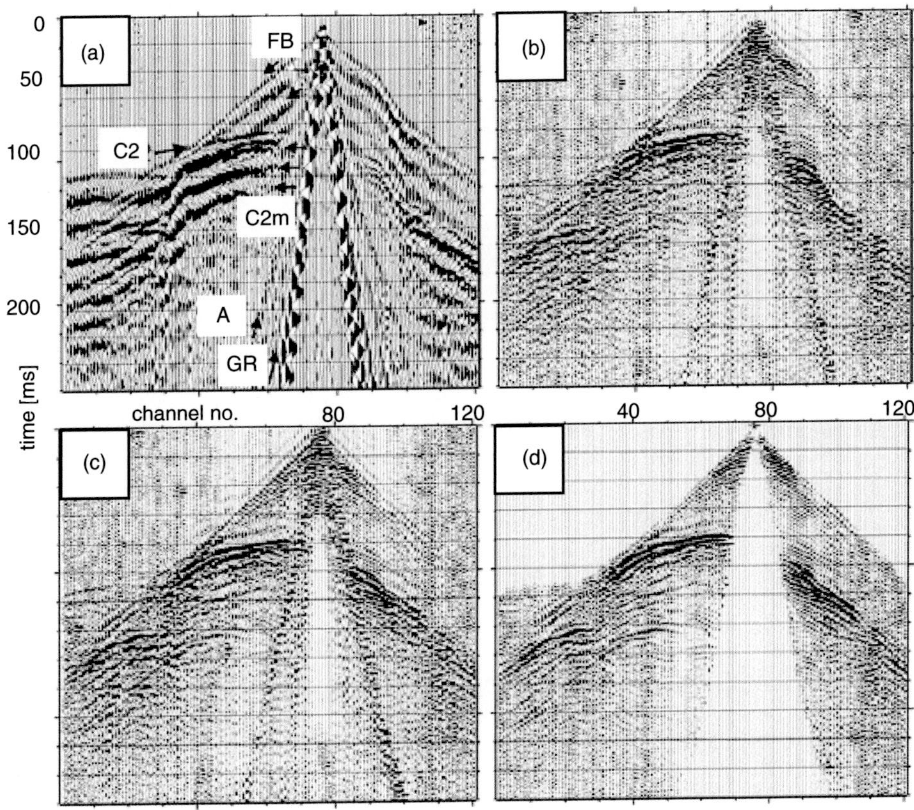

Figure 6.39 Effects of pre-stack processing steps on a high-resolution shot gather from an Alpine fault zone, New Zealand; A = air wave, GR = ground roll, FB = first breaks, C2 = basement reflection, C2M = basement reflection multiples; (a) raw shot gather with automatic gain control (AGC) applied; (b) deconvolution and bandpass filter applied; (c) static (refraction and residual) corrections applied; (d) mutes and τ–p filtering applied; after Kaiser *et al.* (2011).

effects of some of the pre-stack processing steps on a typical shot gather are shown in Figure 6.39. It is easily seen in the figure that the result of the pre-stack processing provides a better definition of the near-surface reflections, along with suppression of source-generated noise.

The post-stack processing included bandpass filtering, followed by 3-D depth migration. The migration algorithm collapsed most of the diffractions that were evident on the unmigrated sections. The interpreted main fault strand (AF) dipping at ~ 80°, along with a subsidiary fault strand (SF), are indicated by the dotted lines in the migrated 2-D section at Figure 6.40, left. The strong reflecting events C1 and C2 are from, respectively, the footwall and the hanging wall of the fault in late-Pleistocene basement. The borehole intersects the basement at ~ 26–30 m depth, in agreement with the location of the basement seismic reflector. The A and B reflectors are due to stratification within the overlying glaciolacustrine and glaciofluvial sediments. The migrated full 3-D volume is shown in

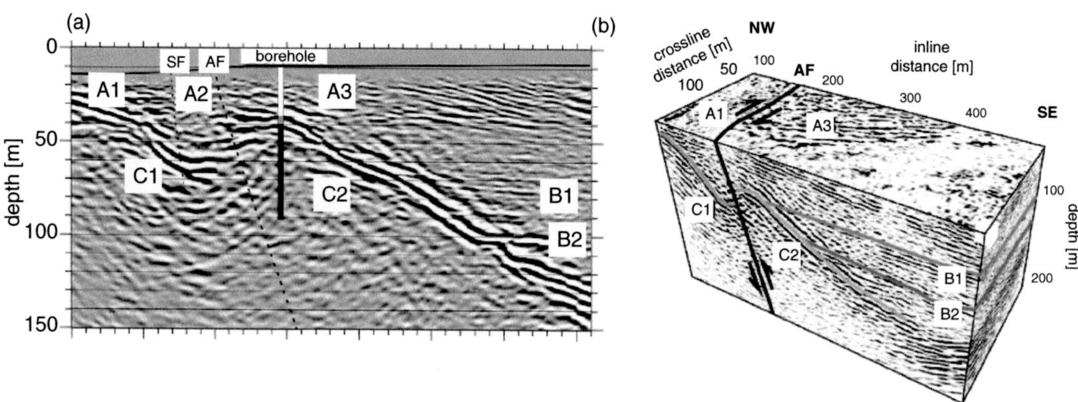

Figure 6.40 High-resolution near-surface reflection data from the Alpine fault zone, New Zealand: (a) a migrated 2-D section extracted from the migrated 3-D volume shown in (b); after Kaiser *et al.* (2011).

(Figure 6.40, right) in which it can be seen that the fault zone is imaged continuously along the strike of the main strand.

6.16 Ray-path modeling

The propagation of seismic waves though heterogeneous elastic media, possibly containing acoustic impedance discontinuities, can be approximated by *ray tracing*. By analogy with geometric optics, seismic wavefront tracking in the high-frequency approximation is described by rays. A ray is a narrow, pencil-like beam that reflects and refracts at material interfaces and bends as it travels in materials characterized by a continuously varying velocity structure. The ray approximation is excellent if the seismic wavelength is small in comparison to the characteristic length-scale of the material heterogeneities. Ray tracing constitutes the forward modeling component of many popular seismic tomographic reconstruction algorithms (e.g. Zelt *et al*, 2006).

Here we provide the classical derivation of ray trajectories $x(t)$ and $z(t)$ in a 2-D acoustic medium (Eliseevnin, 1965). The acoustic wave equation is

$$\frac{\partial^2 p}{\partial x^2} + \frac{\partial^2 p}{\partial z^2} = \frac{1}{c^2}\frac{\partial^2 p}{\partial t^2}, \tag{6.51}$$

where $c(x, z)$ is the acoustic wave speed and p is the pressure field. An acoustic medium can support compressional but not shear waves. Assume that the pressure field is time harmonic, $p(x, z, t) \sim \exp(-i\omega t)$. We can expand the pressure field into a power series in inverse powers of frequency,

$$p(x,z) = \exp(-i\omega\tau)\sum_{m=0}^{\infty}\frac{u_m}{(i\omega)^m} \tag{6.52}$$

for which $\tau(x, z)$ is a traveltime, or *pseudophase*, function and $u_m(x, z)$ is an amplitude function.

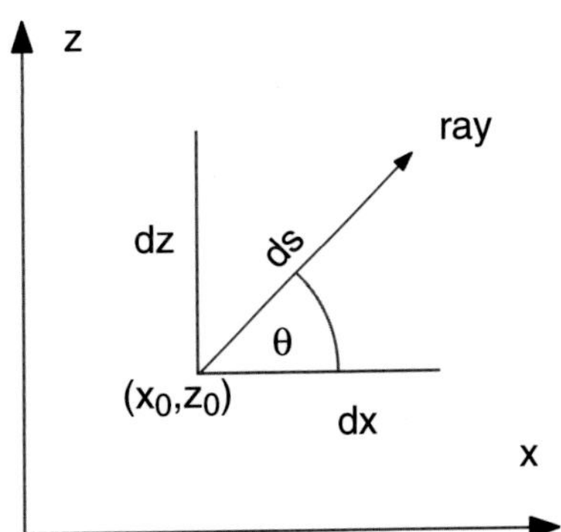

Figure 6.41 A ray emanating from point (x_0, z_0).

The series is rapidly convergent for the high frequencies at which the ray approximation is valid. Inserting Equation (6.52) into (6.51) and keeping the highest-order ($m = 0$) term results in the *eikonal equation*

$$\left(\frac{\partial\tau}{\partial x}\right)^2 + \left(\frac{\partial\tau}{\partial z}\right)^2 - n^2 = 0, \tag{6.53}$$

where $n^2 = 1/c^2$ is the squared wave slowness function.

The eikonal equation describes the spatial distribution of the traveltime $\tau(x, z)$ function in a medium characterized by acoustic wavespeed $c(x, z)$. The interpretation of the traveltime function is that, for any location (x, z) within the medium, the first wave arrives at time $t = \tau$. Hence, solving the eikonal equation provides a method for predicting the first-arriving waveform at any location throughout the medium. To see how the ray trajectories $x(t)$ and $z(t)$ are related to the eikonal equation, consider the ray shown in Figure 6.41. Let s be the distance along the ray such that $c(x, z) = ds/dt$.

From the figure, we can see that $dx = ds \cos\theta$ and $dz = ds \sin\theta$. Then it follows that

$$\frac{dx}{dt} = \cos\theta\frac{ds}{dt} = c\,(x,z)\cos\theta; \tag{6.54a}$$

$$\frac{dz}{dt} = \sin\theta\frac{ds}{dt} = c\,(x,z)\sin\theta. \tag{6.54b}$$

We can use Equations (6.54a, b) to propagate the ray forward from point (x_0, z_0) if we know the angle θ. Once the ray arrives at the new point (x_1, z_1) we need to determine the new ray direction θ. To find an equation for $d\theta/dt$, the eikonal equation (6.53) is re-written as

$$\frac{1}{n^2}\left(\frac{\partial\tau}{\partial x}\right)^2 + \frac{1}{n^2}\left(\frac{\partial\tau}{\partial z}\right)^2 = 1 = \cos^2\theta + \sin^2\theta, \tag{6.55}$$

from which we can identify $\cos\theta = (1/n)\,\partial\tau/\partial x$ and $\sin\theta = (1/n)\,\partial\tau/\partial z$. After some elementary algebra, the details of which can be found in Eliseevnin (1965), the following equation is obtained:

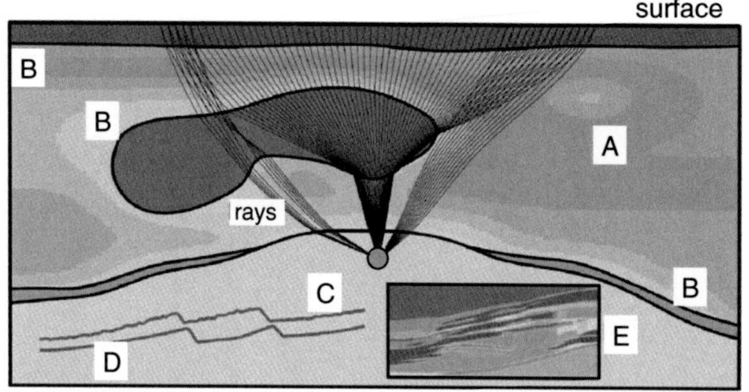

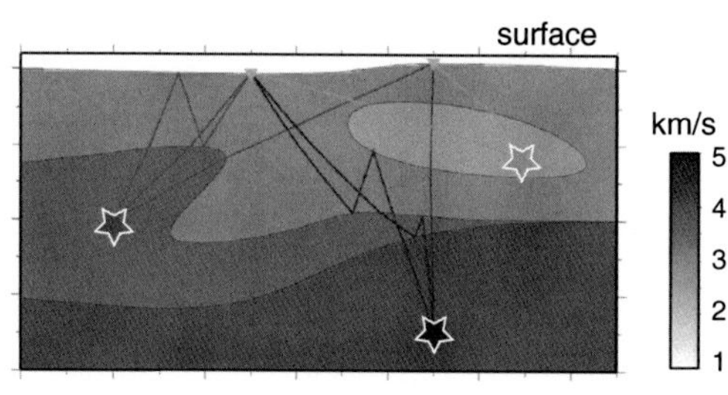

Figure 6.42 (Left) Rays traced from a subsurface source (large dot) through and around a complex-shaped fast-velocity anomaly to the surface (after Gjoystdal *et al.* 2007). (Right) Different ray paths are possible in complicated models, such as this one containing a pinchout, overturned fold, and anomalously fast body (after Hauser *et al.*, 2008). See plate section for color version.

$$\frac{d\theta}{dt} = \frac{\partial c}{\partial x}\sin\theta - \frac{\partial c}{\partial z}\cos\theta. \tag{6.56}$$

Together, Equations (6.54) and (6.56) represent a system of ordinary differential equations whose solutions determine ray paths through an inhomogeneous acoustic medium. A ray-tracing algorithm simply integrates this system of equations through time from $t = 0$ and a given source location (x_0, z_0). Note, however, that the geometry of a ray path is heavily influenced by the initial take-off angle θ_0.

In areas of structural complexity, iterative ray tracing can be used to determine a seismic velocity model that is consistent with the seismic observations. The typical assumption is that reflections mark boundaries between undulating layers of uniform velocity. The essential rule is that the rays obey Snell's law at each interface. Some examples of ray tracing in complex geology are shown in Figure 6.42. Information about the subsurface is obtained only for those regions that are illuminated by rays. Note that, for each of the three subsurface sources shown in Figure 6.42 right, rays that emanate with slightly different take-off angles can follow very different pathways and provide information about very different structures within the model. Also, rays emanating from different sources can arrive at a single point on the surface.

In Figure 6.43, it is shown how ray tracing can be used to understand complex wavefront morphologies, such as a *triplication*, which can develop even for relatively simple structures such as a slow anomaly embedded in a faster host medium. The convoluted shape of the wavefront at time $t + \Delta t$ may be understood by applying Huygens' principle to the wavefront at time t. The three arc segments labeled 1, 2, and 3 comprising the triplication form a characteristic "bowtie" structure. The inset in the figure shows that the signal at the receiver contains three distinct arrivals corresponding to the three ray paths. The first arrival is via ray path 1, which has traversed only the faster host medium. The second and third arrivals are via ray paths 2 and 3, respectively, that have traversed the slower anomalous zone.

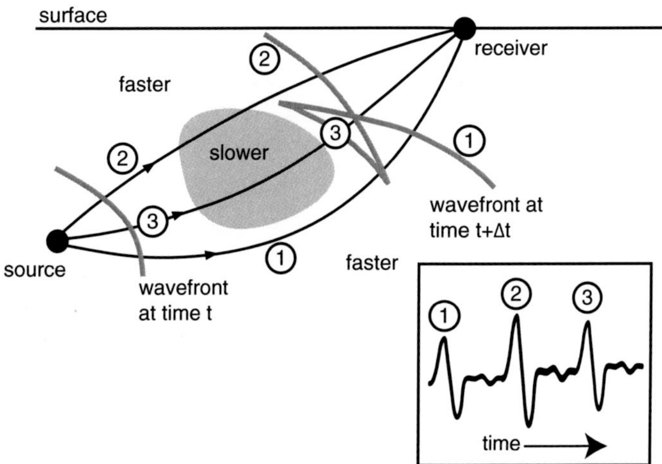

Figure 6.43 Ray paths for a medium containing a slow velocity anomaly showing a triplication. The bottom, top, and middle rays are respectively the first, second, and third arrivals at the receiver. After *Hauser et al.* (2008).

6.17 Illustrated case studies

The following two case studies illustrate recent developments of the seismic-reflection and -refraction techniques for near-surface investigation.

Example. 3-D refraction tomography at a contaminated site.

A 3-D seismic-refraction survey at a Superfund site within Hill Air Force Base in Utah is described by Zelt *et al.* (2006). A long history at this location of using chlorinated solvents as an industrial cleansing agent has led to the infiltration of DNAPL contaminants below ground surface and their pooling at the base of a surficial sand and gravel aquifer of 2–15 m thickness. The water table is at 9–10 m depth. The refraction survey objective was to image, over an area of ~ 0.4 ha, a paleochannel incised into the top surface of a relatively impermeable silty clay layer that underlies the aquifer. A static array of 601 40-Hz geophones connected to data loggers sampling at 1 ms was laid out in 46 parallel lines, each containing either 13 or 14 geophones, with 2.1-m line spacing and 2.8-m station spacing. A total of 596 shots were deployed using a .22-caliber rifle, with each shotpoint nominally located, where possible, within 0.3 m of a geophone. The maximum TX–RX offset distance is 102 m. The resulting shot gathers were minimally filtered using a bandpass filter to remove ground roll and a 60/120-Hz notch filter to remove cultural electrical noise caused by routine base activities. A few of the first-arrival picks, and the survey geometry, were shown earlier in Figure 6.26.

The 3-D regularized tomographic algorithm described by Zelt and Barton (1998) was used to convert the observed first-arrival traveltimes into a subsurface velocity model. The algorithm, which uses ray tracing as the forward module, favors subsurface models that

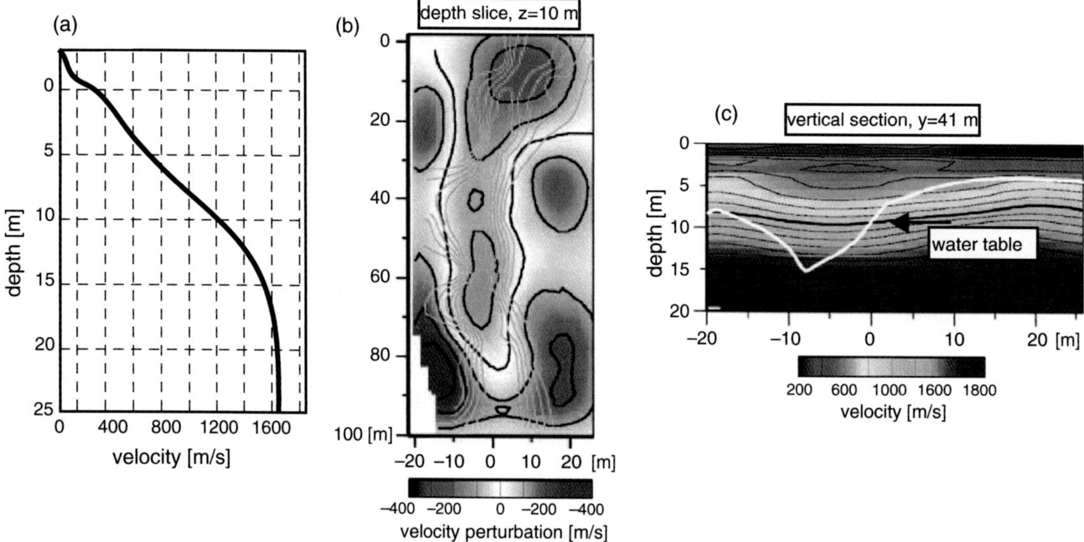

Figure 6.44 3-D refraction tomography at a contaminated site: (a) starting 1-D velocity model; (b) horizontal slice at depth $z = 10$ m through the final preferred tomogram; (c) vertical slice at location $y = 41$ m; light-colored contours in (b) and (c) mark the incised paleochannel inferred from well data. After Zelt *et al.* (2006). See plate section for color version.

contain smooth spatial variations in the velocity structure; i.e. spatially rough velocity models are suppressed. A discussion of tomographic methods and their regularization appcars later in this book in Chapter 12. The starting 1-D velocity model, along with horizontal and vertical slices through the preferred final 3-D tomogram, are shown in Figure 6.44. It is easily seen that a low-velocity zone outlines the shape of the incised paleochannel inferred from drilling and well data. This case study demonstrates both the feasibility and utility of tomographic reconstruction of the near-surface 3-D seismic velocity distribution beneath a contaminated site located at an active military-industrial facility.

Example. Reflection imaging of a buried subglacial valley.

Buried subglacial valleys that form as continental ice sheets retreat are interesting geophysical targets since they often do not have a distinctive geomorphological signature, but they provide a record of climate history and they can host valuable resources such as groundwater, sand and gravel aggregates, and methane gas. Ahmad *et al.* (2009) describe a high-resolution seismic-reflection survey conducted over a subglacial valley in northwestern Alberta, Canada to determine its architecture and to study the hydrological and mechanical processes beneath retreating ice sheets.

The study area is covered by Quaternary glacial deposits left in the wake of Wisconsin glaciation ~ 10 ka. The underlying bedrock is Cretaceous shales ~ 100 Ma, beneath which

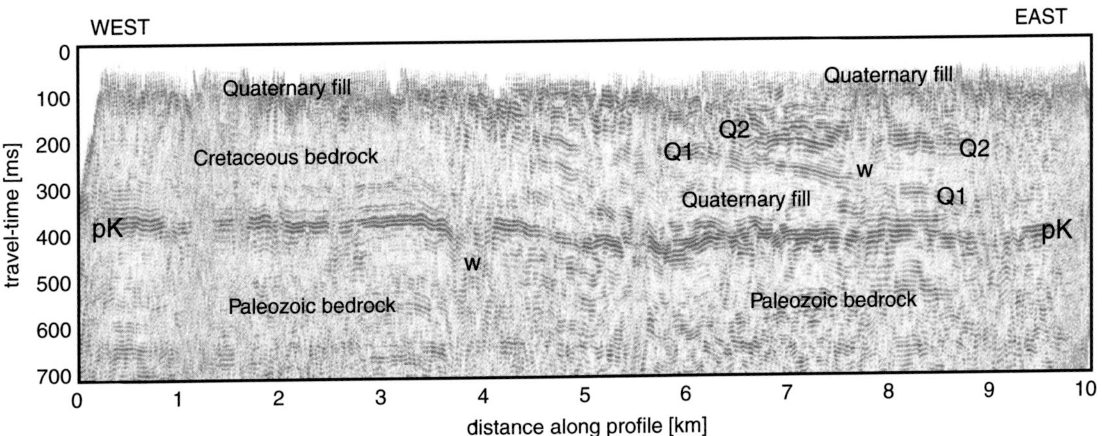

Figure 6.45 Seismic-reflection profile from a buried subglacial valley, Alberta, Canada. See text for details. After Ahmad *et al.* (2009).

are Paleozoic carbonates ~ 340 Ma. The presumably steep-walled valley is filled with ~ 300 m of drift, mainly stratified tills and coarse-grained glaciofluvial and glaciolacustrine sediments. A previously known geophysical log intersected a shallow, high-resistivity (30–100 Ω m) gas-bearing zone at 64–72-m depth.

The objective of the reflection survey was to image the uppermost ~ 350 m. Acquisition parameters included 4-m geophone spacing and 24-m shotpoint spacing along a 10-km profile using a source consisting of a vibrator swept in frequency from 20 to 250 Hz and a 240-channel seismograph. A standard CMP data processing sequence was used with average fold ~ 40. Strong near-surface lateral heterogeneity due to muskeg caused troublesome statics that affected the traveltimes to deeper reflectors. Stacking velocities were determined using a semblance-based velocity analysis. Higher velocities were found to the west that indicated the presence of a thick Cretaceous sequence there. Seismic-refraction analysis and a co-located electrical resistivity tomography (ERT) profile strongly supported this inference.

The final processed reflection profile is shown in Figure 6.45. It shows some washed-out zones marked “w” that extend vertically through the section. These do not contain any reflections and could be associated with free gas saturation. The Cretaceous–Paleozoic unconformity marked “pK” at ~ 300–400 m depth is the most conspicuous reflecting horizon; it is present all along the profile except in the washout zones. The pK reflection horizon appears to undulate but this could be caused by lateral velocity variations in the overlying zones. Lower velocities occur toward the east, associated with thickening of the low-velocity Quaternary fill and the absence of Cretaceous strata. These low velocities tend also to “pull down” the pK reflector. The reflection data poorly image the putative steep valley wall (suggested by the refraction and ERT data) at ~ 3–4 km along the profile. There are some reflectors contained within the Quaternary fill marked as “Q1” and “Q2”. There is a possibility that these might be multiply reflected events but more likely they indicate internal stratification within the Quaternary fill layer.

Problems

1. Show that the law of reflection and Snell's law of refraction may be derived from Fermat's principle of least time, which states that the ray path from a seismic source to a seismic receiver is the one that minimizes the traveltime along the path.
2. Assuming that the TX–RX offset distance x is much less than the layer thickness h, which of course is not always a good assumption in near-surface geophysics, show that the NMO correction is approximately $\Delta T(x) \sim x^2/2T_0V_1^2$.
3. Consider a reflection seismic experiment that includes both diffraction from an edge and multiple reflections. Show that the diffraction hyperbola and the multiple reflections do not align horizontally if an NMO correction based on the primary reflection is applied to each trace. Use the same $x << h$ approximation as in the previous exercise.
4. Assume that the velocity V is known in a reflection experiment over a dipping interface. Derive expressions for down-dip and up-dip traveltimes τ_D and τ_U based on a single source location and a single (moveable) geophone. How might the dip angle φ and depth to the interface h be derived from the two traveltime measurements, assuming that the dip angle φ is small enough that the depth h to the interface is approximately the same for both the up-dip shot and the down-dip shot. What additional measurement, apart from the direct-wave traveltime, should be made if the velocity V is also to be determined?
5. Show that the seismic-refraction cross-over distance x_C, beyond which the head wave arrives earlier than the direct wave, is given by the formula

$$x_C = 2h\sqrt{(V_2 + V_1)/(V_2 - V_1)}.$$

6. Consider a down-dip refraction experiment. (a) Show that the head-wave traveltime curve $T_D(x)$ is given by Equation (6.39), which implies that the down-dip refracted wave moves out with apparent velocity $V_D = V_1/\sin(i_C + \varphi)$. (b) Using a similar analysis, show that the apparent up-dip velocity is $V_U = V_1/\sin(i_C - \varphi)$. (c) Show that the dip angle is given by Equation (6.40).
7. Consider a short pulse consisting of the superposition of two equal-amplitude sinusoidal waves each of the form $\exp[i(\omega t - \beta x)]$ and oscillating at closely spaced frequencies $\omega \pm \Delta\omega/2$ and wavenumbers $\beta \pm \Delta\beta/2$. Show that the pulse consists of *beats*, which move with a group velocity v_g that is related to the phase velocity v_p of the individual sinusoids by $v_g = v_p + \beta dv_p/d\beta$.
8. Show that the bulk modulus, for an elastic body under hydrostatic pressure, is the ratio of the pressure p to the dilatation Δ.
9. Derive Equation (6.7a) starting from Equation (6.6).
10. Show that the refraction traveltime curve $t_2(x)$ for propagation through a two-layer Earth is given by the equation

$$t_2(x) = \frac{x}{V_3} + \frac{2h_2}{V_2}\cos\theta_{C2} + \frac{2h_1}{V_1}\cos\theta_{C1},$$

where the upper layer is characterized by thickness and velocity (h_1, V_1), the underlying layer by (h_2, V_2) and the terminating halfspace by velocity V_3. This equation is the form of a straight-line segment with slope $1/V_3$ and intercept given by the sum of the last two terms.

11. Derive the eikonal Equation (6.53).

7 Seismic surface-wave analysis

Surface-wave-based methods involving active or passive sources are used in investigations spanning a wide range of scales from ultrasonic non-destructive evaluation of civil infrastructure to global seismic imaging of the Earth's mantle. Near-surface geophysical applications with active sources, probing to depths ~ 30 m, are experiencing steady growth (Socco *et al.*, 2010). For the most part, the key information is embedded in high-amplitude, low-frequency Rayleigh waves, i.e. the *ground roll* that is normally regarded as a source of noise in seismic body wave reflection and refraction studies. Typically, surface or interface waves of various types (e.g. Rayleigh, Love, Scholte, Lamb, and Stoneley waves) are guided and highly dispersive. Recognition of these properties drove the development in the 1980s of the spectral analysis of surface wave (SASW) method (Nazarian and Stokoe, 1986) and, later, the multichannel analysis of surface wave (MASW) method (Park *et al.*, 1999). In these and other related techniques, apparent Rayleigh phase velocity versus frequency curves are first constructed, and then inverted to obtain shear-wave depth profiles. The resulting estimates of shear-wave speed in the shallow subsurface can be interpreted in terms of physical properties such as stiffness, liquefaction potential, and moisture content. These properties are of great interest to geotechnical and construction engineers, soil scientists, and others. In addition, the magnitude of ground shaking in response to a nearby earthquake is highly dependent on the subsurface shear-velocity structure.

7.1 Rayleigh waves

Consider a mechanical disturbance within an infinite homogeneous elastic medium. Both compressional (P-) and shear (S-) body waves are generated, as described in the previous chapter. Suppose now the elastic medium occupies only the lower halfspace, such that a free surface is present. In this case there exists also a *Rayleigh wave* (R-wave) solution to the elastic wave equations (Richart *et al.*, 1970). If the medium is heterogeneous, with spatially varying elastic moduli, the R-wave packet is dispersive. The wave packet can be decomposed by Fourier analysis into its individual frequency components. Each frequency component of the wave packet travels at its own characteristic phase velocity. The shape of the phase velocity versus frequency curve is sometimes called the *dispersion characteristic*. Note that an R-wave traveling in a homogeneous elastic medium is not dispersive.

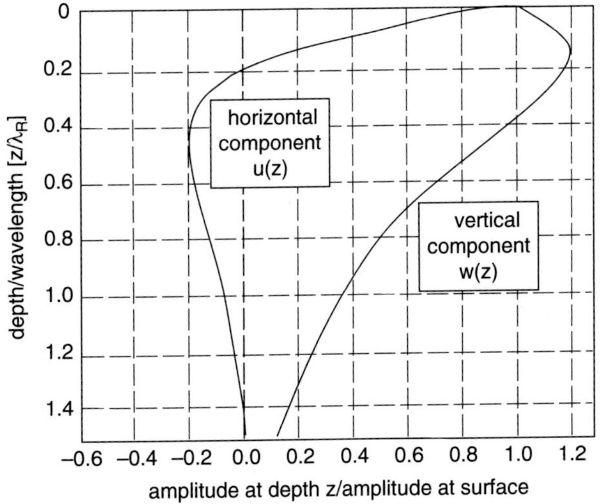

Figure 7.1 Rayleigh-wave displacements as a function of depth. Poisson's ratio $\sigma = 0.25$. After Richart *et al.* (1970).

An elementary R-wave source might consist of a heavy plate vibrating on the surface of an elastic halfspace. The source will radiate a full spectrum of P-waves, S-waves, and R-waves. The elastic energy is typically partitioned into 67% surface waves; 26% shear waves; and 7% compressional waves (Miller and Pursey, 1955). The geometrical damping of the energy in the P-waves and S-waves falls off as $1/r^2$, where r is the distance to the source, since the body-wave energy spreads spherically outward. The energy of the R-wave, which propagates only into a cylindrical region centered on the source on the free surface, falls off as $1/r$. Thus, R-wave amplitudes measured at the free surface are considerably greater than body-wave amplitudes. Note that propagation velocities refer to the wavefront moveout velocities rather than individual particle velocities.

The Rayleigh wave is not a body wave; it is instead guided along the free surface of the underlying elastic medium. The particle motion is retrograde elliptical near the surface, changing to prograde elliptical with increasing depth (Richart *et al.*, 1970). The amplitude of the R-wave decays rapidly with depth, so that at a depth corresponding to one wavelength, the amplitude is reduced to less than 30% of its surface value (see Figure 7.1). The R-wave velocity V_R is always less than the S-wave velocity V_S. For a material with Poisson's ratio $\sigma = 0.25$, for example, we have $V_R \sim 0.92 V_S$.

As will be shown below, the dispersion characteristic of R-waves is sensitive primarily to the S-wave-velocity depth profile. In turn, well-known relationships exist between S-wave velocity V_S and the elastic moduli, such as

$$\mu = \rho V_S^2; \quad E = 2\rho V_S^2(1+\sigma); \tag{7.1}$$

where ρ is the density, μ is the shear modulus, and E is Young's modulus of the material excited by the R-wave. Thus, measurements of the R-wave dispersion characteristic can be used to infer the stiffness of the medium through which it propagates.

7.2 Dispersion

Suppose a vertical motion sensor (geophone) is moved along the surface, away from an idealized source vibrating at a single frequency f, such that the Rayleigh wavelength λ can be determined by observing the distance between successive peaks and troughs in the resulting surface-wave motion. A low-frequency idealized source generates a long-wavelength R-wave. This corresponds to deep sampling of the site since (as shown in Figure 7.2) the depth of penetration of an R-wave scales with its wavelength. Conversely, a higher-frequency idealized source generates a shorter-wavelength R-wave, which corresponds to shallower sampling.

As described earlier, the R-wave velocity is determined by the elastic moduli of the layers which the wave excites. In this way, each of the various frequency components of the R-wave propagating in a multi-layer system does so at its own characteristic phase velocity. This phenomenon is known as *geometric dispersion*. The R-wave velocity as a function of frequency, or its dispersion characteristic, provides information about the elastic moduli of the individual layers within the system. This has turned out to be very useful in civil-engineering applications.

More generally, as discussed in Telford *et al.* (1990), each frequency component in a dispersive wave packet propagates at a phase velocity V that is different from the group

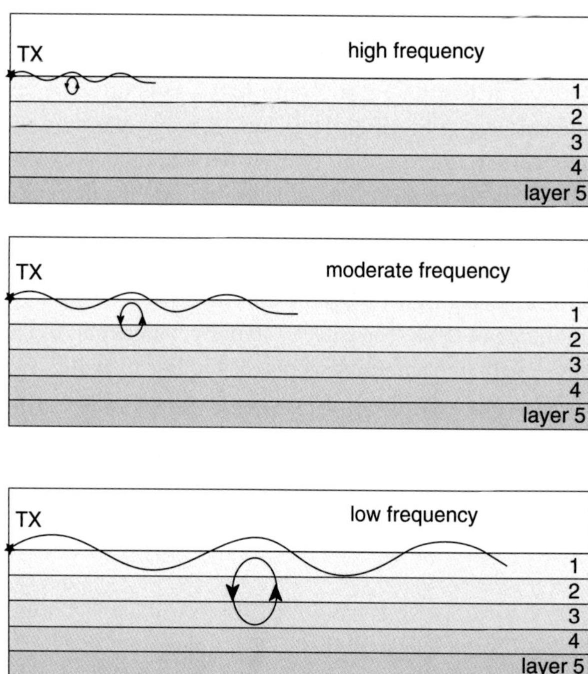

Figure 7.2 Depth of penetration of Rayleigh wave depends on its wavelength and frequency. Here, the low-frequency signal has penetrated into layer 3.

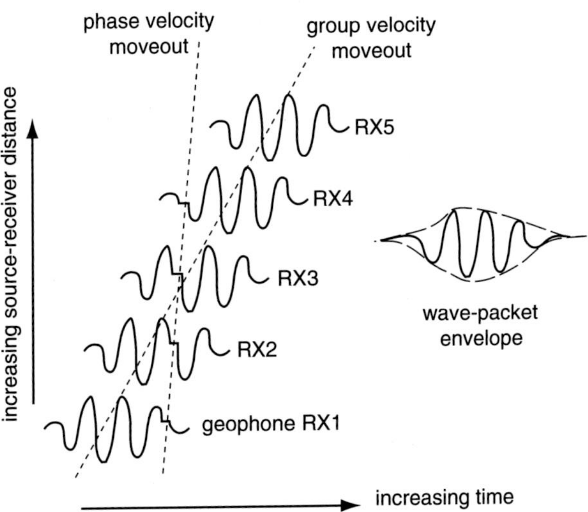

Figure 7.3 Comparison of phase and group velocities within a dispersive wave packet. After Telford *et al.* (1990).

velocity U of the entire packet. An idealized situation is illustrated in Figure 7.3. Notice in this highly stylized example that the high-frequency signal embedded within the lower-frequency wave packet moves out faster than the envelope of the wave packet, such that $V > U$. Generally, the effect of dispersion is to distort a wave packet as it propagates.

7.3 Rayleigh-wave propagation in a multi-layer system

The forward problem of surface-wave propagation in a layered elastic medium due to excitation by a monochromatic source was formulated and solved by Thomson (1950) and Haskell (1953), amongst others, as a classical boundary-value problem of mathematical physics. The important aspects of this solution are outlined by Nazarian and Stokoe (1986); herein I follow their discussion.

Consider the multi-layer elastic system shown in Figure 7.4. The x-direction is horizontal while the z-direction is vertical, positive downward. Each of the $N-1$ layers plus the terminating halfspace has known values of S-wave velocity V_S, P-wave velocity V_P, density ρ, and Poisson's ratio σ. Let these properties within the n-th layer be denoted, respectively, as $(V_{Sn}, V_{Pn}, \rho_n, \sigma_n)$.

Let the horizontal and vertical displacements in the n-th layer be denoted by u_n and w_n, respectively. A pair of elastic-wave potentials (ϕ_n, ψ_n) obeying scalar wave equations are introduced such that

$$u_n = \frac{\partial \phi_n}{\partial z} + \frac{\partial \psi_n}{\partial x}; \tag{7.2}$$

$$w_n = \frac{\partial \phi_n}{\partial z} - \frac{\partial \psi_n}{\partial x}. \tag{7.3}$$

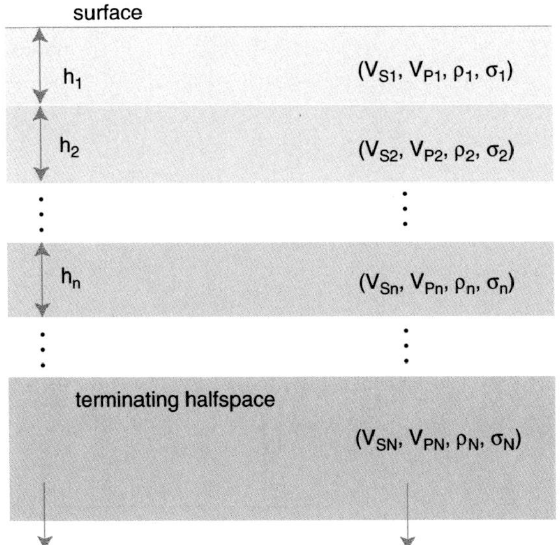

Figure 7.4 A multi-layer elastic system.

The compressional potential ϕ_n in the n-th layer has the general form,

$$\phi_n = U_{Pn}\exp(ir_n z) + D_{Pn}\exp(-ir_n z); \tag{7.4}$$

similarly the shear potential ψ_n is written as

$$\psi_n = U_{Sn}\exp(is_n z) + D_{Sn}\exp(-is_n z). \tag{7.5}$$

The coefficients (U_{Pn}, U_{Sn}) correspond to up-going waves at velocities depending on V_P and V_S, respectively, while (D_{Pn}, D_{Sn}) correspond to their down-going counterparts. The quantities (r_n, s_n) in Equations (7.4) and (7.5) are composite wavenumbers

$$r_n^2 = k_{ph}^2 - k_{Pn}^2; \quad s_n^2 = k_{ph}^2 - k_{Sn}^2; \tag{7.6}$$

where it is recalled that the wavenumber of a wave of angular frequency ω propagating with velocity V is given by $k = \omega/V$. Thus, k_{Pn} and k_{Sn} in Equation (7.6) are the wavenumbers that would correspond to P- and S-waves of angular frequency ω propagating in the n-th layer. The quantity k_{ph} in Equation (7.6) is the wavenumber associated with a wave of frequency $f = \omega/2\pi$ propagating with phase velocity V_{ph}.

The unknown coefficients (U_{Pn}, U_{Sn}) and (D_{Pn}, D_{Sn}) are found by application of appropriate boundary conditions. These include: continuity of normal and shear stress across layer interfaces; continuity of horizontal and vertical displacement across layer interfaces; and the vanishing of normal and shear stress at the free surface. Furthermore, $U_{PN} = 0$ and $U_{SN} = 0$ since there is no up-going wave in the terminating halfspace.

A detailed but straightforward calculation, not reproduced here, reveals that the application of the aforementioned boundary conditions implies the existence of a linear relationship between the surface displacements (u_1, w_1) and the bottom-layer elastic-wave potential coefficients (D_{PN}, D_{SN}) of the matrix form

$$\begin{pmatrix} 0 \\ 0 \\ D_{PN} \\ D_{SN} \end{pmatrix} = \begin{pmatrix} R_{11} & R_{12} \\ R_{21} & R_{22} \end{pmatrix} \begin{pmatrix} u_1 \\ w_1 \\ 0 \\ 0 \end{pmatrix}; \tag{7.7}$$

where the R_{ij} matrices, each of dimension 2 × 2, encapsulate the physics of elastic-wave propagation within the multi-layer system, including the boundary conditions at the material interfaces. The R_{ij} matrices, accordingly, involve the phase velocity V_{ph} and the wave frequency f. Notice that the first two equations in (7.7) can be written as

$$\begin{pmatrix} 0 \\ 0 \end{pmatrix} = R_{11} \begin{pmatrix} u_1 \\ w_1 \end{pmatrix}. \tag{7.8}$$

This equation has a non-trivial solution (u_1, w_1) for the surface displacements if and only if the determinant of R_{11} vanishes, that is,

$$\det R_{11} = 0. \tag{7.9}$$

Equation (7.9) is essentially an implicit equation of the form $g(V_{ph}, f) = 0$. This is sometimes known as the *Rayleigh secular equation* and $\det R_{11}$ is termed the Haskell–Thomson determinant. The equation $g = 0$ constitutes a non-linear constraint on the variables (V_{ph}, f) such that these two unknowns do not vary independently. For a given frequency f, the possible values of phase velocity V_{ph} are fixed by Equation (7.9). In principle, a dispersion characteristic $V_{ph}(f)$ is built up numerically by specifying different values of f and, for each one, varying V_{ph} until Equation (7.9) is satisfied. In practice, special procedures must be implemented to guard against numerical instabilities. Nevertheless, the foregoing analysis has identified the essential physics of surface-wave dispersion. To summarize, given the elastic properties of a multi-layer system, we have sketched how the dispersion characteristic $V_{ph}(f)$ is constructed.

As hinted above, it is often the case that more than one value of V_{ph} satisfies Equation (7.9) for a given frequency f. In fact, multiple solutions to the Rayleigh secular equation typically exist and they can be conveniently organized into a family of dispersion characteristics termed *modes*. An example of modal dispersion characteristics and vertical displacements (at two frequencies, 10 Hz and 20 Hz) for a multi-layered elastic system is shown in Figure 7.5. Each of the modal curves shown in Figure 7.5b describes an independently propagating, dispersive wave packet. The first mode, the curve at the lower left, is termed the *fundamental mode* and it exists at all frequencies. The higher modes have higher phase velocities and they exist only above a cut-off frequency that depends on the mode. At the cut-off frequency, the phase velocity is equal to the highest value of shear velocity in the system, which in this case is 1000 m/s. The fundamental mode does not necessarily dominate; the higher-order modes can sometimes carry a significant fraction of the elastic energy (Socco and Strobbia, 2004).

7.4 Spectral analysis of surface waves (SASW)

The theory described in the preceding section suggests a practical method for testing elastic multi-layer systems. However, such a method would be time-consuming if determination of a velocity–depth profile is required at a signficant spatial resolution,

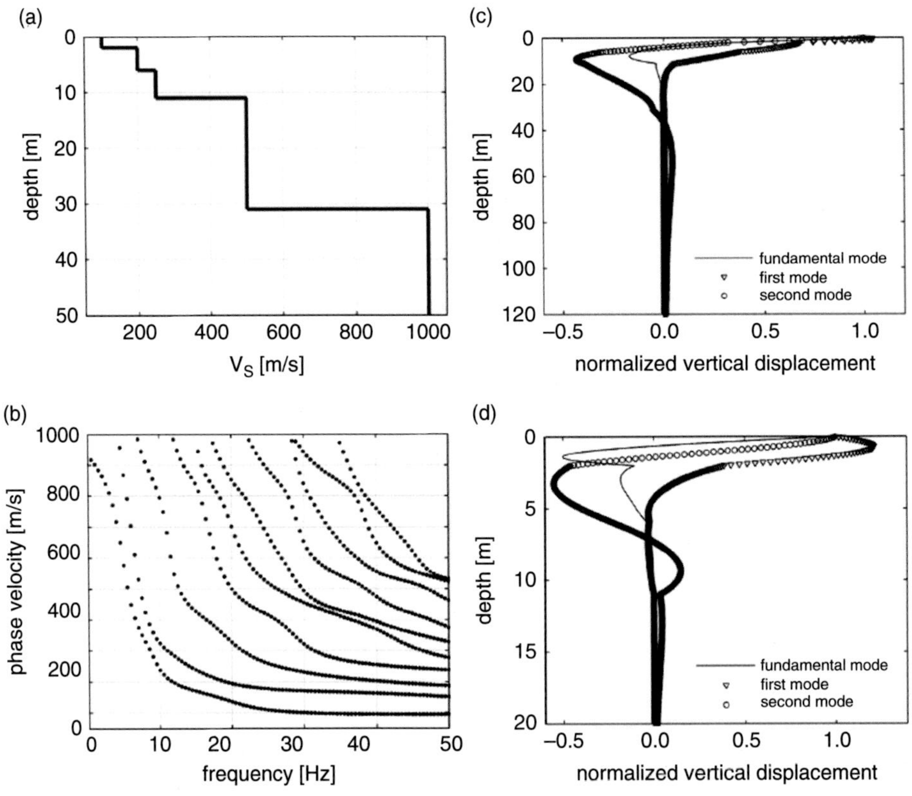

Figure 7.5 (a) A multi-layer S-wave depth profile, and its (b) modal dispersion characteristics $V_{ph}(f)$ and vertical displacements at (c) 10 Hz and (d) 20 Hz. After Socco *et al.* (2010).

since many frequencies would need to be succesively generated by the monochromatic source. It is advantageous to develop a method which can excite and measure at a multitude of frequencies. Such a method could involve the *spectral analysis* of a wave pulse that is generated by an impact at the surface. Spectral analysis of surface waves (SASW) is such a method of analyzing surface-wave propagation to explore multi-layered elastic systems.

A simple field testing setup (Figure 7.6) consists of an impulsive source, a pair of geophones, and a seismograph or spectral analyzer. The desired measurement is that of the time delay between the two receivers of the various frequency components of ground roll arriving from the source. The source must be capable of generating R-waves over a wide range of frequencies at sufficient signal strength to be detected by the geophones.

Suppose the proximal geophone (labeled A) measures the vertical displacement $w_A(t)$ of a transient Rayleigh-wave train generated by the impact source. After stacking the signal to average out incoherent background noise, the frequency spectrum $W_A(\omega)$ can be calculated by applying a Fourier transform. The complex function $W_A\ (\omega)$ reveals the amplitude and phase of the various frequency components that comprise the signal.

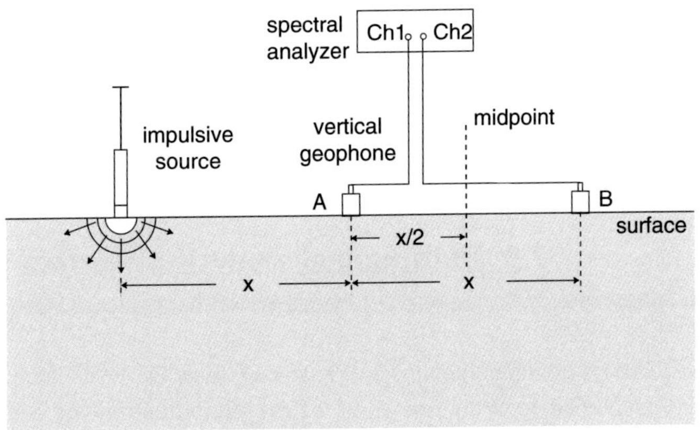

Figure 7.6 Experimental arrangement for SASW tests. After Nazarian *et al.* (1983). Copyright: National Academy of Sciences, Washington, D.C., 1983. Reproduced with permission of the Transportation Research Board.

The power spectrum, defined as $G(\omega) = W_A(\omega)W_A{}^*(\omega)$ where * denotes complex conjugation, determines how the energy of the signal is partitioned into its various frequency components. A cross-spectrum $G_{AB}(\omega) = W_A(\omega)W_B{}^*(\omega)$ can be constructed from the spectra of two different signals $w_A(t)$ and $w_B(t)$, where $w_B(t)$ is the vertical displacement measured by the distal geophone, labeled B. The cross-spectrum amplitude identifies the dominant frequency components that are simultaneously present in both signals, while the phase of $G_{AB}(\omega)$ reveals the relative phase of each of the frequency components present in $w_A(t)$ and $w_B(t)$ (e.g. Wadsworth *et al.*, 1953).

Traveltimes for each frequency component radiated by the source can be extracted from the phase information contained in the cross-spectrum $G_{AB}(\omega)$. The phase difference $\Delta\theta$ at angular frequency ω is a measure of the time lag, or traveltime Δt, for an R-wave component of angular frequency ω to propagate the distance x between the two geophones. A phase difference of $\Delta\theta = 2\pi$, for example, corresponds to a time lag of $\Delta t = T$, where $T = 2\pi/\omega$ is the period of the wave. Hence, the traveltime is related to the phase difference by $\Delta t = \Delta\theta/\omega$. The R-wave phase velocity at frequency f is therefore

$$V_{ph}(f) = \frac{x}{\Delta t} = \frac{2\pi f x}{\Delta\theta}. \tag{7.10}$$

A curve of the R-wave phase velocity can be built up, by this type of spectral analysis, as a function of frequency. This is the sought-after dispersion characteristic, $V_{ph}(f)$. In the SASW method, the R-wave phase velocities are *apparent velocities* since the R-waves sample the layered medium in a complicated fashion; each frequency component does not uniquely probe a separate depth layer.

While SASW testing remains a popular method for geotechnical site evaluation, there are several weaknesses inherent to the procedure. For example, Equation (7.10) shows that the determination of $V_{ph}(f)$ at frequency f depends on the geophone spacing x. Thus, the two geophones must be reconfigured, as many times as necessary, to sample the frequency range of interest. Each re-deployment of the receivers requires a new shot, so that the SASW method soon becomes laborious. Moreover, the noise characteristics in the observed signals can change in a unpredictable manner from shot to shot and from receiver location to receiver location. Noise can be either incoherent, such as the ambient noise from wind, traffic, nearby industrial operations, etc., or it can be coherent, such as the signal-generated noise due to reflected and refracted body waves or surface waves scattered from building foundations, retaining walls, underground storage tanks, etc.

7.5 Multichannel analysis of surface waves (MASW)

The multichannel analysis of surface waves (MASW) technique (Park *et al.*, 1999) was developed to overcome some of the limitations of the SASW method. In this method, normally a swept-frequency vibrator source is deployed into a linear array of geophones. The MASW method is much less time-consuming in the field than the SASW method since the source and geophones need to be deployed only once. The other principal advantage of MASW is that noise sources can often be identified from wavefields that are simultaneously recorded across a number of regularly spaced receivers. The noise is recognized by its frequency content and its moveout across the array. The reduction of noise results in a better signal-to-noise ratio for the Rayleigh-wave train, which ultimately leads to improved accuracy in the estimation of the dispersion characteristic, $V_{ph}(f)$.

In the MASW technique, the ground-roll phase velocity as a function of frequency can be recognized if the recorded wavefields are displayed in a swept-frequency format. This is naturally accomplished if a swept-frequency source is used, but it is also possible to transform recordings from an impulsive source into swept-frequency format using a stretch function (Park *et al.*, 1999). A field example from Kansas using a vibrator source swept from 10 to 50 Hz over a 10-s interval is shown in Figure 7.7. The complete recording is broken into 1.5-s segments. The sweep frequency is shown at the top of each record segment and the recording time is shown at the bottom. The geophone spacing is 1.0 m with minimum TX–RX offset 1.8 m.

Notice that the low-frequency records ($\leq$16 Hz) show a lack of ground-roll coherency, indicating the presence of near-source effects, notably the breakdown of the plane-wave assumption. Similarly, far-field effects such as body waves are evident in the high-frequency records above ~ 40 Hz. Rayleigh-wave phase velocities in the range 250–750 m/s, decreasing monotonically with frequency, can be determined by the linear slopes of the coherent signals in the intermediate frequency range ~ 22–34 Hz.

A general method (Park *et al.*, 1998) is now described for determining multi-modal dispersion characteristics from surface waves generated by an impulsive source. Suppose a shot gather, such as the one shown in Figure 7.8a, is acquired. Let the recorded wavefield be denoted by $u(x, t)$, where x is the TX–RX offset and t is time. A Fourier transform of the wavefield is performed, which obtains $\tilde{u}(x, \omega)$, where ω is angular frequency. The frequency-domain wavefield may be written in the standard form

$$\tilde{u}(x, \omega) = A(x, \omega)\exp[-i\Phi(\omega)x]; \tag{7.11}$$

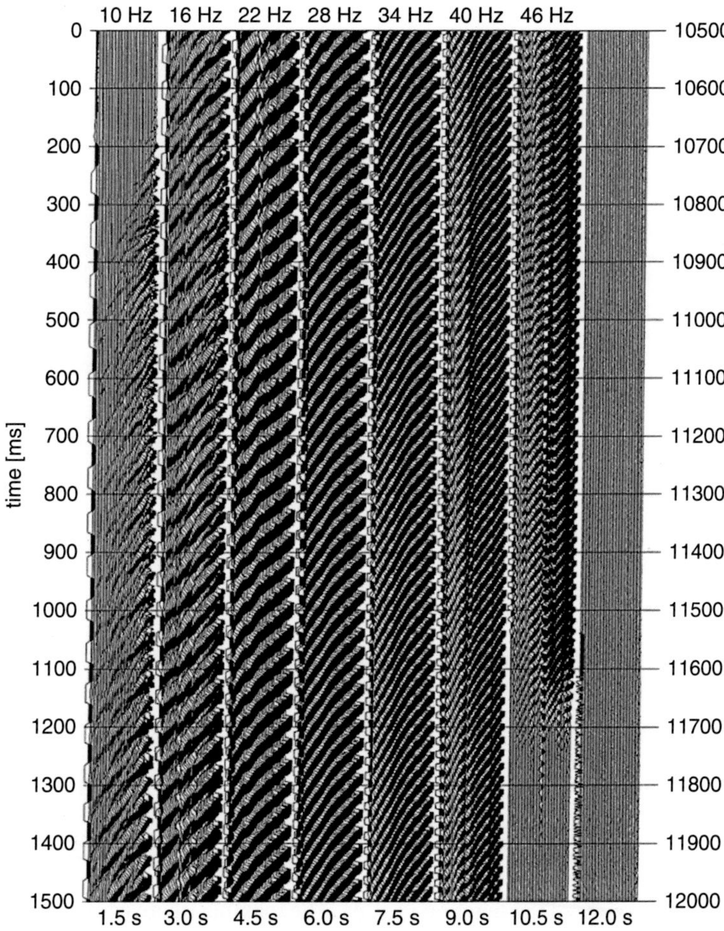

Figure 7.7 MASW swept-frequency wavefields. After Park *et al.* (1999).

where $A(x,\omega)$ is the amplitude spectrum. The function $\Phi(\omega)$ can be regarded as a type of wavenumber spectrum if we identify $\Phi(\omega) = \omega/V_{ph}(\omega)$, with $V_{ph}(\omega)$ being the surface-wave phase velocity at frequency ω.

Next, define the new function $\tilde{v}(\omega, \phi)$, according to the linear transformation

$$\tilde{v}(\omega, \phi) = \int_{x_1}^{x_2} \exp(i\phi x) \frac{\tilde{u}(x, \omega)}{|\tilde{u}(x, \omega)|} dx; \tag{7.12}$$

where (x_1, x_2) are, respectively, the minimum and maximum TX–RX offset on the shot gather. Notice that the transformation (7.12) has the form of a *slant stack* since it involves an offset-dependent phase shift $\exp(i\phi x)$ coupled with an integration over TX–RX offset. The transformation also involves a normalization of $\tilde{u}(x, \omega)$ in the denominator. This is equivalent to a trace normalization, or gain control, to compensate for effects of attenuation and spherical divergence.

Inserting Equation (7.11) into (7.12) yields

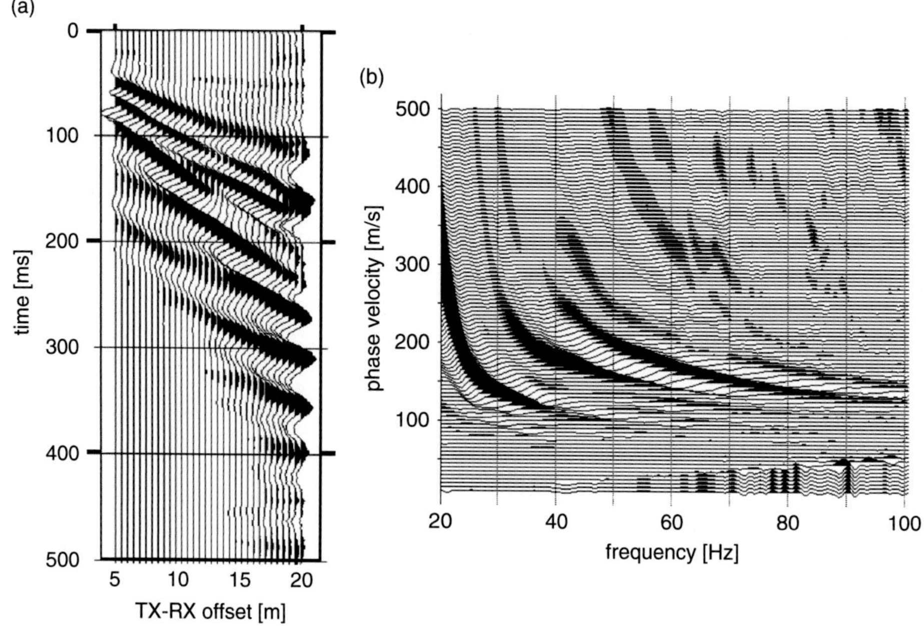

Figure 7.8 MASW determination of dispersion characteristics: (a) shot gather from a test site in Kansas; (b) dispersion image. After Park *et al.* (1998).

$$\tilde{v}(\omega, \phi) = \int_{x_1}^{x_2} \exp(-i[\Phi - \phi]x)\frac{A(x, \omega)}{|A(x, \omega)|}dx. \tag{7.13}$$

It is easy to see from the above equation that the function $\tilde{v}(\omega, \phi)$ attains its maximum when $\phi = \Phi$. This suggests a procedure to vary ϕ, for a given frequency ω, and observe the maxima of the function $\tilde{v}(\omega, \phi)$. The maxima (there could be several) will correspond to the values $\phi = \Phi = \omega/V_{ph}(\omega)$. In this way, the phase velocity $V_{ph}(\omega)$ at frequency ω is determined. Joining together the peak values of $\tilde{v}(\omega, \phi)$ for different values of ω generates the modal dispersion characteristics. The peak values can be visualized on a two-dimensional (2-D) contour plot of $\tilde{v}(\omega, \phi)$, known as the *dispersion image*.

The foregoing method is applied in Park *et al.* (1998) to data from a test site in Kansas. The shot gather acquired using a 10-kg sledgehammer is shown in Figure 7.8a. The resulting dispersion image, plotted by convention as a function of the variables (f, V_{ph}) instead of (ω,φ), is shown in Figure 7.8b. The fundamental and the next two higher modes of R-wave propagation are easily recognized on the dispersion image.

7.6 Inversion of R-wave dispersion characteristics

The objective of active surface-wave geophysical techniques is to determine the shear-wave velocity structure of the uppermost tens of meters. We have already examined the underlying forward theory in which, given a particular $V_S(z)$ depth profile, the dispersion characteristics $V_{ph}(f)$ including fundamental and higher-order modes of wave propagation are found as the roots of the Rayleigh secular equation. We have also examined techniques for extracting the dispersion characteristics from seismic data acquired with a swept-frequency or an impulsive source. It remains now to solve the inverse problem, which involves the back-calculation of a $V_S(z)$ depth profile that is consistent with observations of the fundamental and possibly higher-order modes of $V_{ph}(f)$.

A survey of methods used to solve the surface-wave inverse problem is provided by Socco *et al.* (2010). Typically, the inverse problem (see Chapters 11–13) is formulated as the minimization of a non-linear objective function, which often includes a misfit and one or more regularization terms that can incorporate *a priori* information. The objective function is minimized by varying the underlying S-wave model $V_S(z)$. The search through possible models can be *local*, in which the search is confined to the vicinity of a starting model, or *global*, in which a systematic exploration of the entire model space is attempted.

The misfit term in the objective function normally measures the discrepancy between the observed $V_{ph}(f)$ curve(s) and those that are calculated from the forward theory. The regularization terms weigh against selecting unwanted $V_S(z)$ models that are geologically implausible, contain needlessly large spatial fluctuations, or are far removed from a preferred $V_S(z)$ model established on the basis of some a-priori knowledge about the subsurface. As with all geophysical inverse problems, careful attention must be paid to the effects of uncertainties in the observations on the reconstructed Earth models.

Maraschini *et al.* (2010) have described a new misfit function for surface-wave inversion that does not require the construction of the dispersion characteristics $V_{ph}(f)$. Rather, the new misfit function operates directly on the Haskell–Thomson determinant which, in the notation of Section 7.3, is written as $\det R_{11}$. The misfit function can be written as

$$\chi[\mathbf{m}] = \sqrt{\sum_{i=1}^{D} w_i[\det R_{11}(V_i^{OBS}, f_i^{OBS}; \mathbf{m})]^2}; \tag{7.14}$$

where $\mathbf{m}$ is a $P = 2N - 1$ dimensional vector of model parameters, i.e. S-wave velocities V_{Sn} and thicknesses h_n of each of $n = 1, \ldots, N - 1$ Earth layers along with the S-wave velocity V_{SN} of the terminating halfspace. The quantities (V_i^{OBS}, f_i^{OBS}) for $i = 1, \ldots, D$ in Equation (7.14) are points along the modal curves of the dispersion image; they represent the input data for the inverse problem. The quantities w_i for $i = 1, \ldots, D$ enable each data point to be separately weighted according to its uncertainty or some other criterion. The inverse problem is solved by searching among different model vectors until an optimal vector $\mathbf{m}^*$ is found which minimizes $\chi[\mathbf{m}]$ in Equation (7.14). The advantage of using the misfit function $\chi[\mathbf{m}]$, in lieu of classical formulations of the inverse problem, is that all modes are considered simultaneously and there is no need for a manual identification of the individual modes of propagation on the dispersion image.

Some results of an inversion of synthetic surface-wave data using the new misfit function are shown in Figure 7.9. The data (V_i^{OBS}, f_i^{OBS}) for $i = 1, \ldots, D$ shown as black

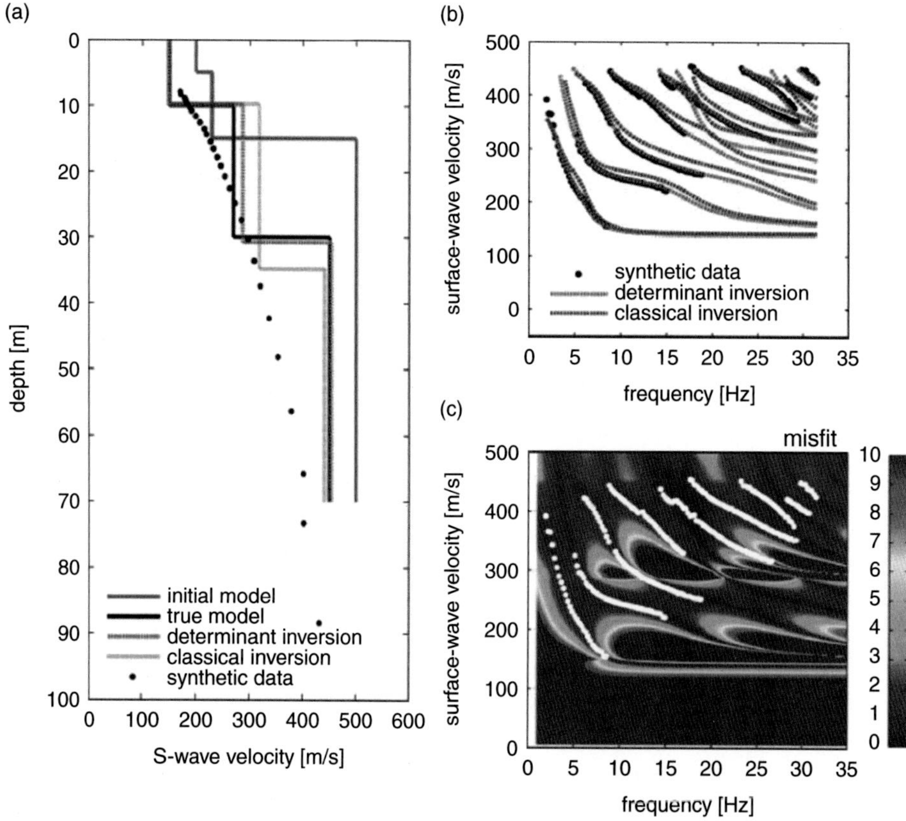

Figure 7.9 Inversion of synthetic surface-wave data using a misfit function based directly on the Haskell–Thomson determinant: (a) S-wave velocity profiles; (b) multimodal dispersion characteristics; (c) dispersion characteristics superimposed on the Haskell–Thomson misfit function. After Marischini *et al.* (2010). See plate section for color version.

dots in Figure 7.9b were generated by solving the forward problem for the "true S-wave velocity model" (black curve in Figure 7.9a) consisting in this case of two layers over a halfspace. Notice that several high-order modes are present in the forward solution. An inversion using the classical misfit function is shown as the blue curve, while the inversion using the Haskell–Thomson determinant misfit function is shown as the red curve in Figure 7.9a. Both inversions started from the same initial model (green curve in Figure 7.9a). Notice that the determinant-misfit inversion result is closer to the true model than the classical-misfit inversion result. The dispersion characteristics for both inversion results, determinant and classical, are compared with the synthetic data in Figure 7.9b. The determinant inversion does a much better job of fitting the higher-order modes. The dispersion characteristics for the determinant inversion result are shown superimposed on a contour plot of the misfit function $\chi[\mathbf{m}]$ in Figure 7.9c and it is seen that the curves indeed trace out zero-lines on the misfit surface, as required. In other words, the white lines shown in the figure are constrained to the blue regions of the contour plot.

7.7 Microtremor and passive studies

Surface-wave methods using active sources are limited in their depth of penetration to the upper 10–30 m. It is recalled that the active-source methods generate surface-wave energy mainly in the frequency range ~ 10–50 Hz. Passive methods probe to greater depths by analyzing ambient noise, termed *microtremor*, which is of generally lower frequencies, down to 2 Hz. Microtremor at most sites is caused by a combination of natural processes and human activities. Unlike the active-source SASW and MASW methods, passive surface-wave methods can offer excellent performance in urban environments with high noise levels. Beginning with fundamental earthquake studies in the 1950s and 1960s, a number of data-processing schemes have since been developed to extract near-surface Rayleigh-wave and Love-wave dispersion characteristics from ambient-noise recordings.

It should be noted that there exists a number of microtremor analysis methodologies (Socco *et al.*, 2010). The spatial autocorrelation (SPAC) method of Aki (1957), for example, has long been used to extract dispersion characteristics from microtremor array data under the assumption that the wavefields are stationary stochastic processes in both time and space. Solid-Earth geophysicists regularly use the SPAC and other passive surface-wave methods to image crust and mantle shear-wave structure on global and regional scales (e.g. Yao *et al*, 2008). Chen *et al.* (2009) used the horizontal-to-vertical (H/V) spectral ratio method, which estimates the transfer function between horizontal and vertical motions at single sites, to compile a map of soft-sediment thickness in support of an earthquake strong-motion assessment for the city of Beijing.

The remaining discussion herein is focused on the *refraction microtremor* (ReMi) technique as described by Louie (2001). In the ReMi method, a linear array is laid out of single-component geophones sensitive to vertical ground motion. A standard refraction configuration can be used. Geophones with small resonant frequencies, such as 4.5 Hz or 8 Hz, are used in order to capture the low-frequency components of the noise. A large geophone spacing, on the order of 10–20 m, is preferred since the maximum depth to which shear-wave structure can be resolved scales with the array length, or *aperture*. The ReMi method probes to ~ 100 m depths for an array length of ~ 200 m, capturing predominantly the fundamental mode of Rayleigh-wave propagation.

As in the MASW method, the essential data-processing step in the ReMi method is a (p,τ) wavefield transformation. However, in this case the slant stack is executed directly in the time domain, followed by a Fourier transformation which converts the wavefield into the (p, f) domain. The sought- after fundamental-mode dispersion characteristic $V_{ph}(f)$ is then extracted from wavefield power images displayed in the (p, f) domain. Louie (2001) introduce a special procedure that takes into account both positive ($p > 0$) and negative slowness ($p < 0$) values. This is necessary since Rayleigh waves can propagate in both forward and reverse directions along the array, reflecting the strong likelihood that the noise sources are distributed over a wide range of azimuths with respect to the alignment of the geophone spread. In the MASW method, on the other hand, the active source is purposely located in the *end-fire* position. In that case, Rayleigh waves will travel across

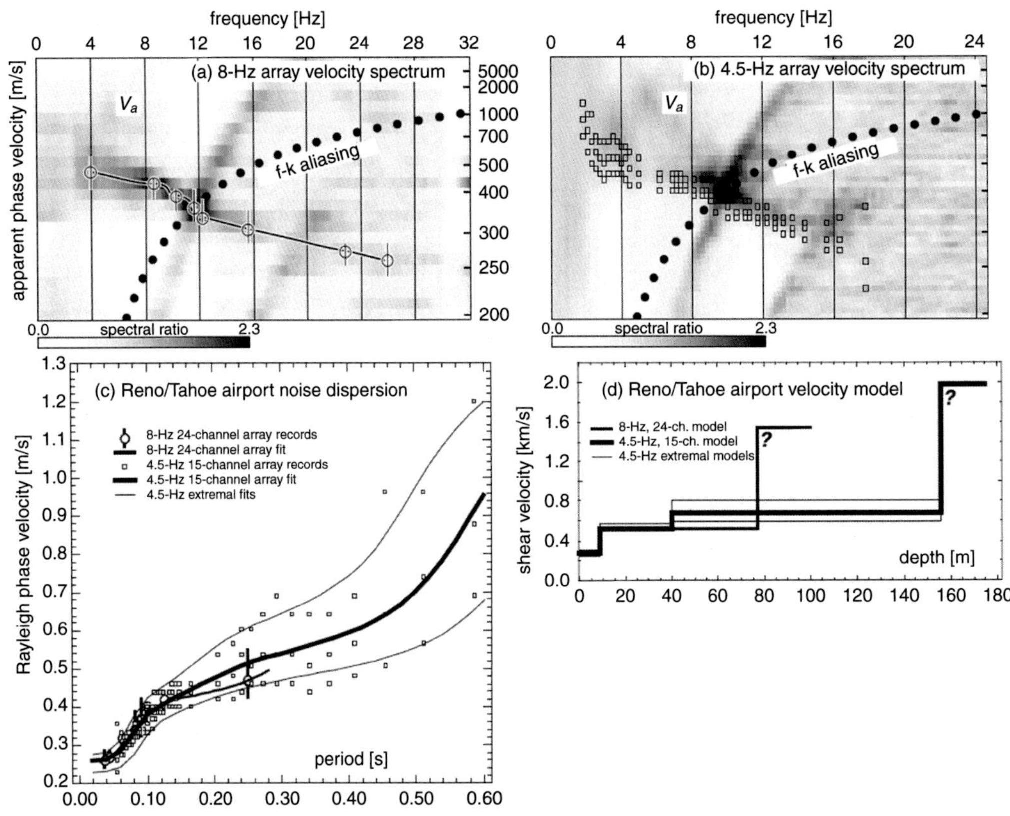

Figure 7.10 ReMi test results from Reno NV international airport showing power spectral ratio images for: (a) the 8-Hz geophone array; (b) the 4.5-Hz geophone array. (c) phase-velocity dispersion characteristics; (d) S-wave velocity depth profiles. After Louie (2001).

the array in one direction only; hence only positive ($p > 0$) slowness values are required for MASW analysis.

The ReMi method was tested by Louie (2001) at a number of locations, including a site near the Reno NV international airport in close proximity to the runways, a busy interstate highway, and a major railway line. Two arrays of geophones were used in two separate experiments: a first array of 24 8-Hz geophones at 15 m spacing, and a second array of 15 4.5-Hz geophones at 20 m spacing. In both cases, noise recordings of ~ 50 s in duration were made.

Wavefield power images in the (p, f) domain are shown in Figure 7.10a for the 8-Hz geophone array and in Figure 7.10b for the 4.5-Hz array. For convenience, the vertical axis is labeled with velocity $V_a = 1/p$ on a non-linear scale but the image is actually plotted on linear–linear (p, f) axes. The actual quantity plotted is a normalized power spectral ratio (for details see Louie, 2001). The important point is that peak values of the spectral ratio correspond to maxima of the power in the noise spectrum. Thus, joining together the points at which the wavefield power image attains its maximum intensity should reveal the fundamental mode of R-wave propagation. These points are shown by the open symbols (circles and squares, respectively) in the images of Figures 7.10a, b. Louie (2001) point out that the peak loci that appear to arc upwards and towards the right of the images are actually (p,τ) artefacts caused by the finite aperture of the sensor array. To the right of the solid dotted line in each image is a region that would be unreliable if an f–k data-processing scheme were to have been employed. The f–p processing scheme does not share this disadvantage.

The dispersion characteristics $V_{ph}(T)$ extracted from the wavefield power images, plotted as functions of period T instead of frequency f, are shown in Figure 7.10c. The heavy solid line is the best overall estimate of $V_{ph}(T)$ while lighter solid lines provide indication of its uncertainty. The results of a standard inversion of the $V_{ph}(T)$ curves to obtain S-wave depth profiles $V_S(z)$ are shown in Figure 7.10d. The S-wave velocity in the uppermost 9 m is well constrained at 280 m/s, increasing to 520 m/s in the 9–40 m depth range. Below that, there is more uncertainty in the shear velocity but it appears to increase with depth. The plot shows that the data acquired with the 4.5-Hz array can provide estimates of shear-wave structure to ~ 160 m depth.

7.8 Illustrated case histories

Example. Shear-wave velocity profiles in a seismic hazard zone.

The New Madrid seismic zone located within the Mississippi embayment of midcontinental United States has long been associated with large intraplate earthquakes. The magnitude of earthquake ground motion in this region is affected by the physical properties, principally the S-wave velocity profile $V_S(z)$, of the underlying soils to depths of hundreds of meters. An SASW analysis was carried out at a number of sites by Rosenblad *et al.* (2010) in order to determine $V_S(z)$. A sledgehammer and a powerful vibrator were used, respectively, to generate high-frequency (up to 40 Hz) and low-frequency (down to 0.8 Hz) energy. Geophone pairs were spaced at < 20 m for the sledgehammer source and at > 20 m for the vibrator source. Dispersion characteristics $V_{ph}(f)$ based on cross-spectral phase measurements were estimated using Equation (7.10) and then inverted using a standard approach to obtain $V_S(z)$ profiles.

The SASW results from three of the Mississippi embayment sites (here labeled as A, B, C) are shown in Figure 7.11. Each $V_S(z)$ profile is compared to a lithological profile estimated from nearby well logs and regional seismic-reflection sections. The fit of the theoretical dispersion characteristic at site C is compared to the experimental $V_{ph}(f)$ curve at the bottom right of the figure. The $V_S(z)$ profiles show soft, near-surface sediment layers in the uppermost 50 m, with S-wave velocity increasing with depth from 200 m/s to 400 m/s. The velocity is roughly uniform and equal to 400 m/s throughout the Upper Claiborne interval. A velocity increase is observed at the top of the Memphis Sand, with the values increasing to ~ 600–750 m/s at 200-m depth. Overall, the SASW-inferred

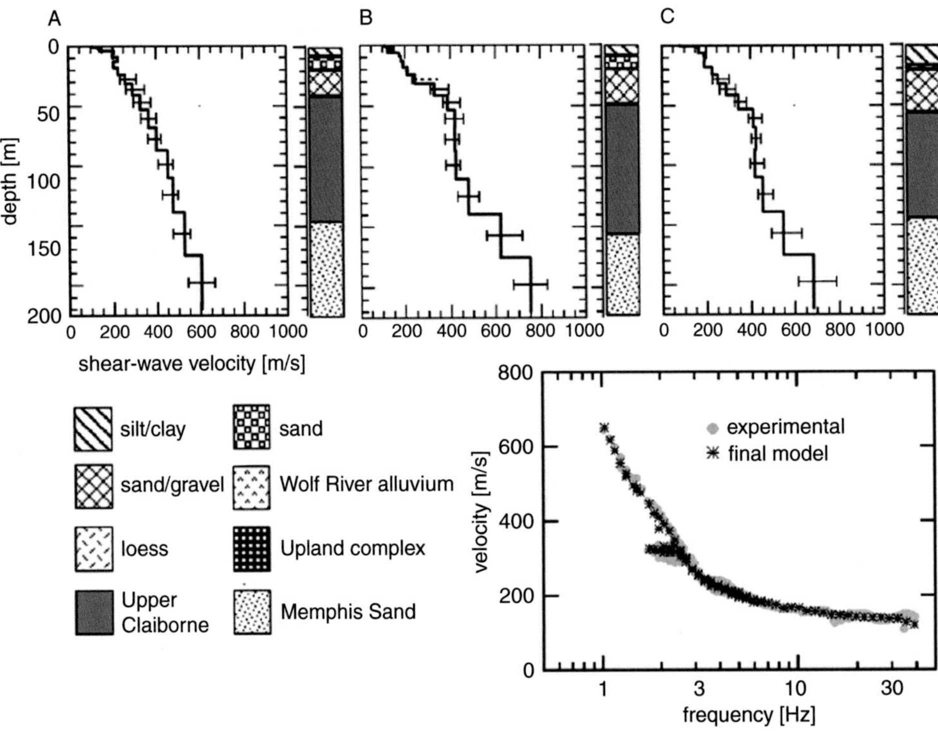

Figure 7.11 SASW analysis in the Mississippi embayment, top row: S-wave velocity profiles and estimated lithology; bottom left: lithology key (see below); bottom right: measured dispersion characteristic $V_{ph}(f)$ at site C compared to the forward response of the S-wave profile at C. Upper Claiborne = Eocene silts and clays, Memphis Sand = coarse- to fine-grained sand, Wolf River = Mississippi River sand/gravel alluvium layers, Upland complex = Pliocene sand and gravel. After Rosenblad *et al.* (2010).

S-wave velocity profiles show an excellent correlation with lithology. This information can be used to improve earthquake hazard assessment across the New Madrid seismic zone.

Example. Surface-wave analysis of volcanic microtremor.

The underground movement of magma and hydrothermal fluids beneath volcanoes often results in a background level of seismic activity known as volcanic tremor. Surface recordings of volcanic-tremor wavefields can be analyzed to extract subsurface shear-velocity structure. Saccorotti *et al.* (2003) report on seismometer array measurements made at the caldera of Kilauea volcano on the island of Hawaii. The SPAC method was used to determine fundamental-mode Rayleigh- and Love-wave dispersion characteristics $V_{ph}(f)$, which were then inverted to obtain S-wave velocity profiles $V_S(z)$.

One of the measurement arrays (Figure 7.12a) consists of a hub seismometer placed at the center of a semi-circular configuration of 30 receivers that are located at regular

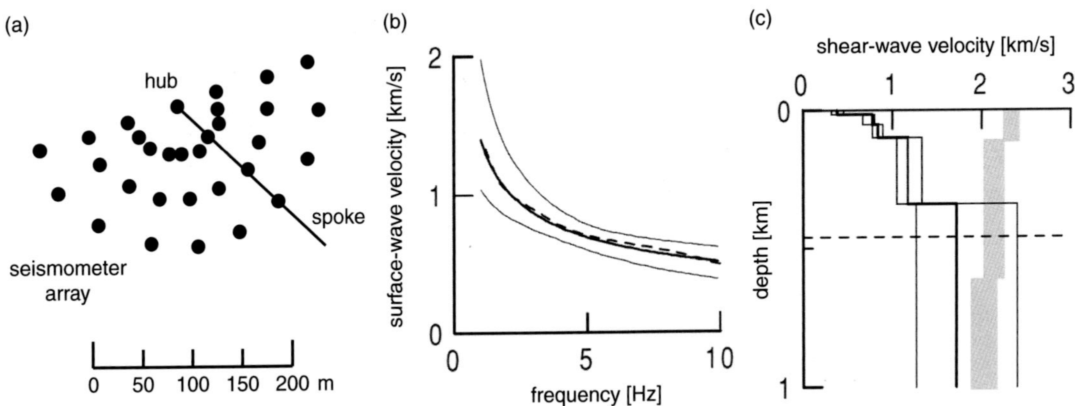

Figure 7.12 SPAC microtremor array analysis from Kilauea volcano, Hawaii: (a) seismometer array; (b) R-wave dispersion characteristics; (c) S-wave velocity profiles; after Soccarotti *et al.* (2003).

intervals along spokes with an angular separation of 20°. The wavefield recorded at the hub is treated as a reference signal. Correlations between the reference signal and the wavefields measured at the other receivers are computed. A semi-circular array geometry is well suited for SPAC analysis since the method involves azimuthal averaging of the correlation data. The dispersion characteristics are extracted by fitting the azimuthally averaged wavefield correlation data to a theoretical expression derived by Aki (1957), which assumes a stochastic, dispersive wavefield that is stationary in both space and time. The fitting procedure is carried out assuming a parametric dispersion curve of the form $V_{ph}(f) = Af^{-b}$. The R-wave result, along with its uncertainty, is shown in Figure 7.12b.

The preferred shear-wave velocity profile $V_S(z)$, obtained by inversion of the dispersion characteristic, is shown as the solid bold line in Figure 7.12c. The thin lines indicate the uncertainty. The shaded gray strips show the result of a previous refraction tomographic inversion at the same location. The shear-wave structure from the Rayleigh-wave analysis and the tomographic inversion clearly show lack of agreement, especially in the upper ~ 400 m. The tomography, however, has very poor resolution in the near-surface. The horizontal dotted line in Figure 7.12c indicates the maximum depth to which the surface-wave inversion is reliable. The dotted line in Figure 7.12b is the forward response of the preferred S-wave velocity profile. Overall, the shear-wave structure beneath Kilauea is consistent with that found at several other volcanoes, i.e. near-surface velocities of ~ 300–600 m/s, increasing to 1.5–2.0 km/s at depths exceeding 300–400 m.

8 Electromagnetic induction

The electromagnetic (EM) induction method, traditionally used for mining, groundwater, and geothermal exploration and geological mapping (Grant and West, 1965; Nabighian, 1988; 1991), is growing in popularity for near-surface geophysical applications (McNeill, 1990; Nobes, 1996; Tezkan, 1999; Pellerin, 2002; Fitterman and Labson, 2005; Everett, 2012). The controlled-source variant of the method utilizes low-frequency (~ 1–100 kHz) time variations in electromagnetic fields that originate at or near the surface and diffuse into the subsurface. The ground-penetrating radar (GPR) technique is distinct from the EM induction method in the sense that the former utilizes a higher frequency range > 1 MHz for which wave propagation, rather than diffusion, is the dominant energy transport mechanism. The diffusive regime is marked by the requirement $\sigma >> \omega\varepsilon$ where σ [S/m] is electrical conductivity, ω [rad/s] is angular frequency, and ε [F/m] is dielectric permittivity. The latter plays no physical role in the EM induction method. Instead, EM induction measurements respond almost entirely to the bulk subsurface electrical conductivity and, in particular, the spatial distribution of highly conductive zones (Everett and Meju, 2005). Instrumentation is readily available, easy to use, and reliable. Electric fields are sensed by pairs of grounded electrodes, i.e.voltmeters, while time-variations of magnetic fields are most commonly sensed using induction coils.

8.1 Introduction

The electrical conductivity structure of the subsurface, as determined from EM induction measurements, can be interpreted in the context of a wide variety of potential targets and application areas, see Table 8.1. Fundamentally, EM methods respond with good sensitivity to fluid type, clay content, and porosity. An overview of the physical and chemical factors that control the bulk electrical conductivity of geomaterials is provided by Gueguen

Table 8.1 Application areas and targets of EM induction geophysics.

Resistive targets	Intermediate targets	Conductive targets
Permafrost zones	Faults, fracture zones	Seawater intrusion
Aggregate deposits	Archaeological structures	Saline and inorganic plumes
Crystalline rock	Precision agriculture	Clay lenses, claypan soils
Caves, karst	Freshwater aquifers	Pipelines, steel drums, UXO

and Palciauskas (1994). The two case histories that follow serve as an introduction to the use of airborne and ground-based EM methods in near-surface investigations.

Example. Airborne EM mapping of a saline plume.

Soil and groundwater salinization negatively affects agriculture, water supplies, and ecosystems in arid and semi-arid regions worldwide. In Texas, especially before the 1960s and the tightening of environmental regulations, subsurface brines produced during oilfield drilling activities were frequently discharged into surface disposal pits which subsequently leaked their contents into the groundwater. Electromagnetic geophysical methods are capable of mapping conductive saline water concentrations, which are typically $\sigma \sim 0.1$–1.0 S/m, that invade geological backgrounds of $\sigma \sim 0.001$–0.01 S/m. Paine (2003) describes airborne EM mapping of such an oilfield brine plume in Texas (Figure 8.1). The plume is associated with a large area barren of vegetation that formed on a Pleistocene alluvial terrace of the Red River in the 1980s.

A helicopter EM system was employed consisting of coplanar horizontal coils and coaxial vertical coils flown at ~ 30 m altitude and operating at several frequencies between 0.9 and 56 kHz. The survey was performed with 100-m flight-line spacing and ~ 3 m along-track station spacing. A map of the 7.2 kHz apparent conductivity (Figure 8.1) clearly reveals the spatial extent of highly conductive ground. Many of the high-conductivity anomalies coincide with known brine-pit locations. The highest conductivity coincides with the barren area, which is bounded to the east by a topographic step to the Permian upland. The electromagnetic geophysical anomalies also find good spatial correlation with total dissolved solid (TDS) analyses based on sampled waters from monitoring wells distributed throughout the study area. The inferred boundary of the brine plume is marked on the map.

Example. Seawater intrusion in a coastal aquifer.

The intrusion of seawater into coastal freshwater aquifers is a serious problem in many places throughout the world. Human activities such as urbanization and the building of large-scale engineering projects such as dams can produce unintended negative effects on the hydrogeolgical conditions of coastal aquifers. Consequently it is important to develop methods that can identify the location of the freshwater–seawater interface. Electromagnetic geophysical methods offer an inexpensive and non-invasive alternative to drilling test boreholes. The electromagnetic detection of seawater is based on its high electrical conductivity relative to freshwater.

A ground-based electromagnetic geophysical survey of the Motril–Salobreña aquifer on the Mediterranean coast in southern Spain is reported by Duque *et al.* (2008). The geological setting is shown in Figure 8.2. The aquifer consists of Quaternary detrital sediments overlying Paleozoic and Mesozoic carbonates and metamorphics. As shown in the figure, a large dam is scheduled for construction along the Guadalfeo River that recharges the aquifer.

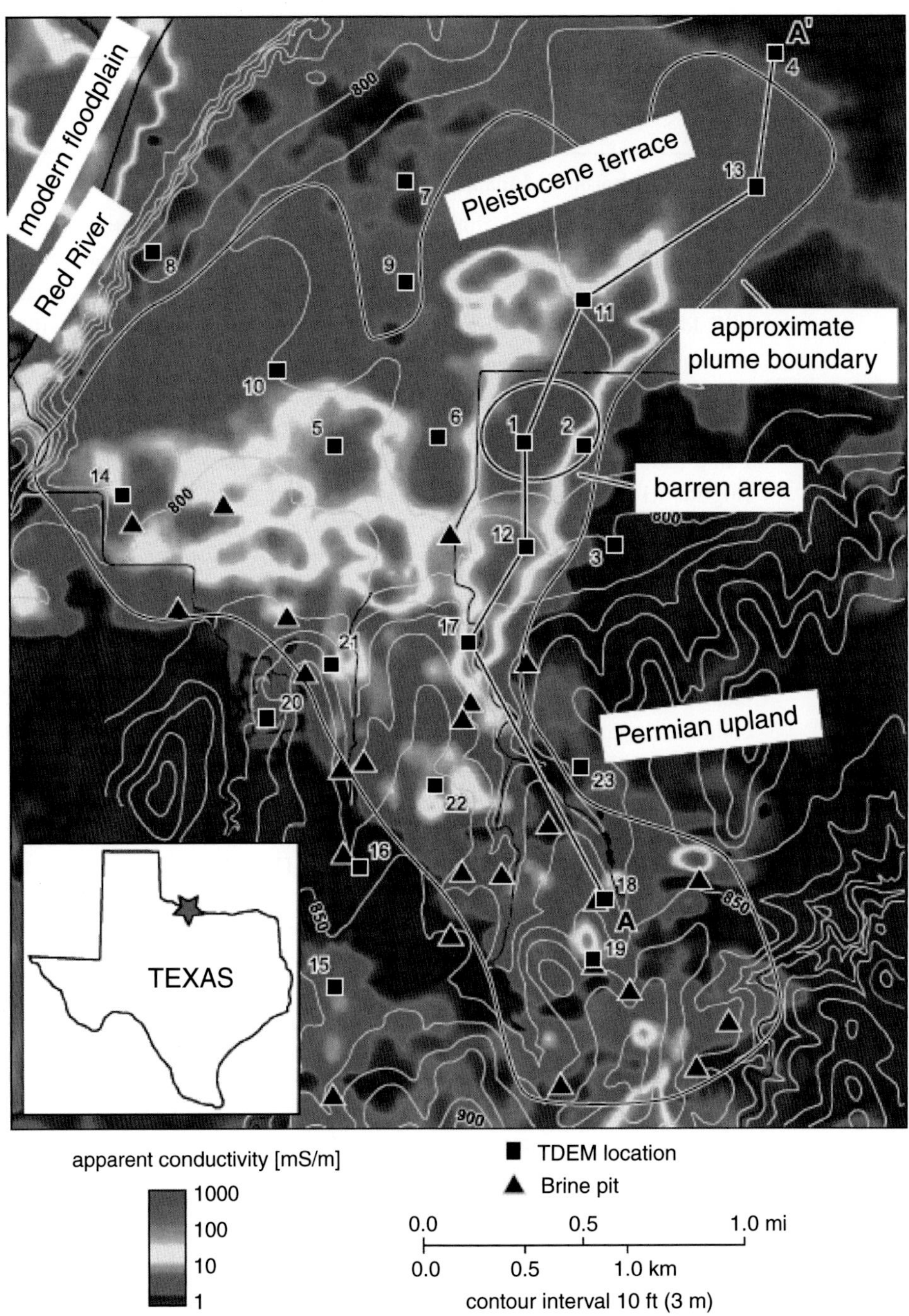

Figure 8.1 Topography and apparent resistivity at 7.2 kHz from airborne EM mapping of a north Texas oilfield brine plume. After Paine (2003). See plate section for color version.

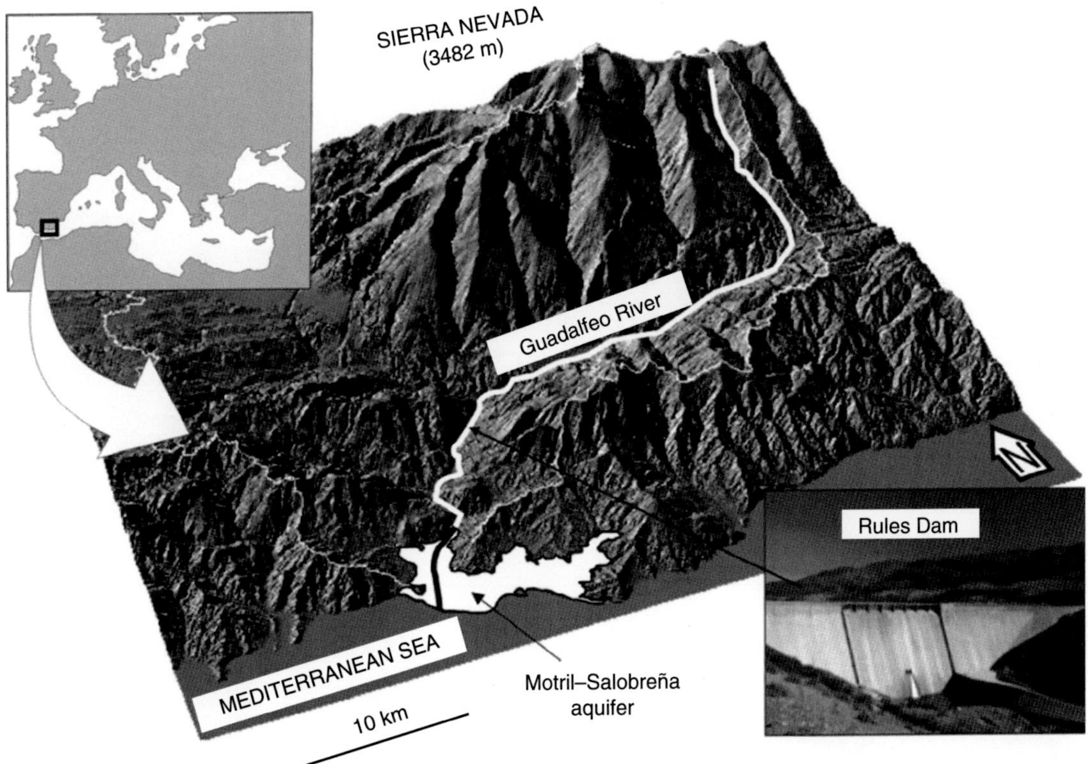

Figure 8.2 Geological setting of the Motril–Salobreña coastal aquifer (the white region) in Spain. After Duque *et al.* (2008).

The time-domain electromagnetic geophysical survey employed a central-loop configuration with transmitter loop sizes 50–200 m and consisted of several coast-perpendicular transects of ~ 2 km in length with station spacing of ~ 100–200 m. Plane-layered inversions of the measured transient responses were performed using a commercial software package; the resulting one-dimensional (1-D) conductivity depth-profiles were "stitched together" to form quasi-2-D depth-sections ready for interpretation. Two representative interpreted depth sections appear in Figure 8.3. The interpretations are based on conductivity assignments of $\sigma \sim 0.2$–2.0 S/m for seawater-saturated sediments; $\sigma \sim 0.003$–0.2 S/m for freshwater-saturated sediments; and $\sigma < 0.003$ S/m for the basement rocks. The location of the basement is further constrained by complementary gravity and borehole data. The results indicate that seawater intrusion into the near-surface aquifer is minor, with the marine wedge extending not more than 500 m inland. However, the position of the seawater–freshwater interface should be monitored carefully in the future owing to the impending urbanization and development of this coastal area. This case study shows that electromagnetic geophysical methods can play an important role in determining the spatiotemporal dynamics of the seawater–freshwater interface.

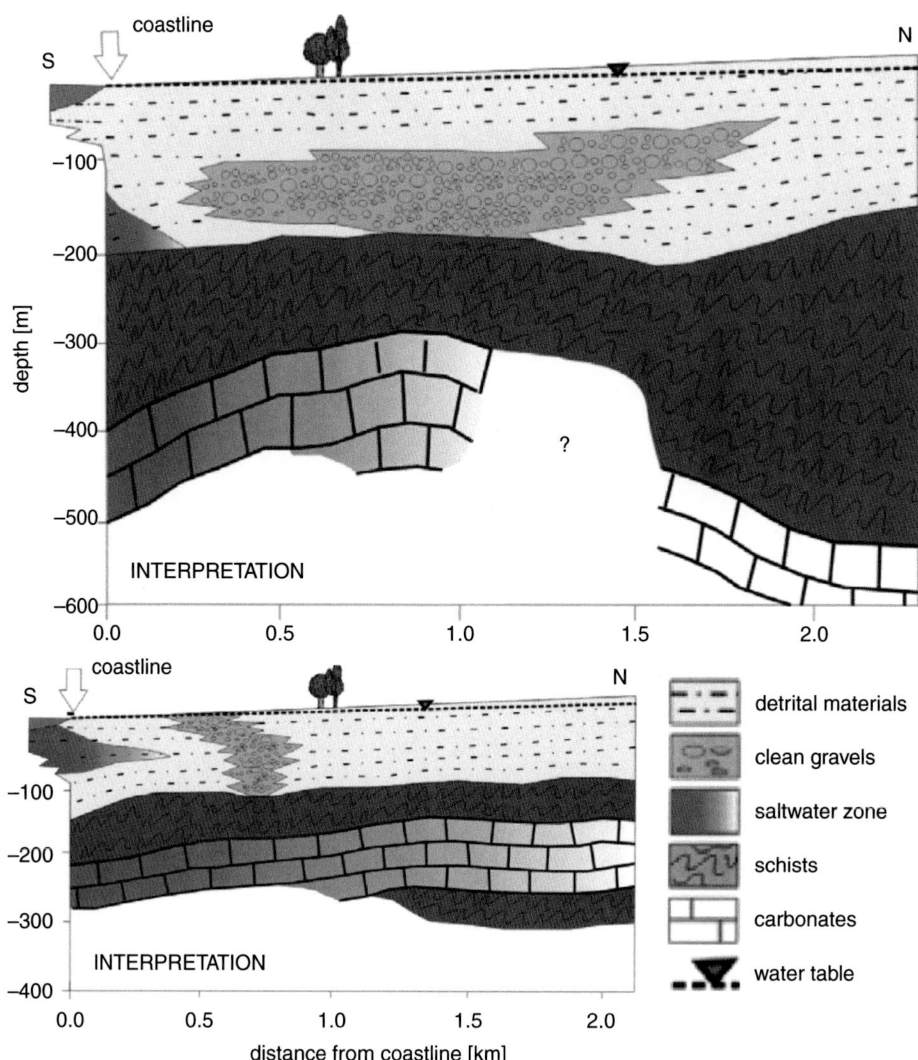

Figure 8.3 Interpretations of central-loop time-domain EM soundings in terms of seawater intrusion (red-colored areas) into the coastal Motril–Salobreña aquifer of southern Spain. After Duque *et al.* (2008). See plate section for color version.

8.2 Fundamentals

The electromagnetic geophysical method is founded on Maxwell's equations of classical electromagnetism. Standard physics textbooks such as Wangsness (1986) and Jackson (1998) are notable for their pedagogical, yet rigorous development of this topic. These texts, however, emphasize wave propagation rather than EM induction. The older books by Stratton (1941), Smythe (1950), and Jones (1964) treat induction in more detail and are highly recommended for advanced study. The reader is also referred to West and Macnae (1991) for a tutorial on the basic physics of induction.

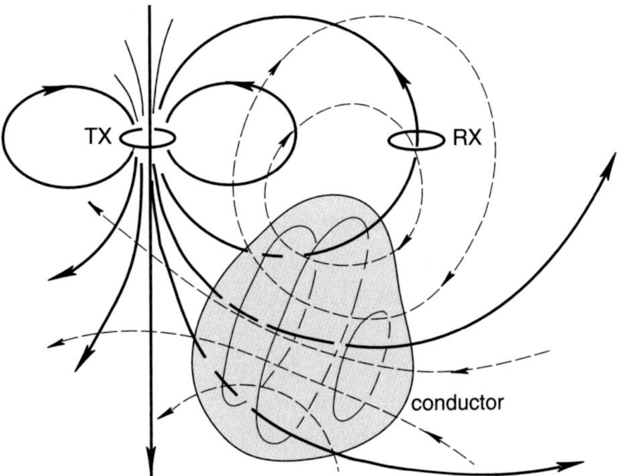

Figure 8.4 A sketch illustrating loop–loop electromagnetic target prospecting. From Grant and West (1965). Reproduced with permission from the McGraw-Hill companies.

Controlled sources used by electromagnetic geophysicists may be either of the *grounded* type, as typified by the electric dipoles used in the resistivity method, or of the *inductive* type in which direct electrical contact with the ground is avoided by using an insulated wire-loop source. In the latter case, the source–ground coupling is achieved entirely by magnetic-flux linkage. The basic principles of inductive loop–loop electromagnetic prospecting for a conductive target are illustrated in Figure 8.4 and explained below.

A time-dependent electric current flowing in the transmitter (TX) coil generates a transient primary field $\mathbf{B}^P(\mathbf{r}, t)$, as shown schematically by the dark solid lines in the figure. A certain number of primary field lines flux through the conductive target, as indicated. This time-varying flux generates an electromotive force which causes eddy currents (light solid lines) of density $\mathbf{J}(\mathbf{r}, t)$ to flow in the conductor. The eddy currents, in turn, generate a secondary magnetic field $\mathbf{B}^S(\mathbf{r}, t)$, as shown by the dashed lines, that is characteristic of the target geometry, and its position and conductivity. Both primary and secondary field lines flux through the RX coil, as shown. Consequently, a voltage is induced in the RX coil containing both the primary signal from the TX and the secondary electromagnetic response of the target. Since the primary signal is known, depending only on the TX–RX configuration, it can be removed, leaving just the unknown target response.

As further aid to understand the EM method, it is worthwhile to recall that the geological medium under investigation, including a conductive target and its host, is considered to be electrically neutral. The investigated volume contains a huge number but roughly equal amounts of positive and negative charge carriers. Some of the charge carriers are mobile and may migrate or drift from place to place within the medium while other charge carriers are essentially fixed in place, bound to lattice atoms or to material interfaces. The EM induction method is concerned only with the mobile charges.

An elementary description of EM induction starts with the Lorentz force $\mathbf{F} = q(\mathbf{E} + \mathbf{v}\times\mathbf{B})$, which is experienced by a mobile charge carrier q moving with velocity $\mathbf{v}$ in an electromagnetic field $(\mathbf{E}, \mathbf{B})$. An amount of work ε is done by the electromagnetic field on the charge carrier as it completes one cycle of an arbitrary closed path L. The quantity ε is called the electromotive force, or *emf.* The emf is not a force per se but rather a voltage or equivalently a potential. Essentially, a voltage develops along any arbitrary closed path L within a conducting body that is exposed to a time-varying **B** magnetic field. The field variations may be naturally occurring but in the controlled-source electromagnetic (CSEM) method, it is the geophysicist who shapes the external time variation with the aid of a transmitter.

The velocity **v** in the expression for the Lorentz force is interpreted as the charge-carrier *drift velocity* $\mathbf{v}_d$. Mobile charge carriers in conducting bodies migrate with an average drift velocity $\mathbf{v}_d$ in response to an applied electric field. The drift velocity satisfies $|\mathbf{v}_d| << c$, where c is the speed of light, the low velocity being due to lattice scattering of the charge carriers as they migrate. The drift velocity of electrons in standard 14-gauge copper wire carrying a 1 A current, for example, is just $|\mathbf{v}_\mathrm{d}| \sim 35$ μm/s (Tipler and Mosca, 2007). The ions of an electrolyte solution are even less mobile. The mobilities of Na^+ and Cl^- ions in seawater at temperature $T = 25$ °C are $m_+ \sim 5 \times 10^{-8}$ m^2/V s and $m_- \sim 8 \times 10^{-8}$ m^2/V s, respectively (Conway and Barradas, 1966). Since $\mathbf{v}_d = m\mathbf{E}$, this implies a very small drift velocity of $\sim 10^{-13}$ m/s in a typical electric field strength of ~ 1 mV/km.

An electric current density is associated with the drift velocity as $\mathbf{J} = nq\mathbf{v}_d$, where n is the volumetric concentration of the charge carrier. The electrical conductivity is related to the density and mobility of the charge carrier by $\sigma = nqm$ (Kittel, 2004; Gueguen and Palciauskas, 1995). Notice that an appreciable induced current does not flow in an insulator such as oil or epoxy since, in these materials, the number density n of mobile charges is negligible. It is the induced drift of mobile charges, acting as a secondary source of electromagnetic field, that generates the electromagnetic response measured by geophysicists.

In addition to the mobile charges that are present in conductive geomaterials, there certainly also exist *bound charges* that are not able to drift freely but nevertheless experience the Lorentz force **F** in the presence of an applied electromagnetic field. The motion of these bound charges leads to several types of *polarization*, foremost of which are atomic and molecular polarization, as earlier discussed in connection with the induced polarization (IP) method. Other types of bound-charge motion are also possible, but none of these are of direct relevance to the CSEM method. In other words, neither the bound charges that are confined to individual atoms, nor the mobile charges that are trapped at material interfaces, make any signficant contribution to the EM induction response. Such motions are capacitive effects; as shown in the next chapter, they are a very important aspect of the ground-penetrating radar (GPR) technique.

8.3 The skin effect

The depth of penetration of the EM induction method is limited by the efficiency of the conversion of the transmitted electromagnetic energy into kinetic energy of the mobilized subsurface charge carriers. The higher the electrical conductivity σ, the greater the efficiency and consequently the smaller the depth of penetration. The well-known *skin effect*

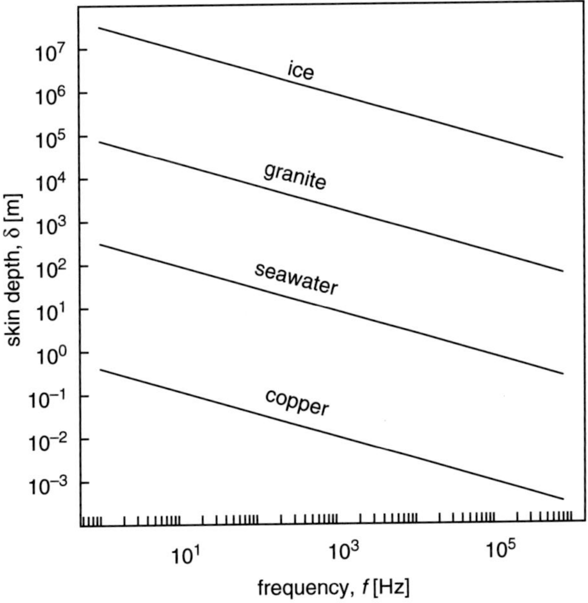

Figure 8.5 Skin depth as a function of frequency for several important non-magnetic geomaterials.

states that normally incident plane-wave signals lose $1/e \sim 0.368$ of their incident amplitude after penetrating one skin depth δ into a medium of conductivity σ, where

$$\delta = \sqrt{\frac{2}{\mu\sigma\omega}}, \tag{8.1}$$

and μ is magnetic permeability, $\omega = 2\pi f$ is the transmitted frequency, and $e \sim 2.71828$ is the base of the natural logarithm, that is, $e = \exp(1)$.

In the EM method, the magnetic permeability is almost always assumed to be equal to its free space value, $\mu = \mu_0$, even in the presence of magnetite-bearing rock formations. An important exception to this rule occurs if compact ferrous metal targets, such as pipelines, steel drums, or unexploded ordnance (UXO), are being investigated (e.g. Pasion, 2007) in which case $\mu \sim 1.05\mu_0$–$50\mu_0$, or even higher. To explore the depth of penetration in some common non-magnetic geomaterials, typical values found in the literature for conductivity can be used: $\sigma \sim 10^{-9}$ S/m (ice); $\sigma \sim 10^{-4}$ S/m (granite); $\sigma \sim 3.2$ S/m (seawater); and $\sigma \sim 10^6$ S/m (copper). The graphs of $\delta(f)$ for these materials appear in Figure 8.5.

For loop-source excitation, the skin effect remains but the field distribution inside the Earth becomes more complicated due to the compact nature of the source. The electric-field intensity associated with diffusion from a loop source into a uniformly conducting halfspace is shown, in the frequency domain, in Figure 8.6a. The contours are shown in a vertical plane passing through the center of loop. Considering the azimuthal symmetry of the problem, the electric-field intensity takes on the appearance of a toroidal "smoke ring." The position of the maximum smoke-ring intensity in terms of the skin depth δ, as given by Equation (8.1), is shown in Figure 8.6b.

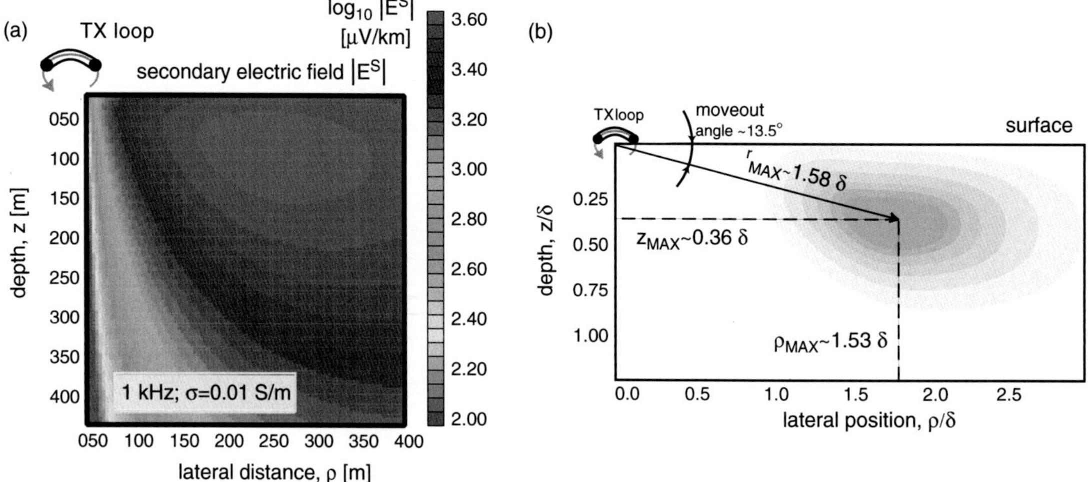

Figure 8.6 (a) Electric-field intensity due to loop excitation in a unform halfspace; (b) position of maximum intensity of the electric field in terms of skin depth. See plate section for color version.

8.4 Inductively coupled *LR* circuits

Considerable insight into the physics of an electromagnetic geophysical prospecting system may be gained by analyzing the behavior of a roughly equivalent system of three inductively coupled *LR* circuits (Grant and West, 1965), as shown below in Figure 8.7. In the equivalent system, the TX loop is modeled as *LR* circuit 1 with resistance R_1 and self-inductance L_1. Similarly, the RX coil is modeled as circuit 2 with its own resistance and self-inductance, respectively R_2 and L_2. The conductive body, which could represent a compact target such as an ore body or a UXO, or even an entire halfspace, is modeled by circuit 3 with (R_3, L_3), as shown.

Let us first examine the primary interaction between the TX loop and the RX coil in free space without the presence of the Earth/target. The *mutual inductance* between the TX and the RX circuits is given by M_{12}. In general, given two loops i and j, the mutual inductance M_{ij} is defined as the magnetic flux that passes through loop j due to a unit electric current flowing in loop i. Thus, mutual inductance is purely a geometric parameter that depends on the relative orientation and location of the two loops with respect to each other. Notice also that $M_{ij} = M_{ji}$. The *self-inductance* L_i is the magnetic flux that passes through loop i caused by unit current flow in the same loop, hence, $L_i = M_{ii}$.

Suppose Φ_j is the magnetic flux through loop j, defined according to

$$\Phi_j = \int_A \mathbf{B}_j \cdot \hat{\mathbf{n}} dA, \tag{8.2}$$

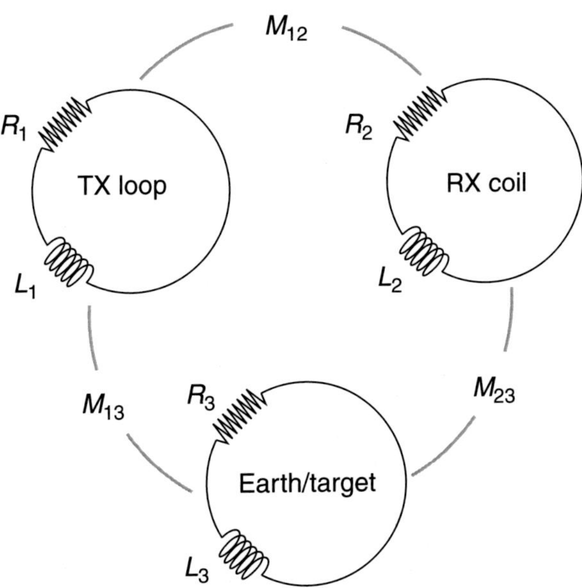

Figure 8.7 Electromagnetic prospecting system conceptualized as inductively coupled *LR* circuits.

where an element of the area A of the loop is denoted by dA and $\hat{\mathbf{n}}$ is the unit normal vector oriented along the axis of the loop. The loop need not be planar, it can be bent or twisted in which case the unit normal vector would depend on location, $\hat{\mathbf{n}} = \hat{\mathbf{n}}(\mathbf{r})$. The magnetic field passing through loop j is denoted as $\mathbf{B}_j$, which presumably is caused by a current I_i flowing in some other loop i. By the definition of mutual inductance, we have

$$\Phi_j = M_{ij} I_i. \tag{8.3}$$

The magnetic, or inductive, coupling of the two loops is embodied in the empirical law generally attributed to Faraday stating that the negative rate of change of flux though a loop equals the voltage V_j induced in the loop, or

$$V_j = -\frac{d\Phi_j}{dt} = -M_{ij}\frac{dI_i}{dt}. \tag{8.4}$$

The above equation reveals that a voltage is induced in loop j if the current flow in loop i is changing with time. Suppose the time variation in the current I_i is harmonic such that $I_i \sim \exp(i\omega t)$ where ω is the angular frequency, then from Equation (8.4) the induced voltage in loop j in the frequency domain is

$$V_j = -i\omega M_{ij} I_i. \tag{8.5}$$

To completely describe the magnetic coupling of two current loops i and j, we need two other empirical laws of classical electromagnetism. The first is a simplified (neglecting radiation effects) version of the Biot–Savart law which, for our purposes, reads

$$\mathrm{B}_j = K_j I_j, \tag{8.6}$$

stating that the magnitude of $B_j = |\mathbf{B}_j|$ of the magnetic field generated by the induced current in loop j is proportional to the strength of the current I_j and completely in phase with it. The constant of proportionality K_j is a geometric factor that depends on the shape, size, and number of turns, of loop j. The second fundamental law is Ohm's law, which, for a general LR circuit in the time domain, is $V(t) = I(t)R + L\partial I(t)/\partial t$, while in this particular time-harmonic case it becomes

$$V_j = I_j[R_j + i\omega L_j]. \tag{8.7}$$

Putting the previous expressions together, we can easily see that the *primary* magnetic field measured by RX loop 2 in Figure 8.6 due to current flow in the TX loop is

$$\mathrm{B}^P = -\frac{i\omega K_2 M_{12} I_1}{R_2 + i\omega L_2}. \tag{8.8}$$

Now, let us add the third loop into the system to incorporate the effect of the conductive Earth or buried target. The voltage induced in the Earth/target loop 3 follows as in Equation (8.5) from Faraday's law and the definition of mutual inductance,

$$V_3 = -i\omega M_{13} I_1, \tag{8.9}$$

to which Ohm's law, Equation (8.7), can be applied to find the induced current I_3 in the ground circuit,

$$I_3 = \frac{V_3}{R_3 + i\omega L_3} = -\frac{i\omega M_{13} I_1}{R_3 + i\omega L_3}. \tag{8.10}$$

The induced current I_3 in the ground in turn generates a secondary voltage signal V_2^S in the RX loop 2 which can be found by another application of Faraday's and Ohm's laws,

$$V_2^S = -i\omega M_{23} I_3 = -\frac{\omega^2 M_{13} M_{23} I_1}{R_3 + i\omega L_3}. \tag{8.11}$$

The secondary induced current I_2^S in the RX loop due to the current flowing in the ground circuit is therefore given by

$$I_2^S = \frac{V_2^S}{R_2 + i\omega L_2} = -\frac{\omega^2 M_{13} M_{23} I_1}{(R_2 + i\omega L_2)(R_3 + i\omega L_3)}. \tag{8.12}$$

Hence, the *total* magnetic field B_2 measured by the RX loop is given by the simplified Biot–Savart law applied to the summation of the primary and secondary induced currents,

$$\mathrm{B}_2 = K_2(I_2 + I_2^S) = -\frac{i\omega K_2 M_{12} I_1}{R_2 + i\omega L_2} - \frac{\omega^2 K_2 M_{13} M_{23} I_1}{(R_2 + i\omega L_2)(R_3 + i\omega L_3)}, \tag{8.13}$$

which reduces to

$$\mathrm{B}_2 = \frac{K_2 I_1}{R_2 + i\omega L_2}\left\{-i\omega M_{12} - \frac{\omega^2 M_{13} M_{23}}{R_3 + i\omega L_3}\right\}, \tag{8.14}$$

or

$$B_2 = B_0\{\Gamma^P + \Gamma^S\}. \tag{8.15}$$

In the above equation, the pre-factor $B_0 = K_2 I_1/(R_2 + i\omega L_2)$ is a known term that depends only on the TX current I_1 and the various geometric properties of the RX loop, which are under the control of the experimenter. The first term in the curly brackets $\Gamma^P = -i\omega M_{12}$ is also known and depends only on the frequency ω and the mutual inductance of the TX and RX loops, both of which are under experimenter control. The second term in the curly brackets $\Gamma^S = -\omega^2 M_{13} M_{23}/(R_3 + i\omega L_3)$ depends on the *unknown* quantities M_{13} and M_{23}, which describe the relative orientation and location of the ground circuit with respect to the TX and RX loops. It also depends on the unknown parameters R_3 and L_3 which contain information about, respectively, the physical properties and the geometry of the ground circuit.

Examining the response of the Earth/target circuit, as a function of TX frequency ω, is important to better understand the EM induction geophysical method. To this end, based on the definition of Γ^P and Γ^S in Equation (8.15), we specify the *normalized response function* as

$$\frac{\Gamma^S}{\Gamma^P} = \kappa \frac{i\omega L_3/R_3}{1 + i\omega L_3/R_3} \tag{8.16}$$

where $\kappa = M_{13}M_{23}/M_{12}L_3$ is a factor that depends only on the geometry of the TX–target–RX configuration. The function Γ^S/Γ^P in Equation (8.16) captures the fundamental behavior of the electromagnetic response of the Earth or a compact buried target while suppressing complexities associated with the TX and RX loop geometries. To examine the frequency dependence of Γ^S/Γ^P we can set $\kappa = 1$ without obscuring the essential physics. A graph of Γ^S/Γ^P versus the normalized frequency, or *response parameter* $\omega L_3/R_3$, is shown in Figure 8.8. The real part

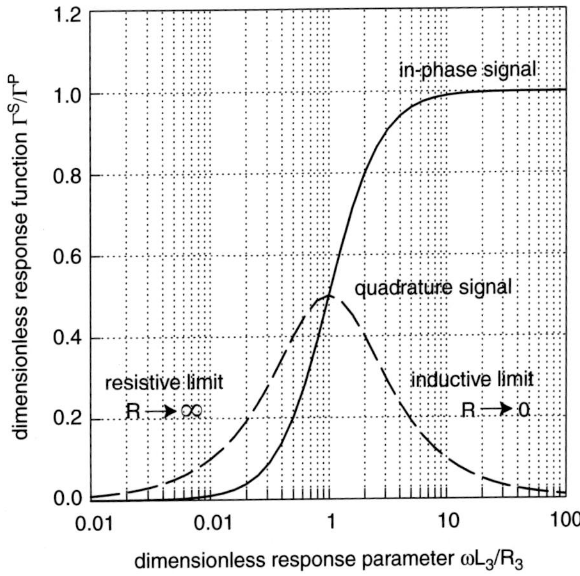

Figure 8.8 Idealized electromagnetic response of a buried target.

of the response function Γ^S/Γ^P is termed the *in-phase signal* since its time variation coincides with that of the TX current. The imaginary part of Γ^S/Γ^P is the out-of-phase, or *quadrature*, signal, so-called since its time variation is exactly 90° out of phase with the TX current.

There are three important frequency regimes in EM prospecting that are apparent from Figure 8.8. The low-frequency regime, in which $\omega L_3/R_3 << 1$, is termed the *resistive limit* and is achieved as the frequency $\omega \rightarrow 0$ or equivalently as the target resistance $R_3 \rightarrow \infty$. The high-frequency regime $\omega L_3/R_3 >> 1$, or *inductive limit*, is achieved as $\omega \rightarrow \infty$ or equivalently $R_3 \rightarrow 0$. Finally, there is an intermediate frequency range at which $\omega L_3/R_3 \sim 1$.

In the resistive limit, as indicated in Figure 8.8, the in-phase signal vanishes and the quadrature signal is small. The low amplitude of the EM response is expected since the target is very resistive and hence induction of an electric current in the ground circuit is poor. Since the response function is of the form $\Gamma^S/\Gamma^P \sim (i\omega L_3/R_3)/(1 + i\omega L_3/R_3)$ it is clear that, in the resistive limit, $\Gamma^S/\Gamma^P \rightarrow i\omega L_3/R_3$. This is a small, purely quadrature signal.

In the inductive limit, the EM response function tends to $\Gamma^S/\Gamma^P \rightarrow 1$ which indicates a large in-phase signal with the quadrature signal becoming very small. The reason that the EM response Γ^S/Γ^P is amost entirely in-phase with the TX current is that there is an additional ~ 90° phase-shifting process that operates at the inductive limit compared to the resistive limit. Specifically, when $R_3 \rightarrow 0$, according to Equation (8.10), the current induced in the Earth target is shifted out of phase with respect to the voltage. In essence, induction is a "*d*/*dt process*." In the inductive limit, there are two additive *d*/*dt* processes at work; one generating the induced current in the target and the other generating the induced current in the RX coil. The superposition of the two *d*/*dt* processes explains why the inductive-limit EM response function is brought back into phase with the TX current.

As further shown in Figure 8.8, the quadrature signal peaks at the intermediate frequency range, $\omega L_3/R_3 \sim 1$. This result indicates that an EM prospecting system can generate a maximal target response, in the quadrature signal, over a specific range of frequencies. The EM quadrature response diminishes at both higher and lower frequencies. Clearly, there is a preferred band of frequencies centered on $\omega \sim R_3/L_3$ that is most useful for detecting the ground circuit. The existence of a preferred band is not surprising when it is recognized that there is a characteristic time for the decay of an electric current flowing in a simple *LR* circuit, namely $\tau = L/R$.

Fitterman and Labson (2005) have analyzed the transient response $V_2(t)$ of a coupled *LR*-circuit system for a step-off in the TX current, $I_1(t) = I_1[1 - u(t)]$, where $u(t)$ is the Heaviside step-on function. They showed that the RX signal $V_2(t)$ consists of a sudden initial rise followed by an exponential decay with time constant τ. A larger time constant, appropriate to describe currents that can flow readily in the Earth circuit, generates a smaller initial amplitude but produces a longer-lived decay of the RX signal. Further discussion of transient EM is deferred to Section 8.6.

8.5 Terrain conductivity meters

A simple electromagnetic geophysical reconnaissance tool is the terrain conductivity meter (McNeill, 1980). This frequency-domain instrument operates as a purely inductive system in which transmitter (TX) and receiver (RX) loops do not make electrical contact with the ground. The two loops are magnetically flux-linked to each other and to the conductive ground as shown schematically by the circuits in Figure 8.7. Two popular terrain conductivity meters are the Geonics EM31 and EM34 instruments (www.geonics.com) shown in Figure 8.9. The EM31 operates at 9.8 kHz and has TX–RX intercoil spacing of 3.66 m. The EM34 has three different intercoil spacings: 10, 20, and 40 m; these operate at, respectively, 6.4, 1.6, and 0.4 kHz. The EM34 can be used with vertical coplanar coils (horizontal dipole, or HD mode), or horizonal coplanar coils (vertical dipole, or VD mode) in which case the coils are laid flat on the ground.

The operating principle of the terrain conductivity meter is based on classical EM induction theory. A time-harmonic current of the form $I(t) = I\sin(\omega t)$ is passed through the TX loop. The primary magnetic field due to the current flowing in the transmitter loop is in-phase with the current and consequently has the form

$$\mathbf{B}^P(\rho, t) = \mathbf{B}_0(\rho)\sin(\omega t), \tag{8.17}$$

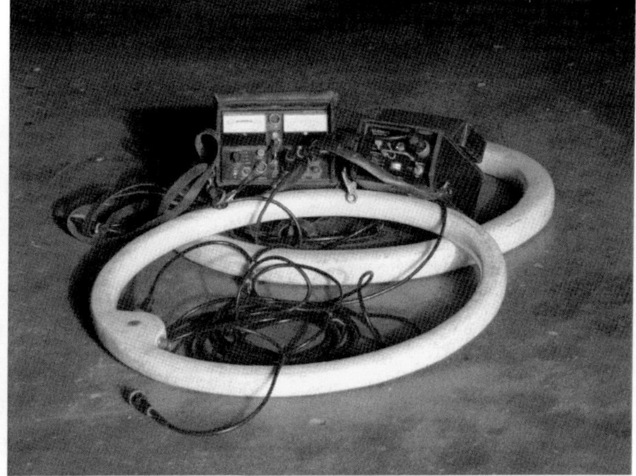

Figure 8.9 Terrain conductivity meters: (left) Geonics EM31; (right) Geonics EM34.

where ρ is the radial distance in a cylindrical coordinate system originated at the center of the TX loop. The conductive ground responds to the imposition of the time-varying primary magnetic flux by establishing a system of electromagnetic eddy currents whose secondary magnetic field $\mathbf{B}^S(\rho, t)$ is organized such that it tends to oppose the *change* $\partial\mathbf{B}^P/\partial t$ in primary flux. This principle was earlier captured in the circuit analogy by Equation (8.4). Essentially, the changing primary flux establishes an *electromotive force* (emf) in the ground. The ground then responds by generating a *back-emf* in an effort to restore the equilibrium that existed before the change occurred in the primary flux. This behavior accords with Lenz's law of electromagnetics, which (like homeostasis) is a specific case of Le Chatelier's principle in chemistry:

> If a system at equilibrium experiences a change of state, the equilibrium will shift in a direction so as to counter-act the imposed change.

The secondary magnetic field then has the form

$$\mathbf{B}^S(\rho, t) = \mathbf{B}_1(\rho)\sin(\omega t + \varphi). \tag{8.18}$$

where $|\mathbf{B}_1| << |\mathbf{B}_0|$, and φ is the phase shift caused by the induced currents which are not completely in-phase with the primary magnetic flux. The magnitude of the phase shift depends on the electrical conductivity of the ground. If the ground is perfectly conducting, $\varphi = 90°$ and the secondary magnetic field is completely out of phase. If the ground is perfectly resistive, $\varphi = 0$ and the secondary magnetic field is completely in-phase. In general, the secondary magnetic field is *delayed* and *attenuated* with respect to the primary magnetic field, indicative of EM induction as an energy-dissipating *d*/*dt* process.

Using the identity

$$\sin(\omega t + \varphi) = \sin\omega t \cos\varphi + \cos\omega t \sin\varphi \tag{8.19}$$

we can decompose the total magnetic field $\mathbf{B}^T(\rho, t) = \mathbf{B}^P(\rho, t) + \mathbf{B}^S(\rho, t)$ into two orthogonal components:

$$\begin{aligned} \mathbf{B}^T(\rho, t) &= \mathbf{B}_0(\rho)\sin\omega t + \mathbf{B}_1(\rho)\sin(\omega t + \varphi) \\ &= \mathbf{B}_0(\rho)\sin\omega t + \mathbf{B}_1(\rho)\sin\omega t\cos\varphi + \mathbf{B}_1(\rho)\cos\omega t\sin\varphi \\ &= [\mathbf{B}_0 + \mathbf{B}_1\cos\varphi]\sin\omega t + [\mathbf{B}_1\sin\varphi]\cos\omega t \end{aligned} \tag{8.20}$$

such that $R = |\mathbf{B}_0 + \mathbf{B}_1\cos\varphi|$ is termed the *real* (*in-phase*) *response* and $Q = |\mathbf{B}_1\sin\varphi|$ is termed the *quadrature* (*out-of-phase*) *response*. The quantities (R, Q) are measured by the EM instrument. The quadrature signal is very small since in most cases,

$$\frac{|\mathbf{B}_1|}{|\mathbf{B}_0|} \sim 10^{-6} = 1\text{ppm}. \tag{8.21}$$

This implies that the strength of the induced current is only a very small fraction of the strength of the primary transmitted current, the latter being typically 1–3 A. The quadrature signal reveals information about the Earth since it vanishes ($Q = 0$) if the conductive Earth is not present. Notice that the known primary signal $\mathbf{B}_0$ can be subtracted to form a secondary in-phase response $R_S = |\mathbf{B}_1\cos\varphi|$. Instead of the real and quadrature components, an EM system can also measure the *amplitude* and *phase* of the ground response. In terms of the quantities (R_S, Q), the amplitude is given by

$$A = \sqrt{R_S^2 + Q^2} = \sqrt{(\mathbf{B}_1 \cos\varphi)^2 + (\mathbf{B}_1 \sin\varphi)^2} = |\mathrm{B}_1|, \tag{8.22}$$

as required, while the phase is

$$\varphi = \tan^{-1}\frac{Q}{R_S} = \tan^{-1}\frac{\mathbf{B}_1 \sin\varphi}{\mathbf{B}_1 \cos\varphi} = \tan^{-1}(\tan\varphi) = \varphi \tag{8.23}$$

as required.

It is demonstrated below that the measured quadrature response Q can be interpreted in terms of the apparent ground conductivity σ_a. An *apparent conductivity* is defined as the conductivity of a homogeneous halfspace that would generate a quadrature response that is identical to the one that is observed. Keep in mind that $\sigma_a = \sigma_a(\omega)$ for an inhomogeneous geological formation since the signal penetration depth depends on frequency, according to the skin effect.

It can be shown that the normalized secondary vertical magnetic field (b_z^S/b_z^P) for coils operating in the VD mode over homogeneous ground, assuming a small horizontal TX loop such that loop radius $a/\rho << 1$, can be simplified to (Wait, 1954; McNeill, 1980):

$$\left(\frac{b_z^S}{b_z^P}\right)_V \approx -\frac{2}{\alpha^2\rho^2}\{9 - [9 - 9i\alpha\rho - 4\alpha^2\rho^2 + i\alpha^3\rho^3]\exp(i\alpha\rho)\}, \tag{8.24}$$

where $\alpha^2 = -i\mu_0\sigma\omega$. Furthermore, at low induction numbers,

$$\mathrm{B} \equiv \frac{\rho}{\delta} \ll 1, \text{or equivalently, } |\alpha\rho| \ll 1, \tag{8.25}$$

such that the TX–RX separation distance is much less than the electromagnetic skin depth (equivalent also to low frequency), it is straightforward to show that Equation (8.24) reduces to

$$\left(\frac{b_z^S}{b_z^P}\right)_V \approx 1 - \frac{i\mu_0\omega\sigma\rho^2}{4}. \tag{8.26}$$

The VD-mode apparent conductivity σ_a is then obtained by re-arranging the out-of-phase component of Equation (8.26),

$$\sigma_a = \frac{4}{\mu_0\omega\rho^2}\left|\mathrm{Im}\left(\frac{b_z^S}{b_z^P}\right)_V\right|. \tag{8.27}$$

Similarly, when the TX and RX coils are oriented in the HD mode, the normalized secondary field is approximated by (McNeill, 1980)

$$\left(\frac{b_z^S}{b_z^P}\right)_H = 2\left\{1 + \frac{3}{\alpha^2\rho^2} - [3 - 3i\alpha\rho - \alpha^2\rho^2]\frac{\exp(i\alpha\rho)}{\alpha^2\rho^2}\right\}, \tag{8.28}$$

and an equation identical in form to Equation (8.27) holds for the HD-mode apparent conductivity.

To summarize, the terrain conductivity meter, e.g. Geonics EM31 or EM34, reads an apparent conductivity that is linearly related to the measured quadrature response by $\sigma_a = 4Q_N/\mu_0\omega\rho^2$ where $Q_N = |\text{Im}(b_z^S/b_z^P)_V|$ or $Q_N = |\text{Im}(b_z^S/b_z^P)_H|$ depending on whether the instrument is being used in the VD or the HD mode.

The above discussion describes the basic operating principle of low-induction-number (LIN) terrain conductivity meters. The meter will fail, or exhibit erratic behavior including displaying negative apparent conductivities, in resistive terrain if the conductivity is sufficiently low such that the quadrature response Q_N falls below the system noise level. Resistive geological materials cannot sustain strong systems of induced currents. The terrain conductivity meter generates stable readings in conductive terrains, for example, if $\sigma \sim 0.001$–0.1 S/m, since in such cases the quadrature response is large compared to the system noise level. The meter may also report negative apparent conductivities in the vicinity of strong lateral heterogeneities, such as might be caused by pipelines or faults, as the assumption of a homogeneous ground that was used to derive Equation (8.24) is not applicable in these cases. Finally, in highly conductive terrains the low-induction number (LIN) assumption (8.25) breaks down and the apparent conductivity σ_a no longer scales linearly with the measured quadrature response Q_N. In such cases, a correction factor such as the one proposed by Beamish (2011) should be applied to better estimate the apparent conductivity.

Measurements using a LIN terrain conductivity meter were taken by geophysics undergraduate students over a buried natural-gas pipeline on Riverside campus at Texas A&M University. The results are shown in Figure 8.10. Data were taken along a profile crossing the pipeline, with a Geonics EM31 instrument resting on the ground. Two VD orientations of the EM31 instrument were used: the in-line (left), or *P-mode*, in which the TX and RX coils are aligned with the profile direction; and the broadside (right), or *T-mode*, in which the TX and RX coils are aligned perpendicular to the profile direction (i.e. parallel to the pipeline). It is easy to see that the pipeline generates a strong, repeatable EM31 anomaly with high signal-to-noise ratio, for both modes. Notice there are some unphysical negative apparent conductivities in the P-mode profile in the vicinity of the pipeline. The striking differences between the P-mode and T-mode signatures reflect the variation in mutual electromagnetic coupling between the TX coil, the subsurface target, and the RX coil caused by differences in the relative geometry of the two TX-RX configurations relative to the target. At the beginning and the end of the profiles, both sets of EM31 readings tend to values of ~ 60–70 mS/m, regardless of the orientation of the coils, reflecting the relatively homogeneous background soil conditions.

A more comprehensive analysis of the EM31 response to a highly conductive target was performed by Benavides and Everett (2005). In that study, which has applications to EM imaging in areas with high cultural noise, a number of steel drill pipes were laid out on the ground in the pattern shown in Figure 8.11a. The VD P-mode and T-mode responses are shown in Figure 8.11b and 8.11c, respectively. The line spacing is 1.0 m and station spacing is 0.25 m. The data were acquired along north–south profiles. It is evident that the pipes generate elongate, bimodal anomalies that are oriented in the direction of the TX–RX coils. The smallest target, the drill pipe collar (item 4), generates only a subtle anomaly in each mode. The anomalies are highlighted if the difference between the T-mode and the P-mode

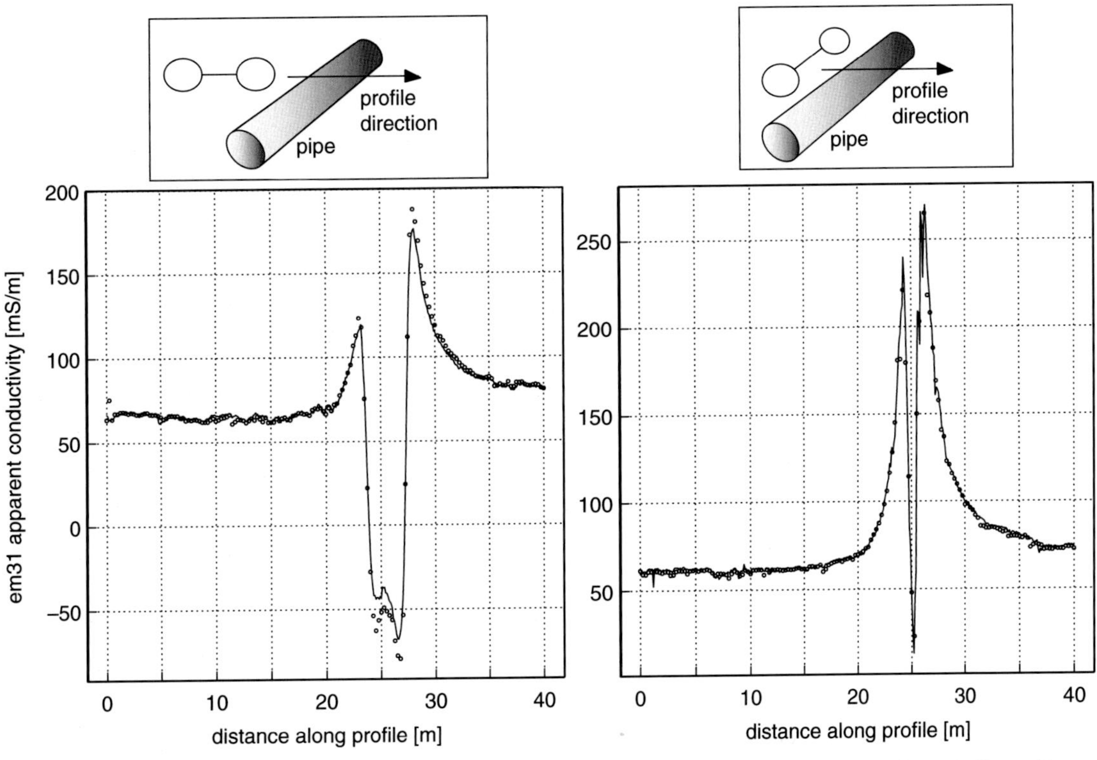

Figure 8.10 EM31 in-line P-mode (left side) and broadside T-mode (right side) signatures of the natural-gas pipeline at Texas A&M Riverside campus. Solid lines indicate original survey with 0.1-m station spacing. Symbols indicate repeat survey with 0.25-m station spacing.

signatures is plotted, as shown in Figure 8.10d. In this presentation, the pipe anomalies exhibit a distinctive quadrupole morphology, while the background soil response remains very subdued. Notice, however, that it is not easy to see from the geophysical signature that the drill pipe (item 5) is aligned at ~ 45° to the data-acquisition grid.

Terrain conductivity meters can also be used for mapping lateral contrasts in near-surface geology. Figure 8.12, left, illustrates EM34 readings taken with 10-m TX–RX intercoil spacing in the P-mode with the coil axes vertical (VD). The profile crosses a known lateral discontinuity juxtaposing a gravel unit with a sand/clay interbedded unit. The purpose of the survey was to delineate the extent of a paleochannel comprised of coarse sediment, in this case, an economic gravel deposit. Generally speaking, fine-grained material with abundant clay is more conductive than clean, coarse-grained sediment. The abrupt transition from low to high apparent conductivity clearly marks the lateral boundary of the gravel paleochannel.

In Figure 8.12, right, the EM34 method (P-mode, VD) is used in its three different TX–RX intercoil spacings (10, 20, and 40 m) to map faults and stratigraphic contacts in a well-characterized consolidated sandstone aquifer in central Texas. The Hickory sandstone, of Cambrian age, is divided into lower, middle, and upper units. It is known that the faults and contacts act as barriers and conduits compartmentalizing groundwater flow in this aquifer.

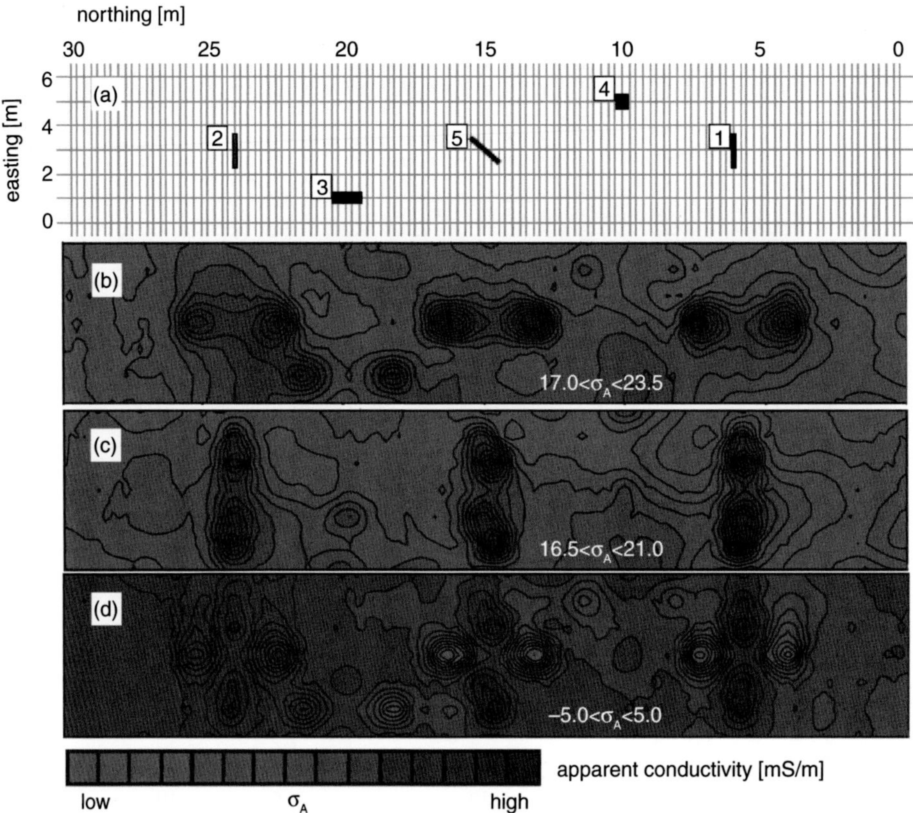

Figure 8.11 EM31 detection of conductive targets. (a) Experimental layout of steel pipes; (b) P-mode apparent conductivity map; (c) T-mode apparent conductivity map; (d) difference of the T-mode and P-mode maps; key: 1, 5 = 155 × 12.4-cm drill pipe, 2 = 150 × 6.4-cm drill pipe, 3 = 75 × 7.5-cm galvanized iron water pipe, 4 = 28 × 7.2-cm drill pipe collar. After Benavides and Everett (2005). See plate section for color version.

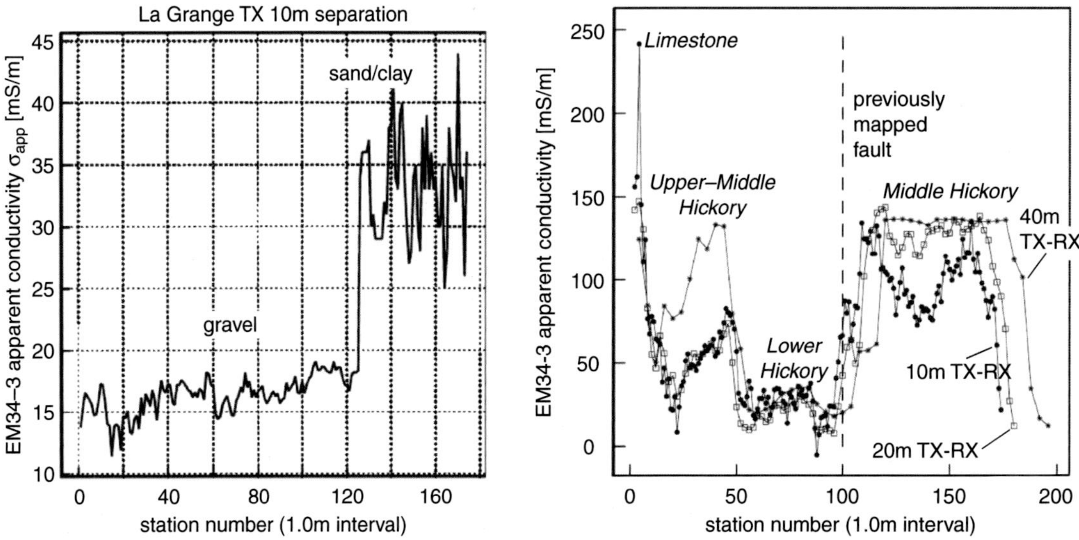

Figure 8.12 EM34 mapping of lateral geological discontinuities: (left) gravel quarry; (right) faulted Cambrian sandstone aquifer. After Everett and Meju (2005); with kind permission from Springer Science + Business Media B.V.

The observed lateral contrasts in apparent electrical conductivity are caused by lateral variations in sedimentary texture. For example, the location of a previously mapped, buried fault is shown by the dotted line. The variations in the EM34 responses between stations 20–50 can be explained in terms of spatial variations in the distribution of clays within the Hickory sandstone formation (Gorman *et al.*, 1998).

8.6 Time-domain EM induction

A physical understanding of *transient* CSEM responses can be obtained by recognizing that the induction process is equivalent to the diffusion of an image of the transmitter (TX) loop into a conducting medium. The similarity of the equations governing electromagnetic induction and hydrodynamic vortex motion, first noticed by Helmholtz, leads directly to the association of the image current with a smoke ring (Lamb, 1994). The latter is not "blown," as commonly thought, but instead moves by self-induction with a velocity that is generated by the smoke ring's own vorticity and described by the familiar Biot–Savart law (Arms and Hama, 1965). An electromagnetic smoke ring dissipates in a conducting medium much as the strength of a hydrodynamic eddy is attenuated by the viscosity of its host fluid (Taylor, 1958; pp. 96–101). The material property that dissipates the electromagnetic smoke ring is electrical conductivity.

The operating principles of an inductively coupled time-domain electromagnetic (TDEM) system are summarized as follows. A typical current waveform $I(t)$ consists of a slow rise to a steady value of I_0 followed by a rapid shut-off as exemplified by the linear ramp shown in Figure 8.13a. Passing such a disturbance through the TX loop generates a primary magnetic field that is in-phase with, and proportional to, the TX current. According to Faraday's law of induction, an impulsive electromotive force (emf) that scales with the negative time rate of change of the primary magnetic field is also generated. The emf drives electromagnetic eddy currents in the conductive Earth, notably in this case during the ramp-off interval, as shown in Figure 8.13b. After the ramp is terminated, the emf vanishes and the eddy currents start to decay via Ohmic dissipation of heat. A weak, secondary magnetic field is produced in proportion to the waning strength of the eddy currents. The receiver (RX) coil voltage measures the time rate of change of the decaying secondary magnetic field, Figure 8.13c. In many TDEM systems, RX voltage measurements are made during the TX off-time when the primary field is absent. The advantage of making off-time measurements is that the relatively weak secondary signal is not swamped by the much stronger primary signal. A good tutorial article on TDEM has been written by Nabighian and Macnae (1991).

During the ramp-off, the induced current assumes the shape of the horizontal projection, or shadow, of the TX loop onto the surface of the conducting ground. The sense of the circulating induced currents is such that the secondary magnetic field they create tends to maintain the total magnetic field at its original steady-on value prior to the TX ramp-off. In this case, therefore, the induced currents flow in the same direction as the TX current, i.e. opposing the TX current decrease that served as the emf source. The image current then

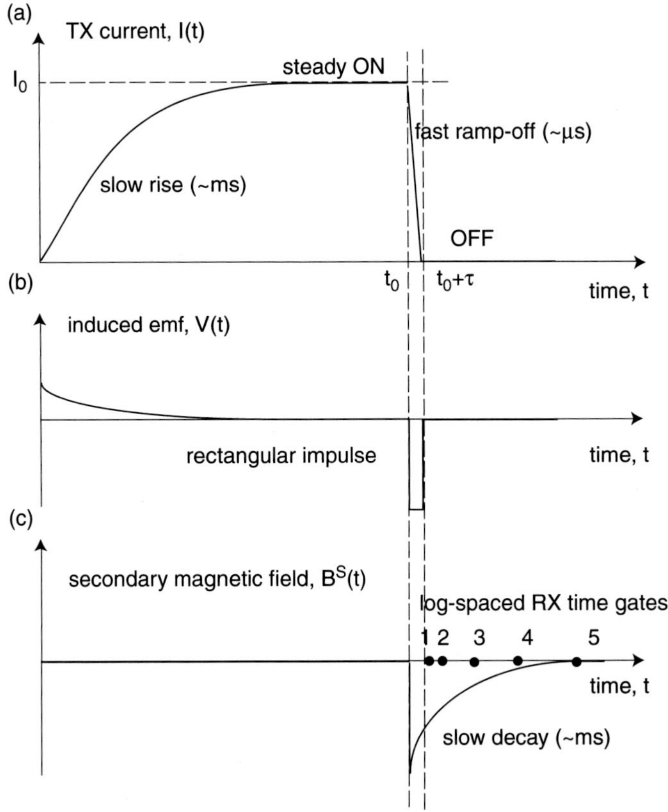

Figure 8.13 (a) A typical TX waveform $I(t)$ with slow rise and fast ramp-off; (b) induced emf $V(t)$ proportional to the time rate of change of the primary magnetic field; (c) the decaying secondary magnetic field $B^S(t)$ due to dissipation of currents induced in the ground.

diffuses downward and outward while diminishing in amplitude. The series of snapshots in Figure 8.14 of the secondary electric field intensity illustrates the transient diffusion of an electromagnetic smoke ring into a uniform halfspace.

In the widely used central-loop sounding method, the RX loop is placed at the center of the TX loop, while in TDEM offset-loop soundings the TX and RX loops are separated by some distance L. As indicated in Figure 8.15, at a fixed instant in time t, the vertical magnetic field $H_z(L)$ at an RX coil due to the underground smoke ring exhibits a sign change from positive to negative as distance L increases. In other words, the vertical magnetic field $H_z(t)$ changes sign from positive to negative as the smoke ring passes beneath a fixed measurement location. It should be clear to the reader that a central-loop measurement of the vertical magnetic field $H_z(t)$ does not exhibit such a sign change.

The "normal moveout" of the sign reversal with increasing TX–RX separation distance L is shown in Figure 8.16. The sharp cusps in the response curves mark the transition between positive and negative flux passing through a RX loop placed at distance L from the TX loop. These solutions correspond to the response of a uniform halfspace of $\sigma \sim 0.1$ S/m, as indicated in the figure legends. Note the late-time responses scale as a power law of

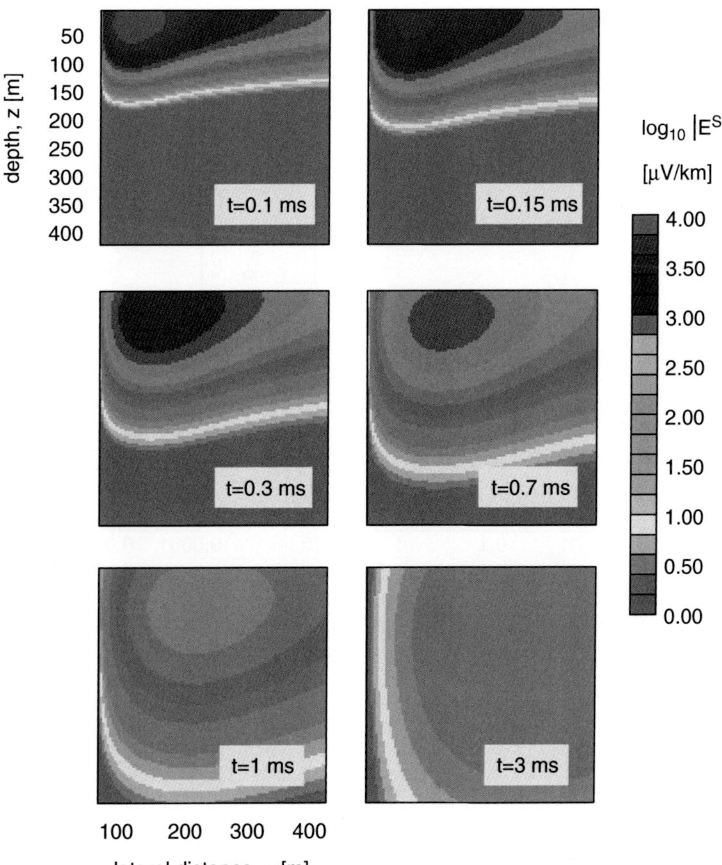

Figure 8.14 Transient smoke-ring diffusion into a uniformly conductive halfspace, $\sigma = 0.1$ S/m. See plate section for color version.

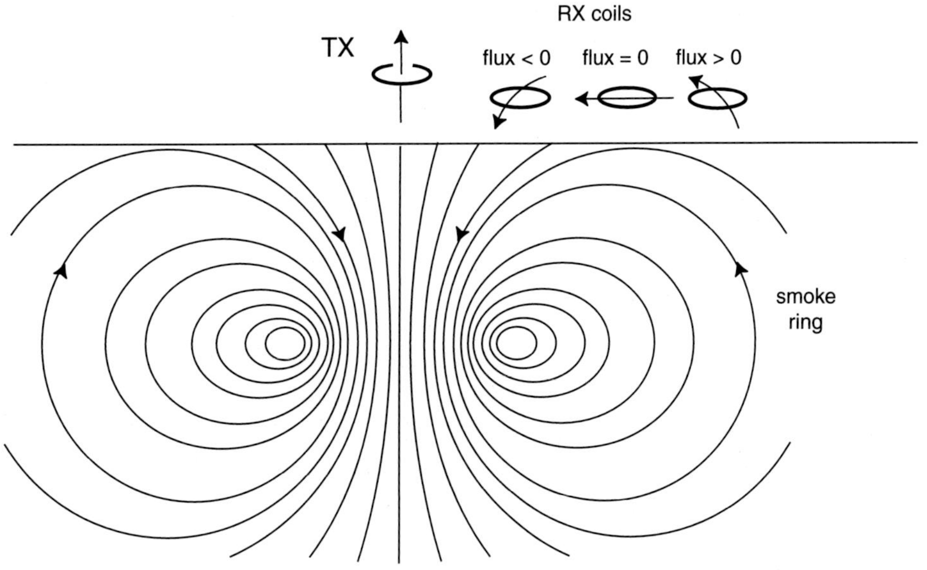

Figure 8.15 The changing sign of magnetic flux from the smoke ring as a function of TX–RX separation. After Nabighian and Macnae (1991).

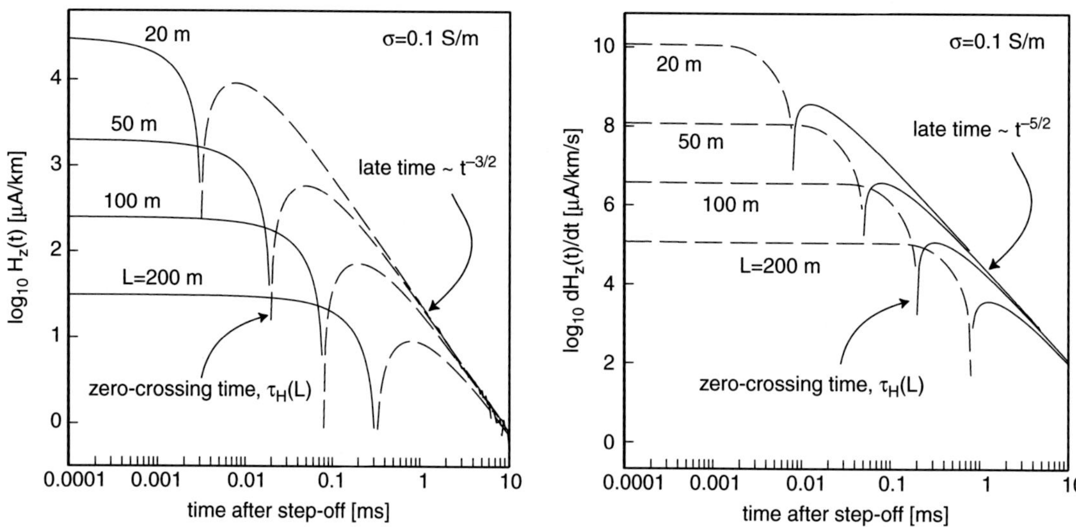

Figure 8.16 (Left) Transient step-off vertical magnetic-field response, horizontal loop source; (right) transient step-off RX voltage response, horizontal loop source.

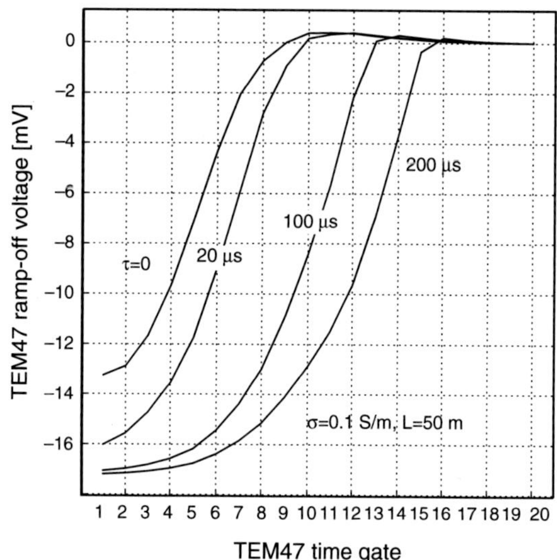

Figure 8.17 Effect of finite ramp-off time on TEM47 response.

the form $t^{-3/2}$ for the magnetic field and, since it is proportional to the time-derivative of the magnetic field, of the form $t^{-5/2}$ for the induction-coil response.

In the ideal TDEM case, step-off transients are analyzed. However, it is impractial in the field to abruptly switch off the current in a finite-sized loop due to self-inductance effects. Consequently, most systems are linearly ramped off over a brief time interval on the order of microseconds. The impact of using a finite TX ramp-off time on the TDEM response is shown in Figure 8.17. For fixed TX–RX separation L, an increase in the ramp-off time τ

serves to delay the response measured at the receiver. The curves in Figure 8.17 are simulated Geonics TEM47 (a popular commercial TDEM instrument) responses that have been computed by convolving the theoretical impulse reponse with a linear ramp-off function of width τ (Fitterman and Anderson, 1987). The curve labeled $\tau = 0$ is the ideal step-off response. The curves are plotted for the logarithmically-spaced times, after initiation of the ramp-off, at which the TEM47 system samples the response.

8.7 Finite-source excitation of a layered Earth

A classic problem of considerable practical importance in near-surface applied geophysics is to determine the electromagnetic response of a plane-layered Earth to finite-source excitation. There are well-known analytic solutions to the Maxwell equations available for finite-sized loops of horizontal and vertical orientation and horizontal grounded sources of finite length; while these are broadly scattered throughout the literature, many are derived in Ward and Hohmann (1988).

In this section we point out some analytic frequency-domain solutions to a few problems of inductive or grounded-source sounding of a layered halfspace. For example, the electromagnetic response of a horizontal loop source of finite radius, like that shown in Figure 8.18, is derived in Morrison *et al.* (1969) and Ryu *et al.* (1970).

From the complete derivation given in Appendix D, the vertical magnetic field recorded by an RX coil placed on the surface $z = 0$ above a homogeneous ground is found to be

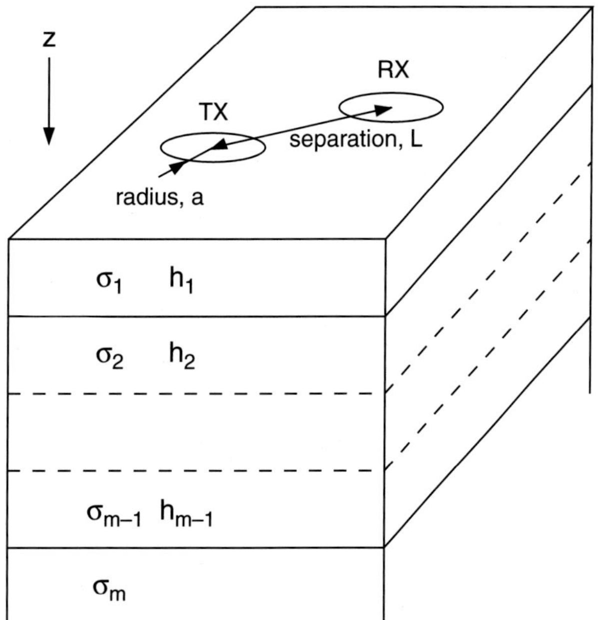

Figure 8.18 Horizontal TX and RX loops on a layered halfspace.

$$b_z(\rho) = \mu_0 I a \int_0^\infty \frac{\lambda^2}{\lambda + i\gamma} J_1(\lambda a) J_0(\lambda \rho) d\lambda, \tag{8.29}$$

where the time-harmonic dependence has been suppressed and now ρ is the TX–RX coil separation. The TX loop radius is a, the TX current is I, and J_0, J_1 are Bessel functions. The parameter $\gamma = \sqrt{\alpha^2 - \lambda^2} = \sqrt{-i\mu_0\sigma\omega - \lambda^2}$ has the dimensions of a wavenumber. A derivation of the solution for the m-layered Earth is left for the reader as an exercise at the end of the chapter.

In most if not all 1-D layered Earth problems, the electric and magnetic fields can be written as Hankel transforms, or integrals over Bessel functions. Various algorithms that can numerically evaluate Hankel tranforms in a fast and reliable manner are readily available. The numerical algorithms are based on convolution, digital filtering, or other specialized quadrature techniques (e.g. Anderson, 1979; Chave, 1983; Guptasarma and Singh, 1997). Transient layered-Earth responses are typically found by inverse Fourier or, occasionally, inverse Laplace transform of the corresponding frequency-domain responses.

The electromagnetic response of a vertical loop of small radius (treated as a horizontal magnetic dipole) over a layered Earth is solved by Dey and Ward (1970). This problem is more difficult to solve than the horizontal-loop problem since it does not possess azimuthal symmetry. Passalacqua (1983) solves the problem of a finite-length grounded horizontal electric dipole deployed over a layered Earth. Finally, Poddar (1983), using a reciprocity principle, develops the solution for a rectangular loop source deployed over a layered Earth. Finally, the article by Spies and Frischknecht (1991) contains a large number of calculated layered-Earth response curves for various combinations of transmitter and receiver coils.

8.8 Plane-wave excitation methods: VLF, RMT, CSMT

To this point, we have been considering electromagnetic geophysical methods that utilize a controlled source of finite extent, most commonly a loop source. The measurements are assumed to be made close enough to the source that the spatial distribution of the primary transmitted field exhibits some curvature. Another class of EM methods has been developed that is based on plane-wave excitation from distant sources. The classical plane-wave method for deep-crustal and upper-mantle exploration is the magnetotelluric (MT) method (Simpson and Bahr, 2005; Chave and Jones, 2012), which is based on excitation by long-period ($f \sim 0.1$ mHz to 300 Hz) geomagnetic disturbances originating principally in the ionosphere at an altitude $h \sim 110$ km, with skin depths $\delta \sim 300$ m to 50 km. The adaptation of plane-wave methods for near-surface investigations has been reviewed by Tezkan (1999). A large number of such methods exists; herein, we focus brief attention on the relevant aspects of the very-low-frequency (VLF), VLF–resistivity (VLF–R), radiomagnetotelluric (RMT), and controlled-source magnetotelluric (CSMT) methods along with a few of their variants.

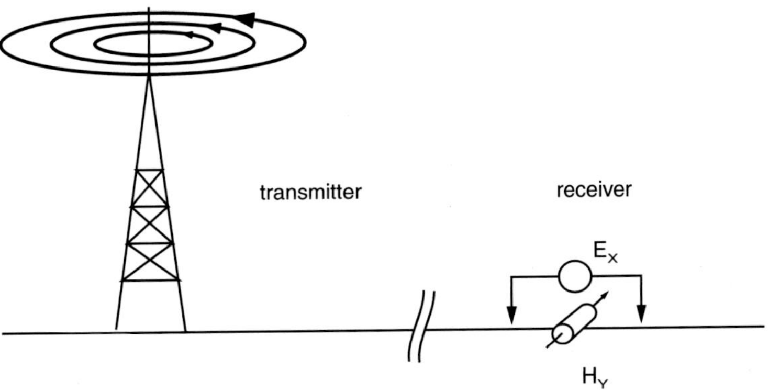

Figure 8.19 The VLF–R method is based on measurements of E_x and H_y signals due to the interaction of transmissions from a distant radio antenna with a variably conductive ground.

The VLF and VLF–R methods take advantage of the electromagnetic field generated by a distant, powerful radio transmitter (the VLF–R configuration is shown in Figure 8.19). Maps showing the locations and global coverage of a number of such transmitters are provided by McNeill and Labson (1991). The carrier signal is typically in the 15–30 kHz frequency range. At great distances from a transmitter, which acts as a vertical electric dipole, the primary field can be approximated as a cylindrical plane wave with non-zero components (E_z, H_ϕ). The transmitted plane wave propagates almost horizontally, arriving to a remote receiver location at grazing incidence.

The small angle of incidence, coupled with the large constrast in electrical conductivity between the air and the ground, results in a refracted plane wave that propagates essentially downward into the conductive Earth. The principal effect, at the surface, of this vertically attenuating plane wave is to generate local, secondary field components such as (E_x, H_z) whose amplitude and phase depends on the terrain conductivity (McNeill and Labson, 1991), and where the $\hat{\mathbf{x}}$ direction is parallel to a line drawn between the transmitter and a receiver deployed at the surface.

In the classical VLF technique (Parasnis, 1997; Sharma, 1997), a single RX coil is tuned to the TX frequency, aligned to the transmitter azimuth, and then rotated up and down until a minimum signal is found. The TE mode is excited if the RX station is located such that the incoming VLF wave propagates along the predominant geological strike direction. If that is the case, then the tilt angle θ at which the minimum signal is found is a measure of the in-phase (real) component of H_z, denoted by $\mathrm{Re}(H_z)$, normalized by the strength of the primary magnetic field H_ϕ,

$$\theta \sim \frac{\mathrm{Re}(H_z)}{|H_\phi|}. \tag{8.32}$$

A distinct tilt-angle anomaly is generated by a lateral discontinuity in terrain conductivity. The tilt angle changes sign over a tabular conductor, as shown in Figure 8.20. A TM mode of excitation is also possible if the transmitter direction is perpendicular to the predominant geological strike. Generally, VLF methods are well suited to detect tabular features such as water-bearing fractures or faults. A number of field examples are presented in McNeill and

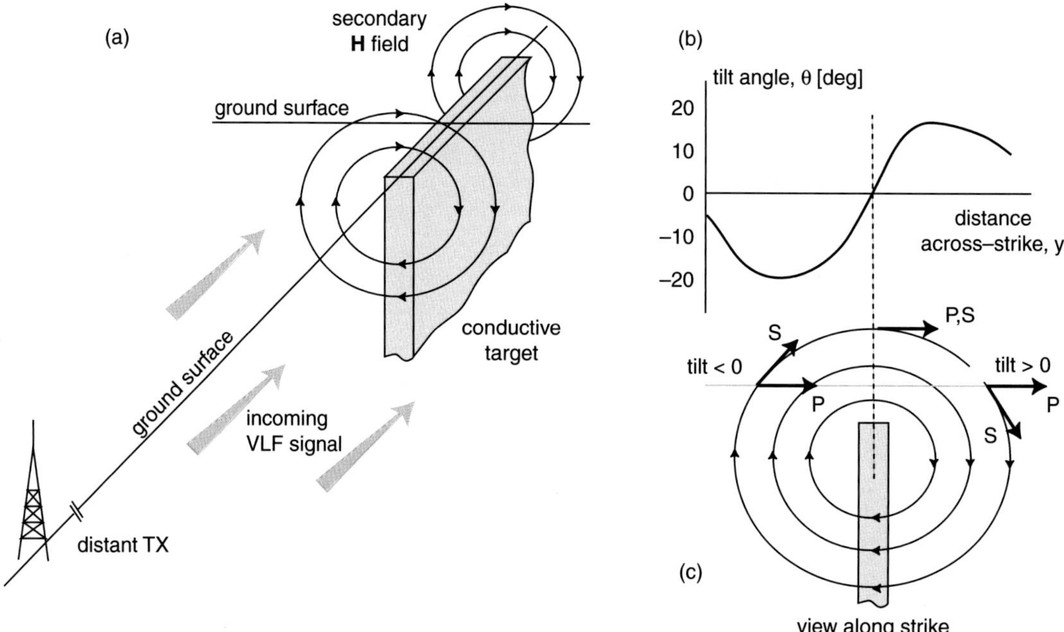

Figure 8.20 VLF (TE mode) exploration of a tabular conductor: (a) the interaction between the incoming plane wave from a distant transmitter and a tabular conductor generates a secondary magnetic field, as shown; (b) the tilt angle as a function of location across the strike of the conductor; (c) explanation of the tilt-angle curve, P = primary magnetic field, S = secondary magnetic field.

Labson (1991). A more recent case study using VLF data to detect faults in a granite quarry is found in Gurer *et al.* (2008).

The quantity measured in the VLF–R method is the ratio of *horizontal* field components E_x and H_y, as suggested in Figure 8.19. The electrodes are normally oriented in the direction of the transmitter. An apparent resistivity ρ_a and phase φ at the observed frequency ω is defined as

$$\rho_a = \frac{1}{\omega\mu_0}\left|\frac{E_x}{H_y}\right|^2; \tag{8.33}$$

$$\varphi = \tan^{-1}\left[\frac{\mathrm{Im}(E_x/H_y)}{\mathrm{Re}(E_x/H_y)}\right]. \tag{8.34}$$

The quantities (ρ_a, φ) are affected by lateral variations in the subsurface conductivity structure and hence contain information about the geology in the intervening region between the transmitter and receiver locations (Beamish, 2000). However, VLF and VLF–R techniques offer poor depth resolution since measurements are made at only one frequency.

Radiomagnetotellurics (RMT) is essentially an extension of the VLF–R method to multiple and higher frequencies, toward the 100 kHz to 1 MHz range (Tezkan, 1999). The RMT method uses simultaneously the transmissions from a number of radio sources

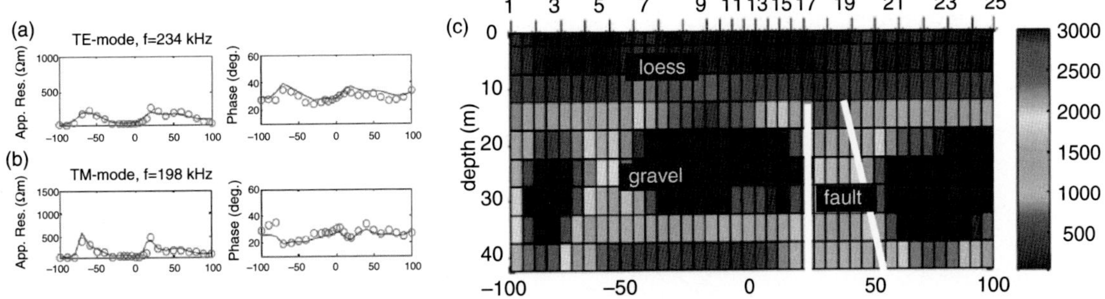

Figure 8.21 RMT survey results from Cologne, Germany: (a) TE-mode apparent resistivity and phase at 234 kHz; (b) TM-mode apparent resistivity and phase at 198 kHz. Blue lines are observed data, while red circles are computed from the resistivity model shown at right; (c) 2-D RMT smooth inversion result. The color bar indicates resistivity ρ [Ωm]. The plot ordinates in all cases is RX location [m]. After Candansayar and Tezkan (2008). See plate section for color version.

distributed over a range of azimuths relative to the survey area. The method is popular in Europe due to the abundance of transmitters and is being used increasingly there and elsewhere for engineering and environmental applications since the skin depths are small, $\delta \sim 5$–10 m. Measurements are made of ratios of horizontal field components and these are converted into apparent resistivity and phase, as in Equations (8.33) and (8.34). The use of multiple frequencies permits the determination of subsurface electrical conductivity profiles with depth. Data from along a profile of measurement sites can be inverted for 2-D subsurface conductivity structure using standard MT inversion routines (e.g. Simpson and Bahr, 2005).

An RMT survey conducted by Candansayar and Tezkan (2008) crossed a near-surface fault outside the city of Cologne, Germany. The geology at the site consists of a conductive wind-blown silt (loess) layer of ~ 15–20 m overlying a resistive gravel unit. The gravel is cut by a steeply dipping, conductive fault of thickness ~ 12–15 m. A total of 25 radio stations spanning the frequency range 10–240 kHz served as the plane-wave sources. The radio stations were located at different azimuths, thus providing both TE- and TM-mode excitation. RMT apparent resistivities and phases were acquired every 5–10 m along a single profile. Representative RMT data are shown in Figure 8.21a, b. The complete dataset was inverted to obtain the 2-D resistivity distribution $\rho(x, z)$ shown in Figure 8.21c. A non-linear, smooth least-squares inversion approach was used; such methods are discussed in Chapter 12. The inversion result shows the moderately conductive fault cutting across the resistive gravel unit.

The controlled-source magnetotelluric (CSMT) methods (Goldstein and Strangway, 1975; Zonge and Hughes, 1991) are similar to the MT and RMT methods except that the source consists of one or more remote grounded dipoles or a large horizontal loop source instead of, respectively, natural ionospheric disturbances or distant radio transmitters (Figure 8.22). The main advantage of a CSMT method is that the geophysicist can control important parameters such as the orientation, location, length/radius, and moment of the transmitter, and the selection of frequencies to be used. A major drawback is that a long, powerful grounded dipole or large horizontal loop can become unwieldy to deploy in the field. As in the RMT method, measurements at selected receiver locations can be made of

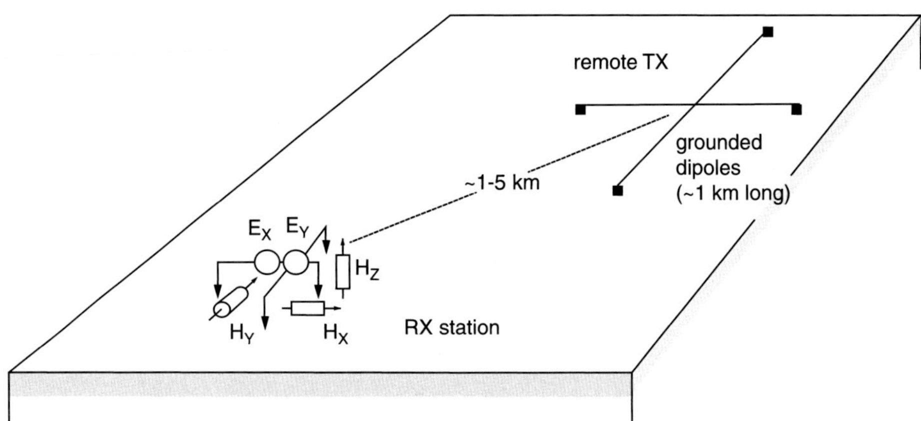

Figure 8.22 A typical tensor CSMT exploration setup with orthogonal TX grounded dipoles and a RX station consisting of surface measurements of 5 EM field components (E_x, E_y, H_x, H_y, H_z).

TE-mode and TM-mode surface impedances that are then converted into apparent resistivity and phase estimates.

In CSMT experiments, the electric and magnetic field receivers should be placed at a sufficient distance from the dipole or loop transmitter such that plane-wave analysis can be used to interpret the data. The plane-wave approximation generally requires that the TX–RX separation distance is ~ 4–20 skin depths in the uppermost medium (Goldstein and Strangway, 1975; Wannamaker, 1997) with the latter value prevailing if the geology consists of conductive sediments over resistive basement. A wide frequency range has been used for CSMT, f ~ 1 Hz–20 kHz, depending on the required depth of signal penetration.

Ismail and Pedersen (2011) combined multi-component (tensor) CSMT (1–12 kHz) and RMT (14–250 kHz) techniques to map the resistivity structure of fracture zones located at ~ 150 m depth within Precambrian crystalline bedrock of Sweden. The RMT data imaged the upper conductive clay layer while the CSMT data successfully probed deeper into the bedrock. A focus of the investigation concerned the applicability of plane-wave, or far-field, techniques since the controlled loop source was located only ~ 300–600 m from the RX profiles. It was found that near-field effects were prominent especially in the low-frequency TM-mode phase responses. These data should be downweighted, or not used at all, if a conventional CSMT 2-D inversion based on a plane-wave source assumption is used. In general, a full 2.5-D (i.e. 3-D-source excitation of 2-D Earth structure) inversion is recommended to interpret data that are strongly influenced by near-field effects (see Figure 8.23).

8.9 Airborne electromagnetics

The airborne electromagnetic (AEM) method has been used for decades in the mining industry as a primary geophysical tool to locate and delineate economic mineral deposits. There have been significant recent developments in near-surface AEM applications, as reviewed by

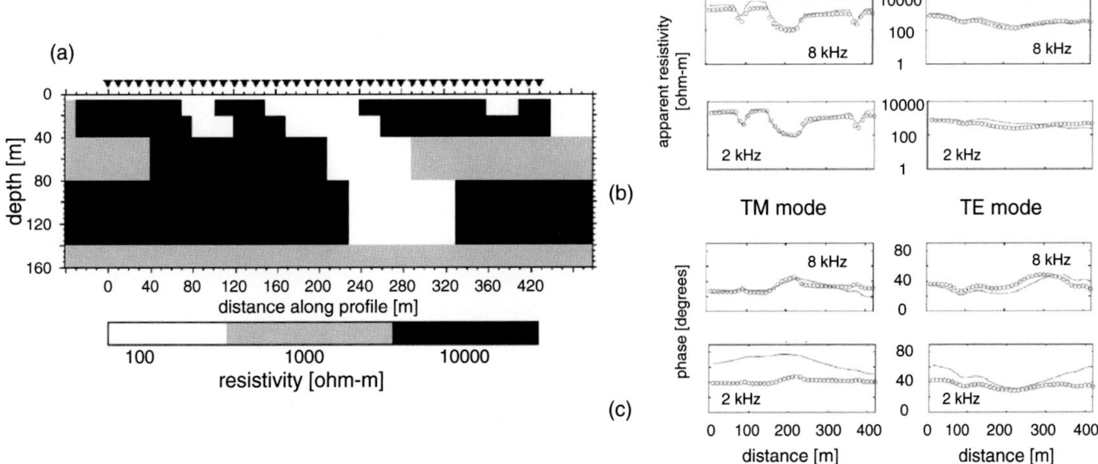

Figure 8.23 Test of the CSMT plane-wave assumption. The 2-D resistivity model in (a) is excited by TE- and TM-mode plane-wave sources and a loop source located 300 m along strike. Computed responses at 2 and 8 kHz, in the form of apparent-resistivity and phase profiles, are shown in (b) and (c), respectively. The circles show the plane-wave responses while the solid lines show the 2.5-D controlled-source responses. As mentioned in the text, the most significant differences appear in the low-frequency TM-mode phase responses. After Ismail and Pedersen (2011).

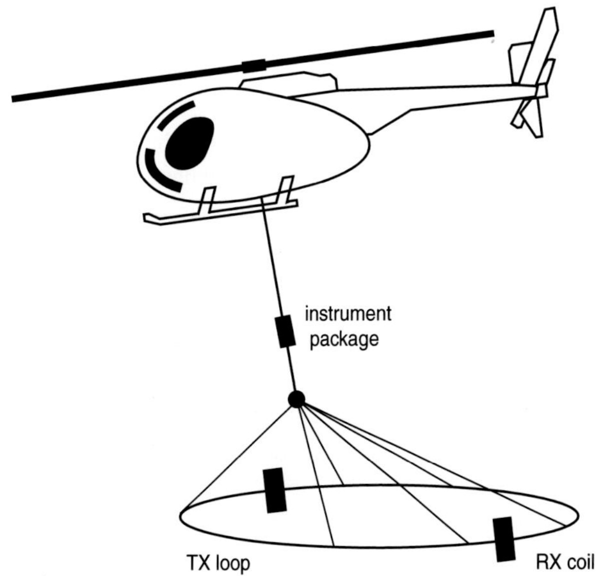

Figure 8.24 A helicopter AEM system.

Auken *et al.* (2006). In most AEM systems a transmitter loop is either mounted on or towed beneath a helicopter or a fixed-wing aircraft and at least one receiver coil is either mounted on or towed behind or beneath the aircraft along with an instrument package sometimes called the bird. A simplified illustration of an AEM system is shown in Figure 8.24.

The operating principles of an AEM system are essentially the same as those of ground-based loop–loop electromagnetic systems, described earlier. Both frequency-domain and time-domain systems are in present use. AEM data processing must take into account a number of important sources of noise including in-flight variations of the altitude and tilt of the mechanical assembly that supports the TX loop and RX coils. The processed data are then usually interpreted based on 1-D layered-Earth forward modeling combined with some type of stable, regularized least-squares inversion (see Chapter 12). In the widely used spatially constrained inversion (SCI) method (Viezzoli *et al.*, 2008), for example, smoothness constraints are enforced that prevent the introduction of flight-line artefacts and other unphysical, sharp lateral changes in the reconstructed subsurface electrical conductivity structure. Examples of the use of AEM systems for hydrogeological investigations are provided at the beginning and the end of this chapter.

8.10 EM responses of rough geological media

The EM method provides valuable non-invasive information about the subsurface electrical-conductivity distribution for a range of applications in hydrology, such as groundwater-resource evaluation, contaminant-transport investigation, and the characterization of aquifer heterogeneity. While traditional approaches treat electrical conductivity as a piecewise smooth medium, perhaps a more efficient and realistic description of conductivity can be attained if the inherent roughness of the underlying geological medium is explicitly recognized. The spatial roughness of fractured rocks, and the attendant implications for subsurface fluid transport, has been carefully analyzed by hydrologists (e.g. Painter, 1996; Bonnet *et al.*, 2001; Berkowitz and Sher, 2005). Rough geology is characterized by patterns of spatial heterogeneity that vary over a wide range of length scales.

To explore the electromagnetic response of such a medium, EM-34 measurements were made on the campus of Texas A&M University in Brazos County, as reported in Everett and Weiss (2002). The geological section consists of floodplain alluvium, mainly interbedded sands, silts, and clays with occasional coarse sand and gravel lenses and paleochannels. An azimuthal apparent-conductivity profile $\sigma_a(\theta)$ was acquired (Figure 8.25a) by fixing the position of the TX loop and moving the RX loop in a complete circle around the TX loop at fixed radius 40 m. The apparent conductivity was read at 512 RX azimuths, so that station spacing is 0.5 m. The data, shown in Figure 8.25a, are highly irregular. The traditional interpretation of the EM response as being caused by a piecewise smoothly varying conductivity structure superposed with random, uncorrelated noise is not supported this dataset. Instead, the responses are examples of fractional Brownian motion (fBm), a class of non-stationary signals that describe self-similar processes with long-range correlations (Mandelbrot and van Ness, 1968). An fBm signal is further characterized by stationary increments.

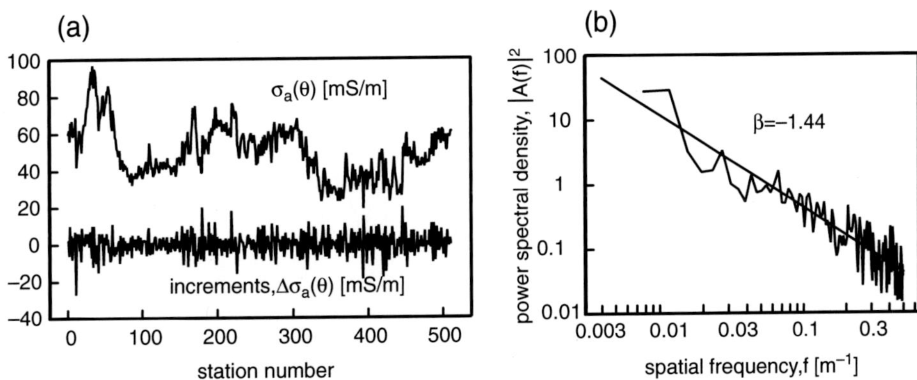

Figure 8.25 (a) EM34 apparent-conductivity measurements from the Brazos River (TX) floodplain; (b) power spectrum of the apparent-conductivity data. After Everett and Weiss (2002).

While the response σ_a in Figure 8.25a is clearly non-stationary, i.e. the mean value fluctuates along the profile, the increments $\Delta\sigma_a(\theta)$, or differences between consecutive data points, appear to be stationary with nearly zero mean over the entire profile. Thus, the apparent-conductivity profile suggests a fractal interpretation of electromagnetic induction in the Earth. This suggestion is further supported by the power spectral density (PSD) function. The PSD of the EM34 σ_a profile is shown in Figure 8.25b. A power-law of the form $|A(f)|^2{\sim}f^\beta$ over 2.5 decades in the spatial wavenumber f is evident. The slope is $\beta = -1.44 \pm 0.06$. A value of $|\beta| > 1.0$ indicates a non-Gaussian fractal signal (Eke *et al.*, 2000). For comparison, the PSD of the increments $\Delta\sigma_a$ has slope $\beta = 0.43 \pm 0.06$ which indicates that the increments are a realization of fractional Gaussian noise, that is, a stationary signal without significant long-range correlation. A pure white-noise process would have a PSD slope of exactly $\beta = 0$.

The foregoing analysis shows that spatial variations in the EM response due to rough geological structure may be characterized by a power-law wavenumber spectrum. The presence of a buried man-made conductive target, with a regular geometry, should generate a departure from a power-law spectrum. This was shown by Benavides and Everett (2005). In their experiment, EM31 apparent-conductivity readings were acquired along a profile of length 152.0 m with 1.0 m station spacing. A buried metal drainage culvert was located where the profile crossed a gravel road. The power spectra of the EM31 data are shown in Figure 8.26. Analyzing the data from the first part of the profile, which contains only the background geology but not the culvert, yields the familiar power-law spectrum (gray). The PSD slope is $\beta = 1.417$, corresponding to an fBm signal. The spectrum shown in black is from the entire profile, including the culvert. It shows anomalously high power in the wavenumber range $f \sim 10$–$40\ \text{m}^{-1}$. This anomalous peak in the power spectrum is indicative of a compact, deterministic signal overlying the purely geologic fBm signal.

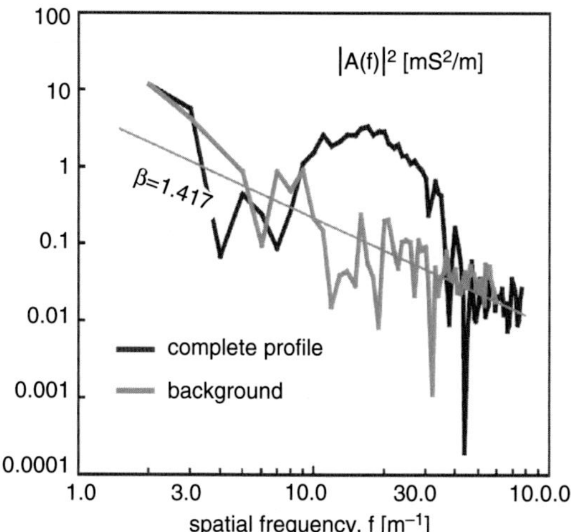

Figure 8.26 The departure from a power-law spectrum due to a buried conductive target. After Benavides and Everett (2005).

8.11 Anisotropy

Near-surface geophysicists now widely recognize that geological formations, as a rule, are electrically anisotropic. Most studies of electrical anisotropy for near-surface (i.e. depths < 100 m) geophysical applications are based on azimuthal DC resistivity methods, as described earlier in Chapter 4. The EM method provides an alternative probe of anisotropic formations that includes the option of using an inductive or a grounded source. The variation of electrical anisotropy with depth can also be explored by using multiple frequencies or else a transient excitation.

In general, a minimum of two scalar parameters is required to characterize electrical anisotropy. These parameters may be termed the along-strike conductivity $\sigma_{||}$ and the across-strike anisotropy $\sigma_{\perp}$. The strike of the anisotropy is then the direction in which the conductivity is different from the isotropic conductivities of the other two directions. A familiar example is the interbedding of sediments, described by an electrical conductivity tensor of the form

$$\sigma = \text{diag}(\sigma_h, \sigma_h, \sigma_v), \tag{8.35}$$

in which the vertical conductivity σ_v differs from the isotropic horizontal conductivity σ_h. Any medium that has two distinct conductivities can be characterized by a *uniaxial* tensor. The reader should recall that the various geological realizations of a uniaxal conductivity tensor were illustrated in Figure 4.12. A *triaxial* conductivity tensor is more general; it is one in which the electrical conductivity is different in all three principal directions, that is $\sigma = \text{diag}(\sigma_x, \sigma_y, \sigma_z)$.

We restrict our attention in this section to horizontal anisotropy for which the electrical conductivity tensor is uniaxial and has the form

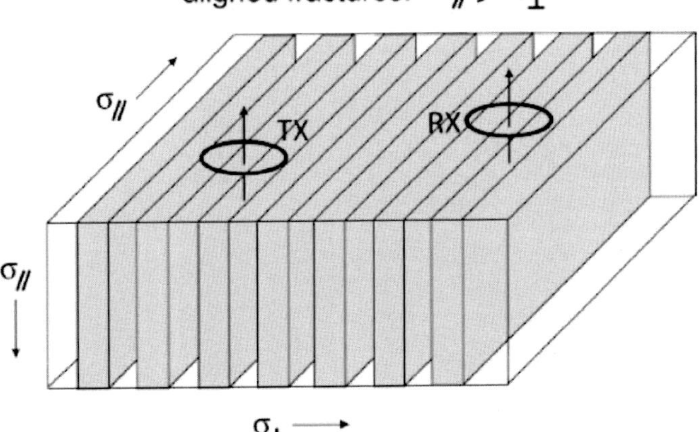

Figure 8.27 Loop–loop electromagnetic sounding over a uniform uniaxial conducting halfspace. After al-Garni and Everett (2003).

$$\sigma = \text{diag}(\sigma_\perp, \sigma_{||}, \sigma_{||}), \tag{8.36}$$

such that the strike is horizontal; specifically, it is in the $\hat{\mathbf{x}}$ direction. The conductivities in the two across-strike directions, $\hat{\mathbf{y}}$ and $\hat{\mathbf{z}}$, are identical. In an early study of such a horizontally anisotropic medium, Le Masne and Vasseur (1981) explored the frequency-domain electromagnetic response of a limestone formation pervaded by parallel vertical fractures. Yu and Edwards (1992) later derived theoretical CSEM responses of uniaxial media characterized by horizontal anisotopy.

Al-Garni and Everett (2003) studied the loop–loop frequency-domain response of an idealized set of vertical fractures for which $\sigma_{||} > \sigma_\perp$ (see Figure 8.27) and they found an electromagnetic paradox of anisotropy in which apparent conductivity σ_a, as defined by Equation (8.27), actually reads higher when the TX and RX loops are aligned across the strike of the anisotropy than when they are aligned along the strike. The paradox is resolved by recognizing that the electromagnetic response is not controlled by the conductivity aligned with TX–RX azimuth, but rather by the conductivity in the direction of the induced current flow beneath the RX loop.

The steeply dipping, weathered, foliated Packsaddle schist formation of the Llano Uplift in central Texas provided the ideal conditions for a field-scale azimuthal TDEM experiment (Collins *et al.*, 2006) to detect horizontal anisotropy. The clay weathering products aligned with the foliations determine the most conductive direction. TEM47 voltage readings were made with a TX loop of 5 m radius held fixed at a central location while the RX coil was moved in a circular path, maintaining a constant TX–RX offset of 40 m. The observed TEM47 responses (Figure 8.28, left) vary with the TX–RX azimuth in the characteristic two-lobed pattern that is predicted by anisotropic forward modeling (Figure 8.28, right). The longer of the two principal axes is aligned at azimuth 146°, precisely parallel to the observed strike direction of the foliations. Isotropic forward modeling predicts that the early-time voltage (the first few time gates) should read higher

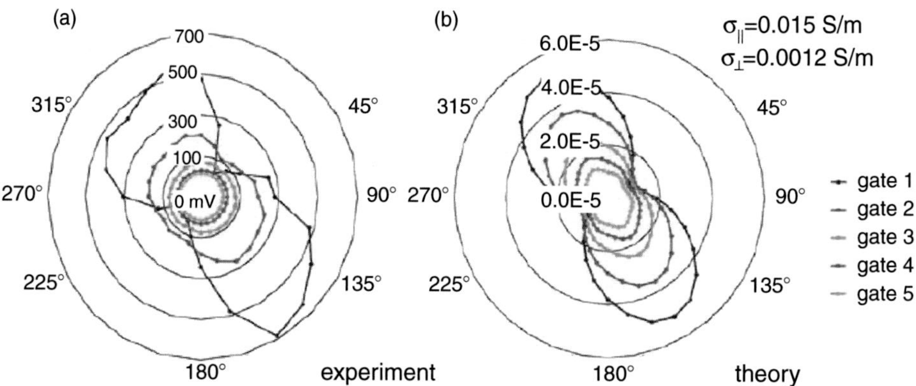

Figure 8.28 (Left) Azimuthal TEM47 voltage readings (first five time gates) over Packsaddle schist, central Texas; (right) calculated response [mV] of the horizontal anisotropic halfspace with $(\sigma_\perp, \sigma_{||})$, as shown; strike direction 146°. After Collins *et al*, (2006). See plate section for color version.

in a resistive medium than in a conductive medium. Naively, it would then hold that the longer principal axis should point in the most resistive direction. However, the paradox of anisotropic responses (al-Garni and Everett, 2003) predicts that the longer axis actually points in the most conductive direction, as observed in the field experiment. The two conductivities found by trial-and-error fit of the theoretical response to the observations are $\sigma_{||} = 0.015$ S/m and $\sigma_\perp = 0.0012$ S/m, which represent reasonable values for resistive metamorphic rocks. Moreover, the presence of the weathering products raises the bulk formation conductivity by more than order of magnitude in the foliation direction.

8.12 Illustrated case histories

Example. Arsenic contamination in Bangladesh.

This case study is an investigation into arsenic concentrations in Bangladesh by Aziz *et al.* (2008). It is well established that groundwater pumped from wells across the Bengal Basin often contains a hazardous (defined herein as > 50 μg/L) level of arsenic. The source of the arsenic is difficult to ascertain in view of the strongly heterogeneous fluvio-deltaic geology of the region. A 25-km^2 survey area was chosen in which $\sim$ 5000 arsenic concentration values from shallow (< 22 m) Holocene aquifers were available.

Loop–loop frequency-domain electromagnetic data were then acquired to discriminate subsurface zones of clays and sands in an attempt to better understand the observed spatial variability of the arsenic concentration data. About 18 500 σ_a readings were made at $\sim$ 4–8-m station spacing along a number of transects, typically 100 m–1 km in length, distributed across the survey area. A spatially continuous σ_a map was constructed by kriging

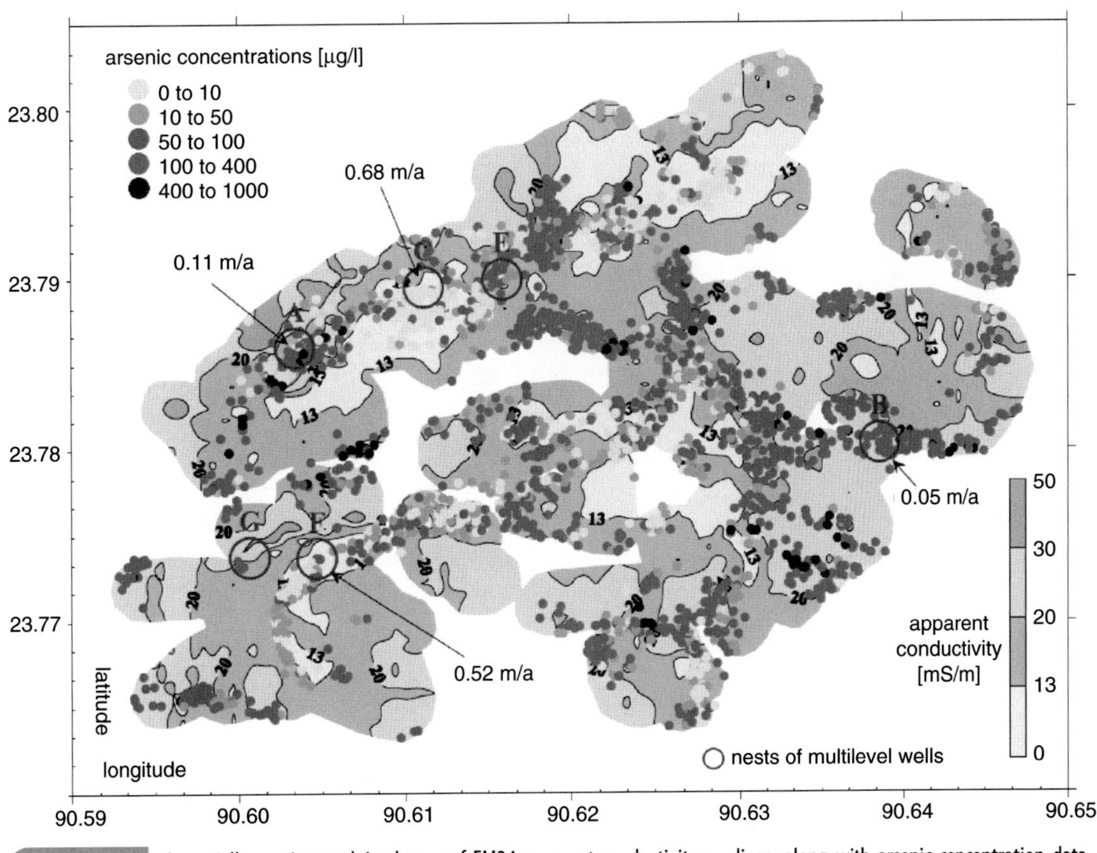

Figure 8.29 A spatially continuous kriged map of EM34 apparent-conductivity readings, along with arsenic concentration data from shallow wells, Bengal Basin. After Aziz *et al.* (2008). See plate section for color version.

(Figure 8.29). It is found that 73% of the wells show non-hazardous arsenic levels in the areas of low $\sigma_a < 10$ mS/m values, which are presumably sand-dominated regions. However, only 36% of the wells show non-hazardous arsenic levels in the high $\sigma_a > 10$ mS/m, clay-dominated regions. The σ_a values were also found to correlate with the depth gradient of arsenic concentration within a well. Overall, the findings suggest that the clay-dominated areas pose higher risk, since groundwater recharge percolating downward through sandy soils tends to prevent arsenic concentrations from rising to hazardous levels.

Example. CSMT at a radioactive waste disposal site.

The CSMT method in the audio frequency range (sometimes termed the CSAMT method) was used by Unsworth *et al.* (2000) to characterize the subsurface hydrological conditions beneath a proposed radioactive waste-disposal site at the Sellafield nuclear facility located

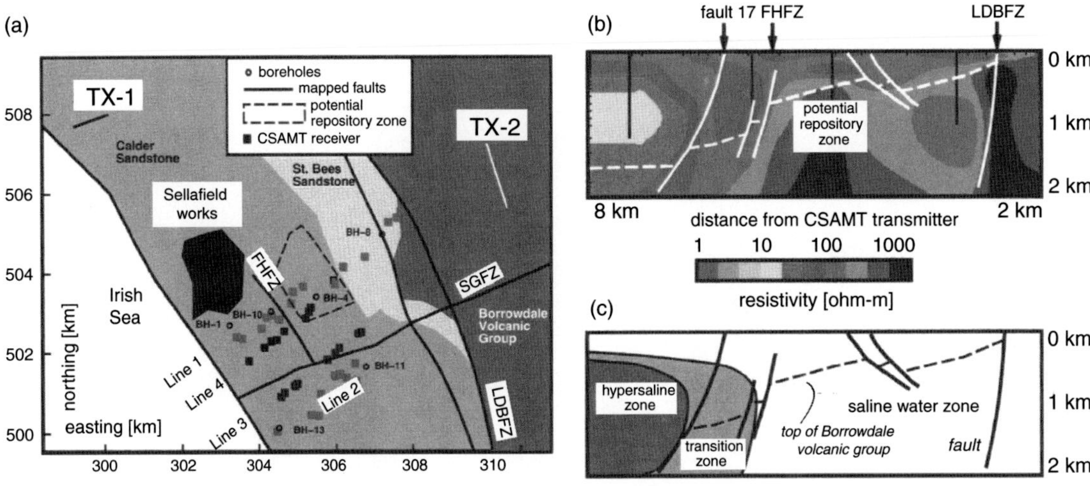

Figure 8.30 CSMT at the Sellafield (UK) nuclear facility: (a) layout of TX grounded dipoles and RX locations with respect to mapped faults and the power plant; (b) 2-D inversion result from Line 1; (c) hydrogeological interpretation. After Unsworth *et al.* (2000). See plate section for color version.

in coastal northwest England. The CSMT data were affected by intense cultural noise due to the infrastructure associated with the power plant. Nevertheless, the apparent resistivity and phase responses were found to be interpretable in terms of a 2-D resistivity section, with strike direction parallel to the coast. The two orthogonal, grounded dipole transmitters TX-1 and TX-2 (frequency range: 1 Hz–4 kHz) are shown in Figure 8.30a, along with the four CSMT RX profiles. The inverted resistivity model for Line 1, which crosses the proposed repository zone, is shown in Figure 8.30b. Known fault locations are overlain for reference. A hydrogelogical interpretation of the resistivity section is shown in Figure 8.30c. The low-resistivity zone (yellow region) is caused by hypersaline groundwater associated with the Irish Sea extending ~ 1 km inland. Groundwater modeling suggests that the brines are likely to remain to the west of the proposed nuclear waste repository but, over geologic time, the site may be negatively impacted by migrating hypersaline waters.

Example. Airborne hydrogeophysics.

A helicopter transient AEM survey with ~ 300 m penetration depth was carried out by Auken *et al.* (2009) over a 190 km^2 area of Santa Cruz volcanic island in the Galapagos archipelago (Figure 8.31). The island has few freshwater resources; a fast-growing population; and unique, pristine ecosystems. Very little, however, is known about the hydrogeology. The acquisition of geophysical data improves this understanding and assists in the development of effective water-resource management strategies. The main exploration target is conductive zones perhaps indicative of water or clay confining layers that are located within the resistive (> 1000 Ω m) volcanic terrain. A 3-D spatially constrained

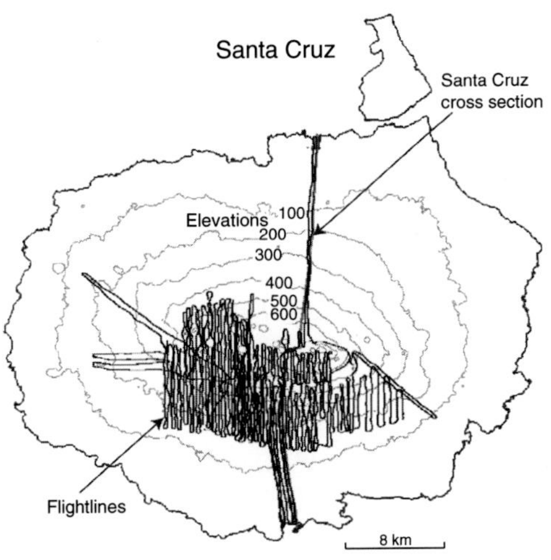

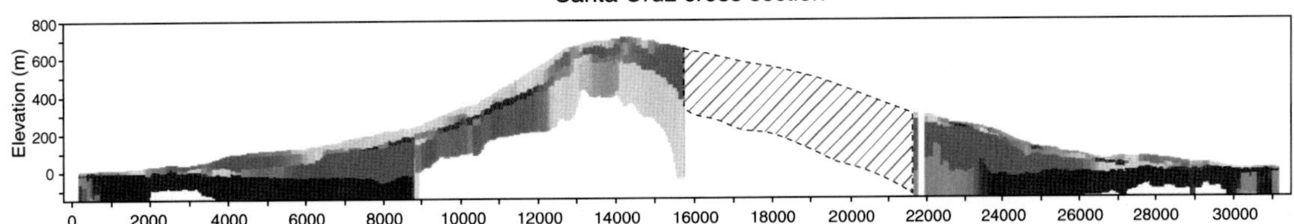

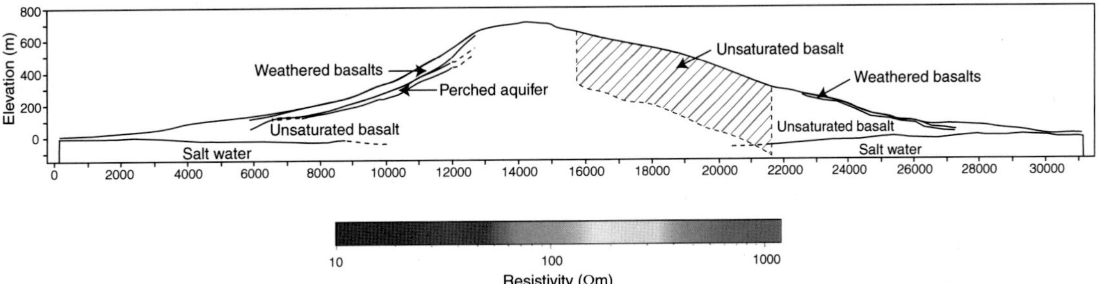

Figure 8.31 (Top panel) SkyTEM flight lines over Santa Cruz island. Contour levels in meters above sea level; (middle panel) resistivity profile along the north–south cross-section; (bottom panel) hydrogeological interpretation of the AEM data, leeward side at right, windward side at left. After Auken *et al.* (2009). Copyright © 2009 Académie des sciences. Published by Elsevier Masson SAS. All rights reserved. See plate section for color version.

series of 1-D local inversions was used to determine the subsurface conductivity structure. The northern leeward side of the island is found to exhibit very low conductivity and thus has a low hydrogeological potential. A perched aquifer at ~ 80–100 m depth may exist on the windward side. The AEM inversions also suggest that intruded seawater underlies the periphery of the island. The geophysical data cannot resolve the presence of possible freshwater lenses above the saltwater.

Problems

1. The analysis by Wait (1954) indicates that the EM response of a small wire loop lying on a homogeneous halfspace is given by Equation (8.24). Verify the approximate relation (8.26) holds at low induction number $\rho/\delta << 1$. Hint: expand the exp[iαρ] term in a power series.
2. The Bessel functions $J_0(x)$ and $J_1(x)$ appear frequently in EM calculations and hence it is important to gain some experience with them. As a first exercise in using Bessel functions, verify the Bessel function identity

$$\frac{\partial}{\partial \rho}\left[\frac{1}{\rho}\frac{\partial}{\partial \rho}\{\rho J_1(\lambda\rho)\}\right] = -\lambda^2 J_1(\lambda\rho)$$

by comparing it with Bessel's differential equation.
3. Show that the vertical magnetic field at distance ρ from a horizontal TX loop of radius a carrying time-harmonic current I and deployed over an m-layered Earth in which the conductivity and thickness of the i-th layer is σ_i and h_i, respectively, is given at the surface $z = 0$ by

$$b_z(\rho) = \mu_0 I a \int_0^\infty \lambda \frac{Z^1}{Z^1 + Z_0} J_1(\lambda a) J_0(\lambda\rho) d\lambda,$$

$$Z^i = Z_i \frac{Z^{i+1} + Z_i \tanh\,(i\gamma_i h_i)}{Z_i + Z^{i+1} \tanh\,(i\gamma_i h_i)},\ \text{for}\ i = 1, \ldots, m,$$

and $Z_i = e_\varphi / b_\rho = -\omega/\gamma_i$ is the intrinsic impedance of the i-th layer.

The above recursion can be developed starting from the m-th or basal layer where $Z^m = Z_m$. Hint: write $e_\varphi = A\ \sinh\,(i\gamma z) + B\ \cosh\,(i\gamma z)$ for the electric field in each layer and use Faraday's law $b_\rho = \partial_z e_\varphi / i\omega$.

9 Ground-penetrating radar

While seismic-reflection and -refraction techniques are commonly employed to map near-surface layers, they do not have the high vertical resolution (detection of subsurface structures with length scales of 1.0 m or less) that is required for many applications. Ground-penetrating radar (GPR) can be a suitable geophysical tool in these situations. The technique is used to detect changes in subsurface electromagnetic impedance via the propagation and reflection at impedance boundaries of an electromagnetic wave generated by a transmitter deployed at the surface or, less commonly, within a borehole. Typical GPR frequencies are in the 10 MHz to 1 GHz range, much higher than the frequencies used in the electromagnetic (EM) induction method (see Chapter 8). The popularity of GPR as a near-surface geophysical technique lies partially in the similar appearance of radar sections to the seismic sections that are familiar to many geophysicists (Figure 9.1). Both seismic reflection and GPR are imaging techniques based on wave-propagation principles but there are important differences; these will be discussed in this chapter. Good overviews of the theory and practice of GPR appear in Davis and Annan (1989), Knight (2001), Neal (2004), Annan (2009), and Conyers (2011).

Example. Perchlorate transport in karst.

The occurrence of the perchlorate ion ClO_4^- in groundwater presents a great risk to human health since perchlorate has long been known to inhibit proper functioning of the thyroid. Beneath the Naval Weapons Industrial Reserve Plant (NWIRP) in central Texas, significant concentrations of perchlorate ions derived from the manufacture of rocket propellant have been detected in groundwater and springs. Hughes (2009) has described a wide-area (~ 500 ha) GPR survey in karst terrain with the goal of mapping subsurface structural features that might be indicative of major pathways for subsurface transport of perchlorate ions. The survey was executed by towing a 50 MHz GPR system for ~ 100 line-km on a sled behind an all-terrain vehicle. The geology consists of a weathered limestone bedrock below a 0–3-m clay overburden. The lateral resolution of the GPR, about 1 m, is far too coarse to detect individual bedrock fractures that are on the order of millimeters in width. The geophysical targets were therefore identified as top-of-bedrock irregularities, such as weathered downcuts of 1–10 m in width and tens of meters in length, that can be spatially associated with the north-northeast-trending regional structure.

A typical GPR profile from the wide-area survey is shown in Figure 9.2a. On the left side of the profile, the main return signal seen at 20–40 ns is interpreted as a reflection from the top of bedrock at 1–2 m beneath the surface. The bedrock signal is lost at the right side

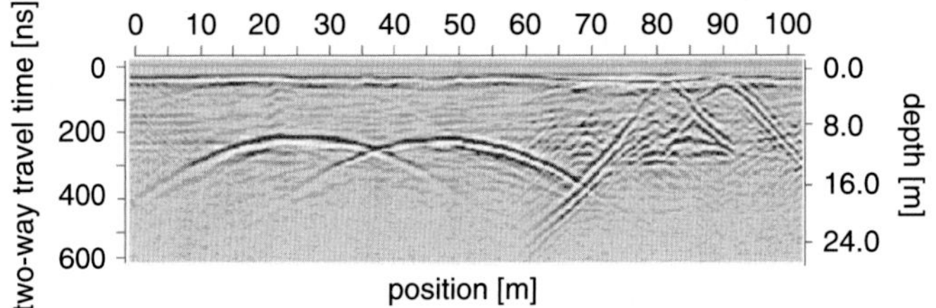

Figure 9.1 GPR section obtained with 50 MHz antennas showing reflection hyperbolas from two road tunnels. The two smaller hyperbolas at the right are caused by scattering from shallow buried objects. After Annan and Davis (1997).

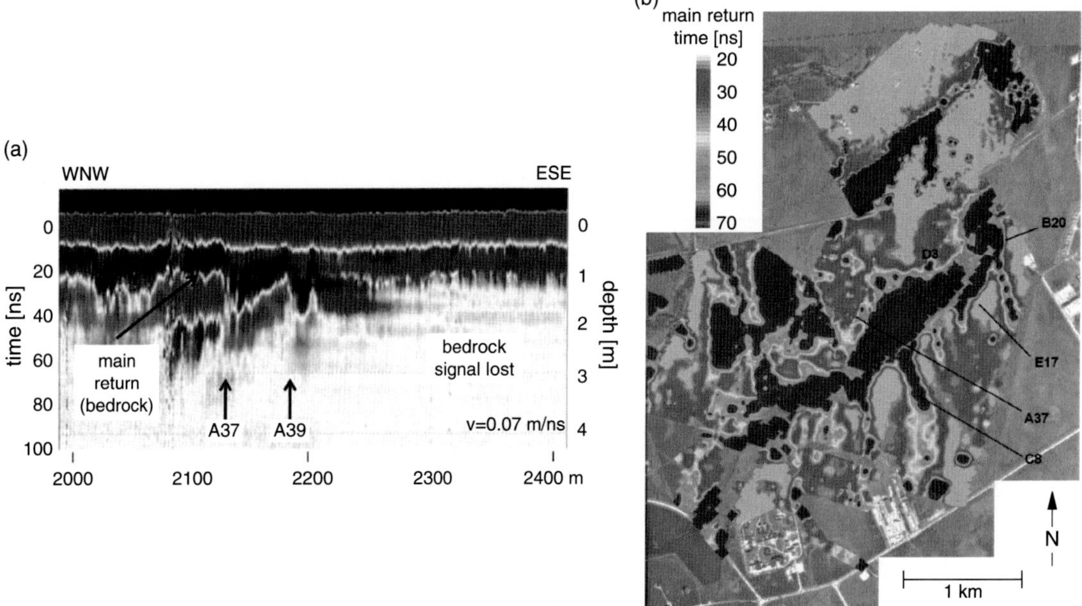

Figure 9.2 (a) Typical GPR section obtained at the NWIRP site, Texas, using 50 MHz antennas; (b) plan-view map of the GPR main returns, interpreted in terms of depth to bedrock. After Hughes (2009). See plate section for color version.

of the profile due to signal attenuation within a deeper clay overburden. The features marked A37 and A39 are interpreted as bedrock lateral discontinuities. A plan-view map of the two-way traveltime to the GPR main return, assembled from all the GPR profiles in the survey area, is shown in Figure 9.2b. This map can be interpreted as a map of depth to bedrock, with the magenta regions corresponding to bedrock highs and the blue regions corresponding to bedrock lows. Notice that the dominant orientation of the GPR-inferred bedrock depth map correlates well with the north-northeast geological structural trend. This lends support to the supposition that GPR can provide valuable information about structural controls on subsurface contaminant transport.

Example. Plastic landmine detection.

Landmines and other explosive remnants of war constitute an enduring and severe environmental hazard in dozens of countries that have previously experienced military conflict. Many landmines are now manufactured that contain little or no metal so that metal detectors, long the primary geophysical tool used by humanitarian deminers, often become ineffective. Alternative geophysical techniques, or combinations of techniques, are currently under intense investigation by various commercial, governmental, and non-profit research organizations around the world. The interested reader is invited to explore the United Nations website www.mineaction.org for further information about all aspects of humanitarian demining.

Metwaly (2007) describes GPR imaging results from a 2.5-m × 6.0-m test site in which five plastic cylinders (Figure 9.3) of similar dimensions (nominal radius 4–5 cm; nominal height 7–13 cm) and construction to anti-personnel landmines were buried up to 22 cm depth in a prepared bed of homogeneous dry sand. It is easy to see from the reflection hyperbolas in the GPR sections that the highest frequency (1.5 GHz) affords the best spatial resolution of the buried objects. The objects labeled 1 and 5 are not well imaged, especially at the lowest frequency, 400 MHz. It may be concluded from this and other studies that plastic, non-metallic landmines may be identified using GPR techniques provided the following conditions hold: (a) there is a sufficiently strong dielectric contrast between the landmine and the host soil; (b) the signal is not overwhelmed by cultural noise (clutter) or shallow subsurface heterogeneities; and (c) the soil electrical conductivity is low, as would likely be the case, for example, in a dry, coarse-grained sedimentary environment with low organic content.

9.1 Fundamentals

In the frequency range 10 MHz $< f <$ 2 GHz, electromagnetic wave propagation in non-magnetic, resistive Earth materials ($\sigma <$ 0.01 S/m) is controlled largely by spatial variations of dielectric permittivity ε in the subsurface. Bound-charge displacement, or *polarization*, is the dominant mechanism although the quasi-free charge migration, or *conduction*, that governs the EM induction technique can play an important role in GPR signal attenuation.

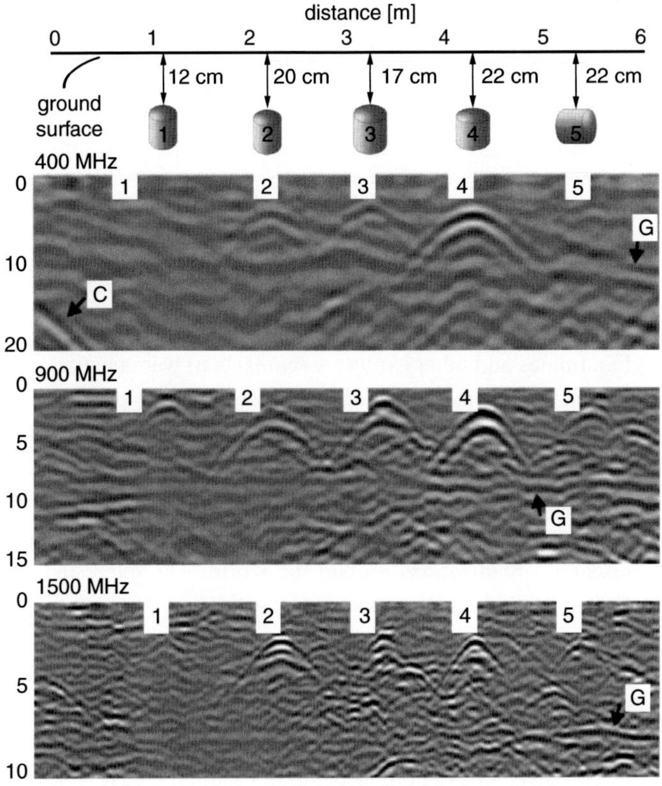

Figure 9.3 GPR detection at three frequencies of plastic-landmine-simulating cylinders buried in homogeneous sand at test site in Egypt; C = unwanted diffraction from an off-site metal sheet; G = reflection from the base of the sand layer; vertical axis is traveltime [ns]. After Metwaly (2007).

The following discussion provides an elementary introduction to the phenomena of atomic, molecular, and interfacial polarization as they are relevant to GPR. Consider an isolated atom consisting of a nucleus of positive charge $+Z$ surrounded by a neutralizing electron atmosphere of charge $-Z$. An applied electric field **E** exerts a force on the electron atmosphere and displaces its charge center, as shown in Figure 9.4a. In essence, the circular orbits of electrons become elliptical (von Hippel, 1954). Similarly, in the presence of an applied field **E**, a polar water molecule will experience a torque which tends to align its *asymmetric* charge distribution into the direction of the applied field, as shown in Figure 9.4b. There is often a small "dielectric loss" caused by some energy that dissipates as the polar molecule rotates, since water is a viscous solvent. The loss term can often be neglected to first order. A non-polar molecule, such as the oxygen molecule O_2, does not exhibit significant molecular polarization due to its more symmetric distribution of positive and negative charges.

In the GPR frequency range, it is the molecular polarization of the water molecule that largely controls the velocity, and hence the reflection, diffraction, scattering, and other aspects of the subsurface propagation of electromagnetic waves. Atomic polarization does not become an important source of polarization until very high frequencies, greater than

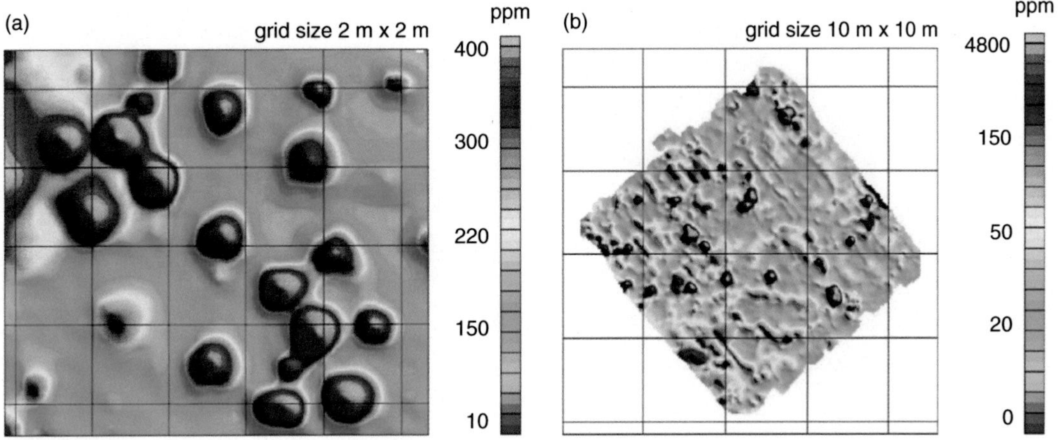

Figure 1.3 (a) EM induction response of metal pipes buried at a test site. (b) EM induction response of unexploded ordnance (UXO) at a live site with highly magnetic soil, Kaho'olawe, Hawaii. From Huang and Won (2003).

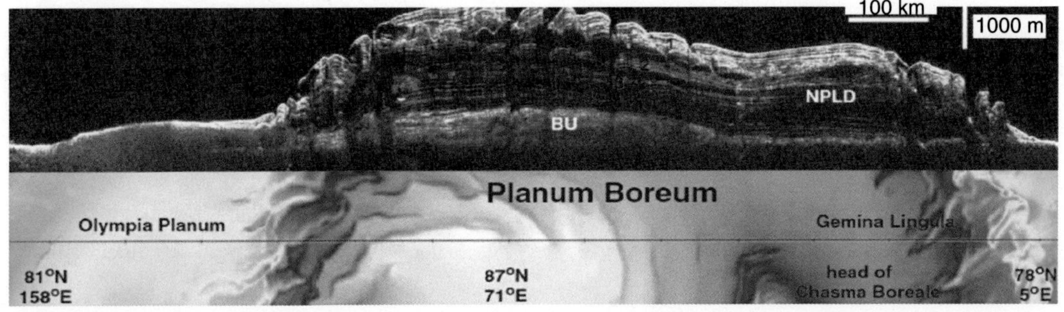

Figure 1.5 Orbiting 85-μs chirp radar image of stratigraphy at the Mars north polar ice cap. From Phillips *et al.* (2008).

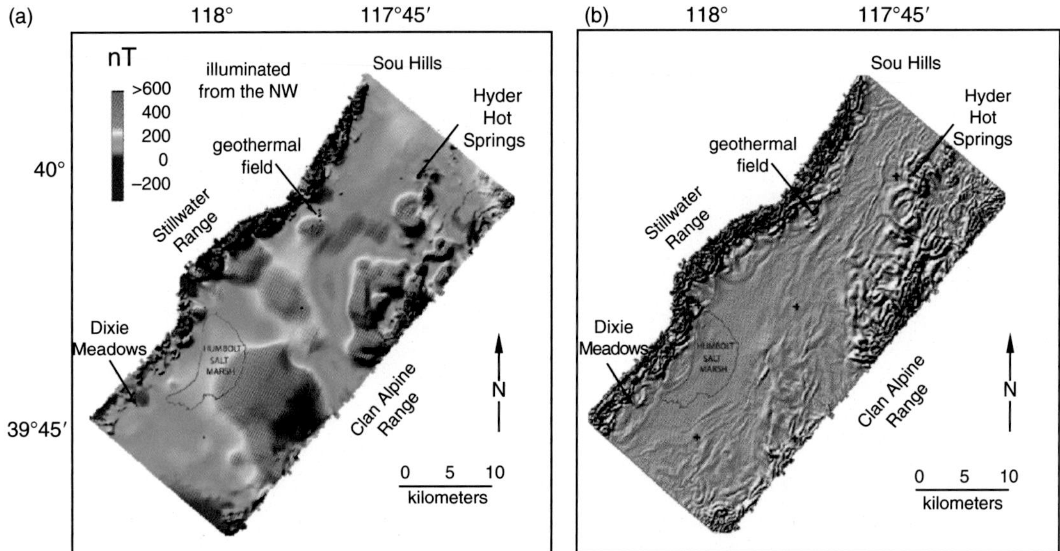

Figure 3.1 (a) Total-field aeromagnetic data after processing in shaded-relief format: Dixie Valley, NV geothermal field; (b) horizontal gradient filter applied to aeromagnetic data to enhance edges and lineaments. After Grauch (2002).

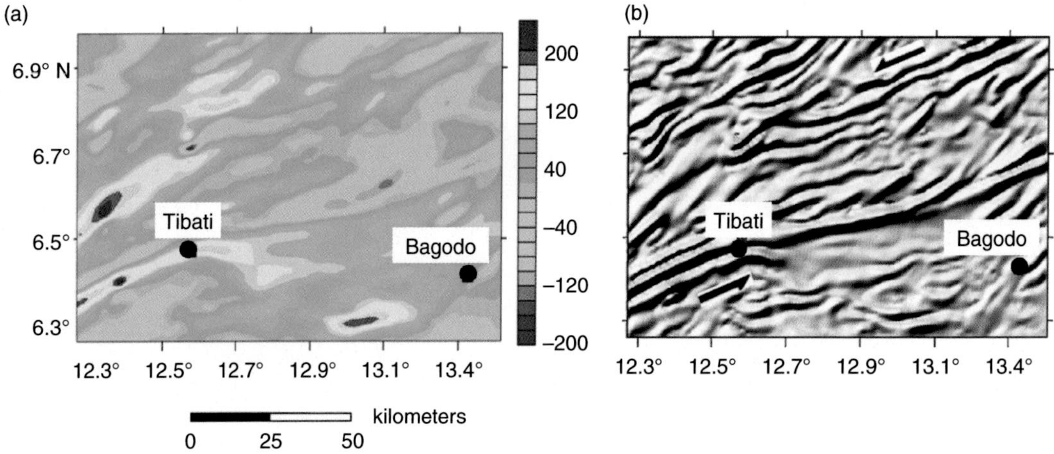

Figure 3.21 (a) Total-field anomaly map in the Adamawa region, Cameroon; (b) shaded-relief map of the total horizontal derivative, *THD*. The area is located in the Foumban shear zone. Black arrows show direction of shearing (from Noutchogwe *et al.* 2010).

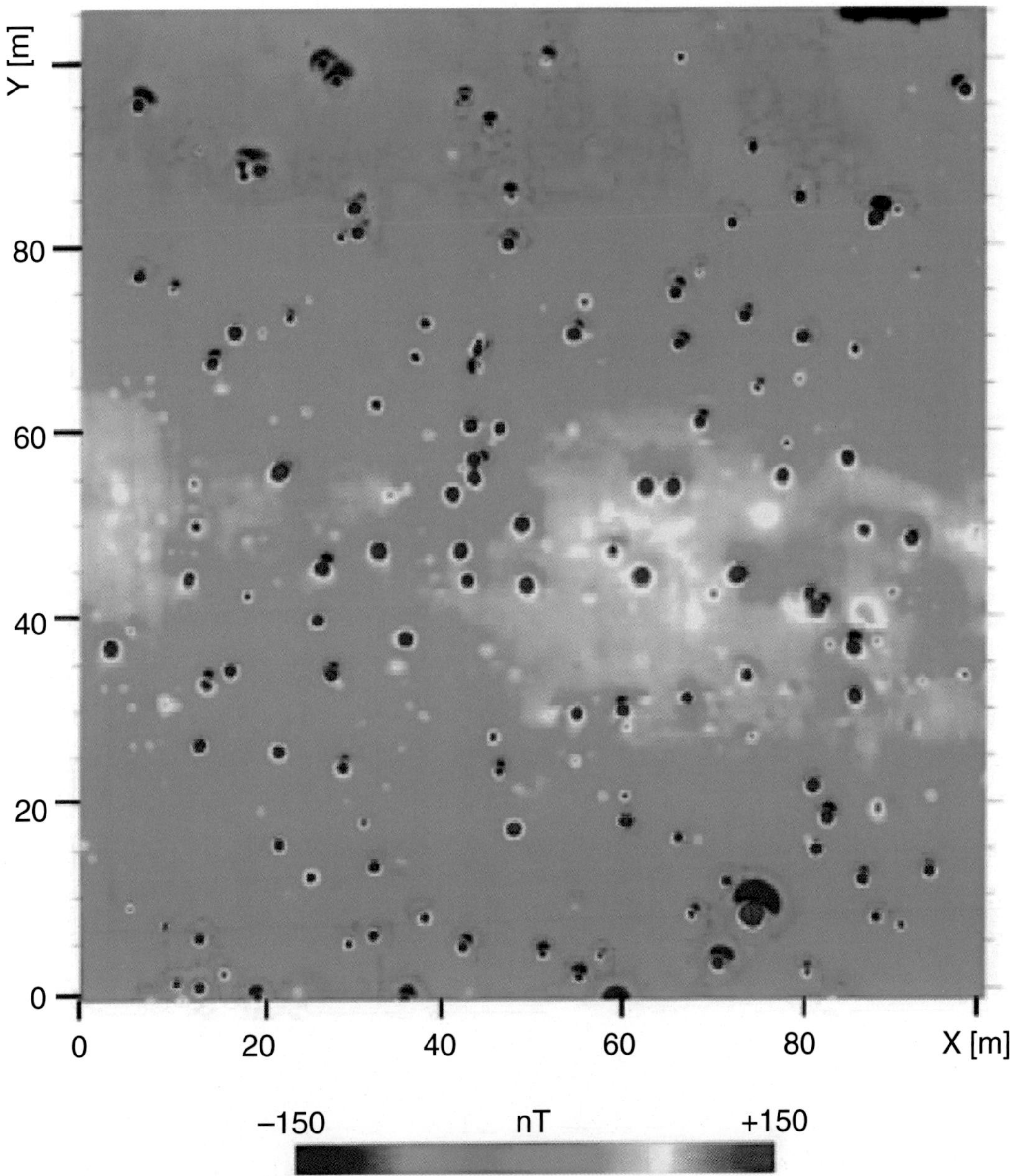

Figure 3.25 Total-field-anomaly map over a live UXO site, Jefferson Proving Ground, Indiana. From Butler (2003).

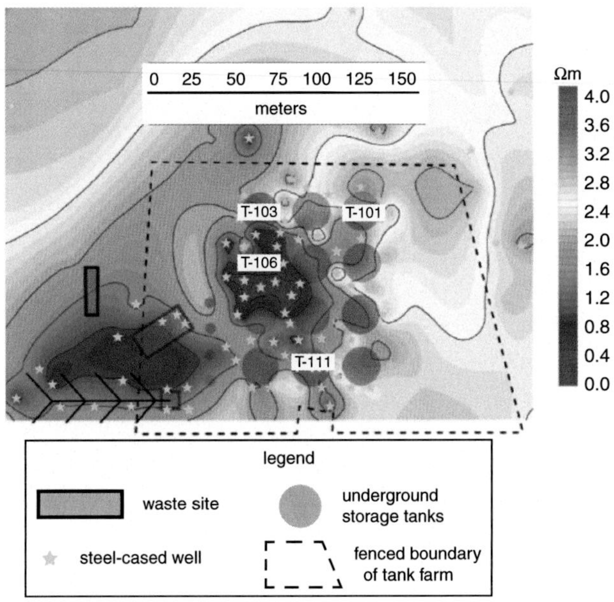

Figure 4.2 WTW resistivity inversion at the Hanford nuclear facility; depth slice at 1.4 m. After *Rucker et al.* (2010).

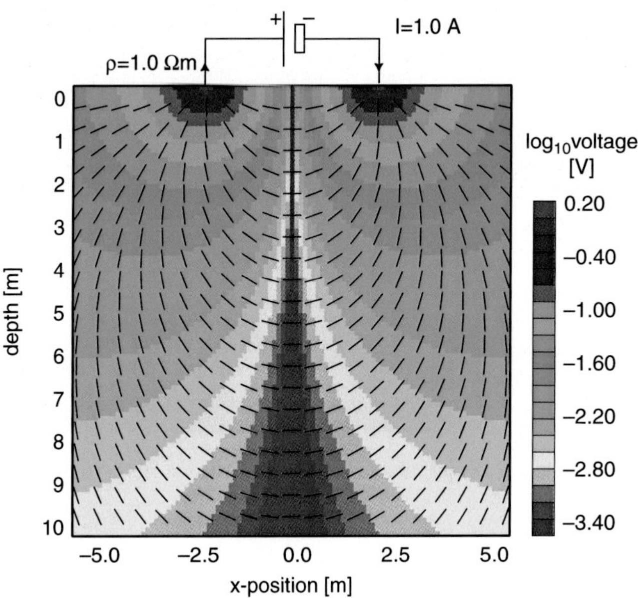

Figure 4.4 Potential and streamlines of electric current for a point source and a point sink of current.

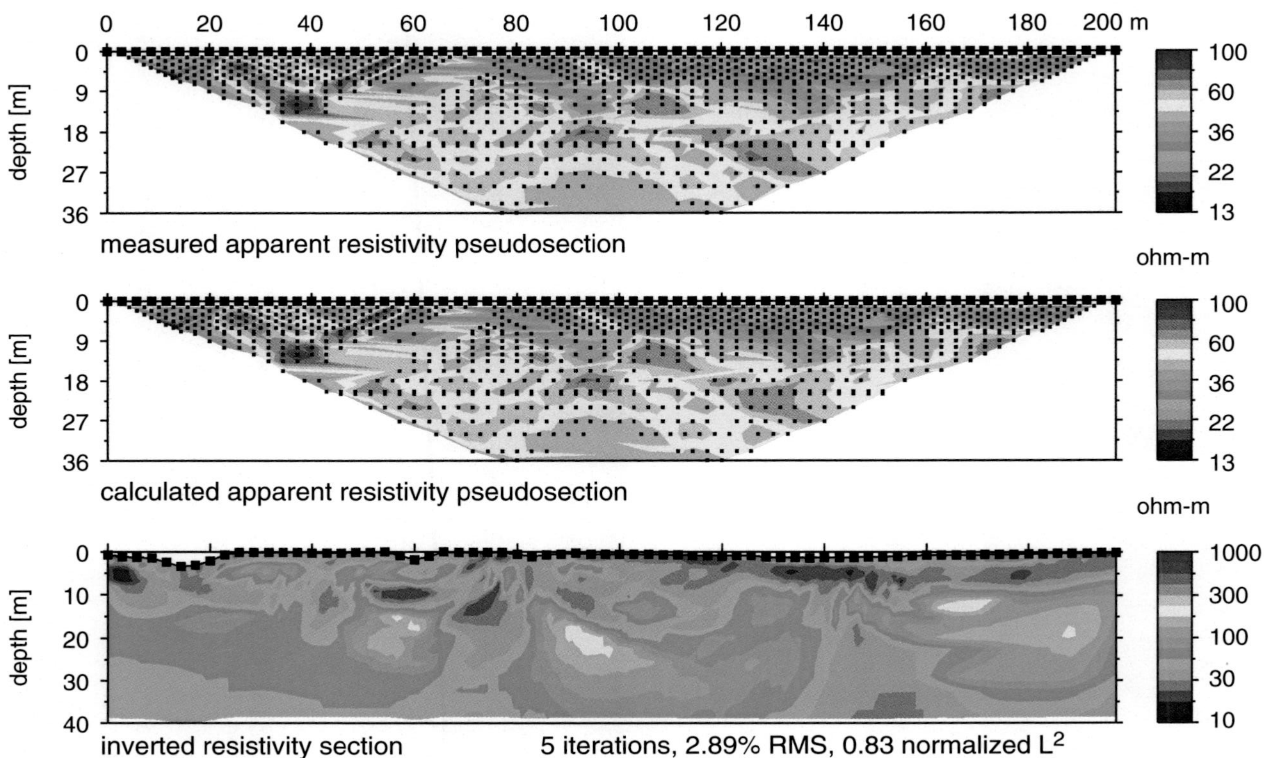

Figure 4.17 Measured apparent resistivity pseudosection (top) for a hybrid Schlumberger-DD electrode configuration, along with the inverted resistivity image (bottom). The calculated pseudosection (middle) is based on solving the forward problem for the resistivity structure shown at the bottom. Note the good match between the measured and calculated pseudosections. See section 4.7 for further details on resistivity inversion.

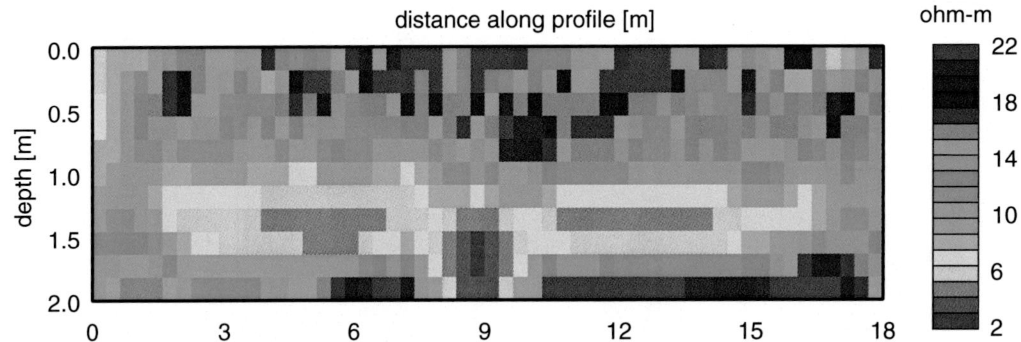

Figure 4.18 Inversion of dipole–dipole resistivity data acquired over a buried pipeline at the Texas A&M Riverside campus. The pipeline is the conductive (*red*) zone at depth 1.5–2.0 m midway along the profile. Electrode spacing is 0.3 m.

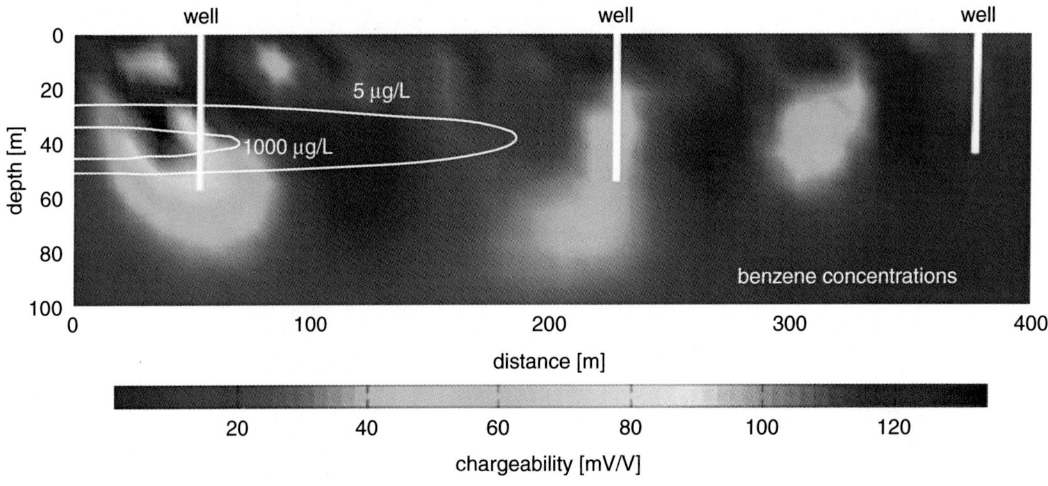

Figure 5.3 IP response of a benzene contaminant plume, Cape Cod, Massachusetts, along with contours of benzene concentration. After Sogade *et al.* (2006).

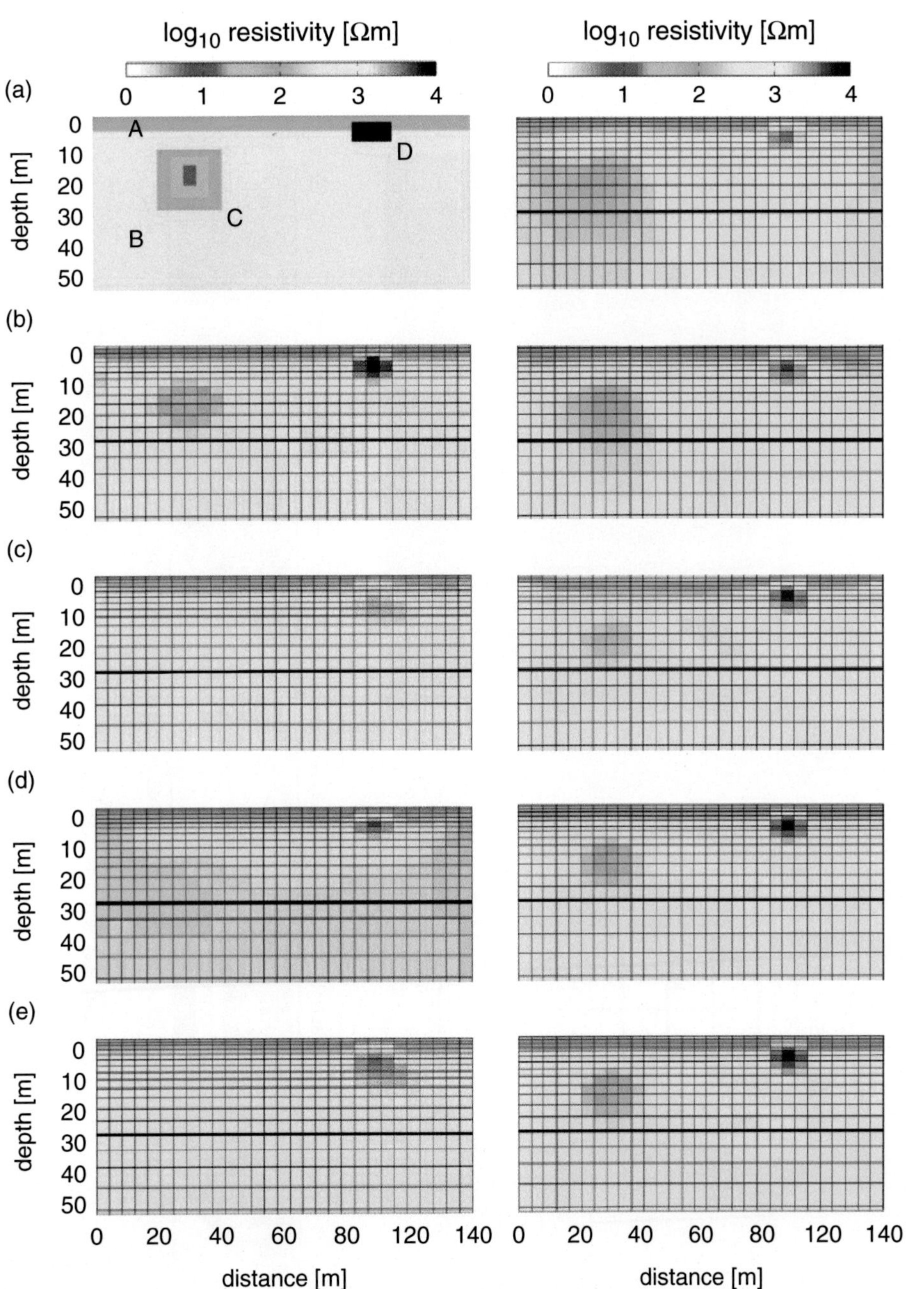

Figure 4.23 (Left panel) Resistivity inversion test, after Stummer *et al.* (2004), (a) starting model. Inversion results using: (b) comprehensive dataset, $D = 51\ 373$; (c) Wenner $D = 135$; (d) dipole–dipole $D = 147$; (e) Wenner/dipole–dipole $D = 282$ electrode configurations. (Right panel) Resistivity inversion based on optimal experimental design, after Stummer *et al.* (2004), (a) $D = 282$; (b) $D = 670$; (c) $D = 1050$; (d) $D = 5740$; (e) $D = 10\ 310$ electrode configurations.

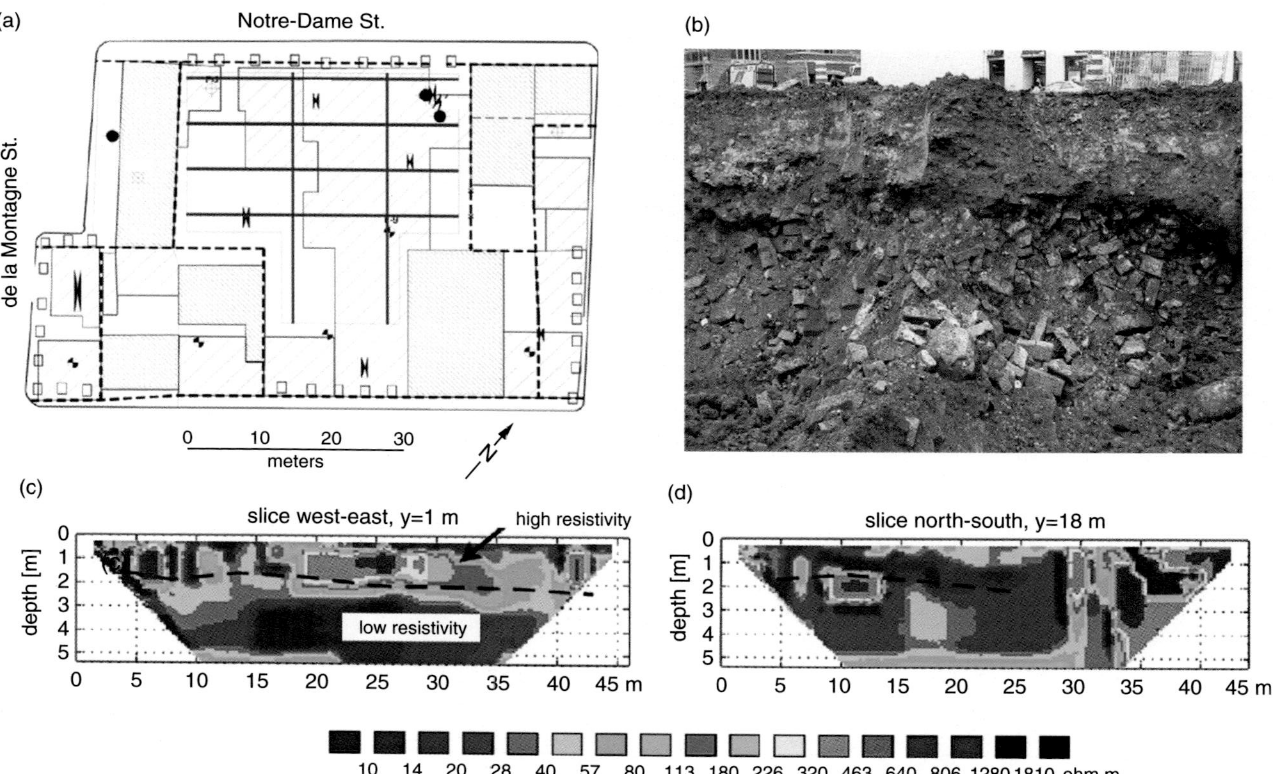

Figure 4.26 (a) Map view of urban redevelopment site, downtown Montreal, Canada. The red lines show ERT profiles. (b) Heterogeneous urban fill containing bricks, concrete, and metal debris. (c) an east–west ERT profile. (d) a north–south ERT profile. The dashed line shows the boundary between the heterogeneous fill and the natural soil, as determined by trench excavations. After Boudreault *et al.* (2010).

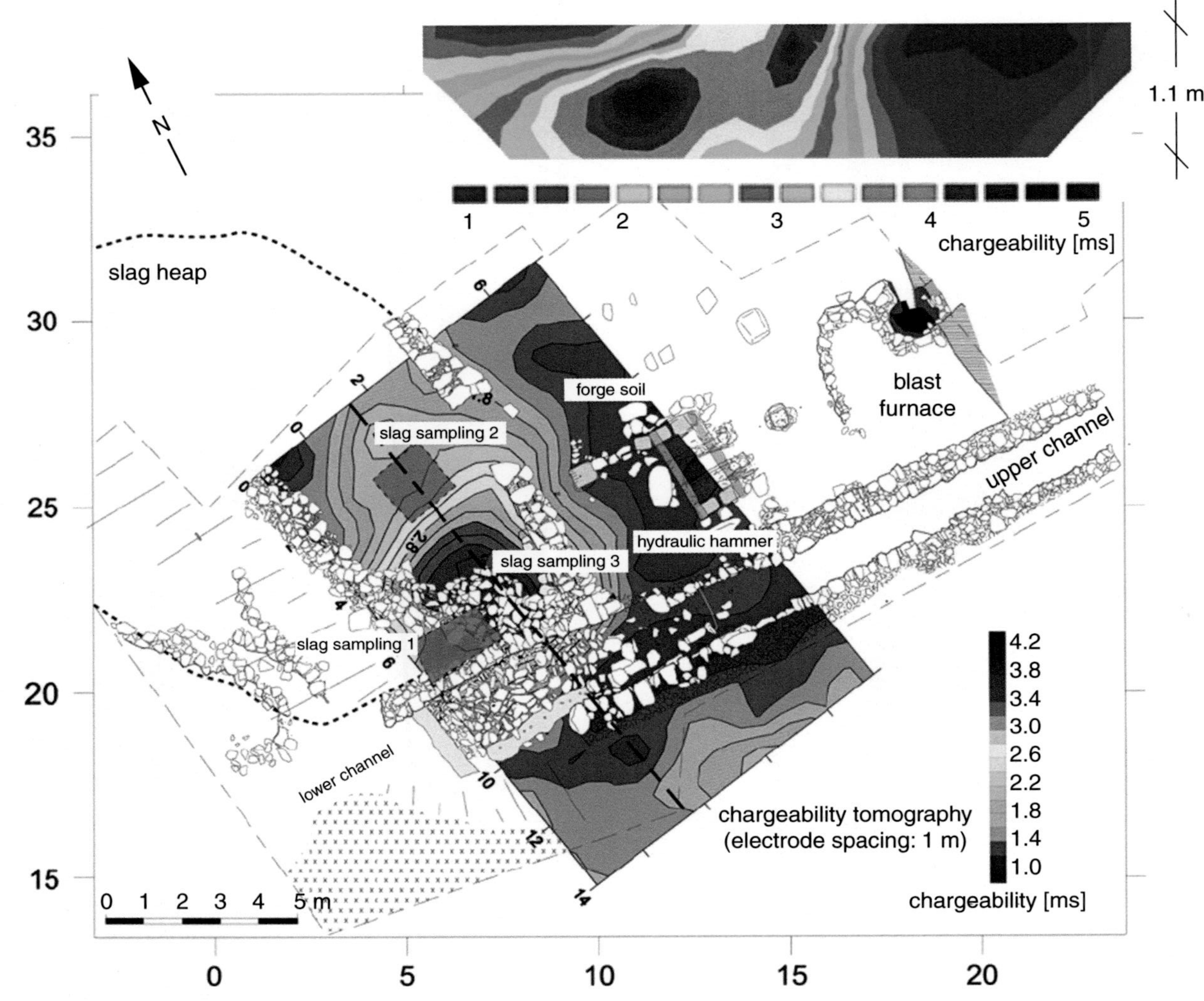

Figure 5.8 Apparent intrinsic chargeability map and vertical cross-section in support of a paleometallurgical investigation. The location of the cross-section corresponds to the N–S dashed line crossing the map. After Florsch *et al.* (2011).

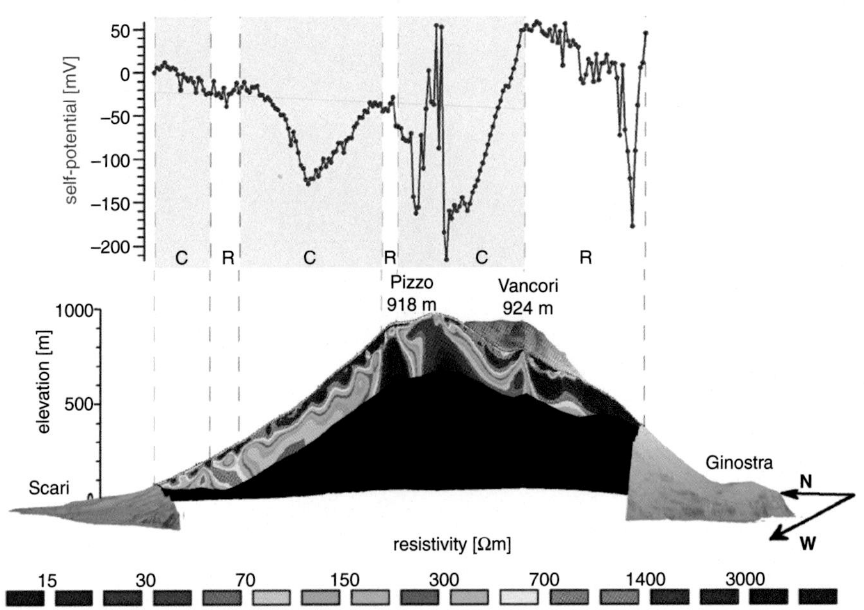

Figure 5.9 SP signals and 2-D electrical resistivity structure at Stromboli volcano, Italy. C = conductive zone; R = resistive zone. After Finizola *et al.* (2006).

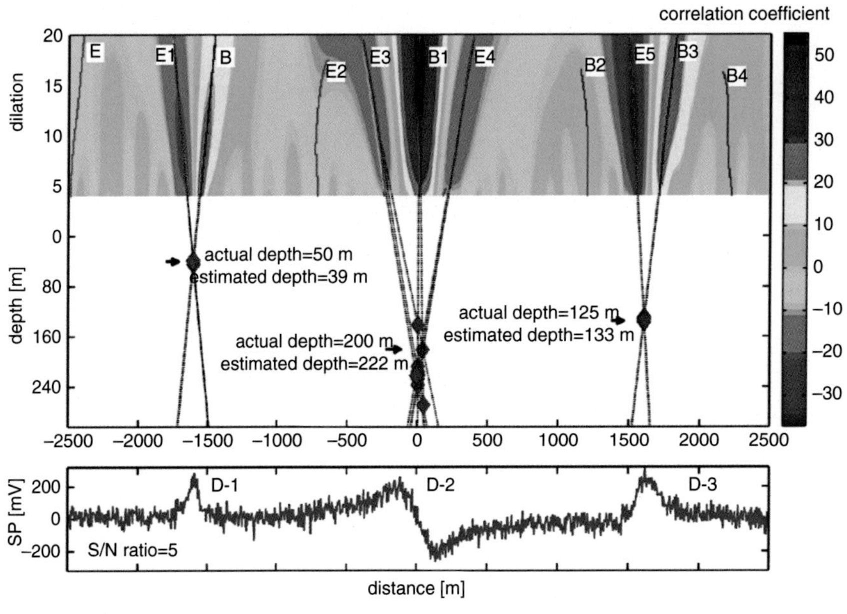

Figure 5.14 Depth determination to SP sources using CWT analysis. After Mauri *et al.* (2010).

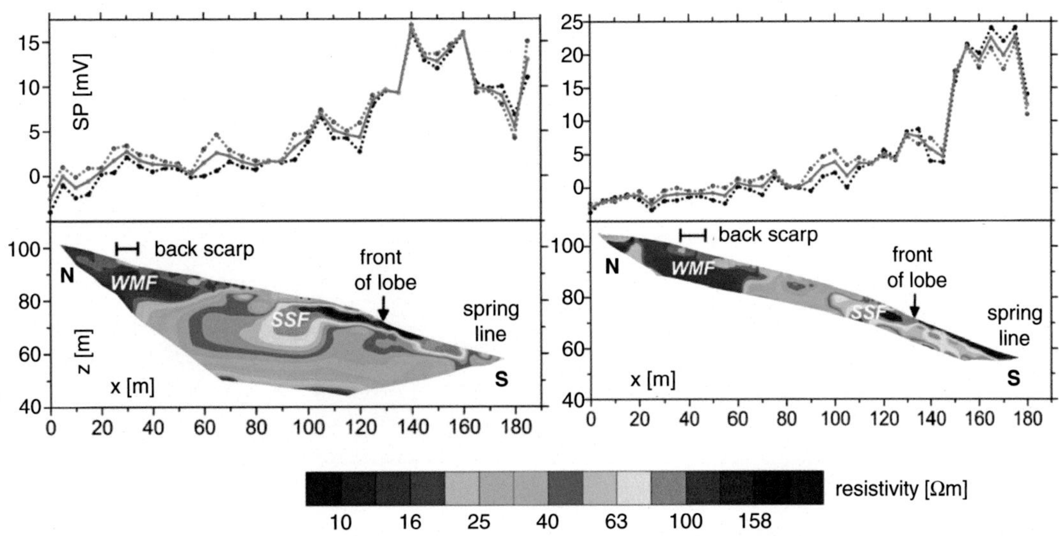

Figure 5.15 SP profiles and co-located 2-D ERT images at an active landslide in northeast England; WMF = Whitby mudstone formation; SSF = Staithes sandstone and Cleveland ironstone formation. After Chambers *et al.* (2011).

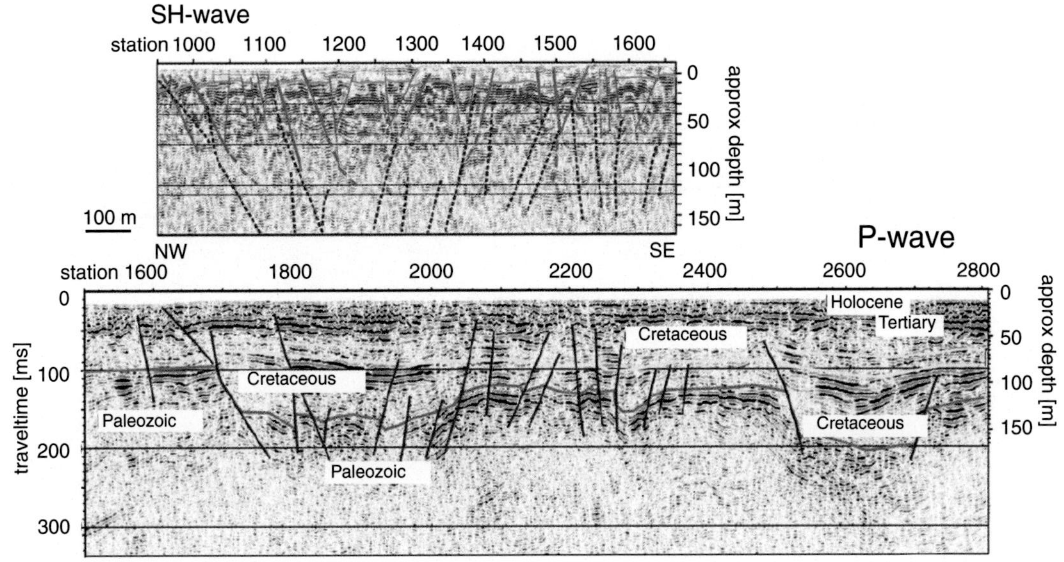

Figure 6.2 SH-wave (top) and P-wave (bottom) seismic profiles in the New Madrid zone of contemporary seismicity revealing reactivated Paleozoic bedrock normal faults. After Bexfield *et al.* (2006).

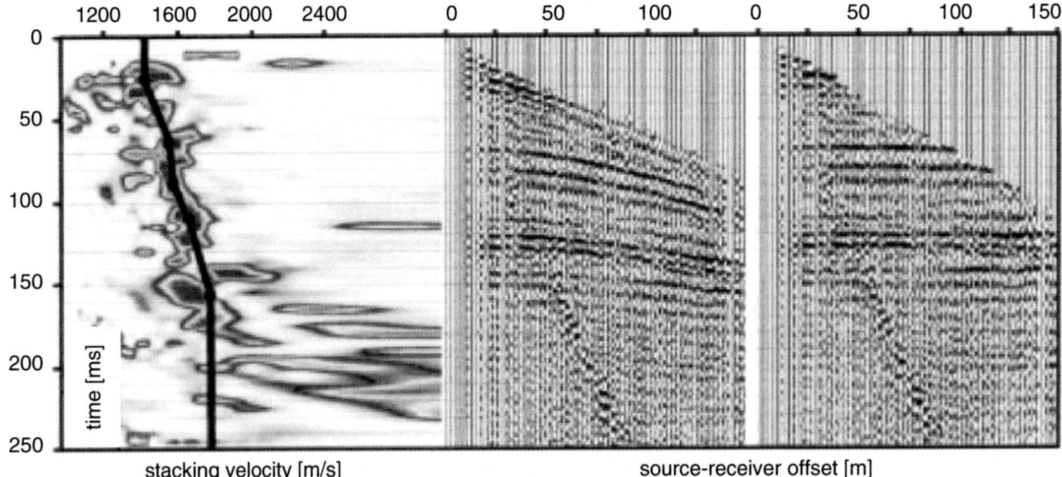

Figure 6.33 (Left) Determination of stacking velocity by passing a continuous curve through regions of high (red) values of semblance; (middle) an uncorrected CMP supergather; (right) the NMO-corrected CMP supergather using the stacking-velocity function in the left panel. After Spitzer *et al.* (2003).

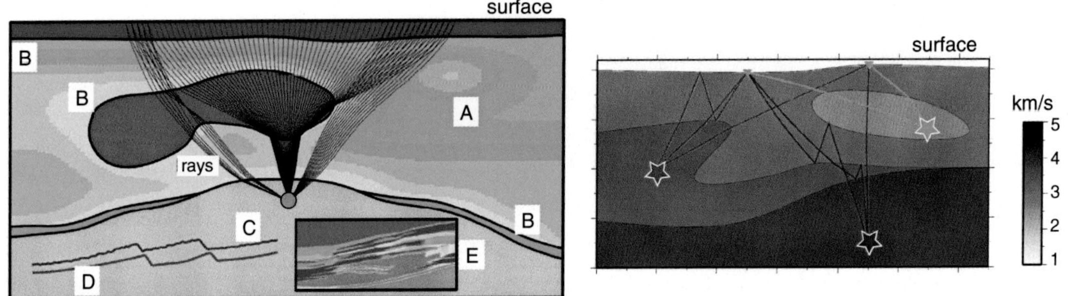

Figure 6.42 (Left) Rays traced from a subsurface source (large dot) through and around a complex-shaped fast-velocity anomaly to the surface (after Gjoystdal *et al.* 2007). (Right) Different ray paths are possible in complicated models, such as this one containing a pinchout, overturned fold, and anomalously fast body (after Hauser *et al.*, 2008).

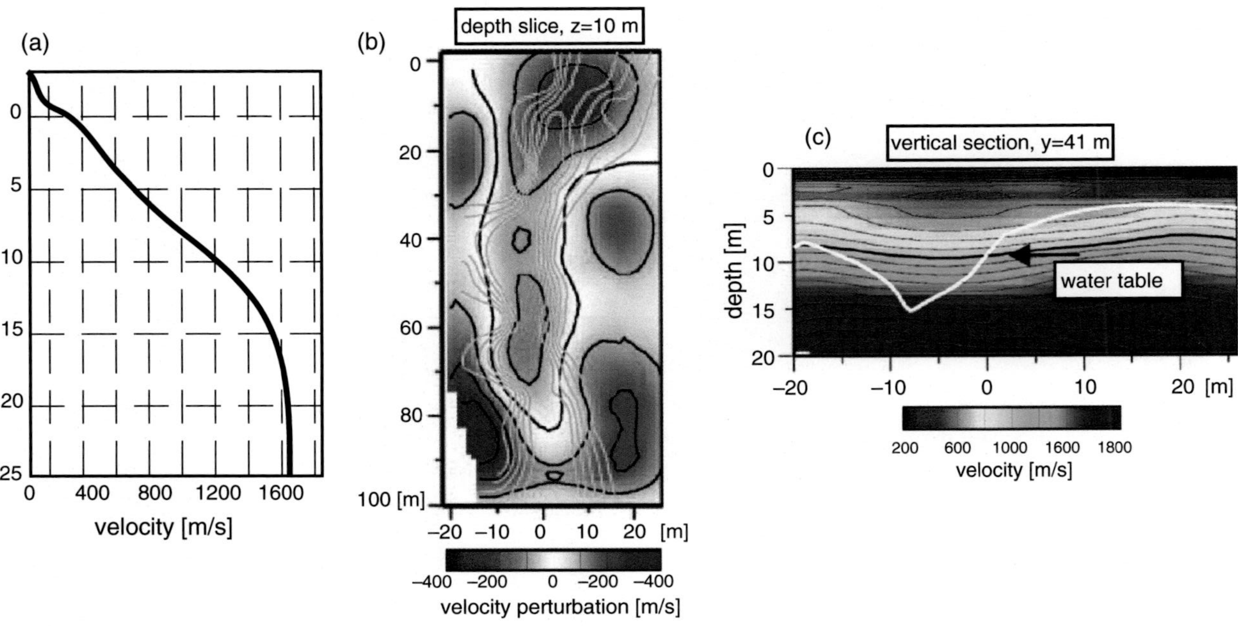

Figure 6.44 3-D refraction tomography at a contaminated site: (a) starting 1-D velocity model; (b) horizontal slice at depth $z = 10$ m through the final preferred tomogram; (c) vertical slice at location $y = 41$ m; green and white contours in (b) and (c) mark the incised paleochannel inferred from well data. After Zelt *et al.* (2006).

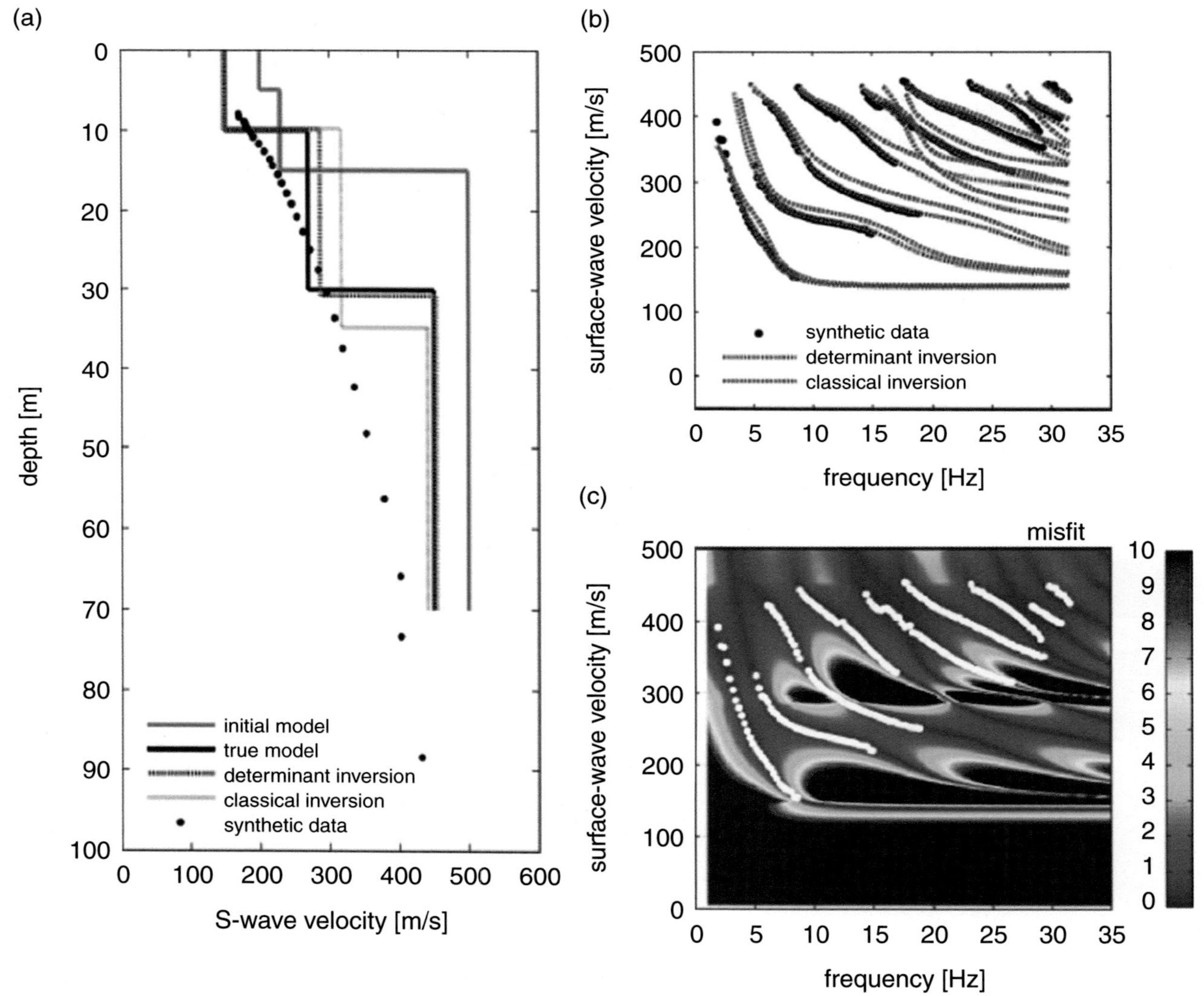

Figure 7.9 Inversion of synthetic surface-wave data using a misfit function based directly on the Haskell–Thomson determinant: (a) S-wave velocity profiles; (b) multimodal dispersion characteristics; (c) dispersion characteristics superimposed on the Haskell–Thomson misfit function. After Marischini *et al.* (2010).

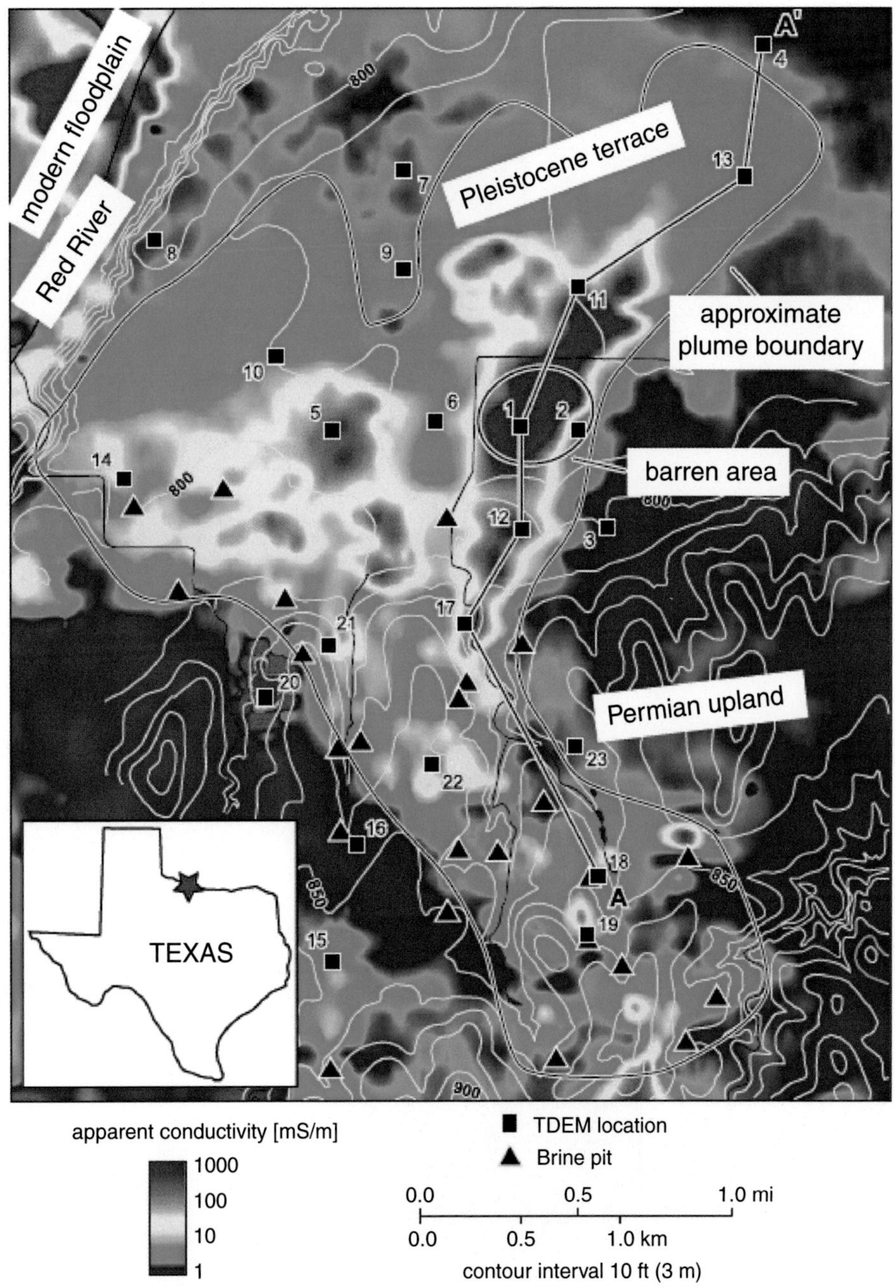

Figure 8.1 Topography and apparent resistivity at 7.2 kHz from airborne EM mapping of a north Texas oilfield brine plume. After Paine (2003).

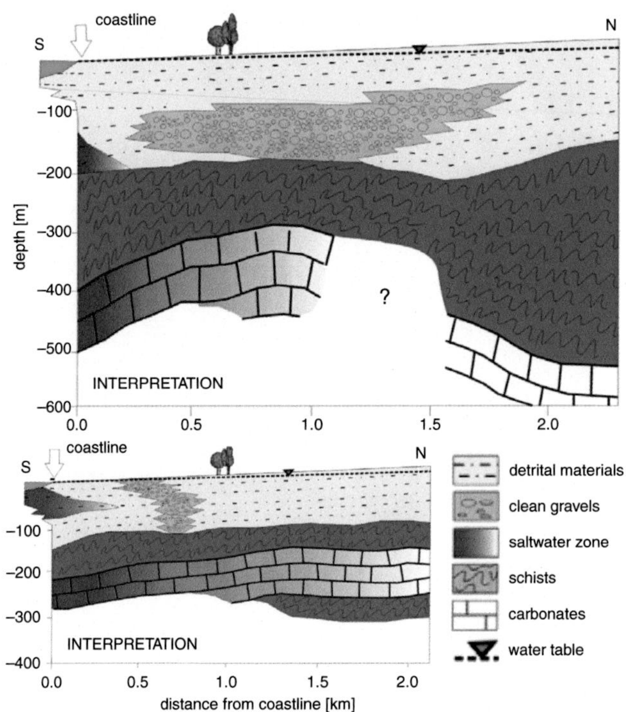

Figure 8.3 Interpretations of central-loop time-domain EM soundings in terms of seawater intrusion (red-colored areas) into the coastal Motril–Salobreña aquifer of southern Spain. After Duque *et al.* (2008).

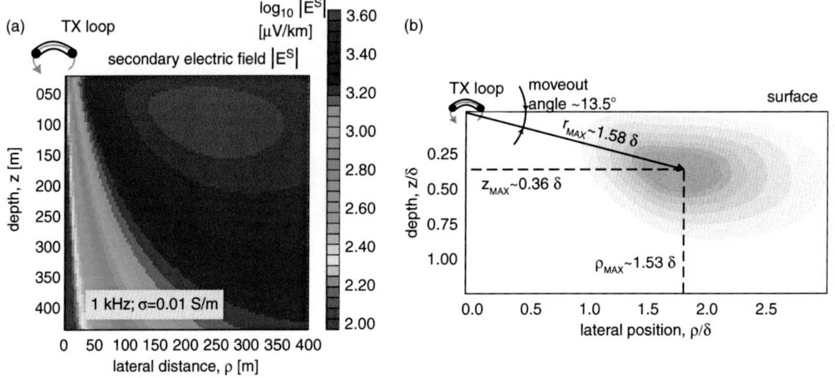

Figure 8.6 (a) Electric-field intensity due to loop excitation in a unform halfspace; (b) position of maximum intensity of the electric field in terms of skin depth.

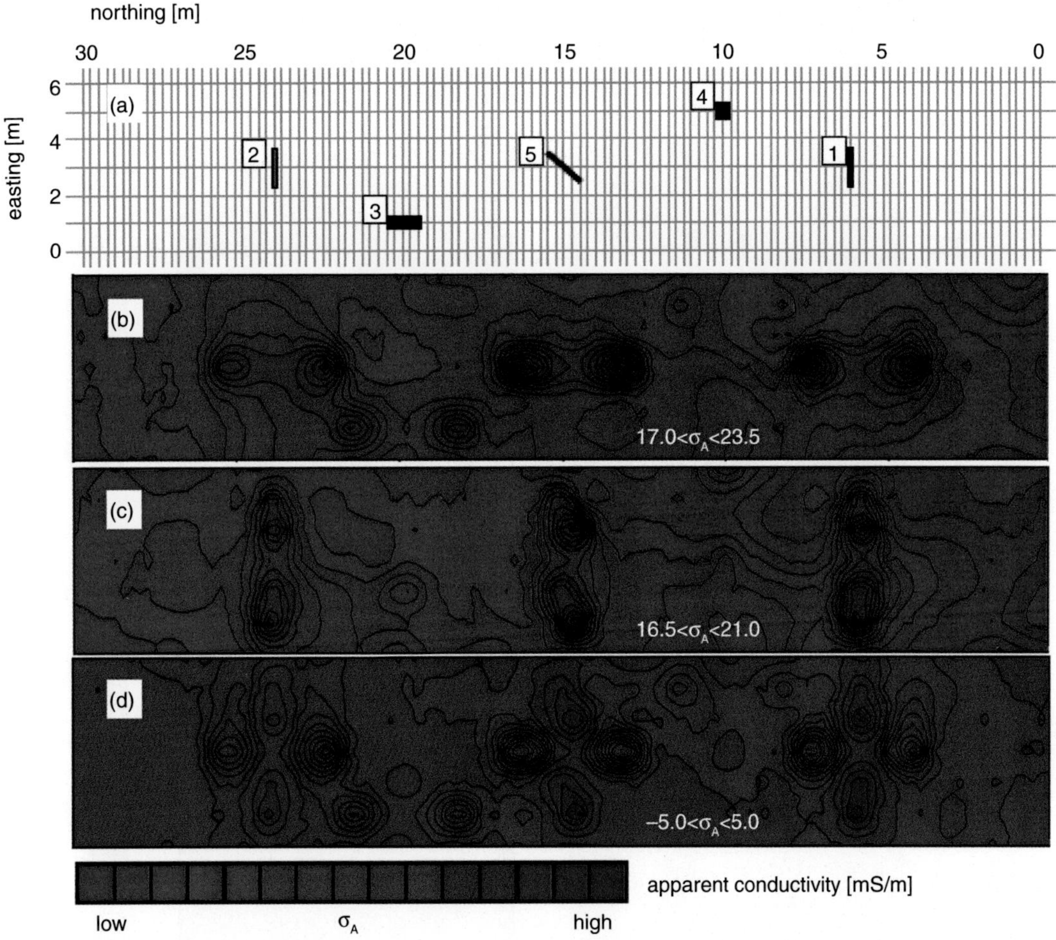

Figure 8.11 EM31 detection of conductive targets. (a) Experimental layout of steel pipes; (b) P-mode apparent conductivity map; (c) T-mode apparent conductivity map; (d) difference of the T-mode and P-mode maps; key: 1, 5 = 155 × 12.4-cm drill pipe, 2 = 150 × 6.4-cm drill pipe, 3 = 75 × 7.5-cm galvanized iron water pipe, 4 = 28 × 7.2-cm drill pipe collar. After Benavides and Everett (2005).

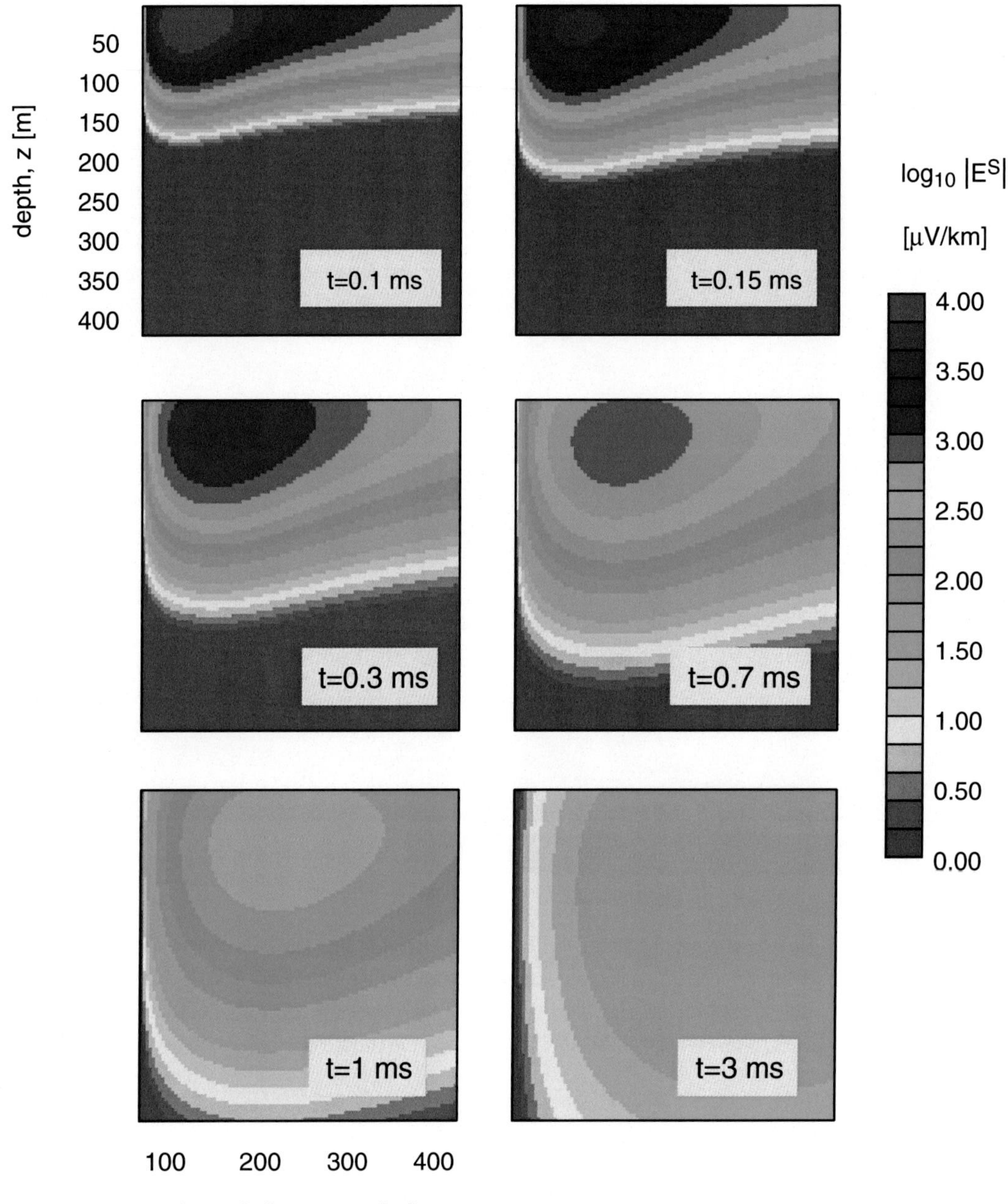

Figure 8.14 Transient smoke-ring diffusion into a uniformly conductive halfspace, $\sigma = 0.1$ S/m.

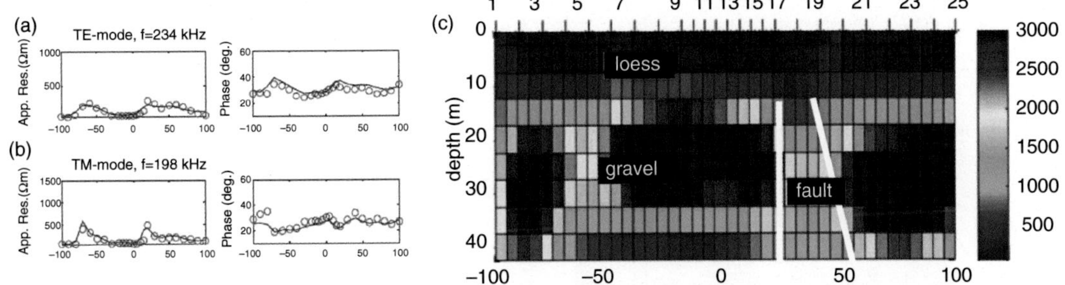

Figure 8.21 RMT survey results from Cologne, Germany: (a) TE-mode apparent resistivity and phase at 234 kHz; (b) TM-mode apparent resistivity and phase at 198 kHz. Blue lines are observed data, while red circles are computed from the resistivity model shown at right; (c) 2-D RMT smooth inversion result. The color bar indicates resistivity ρ [Ωm]. The plot ordinates in all cases is RX location [m]. After Candansayar and Tezkan (2008).

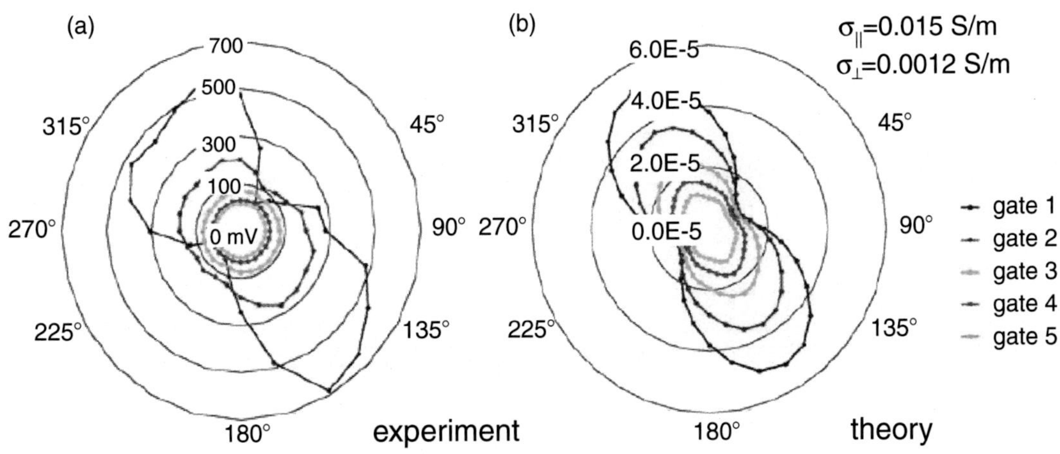

Figure 8.28 (Left) Azimuthal TEM47 voltage readings (first five time gates) over Packsaddle schist, central Texas; (right) calculated response [mV] of the horizontal anisotropic halfspace with $(\sigma_{\perp}, \sigma_{||})$, as shown; strike direction 146°. After Collins *et al*, (2006).

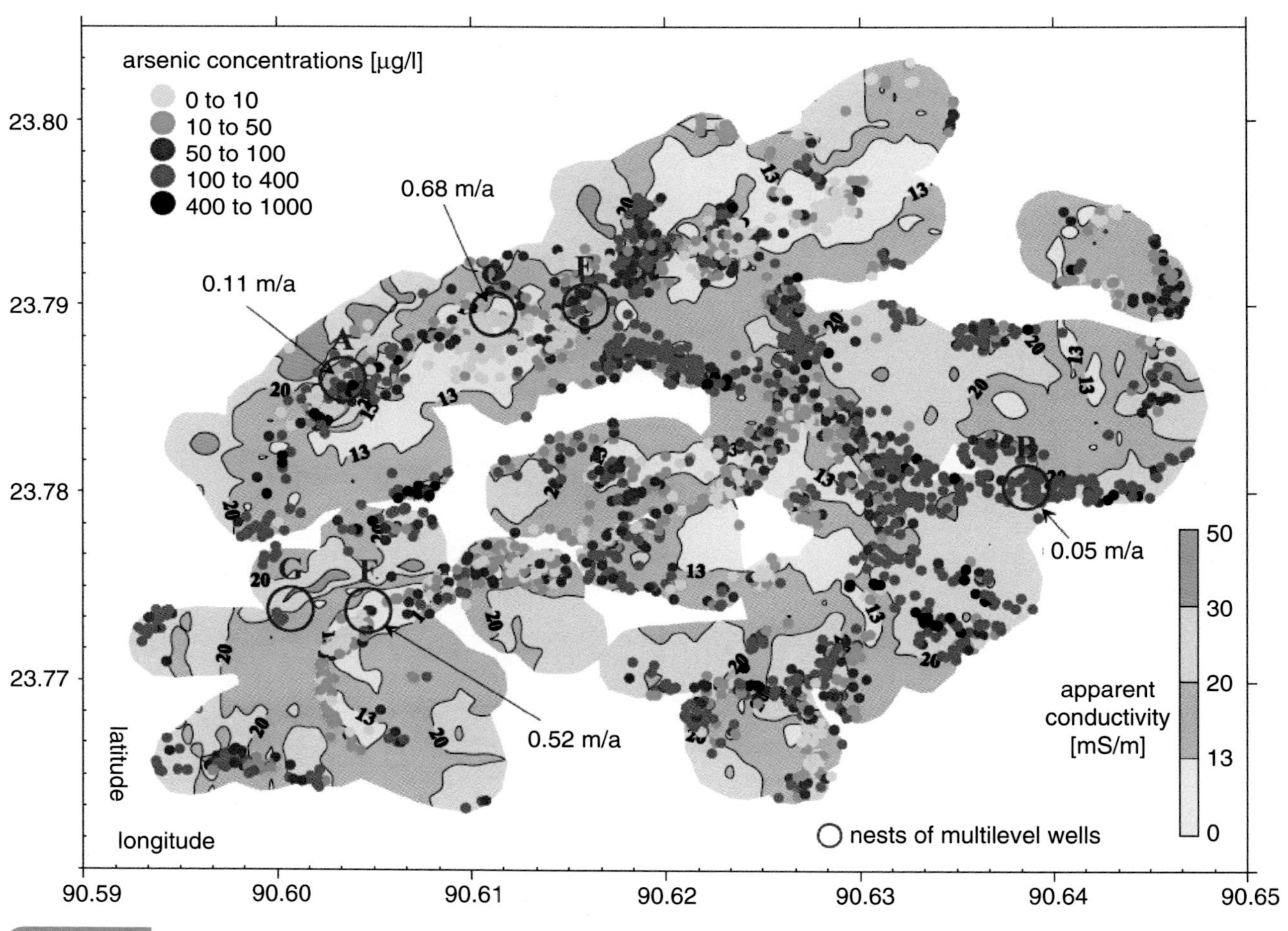

Figure 8.29 A spatially continuous kriged map of EM34 apparent-conductivity readings, along with arsenic concentration data from shallow wells, Bengal Basin. After Aziz *et al.* (2008).

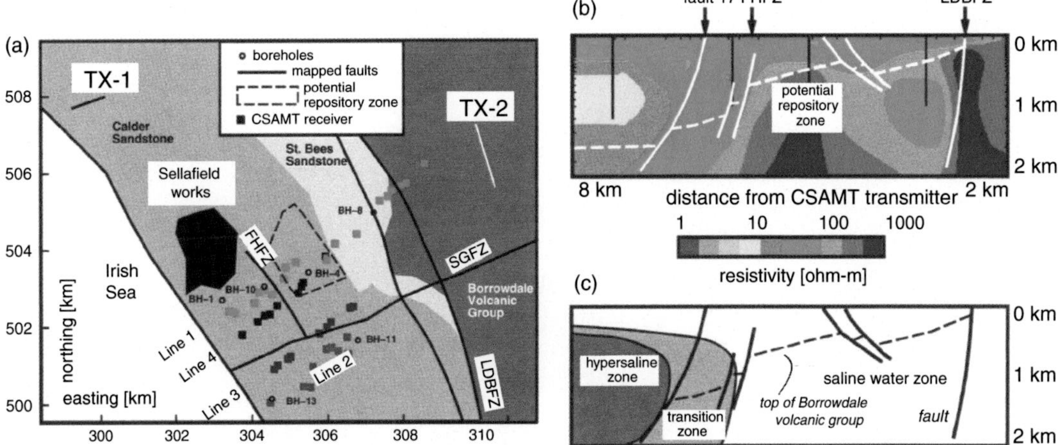

Figure 8.30 CSMT at the Sellafield (UK) nuclear facility: (a) layout of TX grounded dipoles and RX locations with respect to mapped faults and the power plant; (b) 2-D inversion result from Line 1; (c) hydrogeological interpretation. After Unsworth *et al.* (2000).

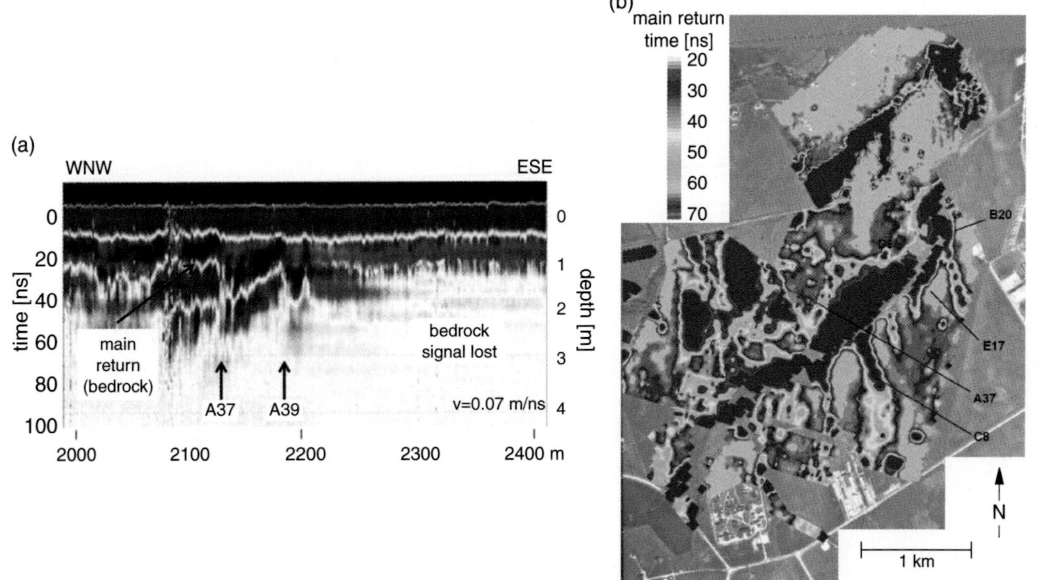

Figure 9.2 (a) Typical GPR section obtained at the NWIRP site, Texas, using 50 MHz antennas; (b) plan-view map of the GPR main returns, interpreted in terms of depth to bedrock. After Hughes (2009).

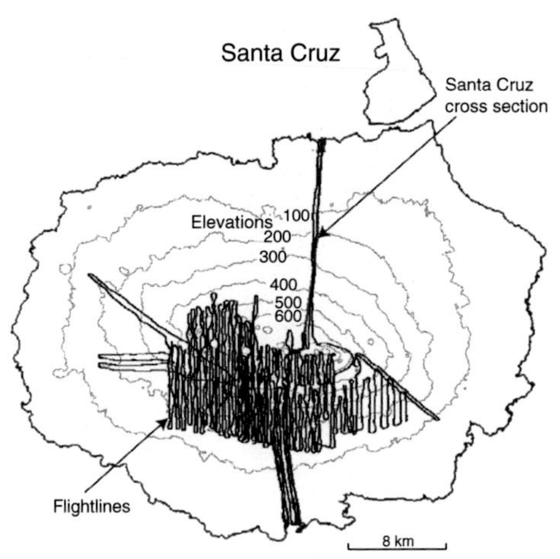

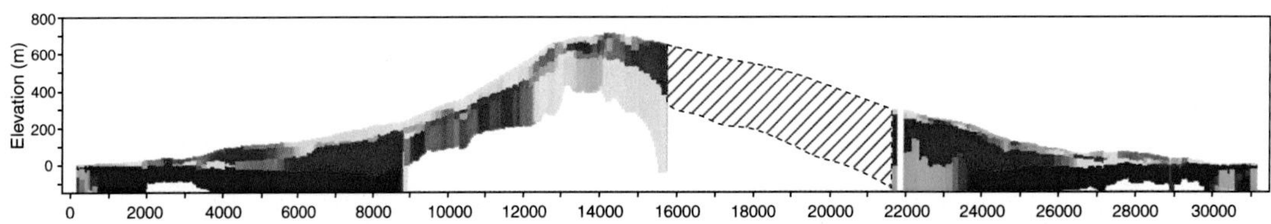

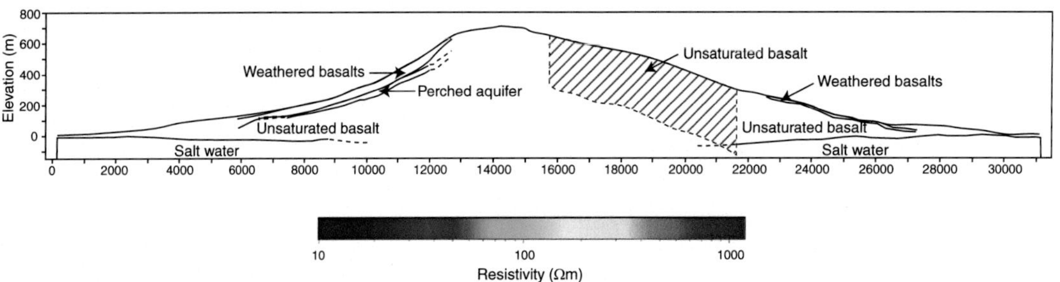

Figure 8.31 (Top panel) SkyTEM flight lines over Santa Cruz island. Contour levels in meters above sea level; (middle panel) resistivity profile along the north–south cross-section; (bottom panel) hydrogeological interpretation of the AEM data. Leeward side at right, Windward side at left. After Auken *et al.* (2009).

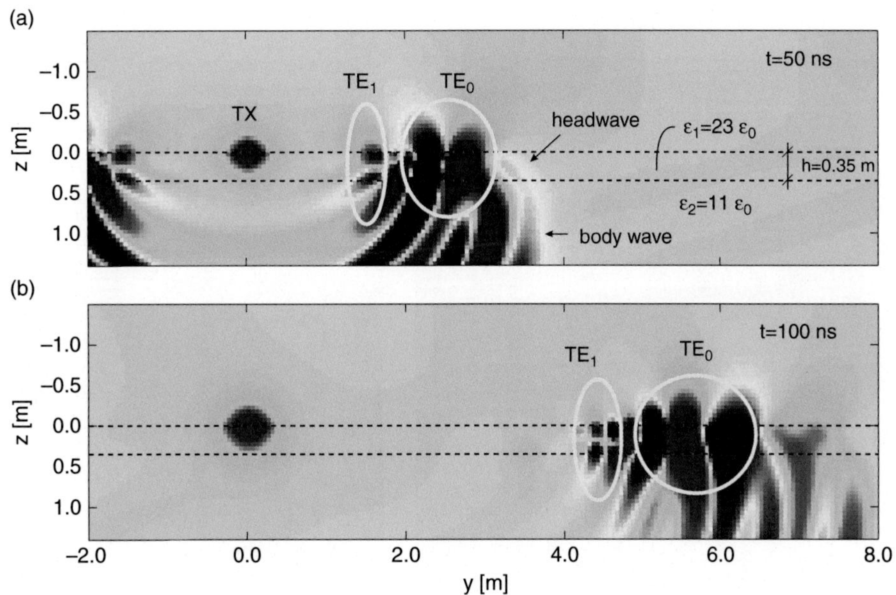

Figure 9.32 Simulations of the electric field E_x within a surface waveguide structure at times (a) $t = 50$ ns and (b) $t = 100$ ns after a pulse is transmitted from the transmitter (TX). The lowest two TE waveguide modes are identified. After van der Kruk *et al.* (2009).

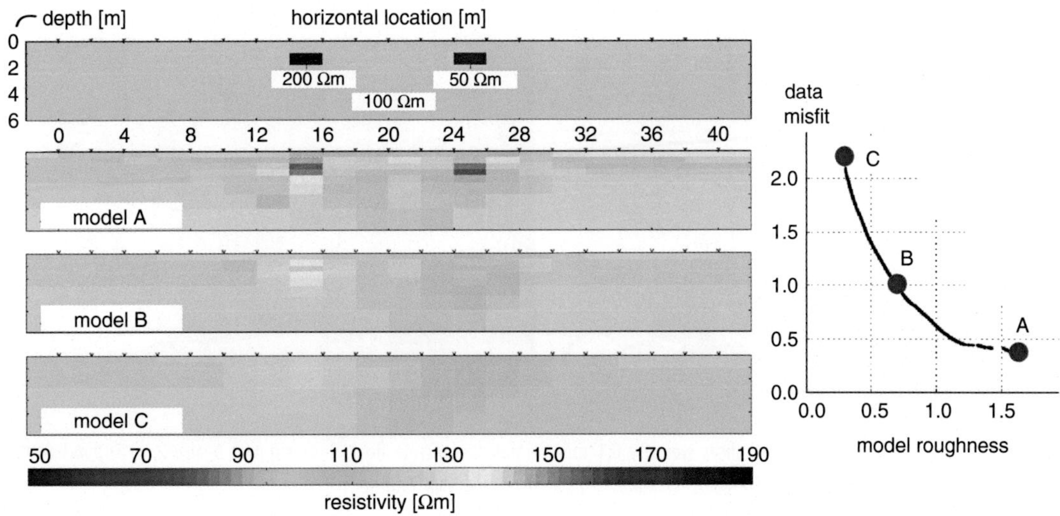

Figure 13.5 Genetic algorithm inversion of 2-D resistivity data: (left, top) synthetic model; (left, bottom three panels) inversion results; (right) the locations of the three models A, B, C on the misfit–roughness trade-off curve. After Schwarzbach *et al.* (2005).

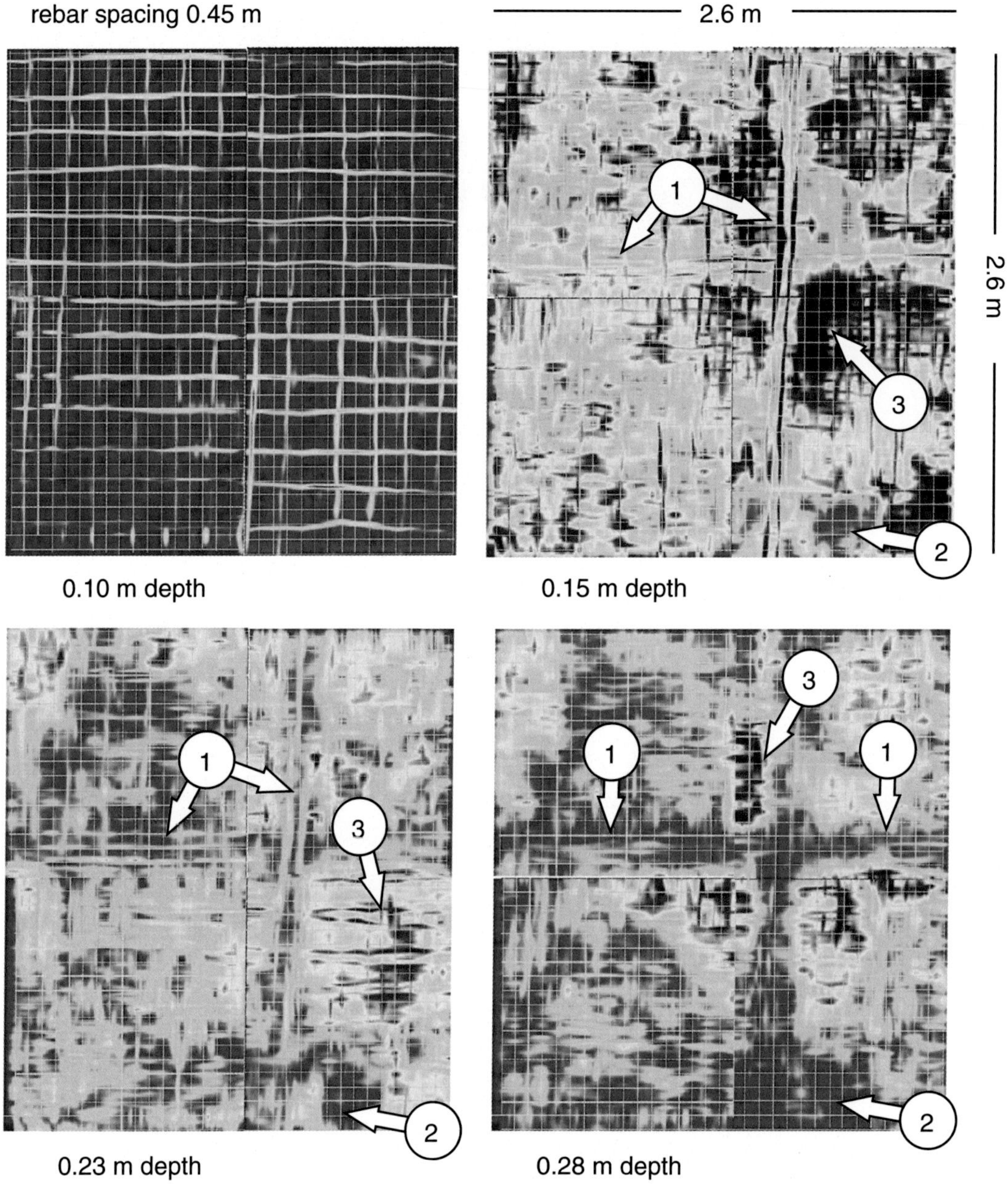

rebar spacing 0.45 m

2.6 m

2.6 m

1

3

2

0.10 m depth

0.15 m depth

1

3

2

1

3

1

2

0.23 m depth

0.28 m depth

Figure 9.34 1.0 GHz radar depth slices, velocity 0.1 m/s, reinforced concrete floor. 1 = support beam; 2 = support pier; 3 = void space; data courtesy Mary Jo Richardson and Wilf Gardner.

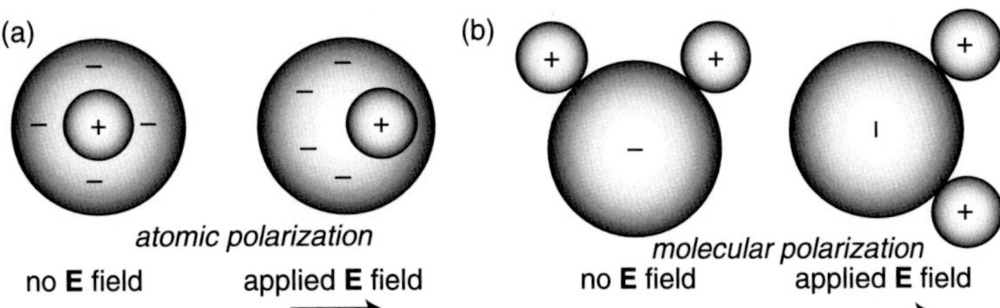

Figure 9.4 (a) Atomic polarization; (b) polarization of the water molecule.

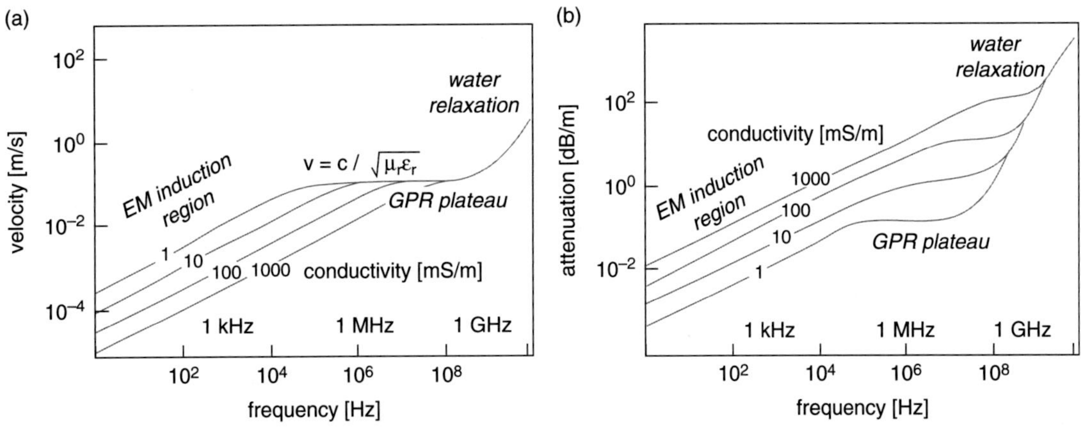

Figure 9.5 (a) GPR wave velocity as a function of frequency; (b) GPR attenuation as a function of frequency. After Davis and Annan (1989).

10 GHz, which is above the normal GPR range. A third type of dielectric polarization, interfacial polarization or the *Maxwell–Wagner effect*, can become important at frequencies less than 100 MHz (Hizem *et al.* 2008). Interfacial polarization at GPR frequencies is caused by electric charges that accumulate at dielectric interfaces when a heterogeneous medium is subjected to an applied electric field.

The electromagnetic wave velocity in a non-magnetic (relative magnetic permeability $\mu_r = \mu/\mu_0 = 1$) medium is given approximately by the formula (Davis and Annan, 1989)

$$v = \frac{c}{\sqrt{\varepsilon_r}}; \tag{9.1}$$

where $c = 3 \times 10^8$ m/s (~ 1 ft/ns) is the speed of light in vacuum and $\varepsilon_r = \varepsilon/\varepsilon_0$ is the relative electrical permittivity, or *dielectric constant*. Sometimes the symbol K is also used for dielectric constant. To first order, in the radar frequency range 10 MHz–2 GHz, the velocity v is independent of both frequency and conductivity. This is evidenced by the "GPR plateau" shown in Figure 9.5a.

For most dry geological materials, such as sand, gravel, and crystalline rock, the dielectric constant varies roughly in the range $3 \leq \varepsilon_r \leq 8$. Water has an anomalously large dielectric constant of $\varepsilon_r \sim 81$ due to the high polarizability of the water molecule in the

Table 9.1 Dielectric constant and radar attenuation of common geological materials at 100 MHz, after Davis and Annan (1989)

Material	Dielectric constant	Attenuation [dB/m]
Air	1	0
Freshwater	80	0.1
Seawater	80	1000
Dry sand	3–5	0.01
Saturated sand	20–30	0.03–0.3
Limestone	4–8	0.4–1.0
Clay	5–40	1–300
Granite	4–6	0.01–1.0

presence of an applied electric field. Thus, water-bearing rocks have significantly higher dielectric constants ($\varepsilon_r \sim 10$–30) than dry rocks of the same lithology. Hydrocarbons such as oil and natural gas have low values of dielectric constants, on the order $\varepsilon_r \sim 1$–2. The dielectric constant and radar attenuation (discussed below) of common geomaterials at 100 MHz is listed in Table 9.1.

The dielectric properties of rocks and soil are generally dispersive at GPR frequencies (West *et al.*, 2003), which implies that permittivity ε is a complex function of frequency, often written as $\varepsilon^*(\omega)$. A primary cause of the dispersion is that, at sufficiently high frequencies, the polarization of atoms and molecules cannot keep pace with the rapid alternations of an applied **E** field. This leads to an out-of-phase component of the polarization that manifests itself as an imaginary, or quadrature, contribution to $\varepsilon^*(\omega)$. Moreover, any dielectric loss caused by viscous dissipation of energy as the water molecules rotate in a rapidly changing **E** field adds to the quadrature part of $\varepsilon^*(\omega)$. At frequencies greater than $\sim 10^{10}$ Hz, well above the GPR frequency range, both the real and imaginary components of permittivity $\varepsilon^*(\omega)$ drop precipitously since water molecules are not capable of responding to such extremely fast fluctuations in the **E** field.

The *attenuation* of a radar wave is given approximately by the formula

$$\alpha \sim 1690 \frac{\sigma}{\sqrt{\varepsilon_r}} \quad [\mathrm{dB/m}]. \tag{9.2}$$

A more general, frequency-dependent formula for attenuation α that accounts for both conduction and various forms of dielectric loss is presented in von Hippel (1954). As shown in Figure 9.5b, attenuation increases with increasing conductivity σ and frequency ω. This accords with the familiar rule that the GPR depth of penetration decreases as the product $\sigma\omega$ increases. Under poor conditions such as wet, clay-rich soils the penetration depth at $\sim$ 100 MHz is roughly 1–2 m. Under better conditions such as dry, clean sands or gravel, the penetration depth at this frequency can be greater than 10–20 m. In pure rock salt, a low-loss ionic solid, penetration depth can be several hundreds of meters (Gorham *et al.* 2002). The occurrence of fine interbedding restricts the penetration depth of radar waves since energy is lost at each reflecting horizon.

The seismic and GPR techniques are somewhat complementary in the sense that poor GPR field conditions (wet clays) are actually good seismic conditions while ideal GPR conditions (dry sands) are unfavorable for the acquisition of high-quality seismic data.

Geophysical imaging using the GPR method is based on the reflectivity of the geological medium under investigation. Suppose a radar pulse is propagated into a non-magnetic ground which consists of a single layer of dielectric constant ε_1 overlying a halfspace of dielectric constant ε_2. The reflection coefficient for a *normally incident* radar plane wave is

$$R = \frac{\sqrt{\varepsilon_1} - \sqrt{\varepsilon_2}}{\sqrt{\varepsilon_1} + \sqrt{\varepsilon_2}}. \tag{9.3}$$

The reflected energy is proportional to R^2. The general case of oblique incidence of a plane wave will be examined later in this chapter.

9.2 Dielectric constant and electrical conductivity

The velocity of a subsurface radar wave depends on the dielectric constant ε_r, as indicated by Equation (9.1). Radar waves reflect off discontinuities, and bend when they encounter spatial gradients in the local dielectric value. Radar waves are attenuated mainly by electrical conductivity σ. Since the two properties (σ, ε_r) affect radar wave propagation in different ways, it is instructive to examine more closely the relationship between them. This will result in a better understanding of the information content of GPR data.

As a conceptual model, following the discussion in von Hippel (1954), consider two oppositely charged parallel-plate conductors like those shown in Figure 9.6a. The parallel plates of area A and separation d serve as a basic model for a capacitor, which is essentially a charge-storage device. The intervening space is filled with vacuum, so that no electrical current can flow between the plates and accordingly they do not discharge. A voltage V is

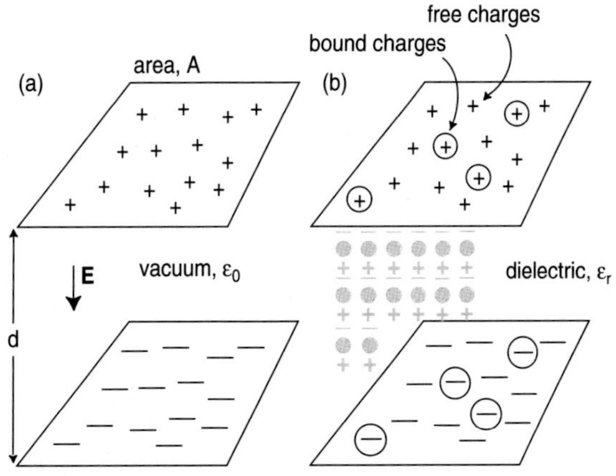

Figure 9.6 A parallel-plate capacitor filled with (a) air; (b) dielectric material.

maintained across the plates by an external battery such that an electric field **E** exists in the $\hat{\mathbf{z}}$ direction, positive downward, as shown.

Let the surface charge density on the two plates be given by $\pm s$ [C/m^2] and the total charge on the plates by $\pm Q$ [C], where $Q = sA$. Now, suppose the space between the plates becomes filled with a dielectric material of relative permittivity ε_r, and zero conductivity $\sigma = 0$, as shown in Figure 9.6b. The electric field **E** now polarizes the dielectric material so that microscopic dipoles form, as shown by the gray symbols in the figure. The dipoles attract and *bind* some of the free charges on the plates. These newly bound charges are the ones that are circled in the figure. Thus, only a portion of the original free charges remain to contribute to the voltage across the plates. The relative permittivity ε_r is then defined as $Q' = Q/\varepsilon_r$ where $\pm Q'$ is the total *free charge* that remains after the dielectric material is added. Note that $Q' < Q$. The free charge density is similarly reduced to $s' = s/\varepsilon_r$. The total bound charge is $Q - Q' = Q(1 - 1/\varepsilon_r)$. Still no current flows across the plate since the added material is non-conductive. The capacitor continues to store electrical energy and the plates do not discharge.

Now suppose the space between the plates is filled by a dielectric material that also has a non-zero electrical conductivity σ. In this case, some of the free charges are able to migrate through the material between the plates, and hence an electric current is created in accordance with Ohm's law $\mathbf{J} = \sigma\mathbf{E}$. The capacitor, in this case, is *leaky* and loses some of its stored electrical energy. The electric field **E** is given, as usual, in terms of the free charge density by $\mathbf{E} = s'\hat{z}/\varepsilon_0$. A new *displacement field* **D** is defined in terms of the *total* charge density as $\mathbf{D} = s\hat{\mathbf{z}} = \varepsilon_0\varepsilon_r\mathbf{E} = \varepsilon\mathbf{E}$. The physical interpretation of the displacement field is explained as follows.

Suppose an external electric current of density **J** [A/m^2] charges the plates. The total surface charge density s increases with time as charges are brought up to the plate by the external current. The increase in charge density, in the case of a pure non-conducting dielectric, is given by $\mathbf{J} = (ds/dt)\hat{\mathbf{z}} = (d\mathbf{D}/dt)$. In the case $\sigma = 0$, all the charges are stopped on the plate and no current is conducted across. Nevertheless, there is a *displacement current* of magnitude $d\mathbf{D}/dt$. If the capacitor is filled with conducting material, some of the charges that are brought up to the plate by the external current **J** can migrate across to the other plate. Therefore, in this case there is also a *conduction current* given by Ohm's law. The total current is then the sum $\mathbf{J} = \sigma\mathbf{E} + d\mathbf{D}/dt$. Sometimes the electric polarization vector $\mathbf{P} = \varepsilon_0\chi_e\mathbf{E}$ is introduced, in which χ_e is termed the *electric susceptibility*, and we write $\mathbf{D} = \varepsilon_0\mathbf{E} + \mathbf{P}$. However, we will find no further occasion to use the electric polarization vector **P**.

The displacement current, and its distinction from the conduction current, is usefully described by the simple parallel-plate model considered above. However, the geophysical situation may be better conceptualized with reference to Figure 9.7 in which an AC voltage $V\exp(i\omega t)$ is applied across a rock mass. The rock has electromagnetic properties (σ, ε_r) and thus both a conduction current $\sigma\mathbf{E}$ and a displacement current $d\mathbf{D}/dt$ will be present. There are both energy-storage and energy-loss mechanisms. It is well known in the history of physics (e.g. Selvan, 2009) that the presence of the energy-storage term, associated with the displacement current, led Maxwell to predict the existence of electromagnetic waves.

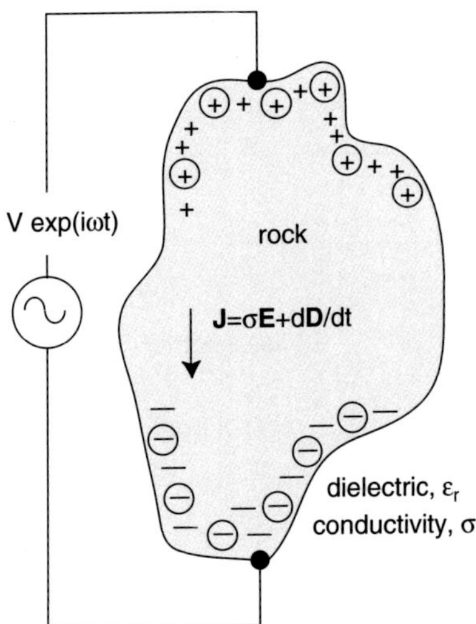

Figure 9.7 Conceptual model of a rock under an applied time-harmonic voltage.

If an AC voltage with harmonic-time variation $\exp(i\omega t)$ is applied to a rock mass, as shown, the total current in the frequency domain becomes $\mathbf{J} = \sigma\mathbf{E} + i\omega\mathbf{D} = \sigma\mathbf{E} + i\omega\varepsilon\mathbf{E} = \sigma^*\mathbf{E}$ where $\sigma^* = \sigma + i\omega\varepsilon$ is a *complex conductivity* that depends on frequency. Neglecting EM induction effects for the moment, the conduction current $\sigma\mathbf{E}$ is in phase with the driving voltage while the real component of the displacement current $i\omega\mathbf{D}$, generated by lossless polarization effects, is out of phase. The dielectric loss term described in the previous section acts, however, mathematically, much like the Ohmic loss term and often in the literature it is seen that both are lumped together into the imaginary component of a complex permittivity $\varepsilon^*(\omega)$. However, it should be remembered that the two phenomena have distinct physical origins: Ohmic loss is essentially the kinetic energy loss of migrating quasi-free charges scattering off lattice ions, while dielectric loss is associated with the rotation of bound charges in a viscous fluid. As such, it is easy to see that the Ohmic term is sensitive largely to connected porosity while the dielectric loss term is more sensitive to total water content, which equals the total (connected and unconnected) porosity in a fully saturated system.

9.3 Dielectric properties of rocks and soils

The electrical properties of porous rocks, including the conductivity σ and the dielectric constant ε_r, are highly sensitive to the pore-scale microstructure and the volume fractions of the solid and fluid phases. The most important factor in determining the dielectric constant ε_r of near-surface geomaterials is the volumetric water content, θ_W. This is because water is characterized by $\varepsilon_r \sim 81$ and air by $\varepsilon_r = 1$ while the dielectric constant of the solid matrix material most commonly falls somewhere close to the range $\varepsilon_r \sim 3$–4. The empirical Topp equation (Topp *et al.*, 1980)

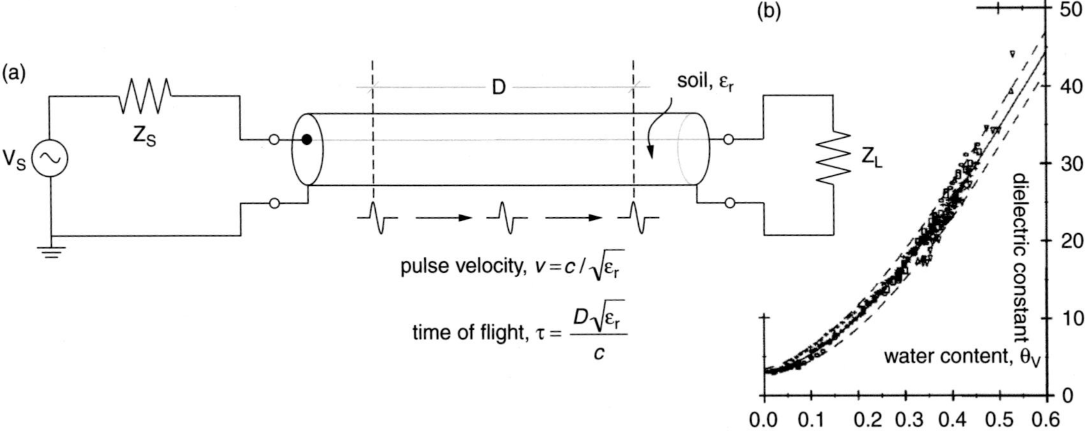

Figure 9.8 (a) Transmission-line method for determining soil dielectric constant; (b) soil dielectric constant as a function of moisture content for different soil types. After Topp *et al.* (1980).

$$\varepsilon_r = 3.03 + 9.3\theta_W + 146.0\theta_W^2 - 76.7\theta_W^3 \tag{9.4}$$

is a widely used formula for expressing the bulk dielectric constant of a soil as a function of its water content. It can be regarded as a GPR equivalent of Archie's law in the sense that it is based on a compendium of laboratory measurements using a wide variety of samples. The method for determining ε_r involves a time-of-flight measurement of an electromagnetic pulse propagating along a coaxial waveguide loaded with the soil sample, as shown in Figure 9.8a. The results from using soils of different types with varying water content θ_W are summarized in Figure 9.8b, in which the curve given by Equation (9.4) and an estimate of its experimental uncertainty are overlain. The Topp equation works well in clays and loams but has less predictive capabilities for organic-rich soils.

The effect of salt content on the bulk ε_r of water can become significant at high values of salinity. In general, ε_r drops with increasing salinity, to values as low as ~ 60 for highly saline pore waters of ~ 100 parts per thousand (ppk) (Hizem *et al.*, 2008). There are three effects at work: (i) more salt by volume implies fewer polarizable water molecules; (ii) water molecules cannot rotate as easily if they are weakly bound to Na and Cl ions; (iii) mobile Na and Cl ions within the pore-fluid electrolyte agitate the water molecules, tending to randomize alignment of the H_2O dipole moments. Hizem *et al.* (2008) also note that ε_r of water drops with increasing temperature T to as low as ~ 45 (at 150 °C) due again to thermal agitation of the H_2O dipole moments.

Many heuristic dielectric mixing rules have appeared in the literature. The heuristic rules, by definition, do not have a firm theoretical basis. The complex refractive index model (CRIM), for example, is based simply on a volumetric averaging of the dielectric constants of the constituents of a composite material (Tsui and Matthews, 1997). The CRIM formula for the dielectric constant ε_r of a partially saturated rock is

$$\sqrt{\varepsilon_r} = \phi(1 - S_W)\sqrt{\varepsilon_0} + (1 - \phi)\sqrt{\varepsilon_1} + \phi S_W\sqrt{\varepsilon_2}, \tag{9.5}$$

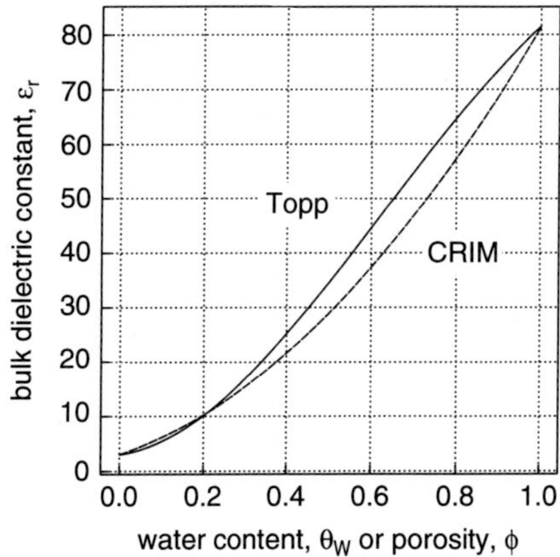

Figure 9.9 Bulk dielectric constant of saturated vesicular basalt as predicted by the Topp and CRIM equations.

where ϕ is the porosity, S_W is the water saturation, ε_1 is the dielectric constant of the matrix material, and ε_2 is the dielectric constant of the pore water. Often, however, CRIM provides unsatisfactory estimates of the bulk dielectric constant since it does not take into account the geometric arrangement of the various solid and fluid phases that make up the composite material (West *et al.*, 2003).

Consider a fully water-saturated vesicular basalt as an idealized system composed of fluid spheres of dielectric constant $\varepsilon_2 = 81.5$ embedded within a solid matrix of dielectric constant $\varepsilon_1 = 3.03$. The bulk dielectric constant ε_r of the composite system, as a function of the volumetric water content, is shown in Figure 9.9 using the Topp equation (9.4) and the CRIM equation (9.5).

A number of more rigorous, physics-based dielectric mixing rules have been developed using an effective medium approach. One of these is the Maxwell–Wagner–Bruggeman–Hanai (MWBH) model (Chelidze and Gueguen, 1999), which grew out of original work by Maxwell in the nineteenth century. The MWBH model treats the composite material as a spatially uniform, concentrated suspension of spherical particles of dielectric constant ε_2 embedded within a host medium of dielectric constant ε_1. The MWBH relationship between the effective dielectric constant ε_r of the composite medium and the dielectric constants of its constituents is

$$\frac{\varepsilon_2 - \varepsilon_r}{\varepsilon_2 - \varepsilon_1}\left(\frac{\varepsilon_1}{\varepsilon_r}\right)^{1/3} = \phi. \tag{9.6}$$

Note that the dielectric constants ε_1 and ε_2 appearing in Equation (9.6) can actually be complex functions of frequency. Robinson and Friedman (2001) have developed a formula that predicts the bulk dielectric constant of mixtures containing n different grain sizes.

The mixing theories described above do not take into account surface electrical-polarization processes that might occur on the interfaces between two components of the mixture. These surface effects are likely to be important factors in determining the effective dielectric constant of actual porous near-surface geomaterials. For example, the Maxwell–Wagner effect

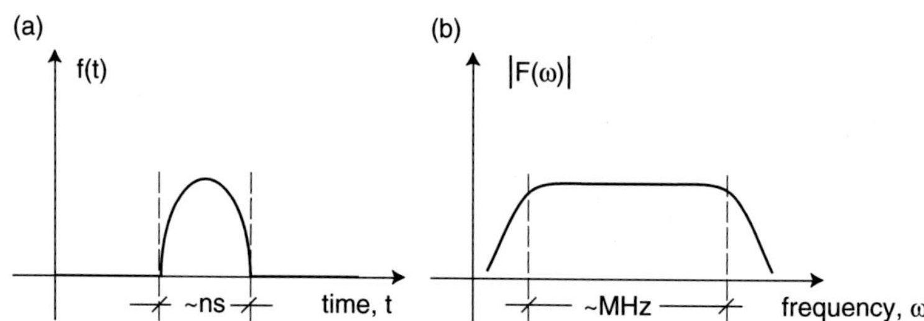

Figure 9.10 (a) Transmitted pulse in the time domain. (b) Broadband amplitude spectrum.

increases with increasing hydrocarbon saturation in a three-phase oil–water–sand mix due to the appearance of additional oil–water interfaces. A good discussion of surface effects on the dielectric properties of porous rocks can be found in Chelidze and Gueguen (1999).

9.4 Resolution

The *resolution* of a GPR system ultimately depends on its capacity to distinguish between two radar returns that are spaced closely in time. The two returns could be due, for example, to the top and bottom interfaces of a buried thin layer. Hence, the resolution is determined by the transmitted pulse width Δt, along with any broadening and distortion of the pulse as it propagates into the subsurface. As the pulsewidth of a given TX decreases, its frequency bandwidth Δf increases (Figure 9.10). The pulsewidth–bandwidth trade-off is a general principle of Fourier analysis, as has been well documented in the seismic literature (Knapp and Steeples, 1986). Therefore, a high-resolution GPR necessarily transmits a broad band of frequencies. Resolution is improved by transmitting at higher frequencies only if the bandwidth is simultaneously increased. This can be achieved in the time domain by narrowing the pulse width. A pulse radar operating in the megahertz range generally has a bandwidth of a similar magnitude and a pulse width of $\Delta t \sim 1$–10 ns.

There is also a trade-off between range and resolution in GPR systems. As was shown earlier in Figure 9.5b, attenuation increases with frequency beyond about ~ 100 MHz. Sharpening the resolution by narrowing the pulse width invariably comes at the cost of a reduction in the depth of penetration at which the subsurface targets may be interrogated. The range–resolution trade-off (Davis and Annan, 1989) places a fundamental limitation on the capability of GPR to image small-scale structures at depth.

9.5 Data acquisition

A standard geophysical surveying geometry for GPR is the common-offset configuration (Figure 9.11a) in which the transmitter and receiver antennas are moved in tandem along a profile while maintaining a fixed separation distance between them. The WARR

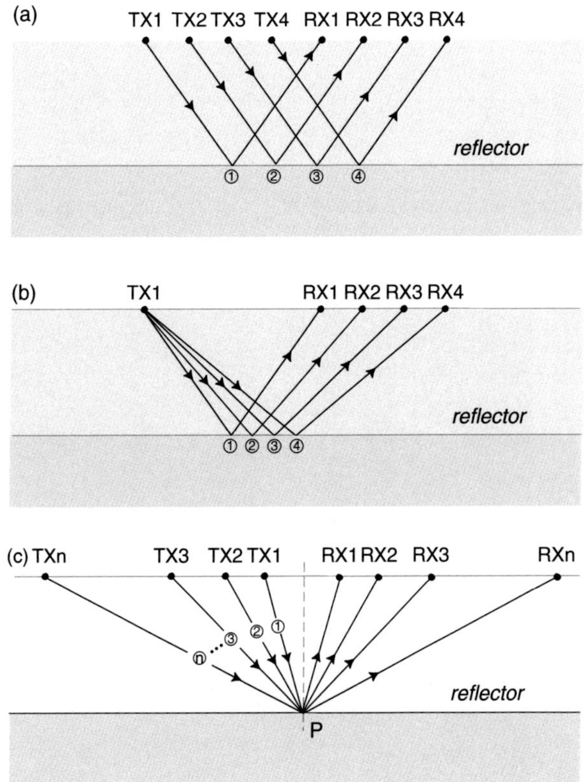

Figure 9.11 (a) Common-offset GPR profiling; (b) wide-angle radar reflection, WARR; (c) common depth-point configuation, CDP.

configuration (Figure 9.11b) is analogous to the deployment of a seismic geophone array except that typically only a single receiver is available and it is stepped out from the fixed transmitter location, as shown. The WARR configuration is not as popular as common-offset sounding since attenuation in the ground is often sufficiently high that good quality data from large TX–RX offsets cannot be acquired. The common depth-point (CDP) configuration is shown in Figure 9.11c. It is often used to estimate the ground radar velocity v by determining the normal moveout of radar return signals, as in the seismic-reflection technique. A new generation of GPR systems, presently under active development and beginning to appear in commercial offerings (e.g. the Sensors and Software product SPIDAR, see www.sensoft.ca/products/spidar/spidar.html), consists of multiple transmitter and multiple receiver antennas working simultaneously. Such systems increase the efficiency of data acquisition in the field, thereby enabling a lower cost and/or an increased scope of a project.

The antennas can be arranged in broadside, in-line or cross-polarized orientations, as shown in Figure 9.12. Due to the vectorial nature of electromagnetic waves, each of the orientations provides a different illumination of buried targets. A discussion on radar target polarization effects appears later in this chapter. A photograph showing in-line 100 MHz GPR data acquisition with 1.0 m TX–RX separation is shown in Figure 9.13.

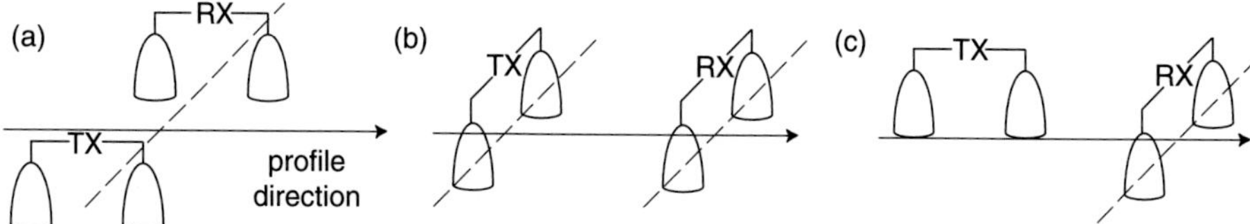

Figure 9.12 GPR antenna orientations: (a) broadside; (b) in-line; (c) cross-polarized relative to the data-acquisition-profile direction.

Figure 9.13 In-line 100 MHz GPR data acquisition with a 1.0-m TX–RX separation distance.

9.6 Basic GPR data processing

A number of standard data-processing steps should be performed after data acquisition in order to transform measured radargrams into a time or a depth section that is ready for advanced processing, qualitative interpretation, and attribute analysis. The standard processing steps described below are recommended in most cases but they do not all have to be performed or in the particular order given. It should be kept in mind that each data-processing step results in a loss or transformation of the information that is contained in the original radar data. Thus, processing should always be performed carefully and, in the final analysis, it becomes a subjective process.

Although it is tempting to apply standard seismic data-processing algorithms (Yilmaz, 2001) to GPR data, significant interpretation errors can be made since the nature of subsurface radar wave propagation is very different from that of seismic wave propagation. As pointed out by Cassidy (2009), GPR signals exhibit considerably more attenuation, dispersion, and scattering from heterogeneities than do seismic signals. The spatial variations in electromagnetic properties of geomaterials are moreover much stronger than variations in

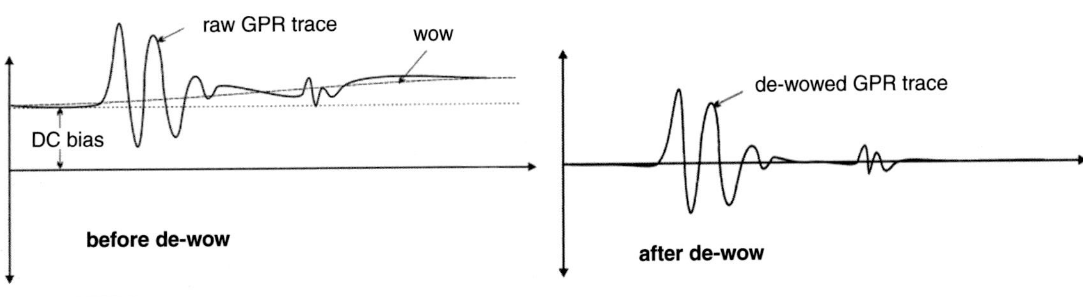

Figure 9.14 Effect of de-wow filter. After Cassidy (2009).

elastic properties; hence, GPR wavefronts generally exhibit more geometric complexity than is found in seismic wavefronts. The GPR waves are also vectorial so that changes in polarization as the wave propagates, reflects, refracts, diffracts, and scatters must be taken into account. Antenna design and ground coupling further contribute to complexities that are specific to GPR wave propagation and these have no exact seismic counterpart.

The first step of GPR data processing is file reconciliation and trace editing. The former is done to bring the datafiles that were actually recorded in the field into the desired survey format. For example, some of the profiles may have been acquired in the reverse direction or acquired out of the correct sequence. Some of the datafiles may be redundant, or there may be some files that are unusable due to equipment malfunctions, poor acquisition protocol, or excessive external noise. Bad traces within a profile may need to be removed. Some profiles may contain extra traces, or too few traces, due to navigation errors. The extra traces can be deleted. The missing traces can be restored by interpolating from neighboring traces. These steps are necessary as most processing algorithms require uniform station spacing. Filtering, as discussed in Chapter 2, may be used at this stage to remove noise spikes and some of the extraneous energy from noise sources whose dominant frequencies lie outside the GPR bandwidth.

An essential GPR processing step to be performed early in the processing sequence, after file reconciliation, trace editing, and preliminary filtering, is the *time-zero correction*. Here, the measured radar traces along a profile are individually shifted along the time axis such that a recognizable feature that is common to each trace, typically the first peak of the earliest arriving pulse train, is aligned to a common temporal datum. Trace misalignment is caused by many factors, including drift in the transmitter or receiver electronics, irregularities in either the cables connecting the transmitter and receiver electronics to the antennas, or in the connectors themselves; or else small, along-profile variations in TX–RX antenna spacing and orientation. The time-zero correction improves the spatial coherency, or cross-trace correlation, of the resulting time section and readies it for further processing.

A *de-wow* low-cut filter should also be applied as one of the early GPR data-processing steps (Cassidy, 2009). "Wow" is the ubiquitous slow variation of the baseline amplitude found in radar traces, and includes any bias, or constant shift, in the baseline amplitude (Figure 9.14, left). The baseline amplitude of a radar trace, at large values of the two-way traveltime, ideally should be a constant zero. Wow variation is caused by the presence of unwanted low-frequency components contained in the spectrum of the transmitted electric field, and also by EM induction effects in the conductive ground. The de-wow filter attenuates these

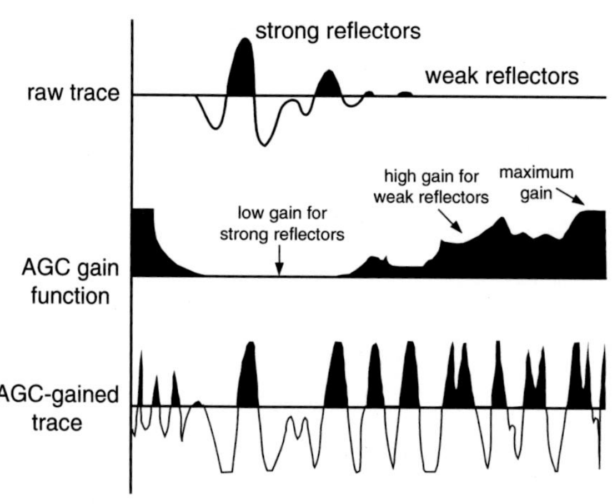

Figure 9.15 The effect of an AGC function applied to a radar trace. (See pulseEKKO 100 User's Guide, www.sensoft.ca)

low-frequency components to produce a zero baseline amplitude. The effects of a de-wow filter are shown in Figure 9.14, right.

As in seismic data processing, *gain control* functions can be applied to radar traces in order to correct for geometric spreading and attenuation of the propagating wavefront, and to equalize the signal returns from all depths. There are many different types of gain functions that could be applied. An automatic gain control (AGC) function, for example, is a simple multiplier that scales with the inverse of the signal strength of the raw trace. The effect of applying an AGC function to a raw trace is shown in Figure 9.15. The objective of applying such an AGC function is to enable visualization of both shallower and deeper reflectors on radar sections at roughly the same display intensity.

A *background-removal filter* involves the subtraction, from each radar trace, of a lateral moving average of the radar amplitudes over a given early-time window. This filter mitigates the unwanted appearance of ground clutter or antenna ringing in displayed radar sections. Ground clutter is the high-amplitude, laterally continuous signal seen in radar sections at early time (Figure 9.16, left). Ground clutter is caused by direct coupling, or cross-talk, between the TX and RX antennas and contains no useful subsurface information. The background removal filter cleans up the early portion of the radar section and sometimes permits a better visualization of very shallow reflectors. A deleterious effect of background filtering, however, is that it can also remove much of the signature of slowly undulating soil horizons and other near-surface geological features, which could include top of bedrock or the water table. Thus, background removal should be used with care or avoided if shallow soil stratigraphic, structural, or hydrostratigraphic mapping is an objective of the GPR survey.

A radar time section processed using a background removal filter and gain control is shown for illustrative purposes in Figure 9.16, right. The ground clutter is largely attenuated and the deeper reflectors are much better imaged in the processed time section.

Elevation measurements should be made along GPR profiles that traverse irregular terrain. A *topographic correction* can then be made to account for the distortion in the acquired radargram due to the along-profile elevation changes. A simplified illustration of a GPR topographic correction is shown in Figure 9.17. Suppose the geology consists of a subhorizontal interface buried at depth D beneath a horst of height h. The radar velocity of the upper

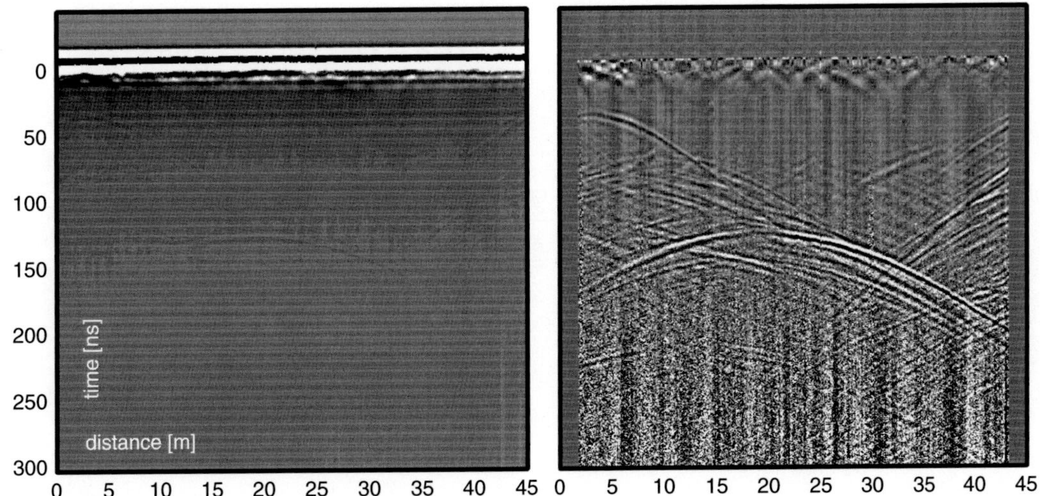

Figure 9.16 500 MHz GPR data. The left panel shows raw traces, while the right panel shows the same data after background removal and gain control.

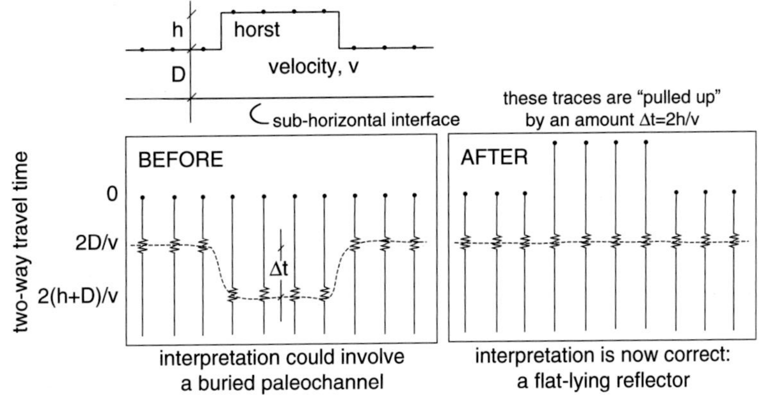

Figure 9.17 Schematic illustration of GPR topographic correction.

unit is known or estimated to be v. An idealized, uncorrected radargram is shown at left. A naive geological interpretation that does not take elevation into account could involve a buried paleochannel, as shown by the dotted line. The topographically corrected radargram is shown at right. The elevation data, combined with the knowledge or estimate of the radar velocity v, are used to "pull up" the traces that were acquired over the horst structure by an amount $\Delta t = 2h/v$. The correct interpretation of a flat-lying radar reflector can now be made.

A field example from Cassidy (2009) is shown in Figure 9.18. The data were acquired over the lobe of a pyroclastic flow. Notice in the topographically corrected section, at right, that the main reflecting horizons (marked as *key basal surfaces*) appear more flat-lying, and are closer to their correct stratigraphic positions. The topographically corrected radargram thus allows a more reliable geological interpretation than does the uncorrected radargram.

A CDP analysis is performed by symmetrically expanding the TX and RX antennas about a common midpoint (Figure 9.11c, bottom). A flat-lying subsurface reflector would exhibit a

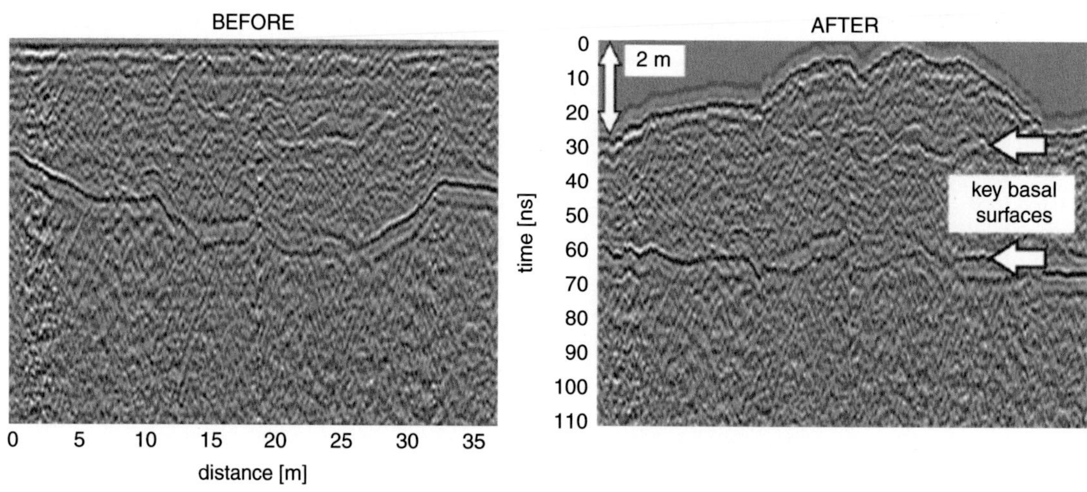

Figure 9.18 225 MHz GPR data. The left panel shows original radargram; right panel, after topographic correction. After Cassidy (2009).

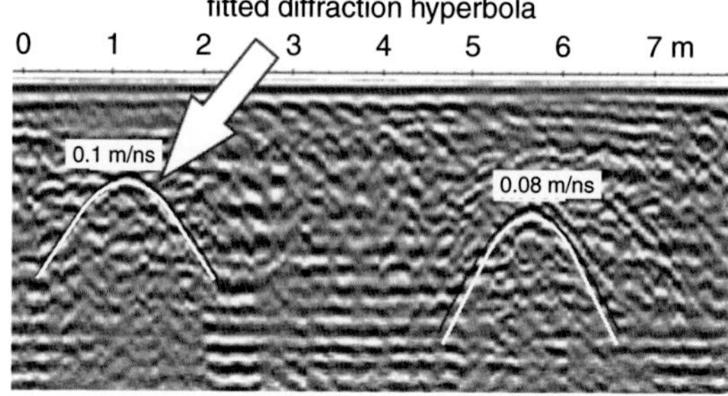

Figure 9.19 Radar velocity analysis by diffraction hyperbola fitting. After Cassidy (2009).

normal hyperbolic moveout in the radar time section, analogous to the previously discussed seismic case (see Chapter 6). As in the seismic case, the subsurface velocity is the one that, when used in an NMO correction, best flattens out the reflector. Alternatively, the subsurface velocity can be estimated by fitting a hyperbolic function to observed diffractions in a radar time section (see Figure 9.19). Once the subsurface radar velocity is known, by either of these methods, the practitioner is enabled to make a *time-to-depth conversion*. Then, one can display a radar depth section rather than a time section. The advantage of a depth section is that the radar reflection horizons presumably appear at their actual depth beneath the surface.

9.7 Advanced GPR data processing

Advanced processing steps are not always required but are briefly summarized here for the convenience of the reader. Most often, the basic processing steps outlined above are sufficient to make a useful geological interpretation. Advanced processing methods include deconvolution, *f–k* filtering, and migration.

Deconvolution is the process of removing the effect of the transmitted source wavelet from the measured radar traces, in an attempt to expose the idealized impulse response of the heterogeneous subsurface. A discussion of convolution and deconvolution, mainly in the context of seismic data processing, appears in Chapters 2 and 11, respectively. Since the source wavelets of GPR are considerably more complex than their seismic counterparts, and the medium is often highly dispersive, deconvolution is more challenging and therefore is used infrequently in GPR data processing.

The *f–k* filter operates as a bandpass filter in both time and space simultaneously. The two-dimensional (2-D) spatiotemporal radar data are first Fourier transformed into the frequency–wavenumber, or (f, k) domain. Then, a pass region in (f, k) is selected. Energy outside the pass region is attenuated and then the data are inverse-Fourier transformed back into the original time–space (t, x) domain. This filter can be used to suppress or enhance dipping reflectors depending on their orientation.

A multitude of seismic migration algorithms have been developed with great success in petroleum exploration geophysics (e.g. Etgen *et al.*, 2009). The main goal of migration is to undo the effects that the finite-velocity wave propagation bestows on measured time sections. Specifically, migration collapses diffractions back to their causative point sources and re-positions dipping events to their correct subsurface locations (see Chapter 6 for further details). Migration requires that the subsurface velocity is known; this is usually estimated in GPR by a common midpoint (CMP) analysis or the diffraction-fitting procedure described above. A seismic migration algorithm however cannot be bodily taken over and applied to GPR data owing to the vectorial nature of electromagnetic wave propagation. Instead, specifically designed *vector migration* algorithms must be developed. These are presently under active investigation (e.g. Streich and van der Kruk, 2007; Streich *et al.*, 2007). Successful schemes must take into full account polarization effects and antenna radiation patterns.

9.8 Electromagnetic plane waves

A practical understanding of the GPR technique is enhanced by a basic knowledge of the underlying theory of electromagnetic wave propagation in conductive media. The classical equations that govern the behavior of electromagnetic (**E**, **B**) fields in source-free regions are the Maxwell equations

$$\nabla \times \mathbf{H} = \mathbf{J} + \varepsilon \frac{\partial \mathbf{E}}{\partial t}; \tag{9.7}$$

$$\nabla \times \mathbf{E} = -\frac{\partial \mathbf{B}}{\partial t}; \tag{9.8}$$

where ε is the dielectric permittivity of the medium, along with the constitutive relations

$$\mathbf{J} = \sigma \mathbf{E}; \tag{9.9}$$

$$\mathbf{B} = \mu \mathbf{H}; \tag{9.10}$$

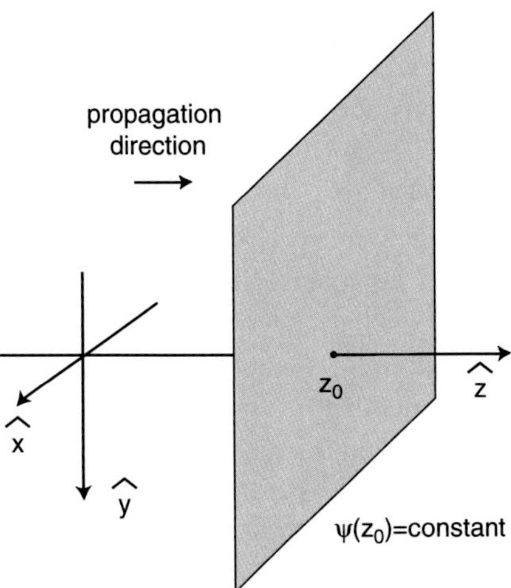

Figure 9.20 Snapshot of a planar wavefront, which moves to the right with velocity v.

where σ is the electrical conductivity and μ is the magnetic permeability of the medium. Following Wangsness (1986), we set $\nabla \cdot \mathbf{J} = 0$ and then eliminate the magnetic field $\mathbf{B}$ from the above set of equations, giving rise to the *vector damped wave equation* for the electric field $\mathbf{E}$,

$$\nabla^2 \mathbf{E} - \mu\sigma \frac{\partial \mathbf{E}}{\partial t} - \mu\varepsilon \frac{\partial^2 \mathbf{E}}{\partial t^2} = 0; \tag{9.11}$$

while there is a similar damped wave equation for $\mathbf{B}$.

Any Cartesian component $\psi = (E_x, E_y, E_z, B_x, B_y, B_z)$ of the electric $\mathbf{E}$ or magnetic $\mathbf{B}$ fields satisfies the *scalar damped wave equation*

$$\nabla^2 \psi - \mu\sigma \frac{\partial \psi}{\partial t} - \mu\varepsilon \frac{\partial^2 \psi}{\partial t^2} = 0. \tag{9.12}$$

It should be noted that in spherical or cylindrical coordinates, the various components of $(\mathbf{E}, \mathbf{B})$ do not satisfy scalar dampled wave equations, as in Equation (9.12), since the unit vectors in these coordinate systems depend on position.

In a non-conducting medium ($\sigma = 0$), such as air to an excellent approximation, the damped wave equation reduces to an ordinary lossless wave equation

$$\nabla^2 \psi - \mu\varepsilon \frac{\partial^2 \psi}{\partial t^2} = 0. \tag{9.13}$$

In the most general situation, the scalar function $\psi = \psi(\mathbf{r}, t)$ is an arbitrary function of position and time. In the specific case of a plane wave, which we can assume without loss of generality to be propagating in the $\hat{\mathbf{z}}$ direction, the function ψ reduces to a planar wavefront form $\psi = \psi(z, t)$ such that for any x, y the function ψ is a constant (see Figure 9.20). Inserting the planar wavefront form into Equation (9.13) reduces it to the one-dimensional lossless wave equation

$$\frac{\partial^2 \psi}{\partial z^2} - \frac{1}{v^2}\frac{\partial^2 \psi}{\partial t^2} = 0, \tag{9.14}$$

where $v = 1/\sqrt{\mu\varepsilon}$ is the wave velocity.

In vacuum, the electromagnetic wave velocity assumes the value $c = 1/\sqrt{\mu_0\varepsilon_0}$. In geological media, the magnetic permeability normally obeys $\mu_r > 1$ and dielectric permittivity similarly obeys $\varepsilon_r > 1$, such that the electromagnetic wave velocity is reduced relative to the speed in vacuum.

Again, closely following Wangsness (1986), the wave equation (9.14) can be conveniently solved using the *separation of variables* technique in which a solution is posited in the form $\psi = Z(z)T(t)$ as a product of a function of z only with a function of t only. Inserting the separable form $Z(z)T(t)$ into the wave equation (9.14) and re-arranging gives

$$\frac{1}{Z}\frac{\partial^2 Z}{\partial z^2} = \frac{1}{v^2 T}\frac{\partial^2 T}{\partial t^2} = -k^2, \tag{9.15}$$

where k^2 is a constant independent of z and t. Thus, the scalar wave equation has separated into two ordinary differential equations

$$\frac{\partial^2 Z}{\partial z^2} + k^2 Z = 0; \tag{9.16}$$

$$\frac{\partial^2 T}{\partial t^2} + \omega^2 T = 0; \tag{9.17}$$

where $\omega = kv$ is the angular frequency. The constant k is termed the *wavenumber*. The general solutions of Equations (9.16) and (9.17) are of the form $Z(z) \sim \exp(\pm ikz)$ and $T(t) \sim \exp(\pm i\omega t)$. The choice of plus or minus sign fixes the phase of the wave relative to the coordinate origin. A plane electromagnetic wave propagating in the $\pm\hat{\mathbf{z}}$ direction has a form such as

$$\psi(z,t) = \psi_0 \exp(ikz - i\omega t). \tag{9.18}$$

Since ψ corresponds to any Cartesian component of the electromagnetic field (**E**, **B**) we can write

$$\mathbf{E}(z,t) = E_0 \exp(ikz - i\omega t)\,\hat{\mathbf{y}}; \tag{9.19}$$

$$\mathbf{B}(z,t) = -B_0 \exp(ikz - i\omega t)\,\hat{\mathbf{x}}; \tag{9.20}$$

where **E** and **B** are not independent of each other but are linked by the Maxwell equations. Note that **E** and **B**, as given by Equations (9.19) and (9.20) are completely *in phase* with each other, that is, their amplitudes wax and wane in tandem when viewed as a function of position or as a function of time. A sketch of the electromagnetic field associated with a propagating lossless electromagnetic plane wave is shown in Figure 9.21a.

The behavior of an electromagnetic plane wave changes fundamentally in a conducting (or lossy) medium, $\sigma \neq 0$. In this case, the one-dimensional scalar damped wave equation becomes

$$\frac{\partial^2 \psi}{\partial z^2} - \mu\sigma\frac{\partial \psi}{\partial t} - \mu\varepsilon\frac{\partial^2 \psi}{\partial t^2} = 0. \tag{9.21}$$

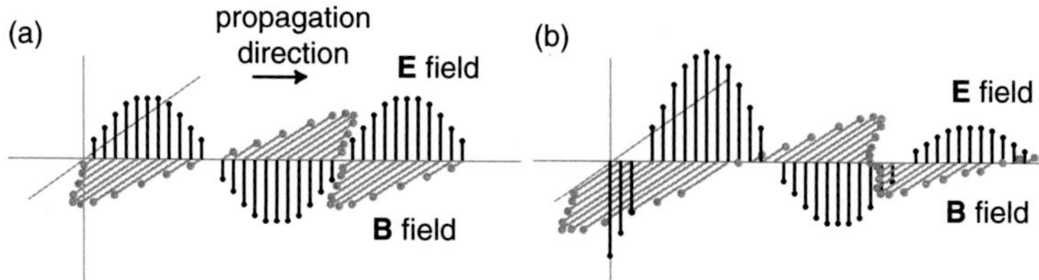

Figure 9.21 Electromagnetic plane-wave propagation: (a) lossless case; (b) lossy case.

Application of the separation of variables technique assuming $\psi \sim \exp(-i\omega t)$ indicates that the separation constant k^2 generalizes to

$$k^2 = \omega^2\mu\varepsilon + i\omega\mu\sigma \tag{9.22}$$

such that the wavenumber is the complex quantity

$$k = \sqrt{\omega^2\mu\varepsilon + i\omega\mu\sigma} \tag{9.23}$$

which can be decomposed into its real and imaginary parts, $k = \alpha + i\beta$. These are readily shown to be, from Equation (9.23),

$$\alpha = \omega\sqrt{\frac{\mu\varepsilon}{2}}\left[\sqrt{1+\left(\frac{\sigma}{\omega\varepsilon}\right)^2}+1\right]^{1/2}; \tag{9.24}$$

$$\beta = \omega\sqrt{\frac{\mu\varepsilon}{2}}\left[\sqrt{1+\left(\frac{\sigma}{\omega\varepsilon}\right)^2}-1\right]^{1/2}. \tag{9.25}$$

Inserting the complex wavenumber $k = \alpha + i\beta$ into Equation (2.18) gives

$$\psi(z,t) = \psi_0\exp(-\beta z)\exp(i\alpha z - i\omega t); \tag{9.26}$$

which can be compared directly with the lossless case, Equation (9.18). An important aspect of electromagnetic plane-wave propagation in a conducting medium is the presence of the *attenuation factor* $\exp(-\beta z)$. In the diffusive regime $\sigma \gg \omega\varepsilon$, the parameter $\beta \to 1/\delta$ where δ is the electromagnetic skin depth introduced in the previous chapter. The oscillating function $\exp(i\alpha z)$ in Equation (9.26) is governed by the real part of a complex k, i.e. Equation (9.24), rather than the real constant $k = \omega\sqrt{\mu\varepsilon}$ as in the lossless case. The electric field $\mathbf{E}$ can be written as

$$E(z,t) = E_0\exp(-\beta z)\exp(i\alpha z - i\omega t). \tag{9.27}$$

In the lossless case, it follows directly from the Maxwell equations that the electric and magnetic fields are linked by the equation $\mathbf{B} = (k/\omega)\hat{\mathbf{z}} \times \mathbf{E}$. Thus, $\mathbf{E}$, $\mathbf{B}$, and $\hat{\mathbf{z}}$ form a mutually orthogonal triad. In the lossy case, the $(\mathbf{E}, \mathbf{B}, \hat{\mathbf{z}})$ vectors remain mutually orthogonal. However, since k is complex, the $(\mathbf{E}, \mathbf{B})$ vectors are no longer in phase with each other, as indicated by the sketch in Figure 9.21b. As shown by Wangsness (1986), the phase difference between $\mathbf{E}$ and $\mathbf{B}$ is Ω where

$$\tan\Omega = \sqrt{1 + Q^2} - Q, \tag{9.28}$$

with $Q = \omega\varepsilon/\sigma$. The appearance of a phase difference between **E** and **B** is a consequence of the non-zero conductivity of the medium.

9.9 Plane-wave reflection from an interface

The electric or magnetic field at a point in space and time has both a magnitude and a direction and is thus a vector. However, the vectorial nature of electromagnetic waves is sometimes ignored by GPR practitioners. Polarization describes the orientation and magnitude of the field vector as the wave propagates and interacts with heterogeneities and buried targets. The polarization characteristics of GPR are useful for defining buried-target properties such as size, shape, orientation, and composition.

Suppose the **E**-field vector is always directed along a certain straight line. In that case, the **E** field is said to be linearly polarized. If the tip of the **E**-field vector instead sweeps out a circle as it propagates, it is circularly polarized. In the general case, the **E**-field vector is elliptically polarized. As described above, the electric **E** and magnetic **B** fields are mutually orthogonal to each other and to the direction of wave propagation. Let the direction of propagation be denoted again by $\hat{\mathbf{z}}$, in which case the electric field can be decomposed into two components

$$E_x(z,t) = E_{x0}\exp(-\beta z)\cos(\omega t - \alpha z - \varphi_X); \tag{9.29a}$$

$$E_y(z,t) = E_{y0}\exp(-\beta z)\cos(\omega t - \alpha z - \varphi_Y). \tag{9.29b}$$

The quantities E_{x0} and E_{y0} are the amplitudes of, respectively, the x and y components of the electric-field vector. Consider now a plane electromagnetic wave propagating in a general direction $\hat{\mathbf{p}}$ in a lossless, non-magnetic medium 1 characterized by dielectric constant ε_1. The wave is obliquely incident, with some incidence angle θ_i, onto a planar interface beneath which the medium 2 is characterized by a different dielectric constant ε_2, as shown in Figure 9.22. The problem is to determine the reflection and refraction coefficients.

The electromagnetic wave is said to be *horizontally polarized* if, as shown in Figure 9.22, the electric-field vector **E** points in the horizontal direction, i.e. the vector **E** lies in a plane that is perpendicular to the vertical plane of incidence. This is also known as the TE (transverse electric) mode of wave propagation. The magnetic-field vector **H** then lies in the vertical plane of incidence, as shown, and is orthogonal to the electric-field vector and the direction of propagation $\hat{\mathbf{p}}$. According to Figure 9.22, we have $\hat{\mathbf{p}} = \sin\theta_i\hat{\mathbf{x}} + \cos\theta_i\hat{\mathbf{z}}$.

The interaction of the incident plane wave with the interface is governed by the fundamental electromagnetic boundary conditions (continuity of the tangential components of the electric and magnetic fields) as well as Snell's law of refraction

$$\frac{\sin\theta_i}{\sin\theta_t} = \sqrt{\frac{\varepsilon_2}{\varepsilon_1}}, \tag{9.30}$$

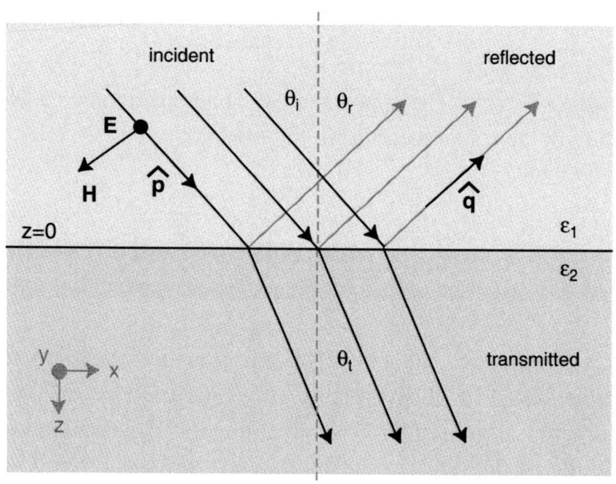

Figure 9.22 A horizontally polarized (TE-mode) plane electromagnetic wave obliquely incident upon a planar interface separating two lossless dielectric media.

and the law of reflection, $\theta_i = \theta_r$. The angle of transmission is θ_t and the angle of reflection is θ_r.

The lossless TE-mode electromagnetic plane wave propagating in the $\hat{\mathbf{p}}$ direction is described by an electric-field vector (of unit amplitude, for convenience)

$$\mathbf{E}_i \sim \exp(ik_1 \hat{\mathbf{p}} \cdot \hat{\mathbf{r}})\hat{\mathbf{y}} = \exp(ik_1[p_x x + p_z z])\hat{\mathbf{y}} = \exp(ik_1[x\sin\theta_i + z\cos\theta_i])\hat{\mathbf{y}}; \tag{9.31}$$

where we have made use in the last expression of the geometric identities $\cos\theta_i = p_z/|\mathbf{p}|$ and $\sin\theta_i = p_x/|\mathbf{p}|$, and $k_1 = \omega\sqrt{\mu_0\varepsilon_1}$ is the characteristic wavenumber of medium 1.

The total electric-field $\mathbf{E}^1$ in medium 1 is the sum of the incident field $\mathbf{E}_i$ and a reflected field $\mathbf{E}_r$ which propagates as a plane wave, according to the law of reflection, in the direction $\hat{\mathbf{q}} = \sin\theta_i\hat{\mathbf{x}} - \cos\theta_i\hat{\mathbf{z}}$. A reflection coefficient R is defined as the ratio of the amplitudes of the reflected and the incident electric fields, $R = |\mathbf{E}_r|/|\mathbf{E}_i|$. We have

$$\mathbf{E}^1 = \mathbf{E}_i + \mathbf{E}_r = \exp(ik_1[x\sin\theta_i + z\cos\theta_i])\hat{\mathbf{y}} + R_{TE}\exp(ik_1[x\sin\theta_i - z\cos\theta_i])\hat{\mathbf{y}} \tag{9.32}$$

where the subscript TE on the reflection coefficient indicates the case of a horizontally polarized incident plane wave. Similarly, we can write the electric-field vector $\mathbf{E}^2$ in medium 2 as a transmitted wave

$$\mathbf{E}^2 = \mathbf{E}_t = T_{TE}\exp(ik_2[x\sin\theta_t + z\cos\theta_t])\hat{\mathbf{y}} \tag{9.33}$$

with transmission coefficient generally defined by $T = |\mathbf{E}_t|/|\mathbf{E}_i|$. It remains now to find the unknown TE-mode reflection and transmission coefficients (R_{TE}, T_{TE}) via application at the interface $z = 0$ of the fundamental electromagnetic boundary conditions.

Enforcing first the continuity of the electric field, which is tangential to the interface since $\mathbf{E}$ is oriented in the $\hat{\mathbf{y}}$ direction, gives

$$\mathbf{E}^1|_{z=0} = \mathbf{E}^2|_{z=0}; \tag{9.34a}$$

$$\exp(ik_1 x\sin\theta_i) + R_{TE}\exp(ik_1 x\sin\theta_i) = T_{TE}\exp(ik_2 x\sin\theta_t).$$

The latter equation must hold for all x along the interface, including $x = 0$. In that case,

$$1 + R_{TE} = T_{TE}, \tag{9.35}$$

which provides the first constraint on the reflection and transmission coefficients. We now enforce continuity of the tangential magnetic-field component, H_x. An expression for H_x in terms of the electric field is obtained from Equations (9.8) and (9.10), keeping in mind the symmetry $E = E_y(x,z)\hat{\mathbf{y}}$,

$$\frac{\partial E_y}{\partial z} = i\mu\omega H_x. \tag{9.36}$$

Thus, continuity of H_x at the interface $z = 0$ is equivalent to

$$\frac{\partial \mathbf{E}^1}{\partial z}|_{z=0} = \frac{\partial \mathbf{E}^2}{\partial z}|_{z=0}. \tag{9.37}$$

Inserting Equation (9.37) into Equations (9.32) and (9.33) yields the second constraint on the unknown set of coefficients R_{TE} and T_{TE},

$$k_1 \cos\theta_i (1 - R_{TE}) = k_2 \cos\theta_t T_{TE}. \tag{9.38}$$

Solving Equations (9.35) and (9.38) for the reflection coefficient R_{TE} yields the final result

$$R_{TE} = \frac{\sqrt{\varepsilon_1}\cos\theta_i - \sqrt{\varepsilon_2 - \varepsilon_1 \sin^2\theta_i}}{\sqrt{\varepsilon_1}\cos\theta_i + \sqrt{\varepsilon_2 - \varepsilon_1 \sin^2\theta_i}}. \tag{9.39}$$

For the special case of normal incidence, $\theta_i = 0$, this expression reduces to a familiar form

$$R_{TE} = \frac{\sqrt{\varepsilon_1} - \sqrt{\varepsilon_2}}{\sqrt{\varepsilon_1} + \sqrt{\varepsilon_2}}, \tag{9.40}$$

which is the result earlier quoted in Equation (9.3).

The case of *vertically polarized*, or TM- (transverse magnetic-) mode wave propagation is characterized by an electric-field vector **E** lying in a plane that is parallel to the vertical plane of incidence. The TM-mode reflection coefficient is, by a similar analysis,

$$R_{TM} = \frac{\kappa\cos\theta_i - \sqrt{\kappa - \sin^2\theta_i}}{\kappa\cos\theta_i + \sqrt{\kappa - \sin^2\theta_i}}, \tag{9.41}$$

where $\kappa = \varepsilon_2/\varepsilon_1$. A graph of the reflection-coefficient amplitudes $|R_{TE}|$ and $|R_{TM}|$ as function of the incident angle θ_i, is given in Figure 9.23, for different values of the underlying dielectric ε_2. Equations (9.39) and (9.41) are known as the Fresnel equations.

Note that both amplitudes $|R_{TE}|$ and $|R_{TM}|$ approach 1.0, corresponding to total reflection, for grazing incident angles of $\theta_i \rightarrow 90°$. Also, as expected, the reflection coefficients tend to vanish as the interface contrast $\kappa \rightarrow 1$, that is, $\varepsilon_2 \rightarrow \varepsilon_1 = \varepsilon_0$. Finally, for the TM mode, it is interesting to note that there exists a range of incident angles $\theta_i \sim 55–70°$ over which there is no reflection for certain dielectric contrasts.

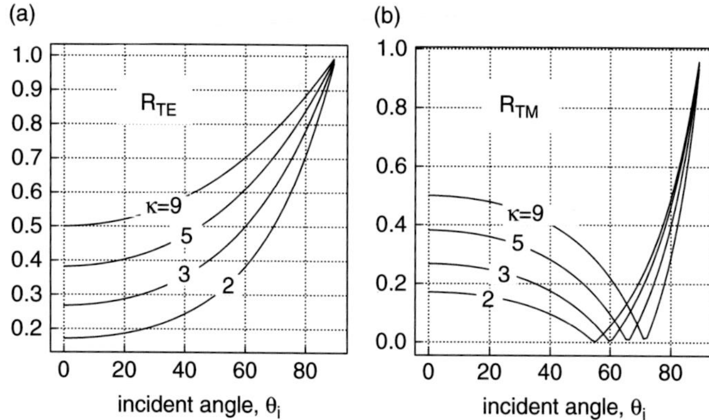

Figure 9.23 Reflection-coefficient amplitude for (a) TE-mode and (b) TM-mode plane waves as a function of incidence angle, with $\varepsilon_1 = \varepsilon_0$.

9.10 Analysis of thin beds

The detection of thin layers has long been of interest to explorationists engaged in seismic imaging of stratified petroleum reservoirs. Near-surface applications such as landslide hazard assessment or fractured-rock aquifer characterization also require imaging of thin subsurface rock layers as these sometimes define planes of slope instability or pathways for subsurface contaminant transport. The classic theoretical work on thin-bed analysis in seismology was performed by Widess (1973) who concluded that a bed as thin as $\lambda/8$ can be resolved, where λ is the dominant wavelength of the probing seismic signal. A more practical limit is suggested however by *Rayleigh's criterion* of $\lambda/4$. Below this limit, reflected signals from the interfaces at the top and bottom of the bed merge together and cannot be separated. Zeng (2009) recently showed that the practical resolution limit is affected by the shape of the source wavelet. Better resolution can be achieved with an asymmetric wavelet compared to a symmetric one.

Resolving a thin layer in GPR is more challenging than its seismic counterpart. The practical resolution limit is only ~ $3\lambda/4$ (Bradford and Deeds, 2006). The reduced capability of GPR to resolve thin layers owes mainly to the highly dispersive nature of the geological medium in which electromagnetic waves propagate. In a dispersive medium, a propagating wavelet rapidly becomes distorted and experiences large phase shifts upon reflection and transmission at material interfaces (Hollender and Tillard, 1998). Thus, it becomes difficult to recognize distinct returns from the top and bottom of a thin bed that have the same or similar shape as the transmitted wavelet. Detailed modeling of electromagnetic wave propagation in dispersive, heterogeneous geomaterials is normally required to determine whether a bed of a given thickness and material type can be resolved.

An important problem in GPR thin-bed analysis is to determine the bed reflectivity, since that can provide important information about the composition of the fill material. For example, the thin bed could be a fracture that contains a hazardous liquid contaminant. Consider the electromagnetic wave incident on a bed of thickness d and permittivity ε_2,

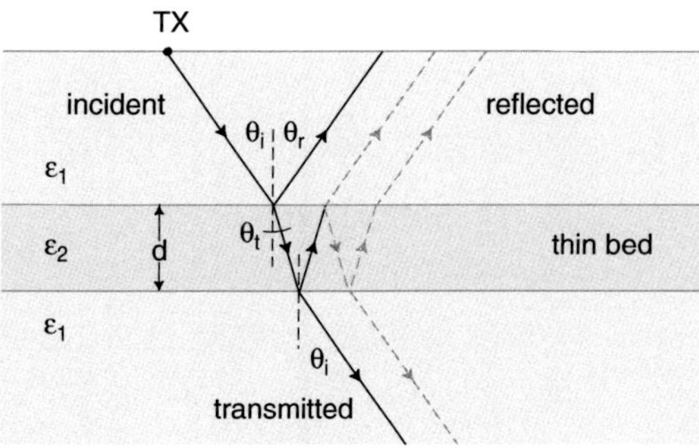

Figure 9.24 GPR reflection and transmission associated with a thin bed.

shown in Figure 9.24. The angles of incidence θ_i, reflection θ_r, and transmission θ_t obey Snell's law, Equation (9.30) and the law of reflection $\theta_i = \theta_r$. The background medium has permittivity ε_1. The reflection and transmission coefficients depend on the polarization of the incident wave, as described in the previous section. Deparis and Garambois (2009) derive the following formula for the reflectivity of the thin layer,

$$R(\omega) = R_{12}(\omega)\frac{1 - \exp[-i\varphi(\omega)]}{1 - R_{12}^2(\omega)\exp[-i\varphi(\omega)]}, \tag{9.42}$$

in which $\varphi(\omega) = 2dk_2(\omega)\cos\theta_i$, and $R_{12}(\omega)$ is either the TE- or TM-mode reflection coefficient (Equation (9.39) or (9.41)), depending on the polarization of the incident wave.

It is easy to see that $R(\omega)$ given in Equation (9.42) reduces to its appropriate values for the limiting cases of a bed of zero thickness, and a bed of infinite thickness. Indeed, as $d \to 0$ it follows that $R \to 0$, which is the value that would be expected if the bed were absent. Also, as $d \to \infty$ one obtains $R \to R_{12}$, which is the correct form for a lower halfspace of dielectric permittivity ε_2.

While the theoretical expression for thin-bed reflectivity (Equation (9.42)) is quite simple, a number of idealizations were used in its derivation. Practically speaking, it is a difficult task to estimate reflectivity from measured amplitudes of GPR reflections (Bradford and Deeds, 2006; Deparis and Garambois, 2009). Factors that were not taken into account in the development of Equation (9.42) include: the shape of the source wavelet; the TX and RX radiation patterns; the coupling of the antennas to the ground; and scattering and reflection losses along the wave-propagation path. Despite these difficulties, Bradford and Deeds (2006) successfully analyzed GPR reflectivity variations with TX–RX offset, using a methodology similar to seismic amplitude versus offset (AVO) analysis (Castagna, 1993).

9.11 GPR antennas

Proper interpretation of GPR images can be modestly enhanced with a rudimentary understanding of the theory of antennas. An *antenna*, for our purposes, is a radiating current element whose dimensions are comparable to an electromagnetic wavelength. The

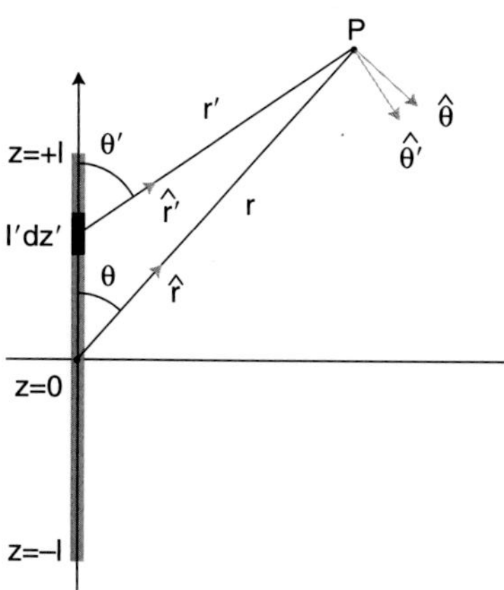

Figure 9.25 Calculation of the electric field of a long-straight-wire antenna carrying an oscillating electric current. After Wangsness (1986).

electric-current distribution, which is established by connecting the antenna to a power supply, is not uniform over the length of an antenna. Following Wangsness (1986), we consider the electromagnetic field radiated by a simple antenna consisting of a straight wire of length $2l$ oriented in free space in the $\hat{\mathbf{z}}$-direction. It is of interest to determine the electric field $\mathbf{E}(r, \theta)$ at some point in the far field, or *radiation zone*, defined by $kr \gg 1$, where $k = \omega\sqrt{\mu_0 \varepsilon_0} = \omega/c$ is the characteristic wavenumber of free space.

An oscillating electric dipole is equivalent to a current element so it is natural to visualize the long, straight antenna as an assemblage of current elements $I'(z')dz'$. We use an unprimed coordinate system to describe the position of an observation point P and a primed coordinate system to describe the position of a source point z', such as the one occupied by the current element $I'\mathrm{d}z'$ indicated in Figure 9.25. Each such current element $I'(z')dz'$ generates an electric field $d\mathbf{E}$, with the total electric field $\mathbf{E}$ being the superposition of the contributions from each of the constituent current elements.

The electric field $d\mathbf{E}$ of an oscillating infinitesimal dipole is well known and following the notation of Wangsness (1986) is of the form

$$d\mathbf{E} = -\frac{i\omega\mu_0 I'(z')dz'}{4\pi r'}\sin\theta' \exp(ikr' - i\omega t)\hat{\boldsymbol{\theta}}' \tag{9.43}$$

where

$$r' = \sqrt{r^2 + z'^2 - 2rz'\cos\theta}, \tag{9.44}$$

and $\hat{\boldsymbol{\theta}}'$ is the unit vector perpendicular to $\hat{\mathbf{r}}'$. The total electric field $\mathbf{E}$ of the long straight antenna is readily obtained by integrating Equation (9.43) over the length of the antenna.

In the radiation zone we have $r \gg \lambda$, where $\lambda = 2\pi/k = 2\pi c/\omega$ is the electromagnetic wavelength in free space. Let us also assume that we are interested in computing $\mathbf{E}$ at a distant point P such that $r \gg 2l$. In that case, for any element z' of the antenna, the relationship $|z'| \ll r$ holds. In that case, Equation (9.44) reduces to

$$r' \approx \sqrt{r^2 - 2rz'\cos\theta} \approx r - z'\cos\theta, \tag{9.45}$$

where use has been made of the binomial approximation $\sqrt{1+x} \approx (1+x/2)$ for $x \ll 1$. We can also use the approximation that $1/r \approx 1/r'$ for distant P, although we cannot likewise assume that $\exp(ikr') \sim \exp(ikr)$. We also use the approximation for distant P that the polar unit vectors are parallel, $\hat{\boldsymbol{\theta}}' = \hat{\boldsymbol{\theta}}$. Putting all this together results in

$$\mathbf{E} = -\frac{i\omega\mu_0}{4\pi r}\sin\theta\exp(ikr - i\omega t)\hat{\boldsymbol{\theta}}\int_{-l}^{+l}\exp(-ikz'\cos\theta)I'(z')dz'. \tag{9.46}$$

This is the general expression for the electric field **E** in the radiation zone for a straight antenna of length $2l$ carrying an arbitrary electric current distribution $I'(z')$. The simplest GPR antenna is a *half-wave* antenna in which the length of the antenna is equal to one-half of the electromagnetic wavelength, that is, $l = \lambda/4$. The simplest electric current distribution for a half-wave antenna is the fundamental mode

$$I'(z') = I_0\cos kz', \tag{9.47}$$

in which the current vanishes at the ends of the antenna and is maximum at the center, $z' = 0$, where it is equal to $I'(z') = |I_0|$. The electric-current distribution given by Equation (9.47) vanishes at the ends of a half-wave antenna.

The electric field **E** for a half-wave antenna is now readily found by inserting the electric-current distribution, Equation (9.47), into the general expression, Equation (9.46). It is straightforward to integrate Equation (9.46) and hence to show that the electric field **E** radiated by a half-wave antenna, in the radiation zone, is given by

$$\mathbf{E} = -\frac{ic\mu_0 I_0}{2\pi r\sin\theta}\cos\left(\frac{\pi}{2}\cos\theta\right)\exp(ikr - i\omega t)\hat{\boldsymbol{\theta}}, \tag{9.48}$$

and that the electric field vanishes at all points P that are aligned with the antenna, i.e. at $\theta = 0$ and $\theta = \pi$. The radiation-zone electric field **E** pattern in free space looks very similar to the electric field of an infinitesimal dipole.

The electric-field distribution produced by a straight-wire antenna is affected substantially by the presence of the dielectric Earth. Consider an infinitesimal electric-dipole antenna lying on an air–soil interface in which the underlying medium is characterized by permittivity $\varepsilon_1 = 4\varepsilon_0$ and conductivity $\sigma_1 \sim 0$, which is appropriate for a dry sandy soil. Suppose the horizontal electric dipole (HED) source is located in the plane of the page and directed to the right, as shown in Figure 9.26, left. The radiated electromagnetic field from this source has been computed using the finite-difference time-domain (FDTD) simulation code of Sassen (2009). The computations reveal that spherical waves propagate radially outward from the source into both the air and the soil. The electromagnetic wave velocities $v = 1/\sqrt{\mu\varepsilon}$ of the two media are different, hence the appearance of boundary waves to ensure continuity of tangential electric and magnetic fields across the interface. The boundary waves include a *head wave* propagating in the soil and an *evanescent wave* traveling in the air. The evanescent wave travels at the velocity of a ground wave and its amplitude decays exponentially with height above the interface. The head wave is the electromagnetic equivalent of a seismic critically refracted wave.

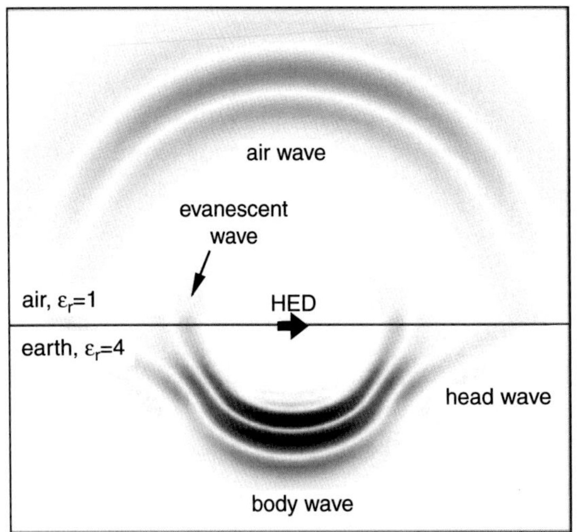

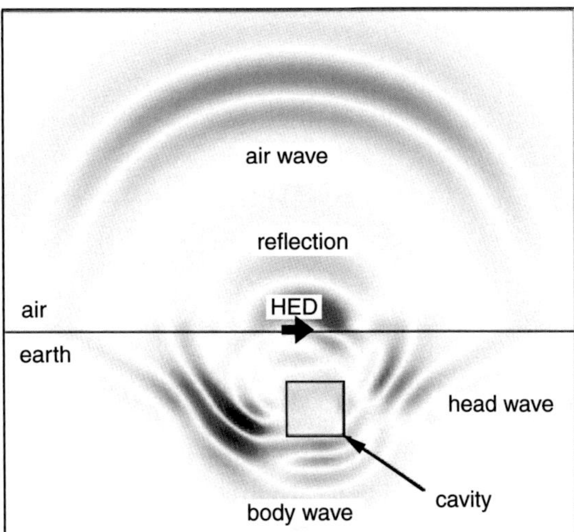

Figure 9.26 (Left) Snapshot of electric fields at $t = 9$ ns into a dielectric halfspace, shown in the vertical plane of the horizontal electric dipole (HED) source (300-MHz Ricker wavelet). (Right) Reflected and diffracted waves due to a 0.3-m^3 buried cavity. Plot dimensions: 2 m $\times$ 2 m. After Sassen (2009).

In Figure 9.26, right, the double-halfspace model is supplemented by a small air-filled cavity located in the ground. In this simulation, a reflected wave appears to be propagating back towards the GPR antenna and then upward into the air. The interaction of the primary double-halfspace field with the cavity also generates a complex pattern of refracted and diffracted body waves in the lower medium, as shown.

9.12 GPR radiation patterns

The previous section has demonstrated numerical solutions for GPR wave propagation in a heterogeneous lossless dielectric medium using the FDTD method. An analytic solution exists for the electromagnetic field radiated by a horizontal electric dipole lying on the interface between two semi-uniform dielectric media characterized by ε_0 (above) and ε_1 (below); it is given in the classic paper by Engheta *et al.* (1982).

The TE- and TM-mode radiation patterns of such a transmitter located on the surface of a ground with dielectric constant $\varepsilon_1 = 3.2\varepsilon_0$ are shown in Figure 9.27, top panel. The TE-mode pattern is the one that is generated when the dipole antenna is directed out of the page. The TM-mode pattern is generated when the dipole antenna lies within the plane of the page. The radiation patterns show, on an imagined vertical plane placed in the far-field in front of the page, how the electric-field strength changes as a function of direction relative to the antenna. The far-field is defined as any distance from the antenna for which $r \gg \max(l, \lambda)$, where l is the length of the dipole and λ is the radar wavelength. The radial axis on the polar plot is a logarithmic measure of relative field strength, relative to the field

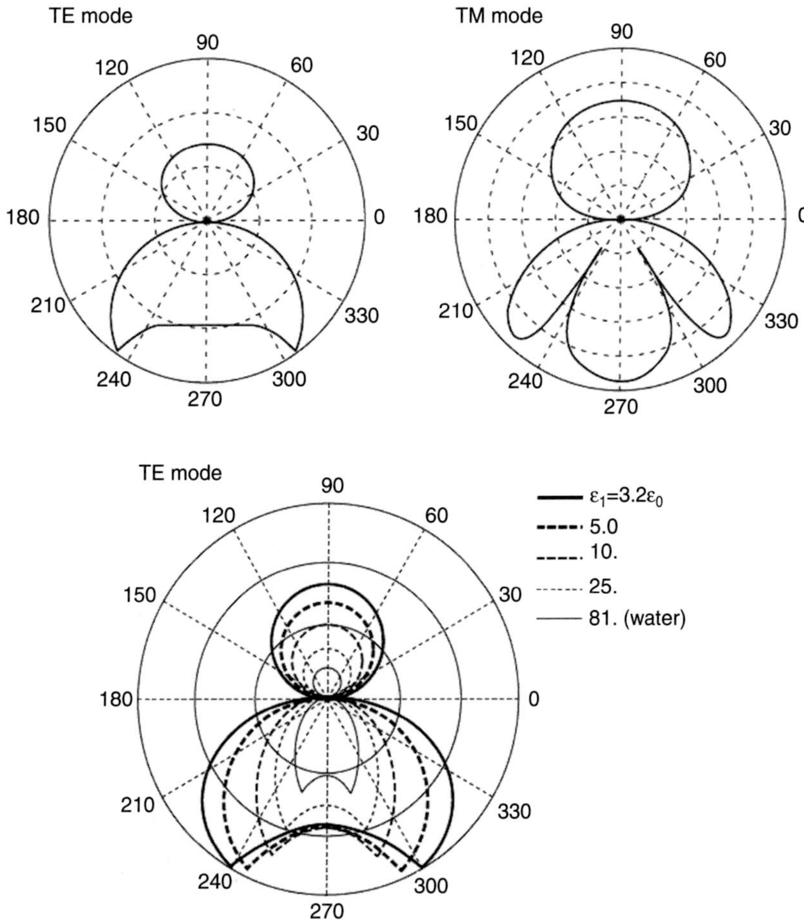

Figure 9.27 (a) Far-field TE- (top left) and TM- (top right) mode antenna directivity; (bottom) TE-mode antenna directivity as a function of ground permittivity. After Annan (2009).

strength at some reference point. For the present discussion, it suffices to note that the higher the field strength in a given direction, the further the radiation pattern extends toward the outer radius of the polar plot in that direction.

Both patterns show that electromagnetic energy is directed preferentially into the ground. Some of the energy, however, is directed uselessly into the air, adding unwanted noise to the GPR measurements. The leakage of signal into the air in some cases can be partially overcome (Annan, 2009) by carefully designing a proper shield to cover the antenna. The TE pattern shows that the greatest depth of signal penetration occurs at azimuths ~ 235° and ~ 305°. The TM pattern exhibits three subsurface lobes, along with two subsurface nulls located at ~ 240° and ~ 300°. The illumination of a buried target by the antenna is efficient if the target is located inside a lobe but inefficient if it is located inside a null region.

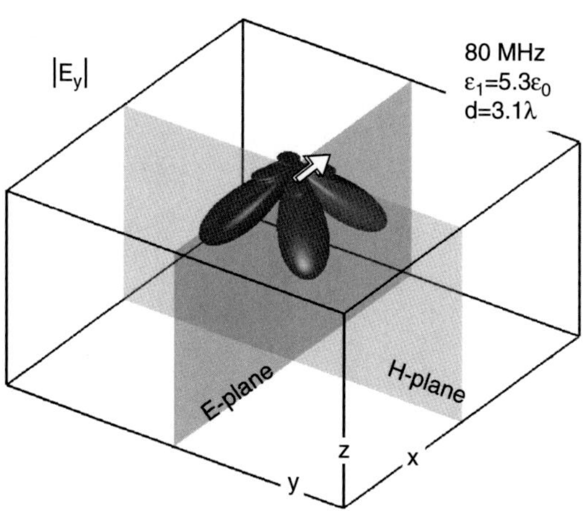

Figure 9.28 Example of a 3-D GPR radiation pattern. After Streich (2007).

The TE-mode radiation patterns that develop as the permittivity of the ground is varied from $\varepsilon_1 = 3.2\varepsilon_0$ to $81\varepsilon_0$ is shown in Figure 9.27, bottom. It can be seen that electromagnetic energy spreads out laterally and vertically from the antenna as the ground dielectric constant decreases. This implies that the antenna scans a wider subsurface area beneath it when the ground dielectric constant is low. In other words, the antenna generally has a larger footprint in dry ground versus wet ground.

A visualization of the three-dimensional radiation pattern from an infinitesimal electric dipole located on a homogeneous dielectric, non-conducting subsurface is shown in Figurc 9.28. A semi-analytic technique developcd by van der Kruk (2001) was used for the calculation. The dipole in this case is oriented in the $\hat{\mathbf{x}}$ direction while the quantity plotted is the strength of the $\hat{\mathbf{y}}$ component of the electric-field vector **E** at a fixed distance 5.0 m (3.1 wavelengths) from the antenna, i.e. in the far-field. The projection of the 3-D pattern onto the "E-plane" would correspond to a TM-mode radiation pattern like the one shown in the previous figure, while a similar projection onto the "H-plane" corresponds to a TE-mode radiation pattern.

9.13 Target polarization

It is of interest to explore the interaction of a transmitted electromagnetic field with a compact subsurface target. The response of a cylinder is discussed here since elongated objects such as buried pipelines are commonly found in environmental and engineering geophysical surveys. Pipes scatter energy into preferential directions depending on the incident radar wave polarization relative to the orientation of the pipe. It is required that enough energy scatters from the target to reach the surface and permit a measurement by the receiver antenna (Radzevicius and Daniels, 2000). Preferential scattering may result in *depolarization* of the incident **E**-field, in which the scattered field has a different direction of polarization than the incident field.

To describe electromagnetic scattering from cylinders, it is intructive to consider the following two situations. The first is a TM-mode excitation in which the incident **E**-field

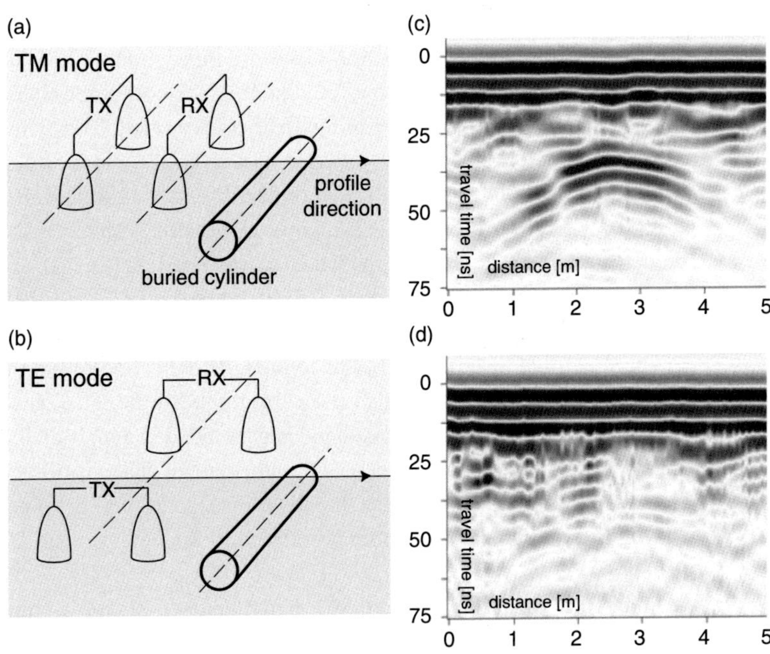

Figure 9.29 GPR survey configurations: (a) TM mode; (b) TE mode. GPR field data (200 MHz) of a pipe buried at 1.5-m depth in a silty sand soil showing the effect of TX and RX orientation relative to target, (c) TM mode; (d) TE mode. After Sassen and Everett (2005).

aligned with the long axis of the cylinder. The second is a TE-mode excitation in which the incident **E**-field is perpendicular to the long axis of the cylinder. As shown in Figure 9.29a, a TM-mode GPR survey is attained in practice when TX and RX dipole antennas are both aligned parallel to the long axis of the cylinder. A TE-mode survey configuration (Figure 9.29b) is attained when both are orthogonal to the long axis. In each case, the survey direction is orthogonal to the long axis.

As noted by Radzevicius and Daniels (2000), to detect the pipe at the receiver antenna, the incident field must couple strongly with the cylinder and cause it to act as an efficient secondary radiator. The resulting scattered field must then contain a significant component that is in alignment with the RX dipole axis, since a RX dipole preferentially responds to that component. Consider first the TM-mode survey configuration. The field that radiates from the TX dipole is polarized mainly in the direction of the dipole axis. The incident field then propagates into the subsurface, with its polarization direction parallel to the long axis of the cylinder. There should follow a strong coupling with the target, since the incident electric field is mainly tangential to the long axis and this component is required by fundamental boundary conditions to be continuous across material interfaces. Since the incident field and the long axis of the target are aligned, there is almost no depolarization. Hence, the cylinder radiates a substantial scattered electric field that is polarized mainly in the same direction as the incident field. Accordingly, as shown in Figure 9.29c, a strong GPR signal is recorded by the RX dipole.

Now consider the TE-mode survey configuration. Here, the TX dipole axis is oriented in a direction orthogonal to the long axis of the cylinder. The incident field, since it is polarized in alignment with the TX dipole axis, should not couple strongly to the target since the tangential component of the field is very small. The cylinder in this case does not act as an efficient secondary radiator. Nevertheless, since induced currents flow most readily along the long axis of the cylinder, any scattered field that does emerge from the cylinder is likely to be polarized mainly along-axis. However, the RX dipole is oriented such that it most efficiently responds to scattered fields that are polarized across-axis. Accordingly, as shown in Figure 9.29d, the cylinder is not detected by the TE-mode experiment.

The foregoing discussion applies to the GPR detection of both metallic pipes and high-dielectric plastic pipes for which the permittivity of the pipe exceeds that of the host soil. However, for low-dielectric plastic pipes embedded in a high-dielectric medium, such as an air-filled PVC pipe in wet sand, a full mathematical analysis of the scattering of cylindrical electromagnetic waves indicates that the GPR signature of the pipe is actually stronger in a TE-mode experiment. The reader is referred to Radzevicius and Daniels (2000) for further details.

Since the capability of a GPR experiment to image a buried elongated cylinder depends on the orientation of the TX and RX dipoles relative to the long axis of the target, it is advisable if time and resources permit to acquire data in both the TE- and TM-mode configurations. A third configuration, crossed-dipoles, in which the TX and RX dipole axes are perpendicular, can also be used profitably to image depolarizing subsurface targets. In fact, Sassen and Everett (2009) describe a *polarimetric* GPR technique for imaging subsurface fractures of unknown orientation. In this technique, data from all three configurations (TE, TM, and crossed-dipoles) are acquired over a 2-D survey grid and, at each measurement point, they are assembled into a 2×2 scattering matrix. The scattering matrix is then rotated to determine which TX–RX dipole alignment would have generated the strongest coupling to the target. The rotation angle determines the strike of the target while the largest eigenvalue of the scattering matrix provides the strength of the coupling.

9.14 GPR guided waves

Near-surface guided waves can be established using ground-penetrating radar if the ground contains a low-velocity layer whose thickness is comparable to an electromagnetic wavelength (Strobbia and Cassiani, 2007; van der Kruk *et al.*, 2009). The analysis of guided waves in such layers can reveal information about soil moisture conditions or the presence of contaminants. In a typical situation, a wet soil layer resides beneath the high-velocity air layer ($v = c \approx 0.3$ m/ns) and above an impermeable bedrock. A soil layer in the vadose zone, by virtue of its high water content, generally has a low velocity. Other situations in which guided waves might be set up include a permafrost layer or a wet-soil horizon overlying dry sand or gravel.

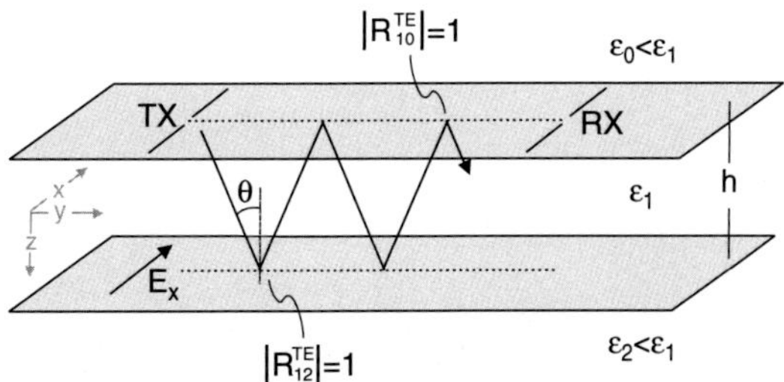

Figure 9.30 TE-mode excitation of a low-velocity GPR waveguide. After van der Kruk *et al.* (2009).

In guided-wave energization, a significant fraction of the transmitted signal from the GPR antenna propagates horizontally at low attenuation within the low-velocity layer. Guided waves are dispersive, that is, the propagation velocity is a function of frequency. Low-frequency components appear to propagate faster than higher-frequency components with the result that the transmitted pulse broadens and becomes distorted with increasing source–receiver separation. Guided waves can dominate GPR reflections and refractions and hence complicate the interpretation of data acquired in the WARR, or variable-offset, mode.

Consider a simple planar waveguide formed by a low-velocity layer $v_1 = c/\sqrt{\varepsilon_1}$ of thickness h, surrounded by two higher-velocity layers $v_0 = c/\sqrt{\varepsilon_0}$ and $v_2 = c/\sqrt{\varepsilon_2}$. For convenience, the GPR antennas are assumed to be oriented as shown in Figure 9.30, such that the TE mode is energized. A TM-mode waveguiding effect is also possible but not discussed here.

As shown in the figure, a wave incident on the lower (v_1, v_2) interface at an angle θ greater than the Snell critical angle is totally reflected so that $|R_{12}^{TE}| = 1$. Similar remarks apply to a wave incident on the upper (v_0, v_1) interface, such that $|R_{10}^{TE}| = 1$. The total internal reflection of this wave is a characteristic feature of ideal waveguiding. Outside the layer in this case, there are only evanescent waves that do not radiate significant energy into the underlying layer (say, bedrock) or overlying layer (say, air). Inside the layer, the total internal reflection implies a greatly reduced attenuation of the horizontally propagating wave. Less efficient, or leaky, waveguides can be formed when the reflection coefficients at the top and bottom are a fraction of unity.

It is shown in Appendix E that a waveguide can support waves propagating in the $\hat{\mathbf{y}}$ direction of the form $E_x \sim \exp[i(k_y y - \omega t)]$. The quantity k_y is the wavenumber in the direction of the wave propagation. However, only certain *modes*, or waves with certain combinations (k_y, ω) of frequencies and wavenumbers, can propagate within the waveguide. In particular, from Appendix E, the waveguide modes must satisfy the non-linear constraint equations

$$\alpha_1 h = \tan^{-1}\left(\frac{\alpha_2}{\alpha_1}\right) + \tan^{-1}\left(\frac{\alpha_0}{\alpha_1}\right) + m\pi; \quad m = 0, 2, 4, \ldots \quad \text{(even)} \tag{9.49}$$

where $\alpha_i = \mu_0 \varepsilon_i \omega^2 - k_y^2$.

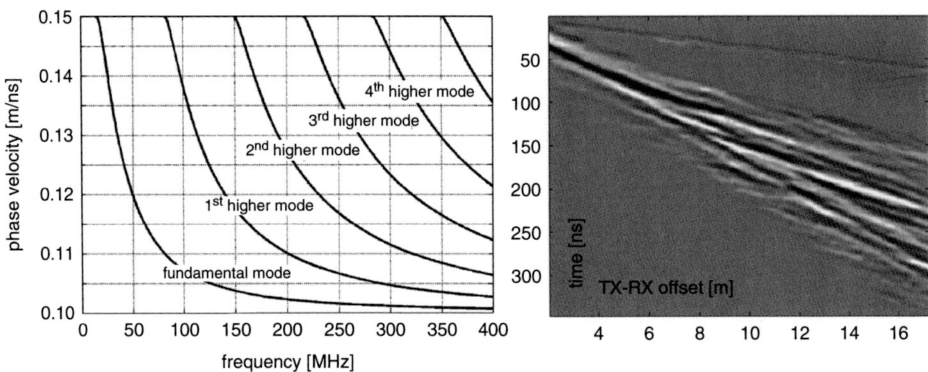

Figure 9.31 (Left) Fundamental mode and first five higher modes in a simple asymmetric waveguide. (Right) Radargram with strongly dispersive guided waves. After Strobbia and Cassiani (2007).

The value $m = 0$ corresponds to the first waveguide mode, the value $m = 2$ represents the first higher mode; $m = 4$ is the second higher mode, and so forth. For each mode m, Equation (9.49) can be viewed as an implicit equation $f(k_y, \omega) = 0$ in the two variables k_y and ω, assuming a waveguide of fixed thickness h and dielectric constants ε_0, ε_1, ε_2. In other words, Equation (9.49) represents, for each mode m, a *non-linear constraint* on the horizontal wavenumber and frequency. For a given frequency ω, the allowed wavenumbers k_y are those that make $f(k_y, \omega)$ vanish. In this way, a *dispersion relation* $k_y(\omega)$ is constructed. Define the waveguide *phase velocity* to be $v(\omega) = \omega/k_y(\omega)$. The phase velocity is a measure of the apparent velocity, within the waveguide, of signals of a given frequency.

An example of the waveguiding effect with $h = 1.0$ m is shown in Figure 9.31, left. Notice that the waveguide signal propagation velocity varies between that of the bedrock (in this case $v_2 = 0.15$ m/ns), at low frequency, to that of the soil ($v_1 = 0.1$ m/ns) at higher frequencies. The fundamental ($m = 0$) mode carries most of the guided energy. Notice that the higher waveguide modes exist only above a certain *cut-off frequency* that depends on m. The dispersive nature of guided waves, that is $v = v(\omega)$, implies that the radar signals spread out with increasing distance traveled along the waveguide; the behavior known as *shingling* is shown in Figure 9.31, right, and it is an effect that is often seen in practical radargrams. Notice that the individual waveguide modes are not easy to identify in the radar image.

Further insight into GPR guided-wave behavior is provided by van der Kruk *et al.* (2009) using the FDTD forward-modeling approach. In Figure 9.32 snapshots are shown of the electric field $E_x(y, z)$ due to transient pulse excitation of a surface waveguide structure. The snapshots reveal trapped electromagnetic energy inside the waveguide and two TE modes can be identified. A leading TE_0 mode has a single-peak amplitude over the vertical range of the waveguide, while a trailing TE_1 mode has a more complicated bimodal signature with alternating positive (red) and negative (blue) peak amplitudes.

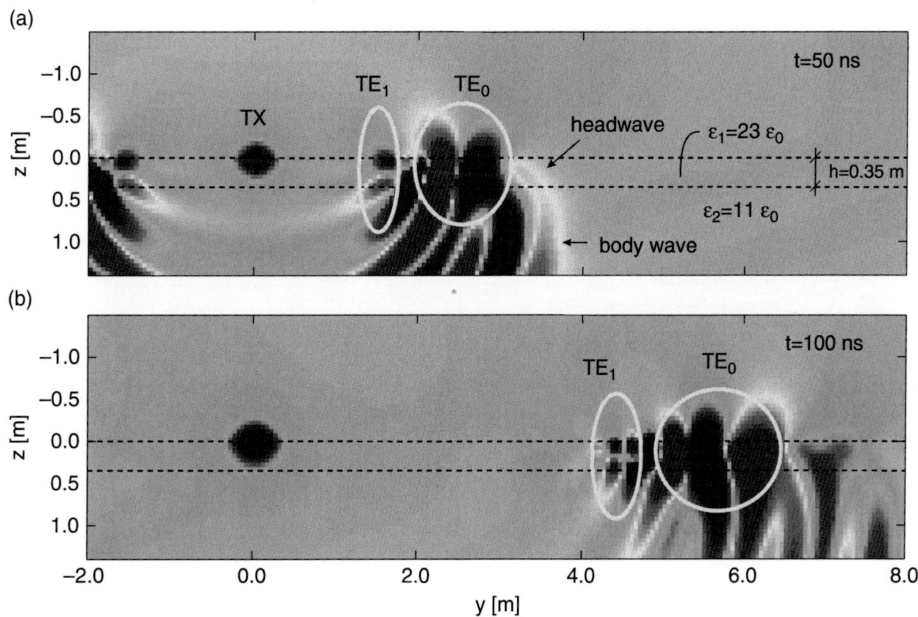

Figure 9.32 Simulations of the electric field E_x within a surface waveguide structure at times (a) $t = 50$ ns and (b) $t = 100$ ns after a pulse is transmitted from the transmitter (TX). The lowest two TE waveguide modes are identified. After van der Kruk *et al.* (2009). See plate section for color version.

9.15 GPR illustrative case histories

Example. Coastal sedimentology.

Ground-penetrating radar has been widely used in sedimentology (Neal, 2004), a discipline that has long relied on seismic sequence stratigraphy as the primary interpretive tool. GPR interpretation is based on identifying radar surfaces and radar facies contained within the sections, combined with ground truth obtained from traditional geological information such as aerial photographs, outcrops, cores, and trenches. The basic assumption underlying a successful GPR interpretation is that bedding structures which cause reflections can be recognized in radar sections and that non-geological reflections are readily identified and either ignored or removed via data processing. As in the seismic techniques, the knowledge and experience of the GPR interpreter remains key to making accurate and reliable sedimentological inferences. There are many potential pitfalls in the data-acquisition, processing, and interpretation stages of both techniques.

In a recent sedimentological application, GPR was used successfully on the Florida Gulf Coast barrier islands to characterize the depth and lateral extents of erosional surfaces and washover deposits associated with hurricane landfalls (Wang and Horwitz, 2007). The interpreted water table is shown in Figure 9.33, as is the base upon which the 2004

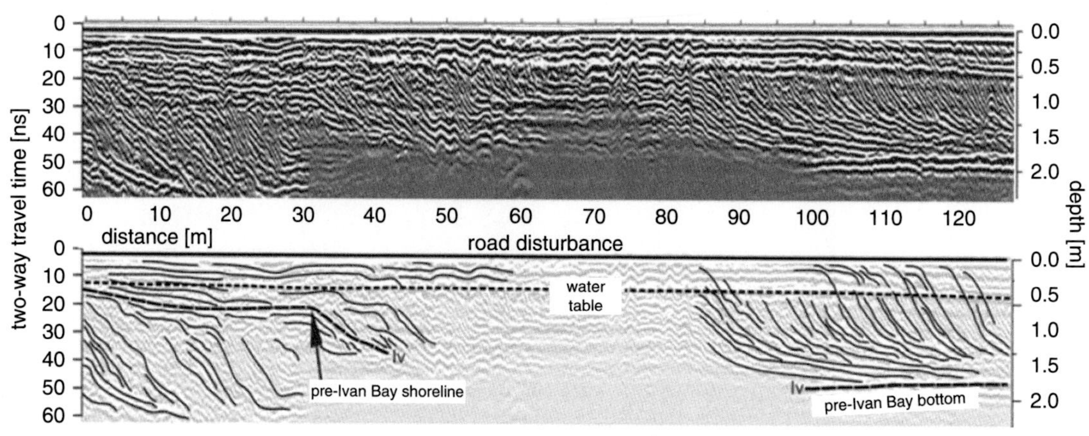

Figure 9.33 (a) 250 MHz radar transect from Florida Gulf coast barrier islands; (b) stratigraphic interpretation; Iv = radar surface interpreted as the shoreline prior to 2004 Hurricane Ivan. After Wang and Horwitz (2007).

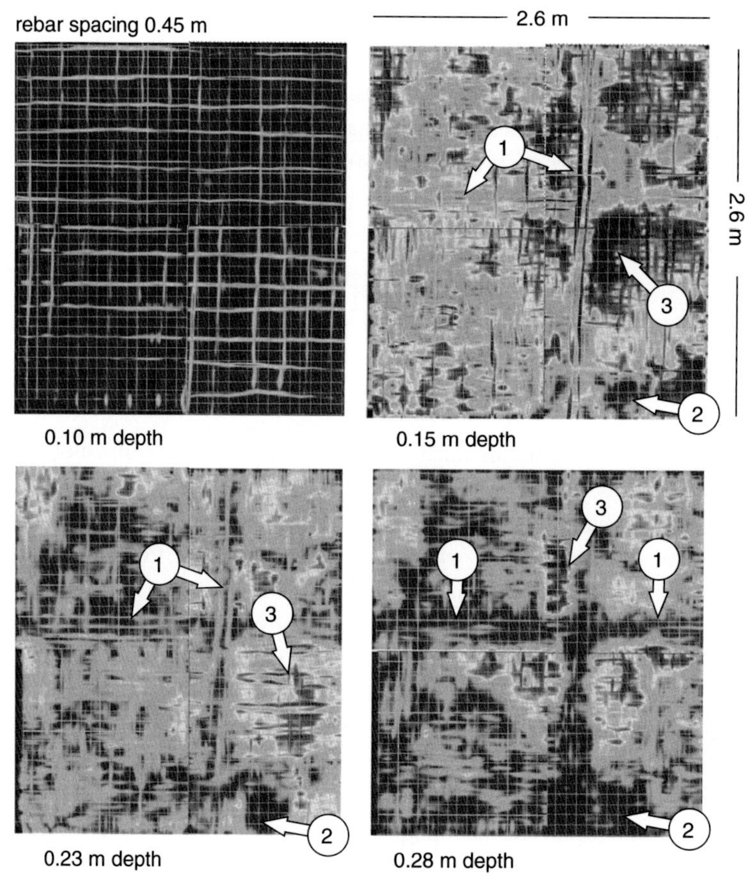

Figure 9.34 1.0 GHz radar depth slices, velocity 0.1 m/s, reinforced concrete floor. 1 = support beam; 2 = support pier; 3 = void space; data courtesy Mary Jo Richardson and Wilf Gardner. See plate section for color version.

Hurricane Ivan washover was deposited. The GPR data proved in this case to be useful in characterizing coastal geomorphological alterations and shoreline-shaping processes that accompany large storms.

Example. Voids beneath reinforced concrete.

In 2006, a residential garage in central Texas built on expansive clay soil shifted ~ 3.8 cm during an extended dry period. A foundation engineer tried to stabilize the garage floor by raising it, installing piers around its perimeter, and injecting grout into the newly created void spaces. Months later, rainfall was observed to infiltrate beneath the raised concrete floor suggesting that the void spaces had not been adequately filled. Further shifting of ~ 2.0 cm occurred over the next three years. A 1.0 GHz GPR survey of the floor with ~ 15-cm station and ~ 15-cm line spacing was then conducted with an objective to determine the extent of the voids. The resulting series of depth slices shown in Figure 9.34 clearly reveals the rebar grid, the original concrete support beams, a support pier installed by the engineer, in addition to a number of putative void spaces. The latter were subsequently confirmed by drilling. The identified void spaces were then filled with pressurized grout in a second attempt at stabilizing the garage floor. This successful case history demonstrates how GPR data were used to guide a small-scale geotechnical remediation project.

Problems

1. Show that electric field **E** and the magnetic field **B** for lossless plane-wave propagation are linked by $\mathbf{B} = (k/\omega)\hat{\mathbf{z}} \times \mathbf{E}$ and hence the three vectors **E**, **B**, and $\hat{\mathbf{z}}$ (the direction of propagation) are mutually orthogonal.
2. Show that the real and imaginary parts of the complex wavenumber $k = \sqrt{\omega^2\mu\varepsilon + i\omega\mu\sigma}$ are given by Equations (9.24) and (9.25).
3. Show that the phase difference between the electric field **E** and the magnetic field **B** is Ω where $\tan\Omega = \sqrt{1+Q^2} - Q$, and $Q = \omega\varepsilon/\sigma$. Does **E** lead or lag **B** when the plane wave is viewed as a function of position, for a fixed instant of time? What about the case when the plane wave is viewed as a function of position?
4. Consider a vertically polarized (TM-mode) plane wave obliquely incident on the planar interface separating two lossless dielectric media. The electric-field vector **E** is parallel to the plane of incidence and orthogonal to the propagation direction $\hat{\mathbf{p}}$. The magnetic-field vector **H** is perpendicular to the plane of incidence and parallel to the interface. Show that the reflection coefficient in this case is given by

$$R_{TM} = \frac{\kappa \cos\theta_i - \sqrt{\kappa - \sin^2\theta_i}}{\kappa \cos\theta_i + \sqrt{\kappa - \sin^2\theta_i}},$$

where $\kappa = \varepsilon_2/\varepsilon_1$. Hint: the solution follows closely the theory for the TE mode given in the text, but pay close attention to the boundary conditions.

5. Show that the electric field **E** radiated by a half-wave antenna, in the radiation zone, is given by Equation (9.48) and that the electric field vanishes at all points P that are aligned with the antenna, i.e., at $\theta = 0$ and $\theta = \pi$.

10 Emerging techniques

This chapter highlights a few of the emerging techniques of near-surface applied geophysics. The discussion is designed to provide the reader with a sense of some of the latest developments in this rapidly growing discipline. The emergent techniques studied here include surface nuclear magnetic resonance, time-lapse microgravity, induced seismicity studies, landmine discrimination, GPR interferometry, and the seismoelectric method. There are many other advances being made, or that will be made in the near future, beyond those described in this chapter; the interested reader is advised to keep watch on the topical journals and conferences.

10.1 Surface nuclear magnetic resonance

The *surface nuclear magnetic resonance* (sNMR) technique is a relatively new geophysical method that can directly sense spatial variations in subsurface water content to depths of ~ 150 m. The sNMR technique holds promise to open new and exciting avenues in groundwater resource investigations. The method is based on the interaction of an applied magnetic field with the magnetic moments of the hydrogen nuclei, or protons, in groundwater. The sNMR concept was first described in a patent by Varian (1962), followed by pioneering field work of Russian geoscientists during the 1970s and 1980s.

Nuclear magnetic resonance in the laboratory was discovered early in the twentieth century by physicists investigating the fundamental properties of electron and nuclear spins (Slichter, 1996; Becker *et al.*, 2007). Practical applications soon followed with the development of medical imaging apparatus (Andrew, 2007) and petroleum well-logging tools (Hurlimann, 2012). Both types of NMR instruments involve a strong permanent magnet that generates a steady field $\mathbf{B}_0$. The role of the magnet is to align the spins of protons contained in a fluid-bearing sample. In the case of medical imaging, the sample is placed inside the permanent magnet. In well logging, the geological formation under evaluation surrounds a magnet that is placed inside the borehole.

In both scenarios, a radio-frequency coil is also employed that produces a short excitation pulse to tip the proton spins by 90° into a plane that is transverse to the direction of $\mathbf{B}_0$. The Larmor resonant frequency $\omega_L = \gamma|\mathbf{B}_0|$ is used, where γ is the gyromagnetic ratio of the proton, a fundamental atomic constant. The proton spin vectors initially precess, at the Larmor frequency, in the transverse plane with a coherent phase. However, over a time scale typically of a few tens of milliseconds, the proton ensemble loses its phase coherence due to random scattering, heterogeneities, and other effects. In ideal systems, the net magnetization signal $M(t)$ decays exponentially according to

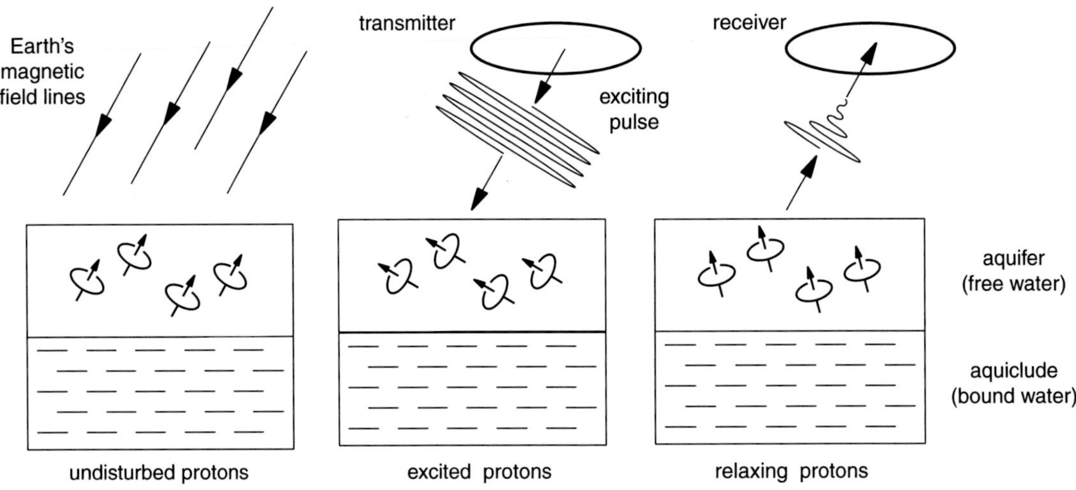

Figure 10.1 Basic physics of the surface NMR technique. After Hertrich (2008).

$$M(t) = M_0 \exp(-t/T_2) \tag{10.1}$$

where T_2 is the relaxation time. In well logging, a measurement of the relaxation time is often interpreted in terms of the pore-size distribution of the sensed geological formation and, with additional assumptions (Kleinberg, 2007), the parameter T_2 can also be related to the permeability of fluid flow.

Weichman *et al.* (2000) substantially modified the basic NMR theory to handle some of the complexities that are particular to geophysical sNMR prospecting. For example, in field geophysical studies, the ambient geomagnetic field is employed in lieu of the permanent magnet. Furthermore, a TX loop of radius ~ 10–100 m is used to generate the excitation signal. However, it must be recognized that the loop is deployed over variably conductive ground and it is located at a substantially remote distance from the subsurface region of investigation. The excitation field $\mathbf{B}_T(\mathbf{r})$ from the transmitter is therefore strongly heterogeneous, unlike the uniform excitation field used in laboratory NMR.

Comprehensive discussions of the sNMR technique are provided in Legchecnko and Valla (2002), Hertrich (2008), and Muller-Petke *et al.* (2011). As illustrated in Figure 10.1, left, in the equilibrium state, the spin axes of protons in groundwater are, on average, aligned with the ambient geomagnetic field. The surface transmitter loop is then switched on (Figure 10.1, middle) and the magnetic field associated with a transient excitation pulse oscillating at the Larmor frequency excites the protons. The strength of the sNMR signal, and the penetration depth of the technique, scales with the pulse moment $q = I\tau$ [As], which is the product of the amplitude I of the transmitted current and the time $\tau \sim 40$ ms during which the current is on. After the transmitter loop current is switched off, the proton spins relax back toward the equilibrium alignment. The sNMR response recorded at the receiver loop, indicated in Figure 10.1, right, consists of the so-called free-induction-decay (FID) signal that is associated with the spin relaxation. It should be noted that the FID signal from protons in bound water, such as found in ice or clay-rich aquicludes, decays too fast to be recorded by sNMR equipment. The method detects chiefly the slower-decaying FID signals from free water such as found in aquifers.

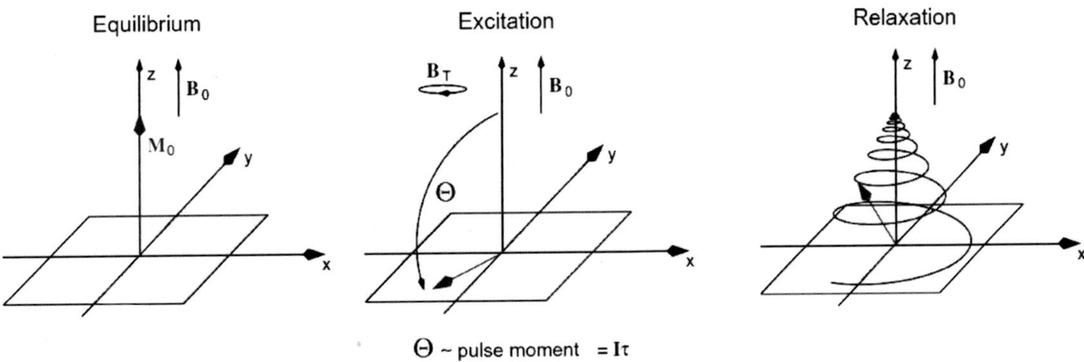

Figure 10.2 NMR free induction decay. After Muller-Petke *et al.* (2011).

Further details of the FID process are shown in Figure 10.2. At left, a net magnetization $\mathbf{M}_0$ of the proton-bearing sample develops due to average proton spin alignment in the direction of the steady, ambient geomagnetic field $\mathbf{B}_0$ of strength 25–55 μT. In the middle panel of the figure, a surface TX coil of radius ~ 10–50 m emits an excitation field $\mathbf{B}_T$ at the Larmor frequency ω_L ~ 1–3 kHz, causing the net magnetization of the sample to re-orientate: the size of the "flip angle" Θ depends on the pulse moment q and the heterogeneity of the subsurface conductivity distribution $\sigma(\mathbf{r})$. At right, after the excitation pulse is extinguished, the net magnetization $\mathbf{M}$ relaxes back into alignment with the geomagnetic field $\mathbf{B}_0$, while emitting an electromagnetic signal that fluctuates at the Larmor frequency and is sensed at the receiver location.

The theoretical development presented in Muller-Petke and Yaramanci (2010) shows that the sNMR relaxation signal $V(q, t)$ depends on a spatially variable sensitivity function $G(\mathbf{r}, q)$ and the subsurface distribution of water content $m(\mathbf{r},T_2^*)$, according to the formula

$$V(q,t) = \int d^3\mathbf{r}\ G(\mathbf{r},q) \int dT_2^*\ m(\mathbf{r},T_2^*)\exp(-t/T_2^*). \tag{10.2}$$

The spatial sensitivity, or kernel function $G(\mathbf{r}, q)$, depends on a number of factors including the transmitted field $\mathbf{B}_T$, the subsurface electrical conductivity distribution $\sigma(\mathbf{r})$, the geomagnetic field $\mathbf{B}_0$, and the magnetic field $\mathbf{B}_R$ that would be measured at the receiver location due to a nuclear spin of unit strength at subsurface location $\mathbf{r}$. The physical interpretation of Equation (10.2) is that the relaxation signal $V(q, t)$ is the aggregate response of all the protons that are excited by the magnetic field of the transmitter. A typical sNMR relaxation signal $V(q, t)$ is shown in Figure 10.3. The spatial distribution of water $m(\mathbf{r},T_2^*)$ is essentially the same as the spatial distribution of the protons, or hydrogen nuclei, since most of the subsurface hydrogen nuclei are present as a constituent of the water molecule H_2O. The dependence of the water content on T_2^* indicates that, in some heterogeneous formations, the FID is better modeled as a multi-exponential process rather than a single exponential decay. For example, if each layer of a multi-layered geological structure is characterized by a different pore-size distribution, then each layer would have its own relaxation-time constant.

An sNMR sounding curve can be built up by transmitting at several different values of pulse moment q. Increasing the pulse moment excites a greater number of protons at greater distances from the transmitter and hence probes a larger volume of the subsurface.

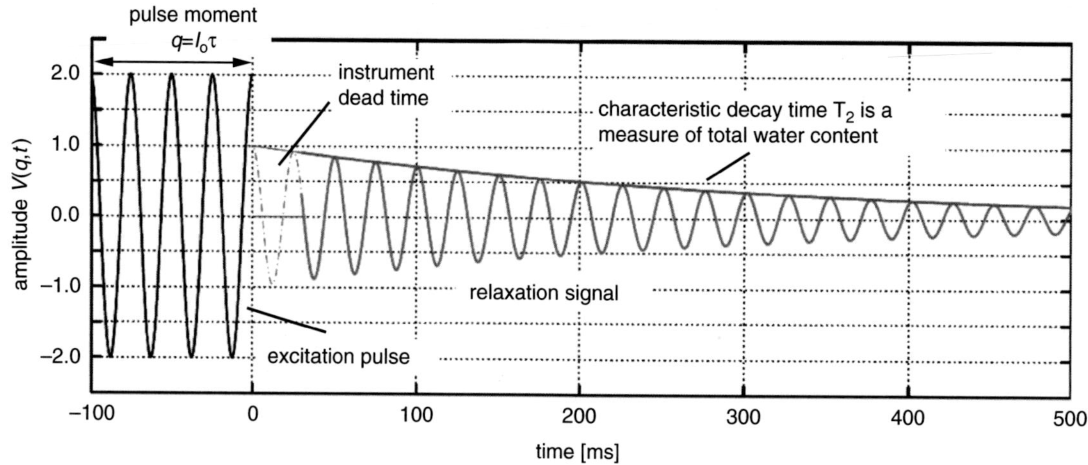

Figure 10.3 NMR relaxation signal $V(q, t)$. After Muller-Petke *et al.* (2011).

In a simplified one-dimensional (1-D) interpretation method for which the FID is modeled by a single exponential, the early-time amplitude of the sNMR relaxation signal E_0 is measured as a function of pulse moment. The resulting $E_0(q)$ sounding curve is then converted into an estimate $m(z)$ of water content as a function of depth only (Legchenko and Valla, 1998). In the more general case, as described by Muller-Petke and Yaramanci (2010), determination of subsurface water content from multi-exponential sNMR signals requires finding the inverse operator of the kernel $G(\mathbf{r}, q)$ in Equation (10.2).

The sNMR method has been applied in recent years by several research groups at a number of field sites to estimate the subsurface water content of aquifers and infer their hydraulic properties. Here, we summarize only the work of Muller-Petke *et al.* (2011) who have provided an interesting verification of the sNMR method at a frozen lake site in the Harz mountains, Germany. The site was selected since the water content of the lake is uniformly 100%, the geology is well known, and its frozen surface facilitated the deployment of the field equipment.

The essential results from the frozen lake investigation are shown in Figure 10.4. The computed 2-D sensitivity function in the top panel displays the sensitive regions of the subsurface at pulse moment $q = 2$ As. Also shown in the figure is the location of the surface TX loop and the lake bathymetry based on GPR soundings. The experimental $E_0(q)$ sounding curve, obtained after a substantial amount of signal processing, is shown in symbols at bottom left. The sounding curve is then converted into a subsurface water content profile $m(z)$ containing sharp layer boundaries. The result is shown at bottom right and it is discovered to be in excellent agreement with the expected geological structure and water content. The $E_0(q)$ sounding curve predicted from the inversion result agrees with the observed sounding curve to within an rms error of 5.8 nV. Note from the $m(z)$ profile at bottom right that the inferred water content of the ice is 0%. This non-detection is a consequence of the fast relaxation of the bound water in the ice layer that occurs entirely within the instrument dead time (the dashed line in Figure 10.3) immediately following TX shut-off. The graywacke below the lake has very small porosity and consequently it also has no NMR-detectable water.

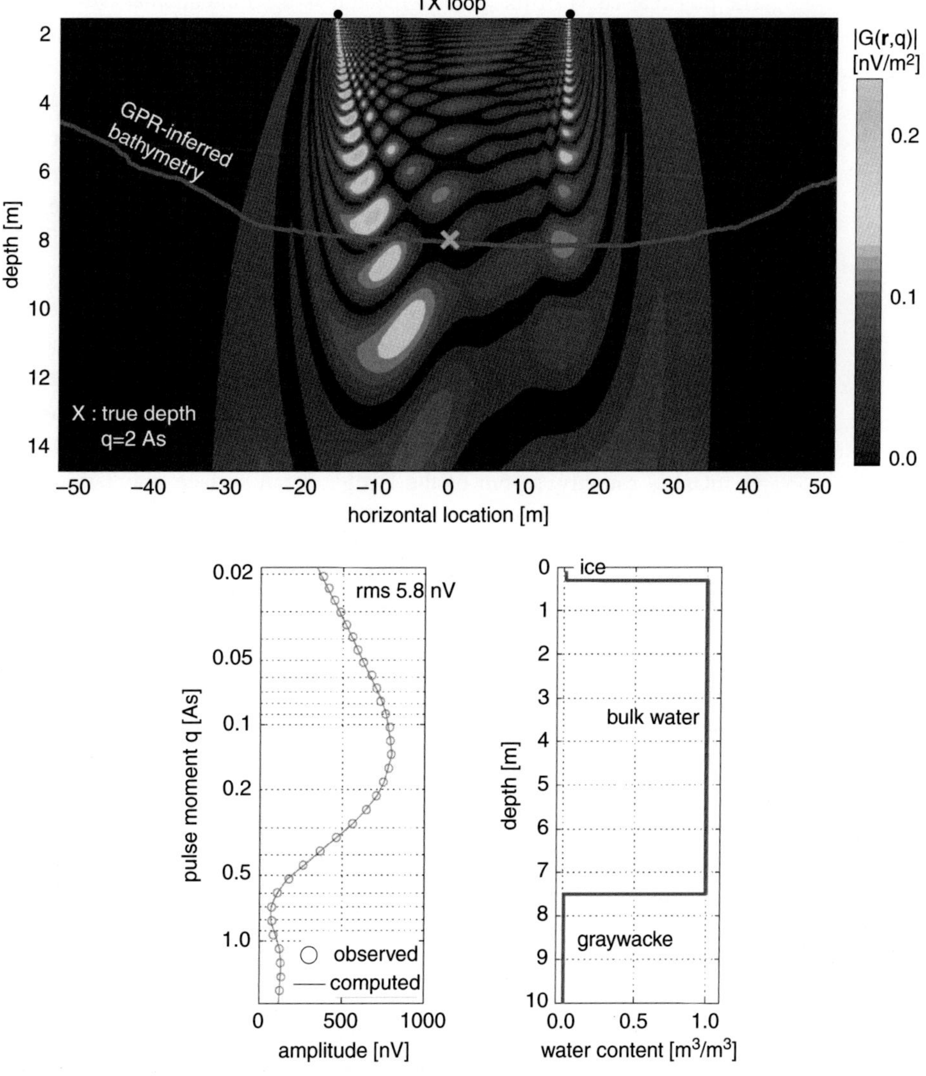

Figure 10.4 (Top) 2-D sNMR sensitivity function, frozen lake site in Germany; (bottom left) sNMR observed $E_0(q)$ sounding curve (symbols) and predicted response (line). The latter is calcuated from the inferred subsurface water content distribution $m(z)$ shown at bottom right. After Muller-Petke *et al.* (2011).

10.2 Time-lapse microgravity

Gravity has a long history as an essential method of applied geophysics in exploration and regional geological mapping contexts. Its value for hydrocarbon and mineral exploration and tectonic studies cannot be over-estimated. Spatial variations in subsurface mass density cause slight perturbations to Earth's gravitational field that are detected by gravity meters.

The physical basis of the gravity method is the fundamental Newton's law of attraction between two point masses. A subsurface body of anomalous density, compared to its surroundings, can be simply regarded as an assemblage of elementary point masses. The gravity effect of a point mass has an elementary and well-known mathematical form. The gravity effect of the anomalous body is then just a linear superposition of the known gravity effects of its constituent point masses. The objective of gravity surveying is to exploit this superposition principle to infer the density, size, shape, location, and burial depth of anomalous bodies from remotely observed gravity readings. A good introduction to the gravity technique, with abundant case histories, is found in Reynolds (2011) while a more comprehensive and mathematical treatment is offered by Blakely (1995).

The gravity method is sometimes used in near-surface (uppermost ~ 30 m) investigations. However, its popularity is tempered by low signal-to-noise levels and the fact that geological interpretation of high-resolution gravity data can be highly ambiguous. While the amplitudes of exploration-scale gravity anomalies are typically ~ 100–500 mGal (1 mGal is roughly 1 ppm of the Earth's mean surface gravitational field), precision to less than ~ 10 μGal (hence the term microgravity) is needed for small-scale investigations and this requires very careful fieldwork and painstaking data reduction. For example, detailed corrections for the gravity effects of individual buildings, bridges, tunnels, earthworks, and other structures should be made if the survey is conducted in a built-up area. This is additional to the standard set of corrections (Blakely, 1995) that account for larger gravity effects due to sensor drift, solar and lunar tides, Earth shape and rotation, survey elevation, local terrain, and sensor motion. Many of the near-surface applications of the traditional gravity method are related to the detection and delineation of underground void spaces such as caverns, tunnels, sinkholes, or crypts.

An emerging area of research in near-surface geophysics is time-lapse microgravity. In this technique, temporal changes in subsurface mass distributions are monitored by comparing microgravity surveys made over the same area but at different times. This technique has been used at the exploration scale to monitor fluid injection into or withdrawal from subsurface reservoirs (Alnes *et al.*, 2008). It has also been used (Branston and Styles, 2003) at the near-surface scale to monitor the development of subsurface void spaces in a built-up area affected by mining-induced ground subsidence, see Figure 10.5.

Measurable time-lapse microgravity signals can also be caused by seasonal changes in aquifer water storage. For example, a simple calculation shows that a 1-m-thick infinite horizontal slab of water produces a gravity signal of 42 μGal. Jacob *et al.* (2010) have made repeat gravity readings over a network of 40 stations spaced ~ 1.5 km apart spanning a ~ 100 km^2 area of the Durzon karst region in southern France. Four gravity surveys were conducted over a two-year period. The observed changes in gravity, after careful data reduction, were then related to changes in aquifer water storage using a mass-balance model that takes into account the recharge due to rainfall and the outflows due to evapotranspiration and discharge at springs. The average gravity change was found to be −12 μGal during the summer season and +13 μGal during the winter season. These values correspond to ~ 0.25-m equivalent water-height fluctuations. Moreover, the strong spatial variability of the gravity changes across the survey network suggest that the water storage in the aquifer is strongly heterogeneous, as would be expected for a karst systems of this type that are dominated by multiscale voids, fissures, and conduits.

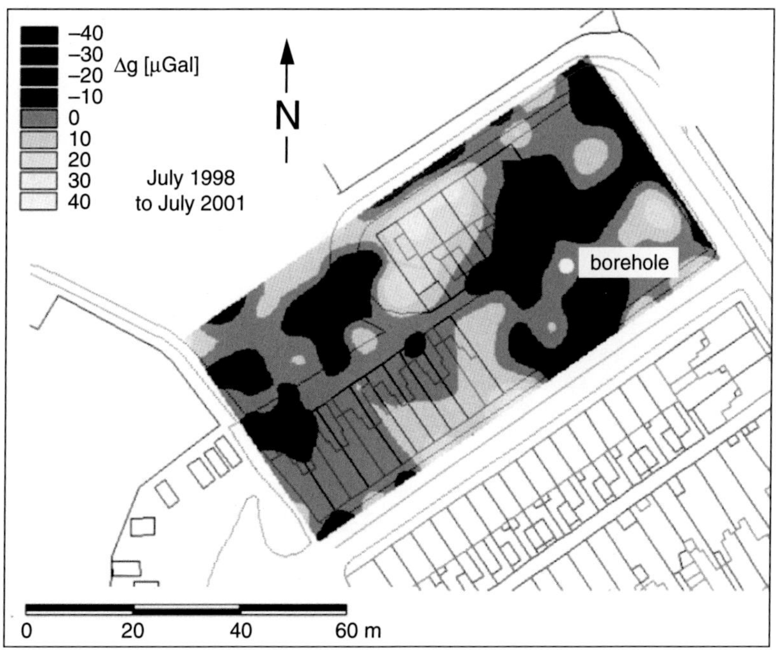

Figure 10.5 Time-lapse microgravity signals over a three-year time interval over an area of ground subsidence caused by subsurface salt extraction, Northwich, UK. After Branston and Styles (2003).

10.3 Induced-seismicity studies

Public attention has recently been called to the generation of moderate earthquakes due to oil- and gas-production activities. While induced seismicity has long been associated with reservoir impoundments, mining, and more recently enhanced geothermal systems and carbon sequestration, a less certain link has been documented between seismicity and routine hydrocarbon-production activities such as hydraulic fracturing and saltwater disposal. In general, there are great difficulties in ascribing seismicity near oil wells to either a natural or an anthropogenic cause. However, in some cases there is little doubt that anthropogenic seismic activity plays a direct role, as in The Netherlands where hundreds of earthquakes at ~ 2–3-km depths are spatially correlated with producing gas fields in the north of the country (van Eck *et al.*, 2006).

Seismicity may be caused by extraction of fluid from a subsurface reservoir (Segall, 1989). According to poroelastic theory, a reservoir should compact as fluid is removed. This can cause both ground subsidence and subsurface faulting. Segall *et al.* (1994) have analyzed the spatiotemporal patterns of seismicity following fluid extraction at the Lacq gas field in southern France. The gas field has experienced ~ 40 mm of ground subsidence between 1967 and 1989. Anthropogenic seismic events were identified to be those that are strongly correlated to the gas field but well separated from regional tectonic events.

Fluid injection into a subsurface reservoir (Fisher and Guest, 2011) may also induce seismicity. As the injected fluid enters microfractures in the rock, the pore pressure increases.

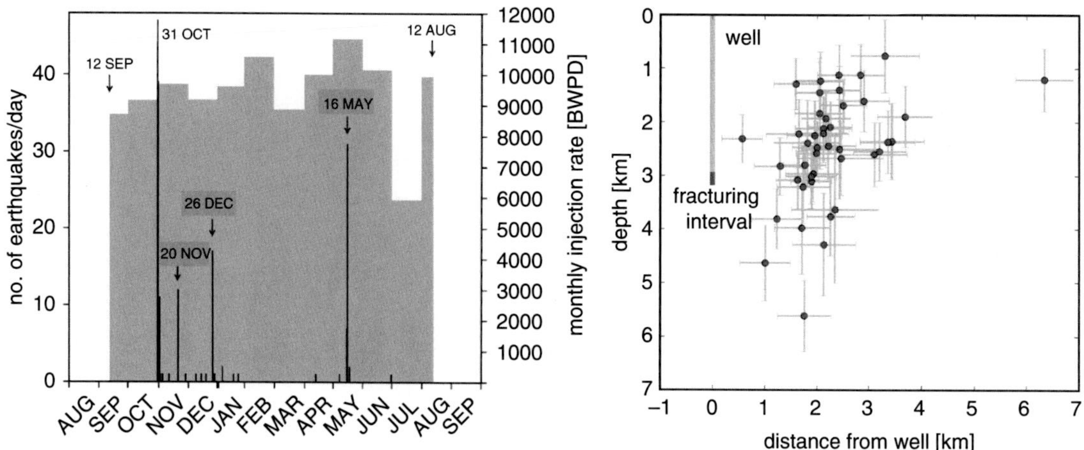

Figure 10.6 Induced seismicity as a result of hydrocarbon-production activities: (left) temporal correlation between seismicity and fluid injection at DFW airport in 2008–2009, after Frohlich *et al.* (2011); (right) spatial distribution of earthquakes, with errors, associated with hydrofracturing of an Eola Field well, Oklahoma, after Holland (2011).

The increase in pore pressure acts in opposition to the confining pressure and thus decreases the effective pressure, bringing rocks closer to failure. The failure envelope is the combination of shear and normal stress that causes rock to fail. Frohlich *et al.* (2011) describe an earthquake sequence at the Dallas–Fort Worth (DFW) airport in Texas between October 2008 and May 2009 consisting of ~ 180 earthquakes of magnitude $M < 3.3$. A detailed analysis of focal mechanisms, hypocentral locations, and onset times suggests the cause to be brine disposal associated with nearby hydrocarbon-production activities. The locations of the induced seismic activity were spatially correlated with the location of a saltwater disposal well. The temporal correlation between fluid injection and DFW seismicity is shown in Figure 10.6, left.

There has also been documented (Holland, 2011) a possible connection between induced seismicity from hydraulic fracturing activity at Eola Field, south-central Oklahoma. A total of 43 earthquakes of size $M < 2.8$ occurred within 36 hours after the onset of initial stages of a fracturing operation. The inferred earthquake positions, with absolute location errors of ~ 300–500 m, show distinct clustering of the seismicity along known fault lines that cut through Eola Field. The pre-existing faults likely provide subsurface fluid pathways. No earthquakes were observed after later stages of fracturing. Consistent with previous studies on induced seismicity at other locations, the earthquakes occurred mainly within 2–3 km of the well in which the fracturing occurred (Figure 10.6, right).

10.4 Landmine discrimination

Landmines are a major environmental concern in many countries that have hosted recent conflict. Much progress has been made in the development of near-surface geophysical technologies to detect mines and to discriminate between hazardous mines and non-

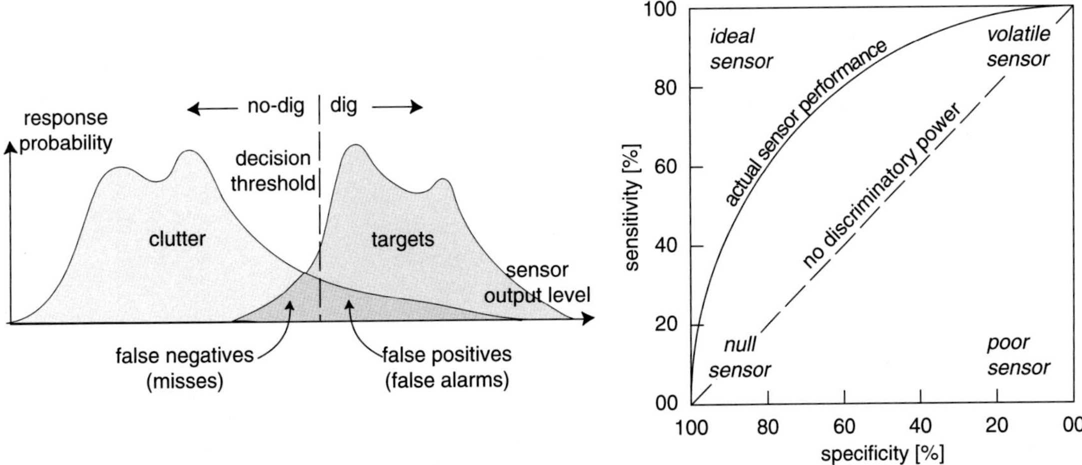

Figure 10.7 (Left) Clutter and target response distributions, and the definitions of false negatives and false positives. (Right) Standard receiver operating characteristic (ROC) curve.

hazardous clutter. A number of systems are under active development; they typically contain multisensor hardware configurations (e.g. Sato and Takahashi, 2012) including GPR, electromagnetic induction, and sometimes supporting instrumentation such as infrared cameras or nuclear quadrupole resonance detectors. The currently fielded systems achieve a high detection rate but they produce too many false alarms and generally show variable performance in difficult or hostile environments. It should be mentioned that, for the purposes of this discussion, a "landmine sensor" consists of the actual hardware, plus the accompanying signal-conditioning, processing, and interpretation software.

A critical task in any landmine clearance operation is to determine a decision threshold, or simply put, to specify the sensor output level that should trigger a dig decision. The situation is illustrated in Figure 10.7 (left) in which it is assumed that a hypothetical sensor has some discriminatory power. In other words, it is supposed that actual mines (targets) generally produce a higher sensor output level than non-mines (clutter). A threshold (the vertical dotted line) is then chosen such that the decision to dig is made only if the sensor output level exceeds the threshold. Notice in this case that some of the clutter items will be misclassified as targets; these are the false positives, or false alarms. Notice also that some of the target items will be missed; these are false negatives and it means that some hazardous landmines are left in the ground. The goal is to determine a threshold that detects almost all the mines while producing few, if any, false alarms.

The receiver operating characteristic (ROC) curve (Metz, 1978) shown in Figure 10.7 (right) describes the performance of a hypothetical sensor. On the vertical axis is its sensitivity; this describes the probability that a buried object is detected. On the horizontal axis is its specificity; this describes the probability that the detected object is correctly classified as either a mine or clutter. As shown in the figure, the ideal sensor plots at the top left: it has high sensitivity (all buried objects are detected) yet high specificity (all detected objects are correctly classified). A poor sensor plots toward the bottom right: it has low sensitivity (most of the buried objects are missed) and low specificity (the objects it does detect are incorrectly

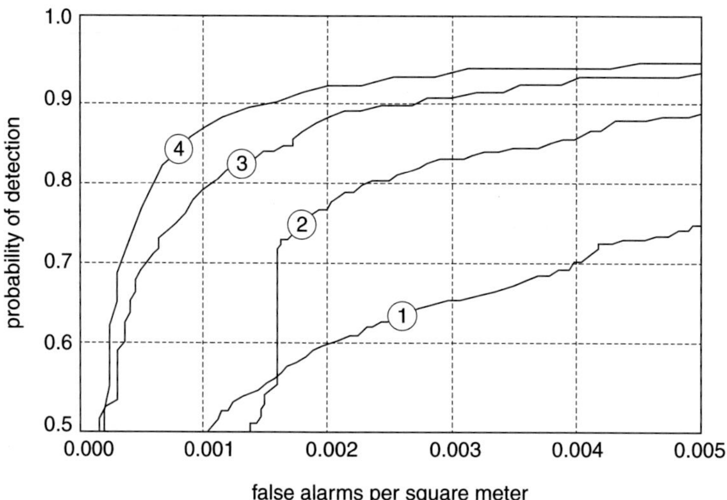

Figure 10.8 Landmine-discrimination ROC curves based on analysis of GPR array data. The discrimination algorithms are: 1 = spectral correlation; 2 = Gaussian–Markov random field; 3 = edge histogram descriptor. ROC curve 4 is the result obtained by fusing six different discimination algorithms. For details, see Frigui *et al.* (2012).

classified). The null sensor at the bottom left can correctly classify buried objects but unfortunately misses most of them. The volatile sensor, upper right, detects everything but is not able to make a correct classification. A sensor with little if any discriminatory power plots close to the dashed line in the figure; as its sensitivity increases, its specificity goes to zero and the dig decision becomes essentially a coin flip. The performance of an actual sensor that has some discriminatory power will plot above the dashed line. As its sensitivity increases (for example, by decreasing the decision threshold), its specificity slowly decreases (it begins to generate false alarms). This behavior traces out its ROC curve. The objective of the landmine-sensor designer is to push the ROC curve as far as possible toward the top left of the plot.

The performance of several landmine-discrimination algorithms has been explored by Frigui *et al.* (2012) using a comprehensive database of ~ 1600 anti-tank-mine encounters from different geographic regions under varying soil types and moisture conditions. The geophysical sensor is a NIITEK vehicle-mounted 24-channel wideband pulse GPR array (Hintz, 2004). A volatile pre-screener algorithm with high sensitivity but low specificity is applied in an intial step to winnow the huge volumes of GPR data that are required to be analyzed. The resulting alarms, many of them false, in the reduced dataset were then analyzed using the discrimination algorithms. The latter may be regarded as different kinds of pattern-recognition methods. The essential idea is to find close matches between measured and either empirical or theoretical GPR signatures of mines. The ROC curves for three of the algorithms (labeled 1, 2, and 3) are shown in Figure 10.8. It is clear that algorithm 3, the edge histogram descriptor, shows the best performance.

Landmine discrimination is a highly complex pattern-recognition task and long experience has shown that no single algorithm can outperform all others under all environmental conditions. Many of the discriminators do however provide complementary information. A great amount of research is currently being undertaken to try to take advantage of the strengths of individual discriminators and to overcome their weaknesses. Some success has been achieved by optimally combining, or *fusing*, the results of several algorithms. The fusion process may be regarded as a kind of consultation of multiple experts who then make an informed group decision. The "voting" procedure used to obtain the group decision is largely heuristic but it characterizes the different fusion methods. The ROC curve labeled 4 in Figure 10.8 is a representative one obtained by fusing six different discrimination algorithms (Frigui *et al.*, 2012). It is clear that the fused discriminator outperforms any of the single discriminators applied separately.

10.5 Passive GPR interferometry

Passive seismic interferometry is a technique that has developed over the past several years enabling seismologists to extract information about Earth's subsurface velocity structure by cross-correlating long time series of seismic waveforms recorded at two or more stations (e.g. Shapiro *et al.*, 2005; Schuster, 2010; Nicolson *et al.*, 2012). The essentials of this type of interferometry can be introduced by consideration of the scalar wave equation.

Schuster (2009) derives the basic reciprocity equation for the scalar Helmholtz equation in a domain D

$$(\nabla^2 + k^2)G(\mathbf{x}, \mathbf{x}_A) = -\delta(\mathbf{x} - \mathbf{x}_A) \tag{10.3}$$

where $k = k(\mathbf{x})$ is a material property, with $\mathbf{x}$ the three-dimensional position vector, and the right side represents an impulsive source applied at $\mathbf{x} = \mathbf{x}_A$ *inside* the domain D. The solution $G(\mathbf{x}, \mathbf{x}_A)$ to Equation (10.3) is a scalar Green's function. Suppose $G_0(\mathbf{x}, \mathbf{x}_B)$ is the scalar Green's function due to a second impulsive source applied at $\mathbf{x} = \mathbf{x}_B$ *outside* the domain D. It is shown by Schuster (2009) that the Green's functions are connected by

$$G(\mathbf{x}_B, \mathbf{x}_A) = \int_{\partial D} [G_0(\mathbf{x}_B, \mathbf{x})\hat{\mathbf{n}}\cdot\nabla G(\mathbf{x}_A, \mathbf{x}) - G(\mathbf{x}_A, \mathbf{x})\hat{\mathbf{n}}\cdot\nabla G_0(\mathbf{x}_B, \mathbf{x})]d^2\mathbf{x} \tag{10.4}$$

where ∂D is the bounding surface of domain D and $\hat{\mathbf{n}}$ is its unit outward normal vector.

The function $G(\mathbf{x}_B, \mathbf{x}_A)$ is the Green's function evaluated at $\mathbf{x}_B$ due to the source located at $\mathbf{x}_A$. The physical interpretation of Equation (10.4) is that the signal measured at location $\mathbf{x}_B$ outside the domain D due to an elementary source located at $\mathbf{x}_A$ inside the domain D equals the cross-convolution of the signals measured at these two locations that would be caused by a continuum of sources distributed on the boundary ∂D. The situation is illustrated in Figure 10.9.

Electromagnetic interferometry is also based on a principle of reciprocity but involves vectorial Green's functions. A general reciprocity theorem for electromagnetics is presented in Cheo (1965). Specifically, the theorem states that the electromagnetic field due to an elementary electric- or magnetic-dipole source is unchanged if the locations of the source and the measurement point are switched. The elementary reciprocity relationships are illustrated in Figure 10.10 and can be written in the frequency domain as Green's functions relationships (Slob and Wapenaar, 2008)

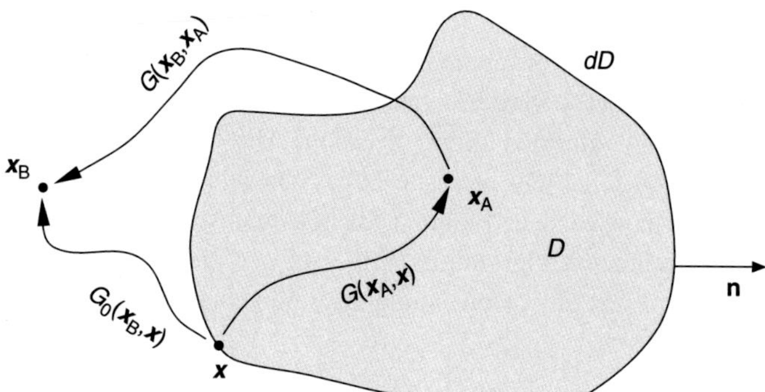

Figure 10.9 The basic interferometry configuration involves a continuum of sources on the boundary of domain D. After Slob and Wapenaar (2008).

$$G_{kr}^{EE}(\mathbf{x}_B, \mathbf{x}_A, \omega) = G_{rk}^{EE}(\mathbf{x}_A, \mathbf{x}_B, \omega); \tag{10.5a}$$

$$G_{kr}^{EM}(\mathbf{x}_B, \mathbf{x}_A, \omega) = -G_{rk}^{ME}(\mathbf{x}_A, \mathbf{x}_B, \omega); \tag{10.5b}$$

$$G_{kr}^{MM}(\mathbf{x}_B, \mathbf{x}_A, \omega) = G_{rk}^{MM}(\mathbf{x}_A, \mathbf{x}_B, \omega). \tag{10.5c}$$

The notation $G_{kr}^{EM}(\mathrm{x}_B, \mathrm{x}_A, \omega)$ in Equation (10.5b), for example, refers to the k-th component of the electric field measured at point B due to an r-directed, time-harmonic magnetic dipole source located at point A. The superscript E stands for "electric" and the superscript M stands for "magnetic." Each of the reciprocity relations (10.5a–c) is shown pictorally in a row of Figure 10.10, and the physical interpretation of each is indicated in thc figure caption.

Both active-source and passive types of electromagnetic interferometry are possible. Controlled-source electromagnetic interferometry is discussed by Hunziker *et al.* (2012); the technique retrieves the subsurface impulse response from measurements made at an array of receivers after decomposing the recorded signals from an active source into their upward and downward decaying parts. Slob and Wapenaar (2008) provide, on the other hand, an excellent tutorial overview of passive electromagnetic interferometry as it applies to GPR imaging of the subsurface. This technique supposes the existence of random, spatially dense but uncorrelated noise sources located above Earth's surface. Recordings of this noise are then made at two receiver locations and these are cross-correlated to identify the Green's function, or impulse response, between the two measurement points. The result is equivalent to that of actually deploying an impulsive source at one location and measuring the response at the other.

Once the Green's function has been determined, it may be used in a standard electromagnetic imaging method to determine the subsurface electromagnetic properties.

A possible approach to passive GPR interferometry is outlined in Slob and Wapenaar (2008). Suppose there exists wide-band, pervasive electromagnetic noise sources in or above the atmosphere. Then, as shown in Figure 10.11, we can consider a semi-infinite domain D in which the upper surface ∂D_1 is located above the highest receiver location and the lower surface ∂D_2 is just above the surface of the Earth.

The objective is to determine the electric and magnetic Green's functions $G(\mathbf{x}_B, \mathbf{x}_A)$, which can be regarded as the signal measured at location $\mathbf{x}_B$ due to an impulsive dipole source located at $\mathbf{x}_A$. There is no actual dipole source located at $\mathbf{x}_A$; instead, there is

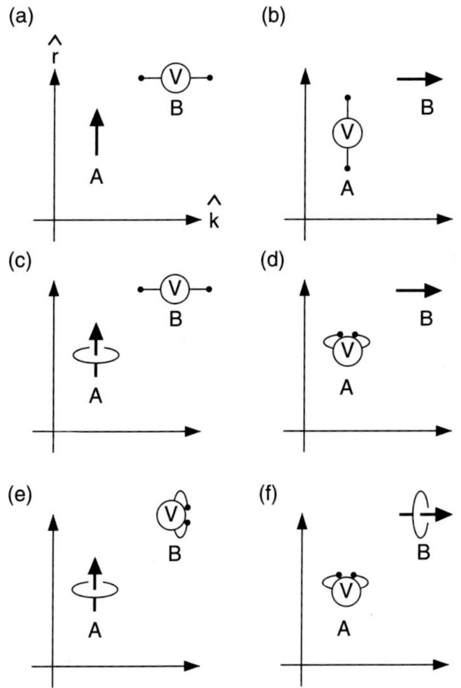

Figure 10.10 Electromagnetic reciprocity relationships: (a, b) the *k*-th component of the electric field measured at B due to an *r*-directed electric-dipole source located at A is equal to the *r*-th component of electric field measured at A due to a *k*-directed electric-dipole source located at B; (c, d) the *k*-th component of the electric field measured at B due to an *r*-directed magnetic-dipole source located at A is equal to the negative of the *r*-th component of magnetic field measured at A due to a *k*-directed electric-dipole source located at B; (e, f) the *k*-th component of the magnetic field measured at B due to an *r*-directed magnetic-dipole source located at A is equal to the *r*-th component of magnetic field measured at A due to a *k*-directed magnetic-dipole source located at B.

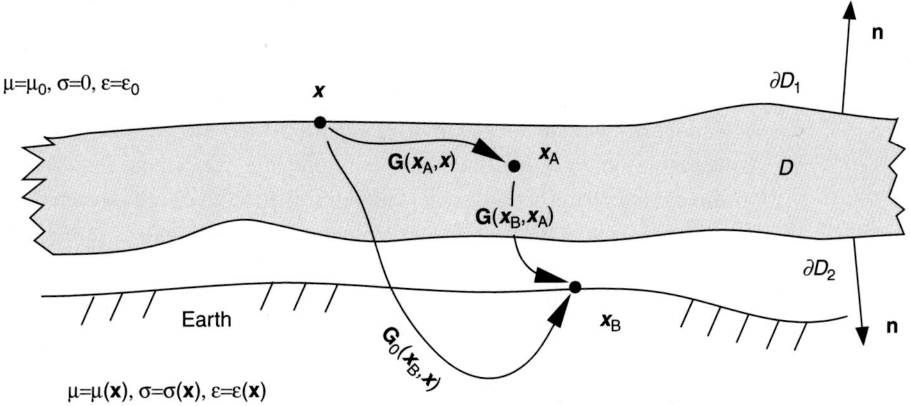

Figure 10.11 The passive GPR interferometry configuration involves a continuum of sources on the upper boundary of domain *D*. After Slob and Wapenaar (2008).

presumed to be a spatial continuum of noise sources located on points $\boldsymbol{x}$ of the boundary ∂D_1. Measurements of this noise are then made at the two receiver locations $\mathbf{x}_A$ and $\mathbf{x}_B$ shown in the figure. Suppose the noise sources are spatially uncorrelated but that each one has the same power spectrum $S(\omega)$. In that case, Slob and Wapenaar (2008) find expressions for the functions $G(\mathbf{x}_B, \mathbf{x}_A)$ such as

$$G_{rk}^{EE}(\mathbf{x}_B, \mathbf{x}_A, \omega)S(\omega) \approx -\Big\langle E_r(\mathbf{x}_B, \omega)E_k(\mathbf{x}_A, \omega)\Big\rangle \tag{10.6}$$

where the brackets $\langle\cdot\rangle$ denote expectation value. The interpretation of Equation (10.6) is that a cross-correlation of electric-field noise measurements made at locations $\mathbf{x}_A$ and $\mathbf{x}_B$ are sufficient to yield the electric–electric Green's function between these two points. Similar equations can be derived to estimate the electric–magnetic and the magnetic–magnetic Green's functions.

10.6 Seismoelectric coupling

Electromagnetic fields may be generated when seismic waves propagate through a heterogeneous porous medium. It has long been speculated that the attendant coupling of seismic and electric fields could provide the basis for a useful geophysical prospecting tool. Although many relevant experiments at petroleum-exploration length scales have been carried out over the past decades (e.g. Thompson and Gist, 1993), researchers now recognize that seismoelectric effects can also be applied to near-surface investigation of aquifers.

The physical mechanism responsible for seismoelectric coupling is widely believed to be a mechanically induced displacement of electric charges in the electric double layer that resides at pore-scale solid–fluid interfaces. The double layer (Figure 10.12), described earlier in Chapter 5, consists of two parallel layers of ions, a surface layer and a diffuse layer that extends into the fluid.

A rigorous theoretical description of the coupling of elastic and electromagnetic waves in porous media has been developed (Pride, 1994). The mathematical details are beyond the scope of the present discussion. Nevertheless, in a first type of seismoelectric conversion, fluid pressure gradients develop as a seismic compressional wave propagates through a poroelastic medium. As shown in Figure 10.13, the pore fluid flows out of zones of compression into the adjacent zones of rarefaction. Since pore fluids are electrolytes, some of the excess charges resident in the outer diffuse layer are advected with the fluid. The resulting *streaming current* creates a charge separation and, accordingly, an *electrokinetic* potential gradient ∇V will appear across the adjacent zones of compression and rarefaction. This potential gradient fluctuates at the seismic frequency and is termed the *coseismic signal* since it is carried along with the seismic wave.

To a first approximation, the causative fluid pressure gradient ∇P and the consequent electokinetic potential gradient ∇V are proportional such that

$$\nabla V = C\nabla P \tag{10.7}$$

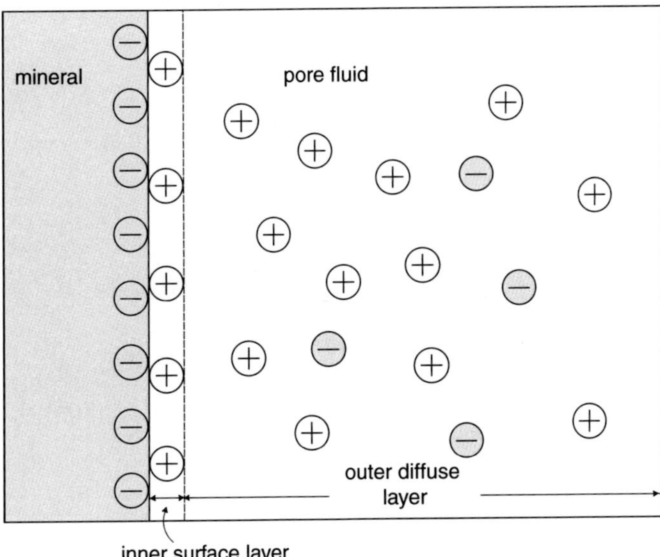

Figure 10.12 The electric double layer that develops at interfaces between charged mineral surfaces and pore fluid electrolytes.

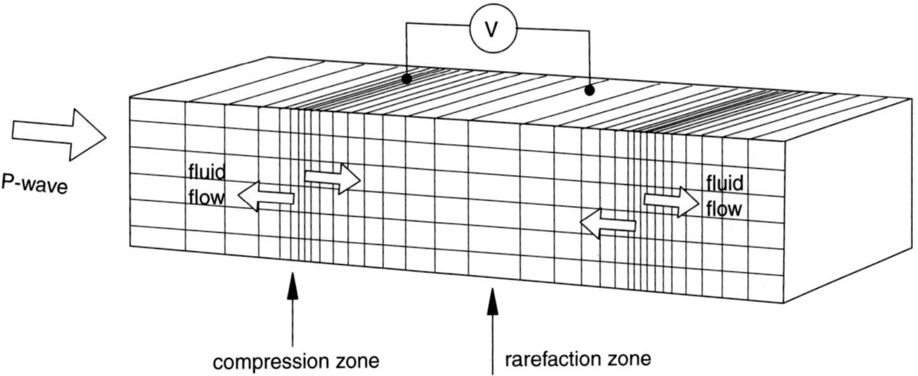

Figure 10.13 The physical basis of coseismic signals. The electrolytic fluid flow causes an electrokinetic voltage to appear across the zones of compression and rarefaction.

with laboratory studies indicating that C can range from – 12 to 350 mV/atm depending on the fluid and rock properties. A theoretical treatment indicates that C varies as

$$C = \varepsilon\zeta/\eta\sigma \tag{10.8}$$

where ε is the fluid dielectric constant, ζ is the *zeta potential*, η is the fluid viscosity and σ is the pore-fluid electrical conductivity.

A series of field experiments was performed by Garambois and Dietrich (2001) to detect seismoelectric conversions. The results are shown in Figure 10.14 and they clearly show coseismic signals. At left is the seismic response acquired with a sledgehammer source ($N = 20$ stacks) as recorded by geophones, while at right are seismoelectric signals acquired with the same source (except $N = 120$ stacks) and using pairs of 1.0-m-long electrodes aligned

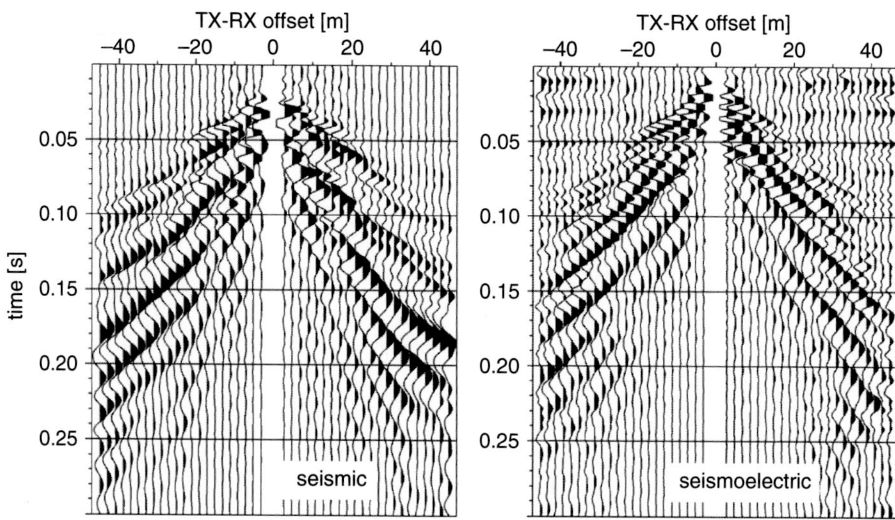

Figure 10.14 Seismoelectric effect: coseismic signals. After Garambois and Dietrich (2001).

with the profile direction. Both datasets were bandpass-filtered between 1–200 Hz. The electric-field data were further filtered to remove the power line noise of 50 Hz and its odd harmonics. The seismic and seisomelectric datasets look very similar with both showing high-frequency first arrivals followed by the signature of lower-frequency, dispersive surface waves.

A second type of electrokinetic conversion, termed the *interface effect*, occurs at an interface separating two different values of the coupling coefficient *C*. The conversion is largely contained within the first *Fresnel zone* of the down-going seismic pulse directly beneath the shotpoint, as shown in Figure 10.15. The Fresnel zone can be regarded as a footprint of the seismic source; as earlier discussed in Chapter 6.

The coupling at the Fresnel zone generates an electromagnetic wave that provides almost instantaneous arrivals at the surface electrode array (Haartsen and Pride, 1997). The seismoelectric conversion can be regarded as the appearance of an extended vertical-electric-dipole (VED) source in the Fresnel zone (as shown in the figure) whose electric **E** field may be computed using standard electromagnetic theory. The maximum **E**-field response occurs at an offset equal to ~ 0.5 the depth to the source. This rule of thumb governs the ideal placement of electrodes for maximum sensitivity to a freshwater interface at a given depth. Expected signal levels for ~ 10–20-m soundings are in the micro- to millivolt range. The interface effect can be seen in Figure 10.14 (right) as the early arrivals at $t = 0.01$–0.07 s whose arrival time is independent of TX–RX offset.

An experiment to determine whether seismoelectric profiling can be used to map the water table and other subsurface interfaces was carried out over a sandy aquifer in Western Australia by Dupuis *et al.* (2007). The seismoelectric stacked section shown in Figure 10.16 (bottom) has been processed to highlight the contributions from the interface effect. In particular, the coseismic signals were muted and the polarity of traces from opposite sides of shotpoints were reversed prior to stacking. Interface signals are expected to have opposite polarities on either side of the shotpoint if they are well represented by an effective vertical electrical dipole at the Fresnel zone. The event labeled 1 is interpreted to be the water table

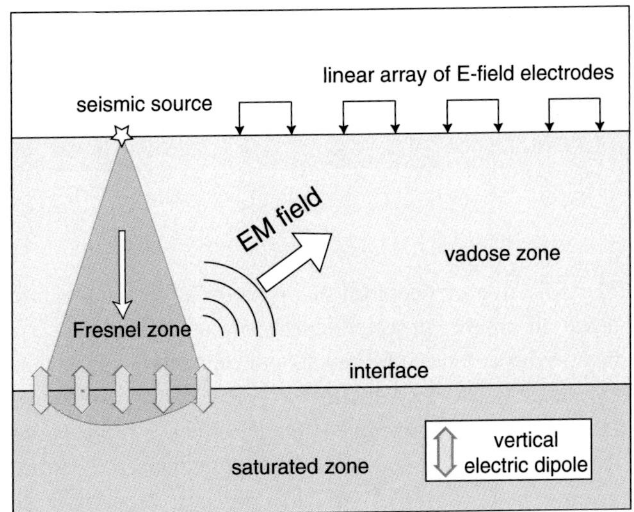

Figure 10.15 The physical basis of seismoelectric interface signals.

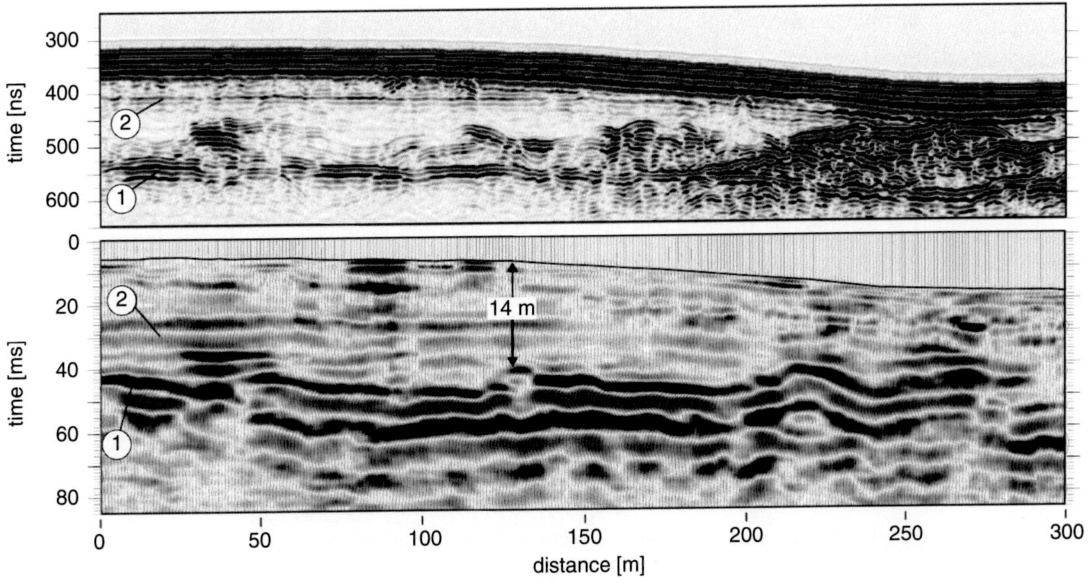

Figure 10.16 50-MHz GPR profile (top) and stacked seismoelectric section (bottom) imaging the vadose zone of a sand aquifer. After Dupuis *et al.* (2007).

at ~ 14 m depth. The event labeled 2 is a shallow water-retentive layer at ~ 6–7 m depth. These two events on the stacked seismoelectric section appear also on a co-located GPR profile (Figure 10.16, top). The study shows that seismoelectric data can detect subsurface interfaces that cannot be resolved by conventional seismic-reflection and -refraction data. Moreover, the seismoelectric method offers greater penetration depths than the GPR method.

11 Linear inversion

The objective of geophysical inversion is to acquire information about Earth's interior based on indirect measurements made at or near the surface. Treatments of geophysical inverse theory have appeared at both elementary and advanced levels in recent monographs by Menke (1984), Parker (1994), Gubbins (2004), Tarantola (2004) and Aster *et al*. (2012). Inverse theory is applicable if there exists a *forward-modeling* algorithm that predicts the response of an assumed model of Earth structure. The forward problem may be stated as: given a model $\mathbf{m}(\mathbf{r})$ of an Earth physical property, calculate the response $\mathbf{u}(\mathbf{r})$ at the measurement locations. Examples of pairs of physical properties and responses are shown in Table 11.1. The calculation of $\mathbf{u}(\mathbf{r})$ sometimes can be carried out analytically for sufficiently simple Earth models, or with a numerical technique such as the finite-difference or finite-element method if the Earth model is too complex to admit an analytic treatment.

11.1 Introduction

Let us suppose that we can describe the essential characteristics of the portion of Earth under geophysical investigation using a finite number of P parameters. For example, we might be trying to interpret electromagnetic measurements in terms of a layered Earth with $(P + 1)/2$ electrical conductivities and $(P - 1)/2$ layer thicknesses, assuming P is odd. This scenario and several others are illustrated in Figure 11.1. In all cases, the model parameters can be assembled into a *model vector* $\mathbf{m} = (m_1, m_2, \ldots, m_P)^T$ where superscript T denotes transpose. The *inverse problem* is to estimate these parameters from geophysical data, in this case electromagnetic-field measurements.

Let D be the number of data, arranged as the *data vector* $\mathbf{d} = (d_1, d_2, \ldots, d_D)^T$. Each data point d_i has an associated error e_i which is estimated by the geophysicist based on the best available knowledge of the instrumentation precision and accuracy, the effects of

Table 11.1 Examples of geophysical forward-model–response pairs

Technique	Physical property (model)	Measurement (response)
Gravity	Density $\rho(\mathbf{r})$	Vertical gravity $g_z(\mathbf{r})$
Magnetics	Magnetization $\mathbf{M}(\mathbf{r})$	Magnetic field $\mathbf{B}(\mathbf{r})$
Seismology	P-wave velocity $V_p(\mathbf{r})$	Traveltime $\tau(\mathbf{r})$
Electromagnetics	Electrical conductivity $\sigma(\mathbf{r})$	Electromagnetic field $\{\mathbf{E}(\mathbf{r}, t), \mathbf{B}(\mathbf{r}, t)\}$
Heat flow	Thermal conductivity $k(\mathbf{r})$	Temperature $T(\mathbf{r})$

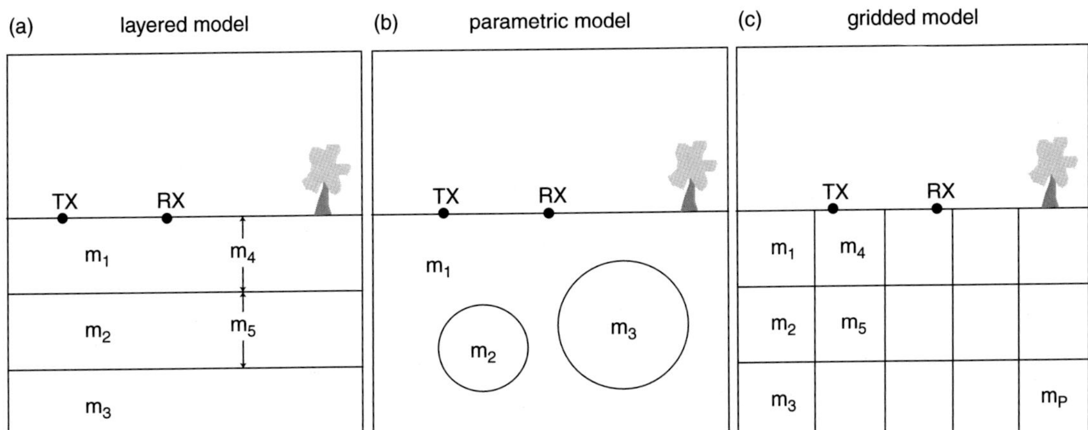

Figure 11.1 Examples of geophysical model vectors, m: (a) layered model with $P = 5$; (b) a structural model consisting of a few parameters, $P = 3$; (c) a voxel-based model consisting of many parameters, with large P.

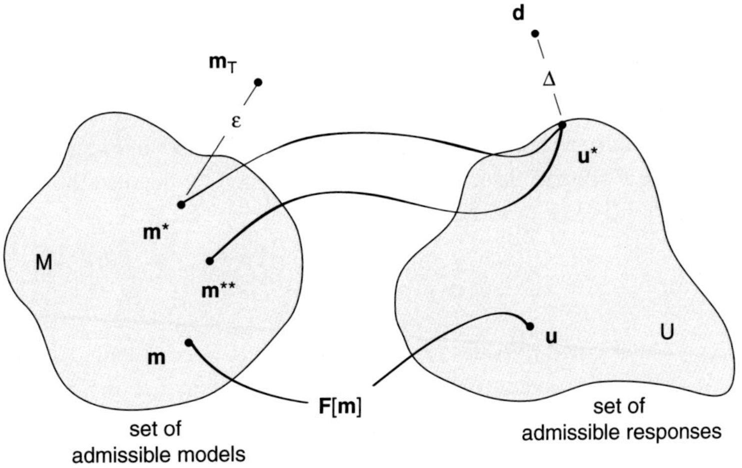

Figure 11.2 The structure of a geophysical inverse problem.

environmental conditions, reading errors, repeatability, etc. The data errors are assembled into the *data-error vector* $\mathbf{e} = (e_1, e_2, \ldots, e_D)^T$.

The essential structure of a geophysical inverse problem is shown in Figure 11.2. The set M represents the entire domain of Earth models that the geophysicist can analyze with the available forward-modeling capabilities. The set M is called the *admissible set* of models. The inverse problem is effectively a search over the models in M, looking for one that optimizes a prescribed criterion. An optimal model $\mathbf{m}^*$ is a solution to the inverse problem and it necessarily belongs to M.

It is assumed that the geophysical forward problem can be written as a stable, possibly non-linear mapping $\mathbf{u} = \mathbf{F}[\mathbf{m}]$ that transforms an Earth model $\mathbf{m}$, selected from M, into a unique response vector $\mathbf{u}$. The mapping $\mathbf{F}[\mathbf{m}]$ is a mathematical operation that presumes to capture the essential physics relating an Earth model to its geophysical response.

The response vector **u** is typically measured as, or transformed into, a set of D real numbers; it is also called the model-prediction vector. The response vector **u** necessarily resides in U, the set of all possible responses. The set U, a subset of $\boldsymbol{R}^D$, is termed the range of the forward operator $\mathbf{F}[\mathbf{m}]$.

For every admissible model **m** in M, there is presumed to exist a *unique* response vector **u** in U which can be constructed by solving the forward problem, i.e. by evaluating the functional $\mathbf{F}[\mathbf{m}]$. This is indicated by the path $\mathbf{m}\rightarrow \mathbf{F}[\mathbf{m}]\rightarrow \mathbf{u}$ shown in the figure. The calculated response vector **u** is then typically compared to the vector of measurements $\mathbf{d} \pm \mathbf{e}$. Both the data vector **d** and the measurement error vector **e** reside in $\boldsymbol{R}^D$. Notice, however, that the vector **d** almost certainly lies outside U since it is a physical measurement generated by the actual, or "true" Earth $\mathbf{m}_T$. The latter, due to its enormous complexity, almost certainly lies outside M. Thus, both the true Earth $\mathbf{m}_T$ and its actual response **d** are outside both the domain and the range of the forward operator $\mathbf{F}[\mathbf{m}]$. Moreover, the forward mapping operator $\mathbf{F}[\mathbf{m}]$ very likely represents an oversimplification of the actual, and to some extent unknown, physical processes that connect the actual Earth $\mathbf{m}_T$ to the measured data **d**.

Perhaps the simplest geophysical inverse problem involves looking for a model which minimizes some measure of the data misfit. Specifically, a model $\mathbf{m}^*$ is sought in M which generates that member of U, namely $\mathbf{u}^*$, lying as close as possible in $\boldsymbol{R}^D$ to the data vector **d**. Let a measure of the distance in R^D between **d** and a vector **u** in U be termed the misfit Δ, as illustrated for the specific case $\mathbf{u} = \mathbf{u}^*$ in Figure 11.2. Typically, the misfit Δ is the Euclidean distance, optionally weighted by the measurement errors, given by

$$\Delta = \sqrt{\frac{1}{D}\sum\nolimits_{i=1}^{D}\left[\frac{u_i - d_i}{e_i}\right]^2}. \tag{11.1}$$

Other criteria besides misfit minimization are commonly used to formulate a geophysical inverse problem. For example, an optimal model $\mathbf{m}^*$ may be defined as one whose response $\mathbf{u}^*$ provides an acceptable fit to the data vector **d** while satisfying, as nearly as possible, a number of side constraints on the model parameters. The imposition of model-parameter constraints effectively reduces the size and alters the topology of the admissible set M, but does not change the essential structure of the inverse problem.

It is critical to note that geophysicists can never determine the *true*, immeasurably complex Earth $\mathbf{m}_T$, for several reasons: (a) the true structure $\mathbf{m}_T$, being immeasurably complex, lies outside the set M and hence is not an admissible model; (b) an immeasurably complex Earth requires a very large number of parameters to describe, far more than could be handled even by the largest supercomputers; (c) the forward operator **F** acts only on members of M and, even if it were properly defined for the true structure $\mathbf{m}_T$, would not generate the observed data vector **d** due to extraneous noise in the measurements; (d) the inverse mapping $\mathbf{m}^* = \mathbf{F}^{-1}[\mathbf{d}]$ between the data vector **d** and the optimal model $\mathbf{m}^*$ is inherently unstable and non-unique (see below); and (e) the data are necessarily sparsely sampled and hence inadequate to resolve the full structure of $\mathbf{m}_T$.

As suggested in the foregoing paragraph, geophysical inverse problems are unstable, non-unique, and oftentimes intractable. If small changes to the data vector **d** bring about large changes in the optimal model vector **m***, the inverse problem is said to be *unstable*. *Non-uniqueness* lies in the fact that many different models **m** have essentially the same response **u**. For example, it is seen in Figure 11.2 that the two distinct models **m*** and **m**** are both valid solutions to the inverse problem since both generate the identical response **u***. Finally, note that a solution to the inverse problem *may not exist*; that is, there might be no model **m** in M whose response **u** fits the data vector **d** to within a prescribed acceptable level of misfit, i.e. $\Delta < \Delta_{TOL}$, where the subscript TOL indicates "maximum tolerable."

In general, geophysical inverse problems are ill-posed in the sense that they must be formulated as optimization problems, rather than uniquely solveable problems. Optimal solutions are then found by application of inductive reasoning or statistical inference. Forward problems, by constrast, are well-posed, uniquely solveable, and their solution is found by deductive reasoning. In other words, if **F**[**m**] encapsulates the geophysicist's belief about the Earth and how it responds to geophysical probing, the forward problem answers the well-posed question: "What does my belief imply about the evidence?" while the inverse problem addresses the more challenging question: "How does the evidence shape my belief?" Stated yet another way, the forward problem is to calculate frequencies from known or hypothesized probability distributions while the inverse problem is to estimate unknown probability distributions from observed frequencies (Johnson, 1932).

11.2 Linear-parameter estimation

There are certain geophysical problems in which each data point is presumed to be a *linear* function of the model parameters. A linear forward problem can be written in matrix form $\mathbf{d} = A\mathbf{m} + \Delta$. The matrix A is a $D \times P$ *rectangular* matrix of known coefficients that approximates the physics connecting the model parameters to the response vector. In other words, the matrix–vector product $A\mathbf{m}$ is a linear version of the non-linear forward-mapping operator **F**[**m**], such that $\mathbf{u} = A\mathbf{m}$. The matrix A does not depend on the model vector **m** or the data vector **d**. The vector Δ, which describes the discrepancy between the response vector **u** and the data vector **d**, can be regarded as the modeling error. In other words, the vector Δ is present due to the inadequacy of the forward modeling operator ($A\mathbf{m}$ in the linear case or **F**[**m**] in the non-linear case) to fully describe the true Earth and all the physical processes that connect the true Earth to the observations. Notice that the modeling error vector Δ is not the same as the measurement error **e** introduced in the previous section. The latter, as shown in Equation (11.1), is used simply to weight the data.

For example, in gravity prospecting, a linear forward problem can be defined in which the elements of matrix A are based on Newton's gravitational law. Specifically, matrix element A_{ij} describes the gravity field at observation point i due to an elemental volume of mass located at source point j. The i-th row of matrix A is called the *data kernel*. It describes how a gravity observation at point i depends on the various parameters contained in the model vector **m**. In fact, each data point d_i determines a linear constraint on the model, summarized by the corresponding data kernel. Thus, the number of linear-constraint equations is equal to D, the total number of rows in the matrix A.

There are three possible relationships between the number of unknown model parameters and the number of known data points. If $D = P$, there are equal numbers of unknowns and equations and it may be possible to find a unique, stable solution for $\mathbf{m}$. In that case, the inverse problem is considered to be *well-determined*.

If $D > P$, there are more constraints on the unknowns than unknowns themselves and the inverse problem is *over-determined*. In this case, unless some of the constraints are redundant, there is no model vector $\mathbf{m}$ that can satisfy all of the constraints exactly. A number of distinct model vectors may then be constructed each of which approximately satisfy the constraints to within a prescribed tolerance Δ.

Finally, if $D < P$ there are more unknowns than constraints and in this case the inverse problem is *under-determined*. There is not enough information in the comparatively sparse data set to uniquely determine the larger number of parameters contained in the model vector $\mathbf{m}$. Again, a number of model vectors may be constructed which approximately satisfy the constraints to within the tolerance. In this case, some linear combinations of the model parameters may be poorly resolved while other linear combinations are better resolved. The poorly resolved combinations of model parameters could take on arbitrary, unphysical values that negatively affect the interpretation but have little or no impact on the response vector. Accordingly, an inverse problem exhibiting this property is classed as unstable.

11.3 Least-squares solution

A well-determined linear inverse problem sometimes leads to a unique and stable solution for the model parameters. In the over-determined case however, in which $D > P$, a unique solution that exactly satisfies the data is practically impossible unless by extreme serendipity. The *least-squares* approach and its variants (Lines and Treitel, 1984) minimizes a weighted sum of squares of the discrepancies between each computed response and the corresponding measured value.

From the linear forward problem $\mathbf{d} = A\mathbf{m} + \Delta$, the components of the vector Δ can be written as

$$\Delta_i = d_i - \sum_{j=1}^{P} A_{ij} m_j, \quad i = 1, \ldots, D. \tag{11.2}$$

For now we ignore the observation error $\mathbf{e}$ and assume the data are equally weighted. The sum of the squares of the errors is the scalar product $\Delta^T \Delta$, which is the quantity to be minimized. Specifically, the model parameters m_j for $j = 1, \ldots, P$ are adjusted; their best values are considered to be those for which $\Delta^T \Delta$ attains its lowest value. This is the essential characteristic of the most elementary least-squares approach.

A necessary condition on the optimizing set of model parameters is that, for each $k = 1, \ldots, P$ we must have $\partial \Delta^T \Delta / \partial m_k = 0$. This is the multivariate equivalent to the statement that a minimum of a univariate function $f(x)$ occurs at the point x^* such that $\partial_x f(x^*) = 0$, where we have used the notation

$$\partial_x f(x^*) = \left.\frac{\partial f}{\partial x}\right|_{x=x^*}.$$

In P-dimensional space, the minimum of the multivariate function $f(\mathbf{x}):R^P \to R^1$ occurs at the point $\mathbf{x}^* = (x_1{}^*, x_2{}^*, \ldots, x_P{}^*)$ such that each partial derivative $\partial_x f(x_k{}^*) = 0$ vanishes, for $k = 1, \ldots, P$. Applying the set of necessary conditions $\partial \Delta^T \Delta / \partial m_k = 0$ for $k = 1, \ldots, P$ (which can be written as $\partial_{\mathbf{m}} \Delta^T \Delta = 0$) leads to a set of linear equations $A^T A \mathbf{m} = A^T \mathbf{d}$ (see problems at the end of this chapter) called the *normal equations*.

The solution to the normal equations is

$$\mathbf{m}^* = (A^T A)^{-1} A^T \mathbf{d}. \tag{11.3}$$

It can be shown that the matrix $A^T A$ is positive semi-definite which means that its eigenvalues are all positive or zero. The $P \times D$ rectangular matrix $(A^T A)^{-1} A^T$ is called the least-squares inverse of matrix A.

11.4 Example: near-surface magnetization

While most practical inverse problems in near-surface applied geophysics are non-linear, an instructive linear inverse problem can be formulated that involves the interpretation of magnetic-anomaly data. A related computer assignment designed to familiarize the reader with some of the various linear-inversion algorithms discussed herein is provided at the end of this chapter.

The magnetics method was described in Chapter 3 of this book but we recall the essential details here. The relevant forward problem is: given the magnetization $\mathbf{M}(\mathbf{r}')$ of a subsurface magnetized body, determine at some remote locations the total field anomaly $T(\mathbf{r})$. Notice that primed position vector $\mathbf{r}'$ refers to a location occupied by the source while the unprimed position vector $\mathbf{r}$ refers to an observation point occupied by a magnetometer.

Recalling the discussion in Chapter 3, the magnetic field of an elementary dipole is (Blakely, 1995)

$$\mathbf{B}_0(\mathbf{r}) = \frac{\mu_0}{4\pi} \nabla \left[\mathbf{m} \cdot \nabla \frac{1}{|\mathbf{r} - \mathbf{r}'|} \right], \tag{11.4}$$

where μ_0 is the magnetic permeability of free space and the gradient operator is $\nabla = \hat{x}\partial_x + \hat{y}\partial_x + \hat{z}\partial_z$. The magnetic field $\mathbf{B}(\mathbf{r})$ at some point outside the magnetized body can be regarded as the superposition of contributions from each of its constituent dipoles, namely,

$$\mathbf{B}(\mathbf{r}) = \frac{\mu_0}{4\pi} \nabla \int_V \left[\mathbf{M}(\mathbf{r}') \cdot \nabla \frac{1}{|\mathbf{r} - \mathbf{r}'|} \right] d\mathbf{r}', \tag{11.5}$$

where V is the volume occupied by the body. Note that the integration is over the primed (source) coordinates. If a total-field magnetometer is used, the actual quantity measured is

$|\mathbf{B} + \mathbf{B}_{REF}|$ where $\mathbf{B}_{REF}$ is the ambient geomagnetic field. The *total-field anomaly* T is then defined as $T = |\mathbf{B} + \mathbf{B}_{REF}| - |\mathbf{B}_{REF}|$. Since the magnitude of the ambient geomagnetic field is always much larger than the magnetic field of the body, it was shown in Chapter 3 that

$$T(\mathbf{r}) = \frac{\mu_0}{4\pi}\mathbf{B}_{REF}\cdot\nabla\int_V\left[\mathbf{M}(\mathbf{r}')\cdot\nabla\frac{1}{|\mathbf{r}-\mathbf{r}'|}\right]d\mathbf{r}', \tag{11.6}$$

where $\mathbf{B}_{REF}$ is the *direction* of the ambient geomagnetic field. Next, we make the simplification that the subsurface block is magnetized in the direction of the present-day geomagnetic field, i.e. there is no remanent magnetization remaining from previous geomagnetic polarity reversals or from previous tectonic rotations. This implies that $\mathbf{M}(\mathbf{r}') = M(\mathbf{r}')\mathbf{B}_{REF}$ and the total-field anomaly is then written

$$T(\mathbf{r}) = \frac{\mu_0}{4\pi}\mathbf{B}_{REF}\cdot\nabla\int_V\left[M(\mathbf{r}')\mathbf{B}_{REF}\cdot\nabla\frac{1}{|\mathbf{r}-\mathbf{r}'|}\right]d\mathbf{r}'. \tag{11.7}$$

This equation has the form of a Fredholm integral equation of the first kind

$$T(\mathbf{r}) = \int_V M(\mathbf{r}')A(\mathbf{r},\mathbf{r}')d\mathbf{r}' \tag{11.8}$$

in which it is evident that the measured quantity $T(\mathbf{r})$ is a *linear function* of the Earth physical property $M(\mathrm{r}')$. The function $A(\mathrm{r},\mathrm{r}')$ is the data kernel that contains the physics of the forward problem.

There are many ways to convert Equation (11.8) into a practical inverse problem. Here, we divide the near-surface region into two-dimensional (2-D) uniformly magnetized polygonal blocks with vertical sides, as shown in Figure 11.3. There are a total of P blocks; the j-th block has magnetization M_j, nominal width w_j, nominal height h_j and the depth to its top left corner is z_j. There are a total of D data points; the i-th data point is denoted by T_i.

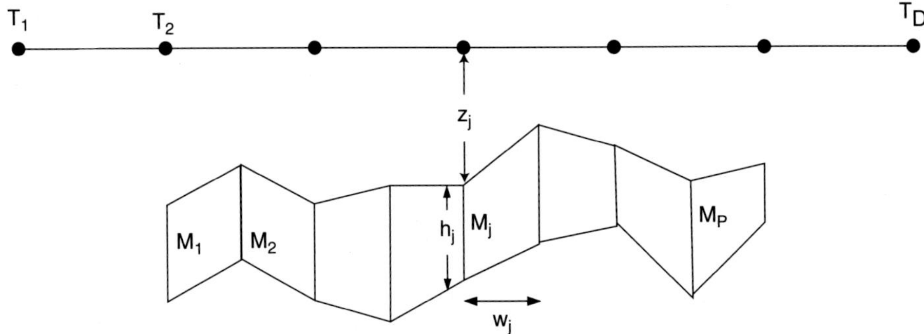

Figure 11.3 A linear inverse problem: near-surface magnetization from total-field anomaly data.

Discretization of the integral in Equation (11.8) leads to the linear system

$$T_i = \sum_{j=1}^{P} A_{ij} M_j, \text{ for } i = 1, \ldots, D, \tag{11.9}$$

where A_{ij} is described below. Equation (11.9) is in the form of a linear inverse problem $\mathbf{d} = A\mathbf{m} + \Delta$ if we add a modeling-error term Δ to account for factors such as the idealized parameterization of near-surface magnetization, and the simplifying assumptions that went into the forward physics. The inverse problem is then to determine optimal model parameters M_j, $j = 1, \ldots, P$ that are most consistent, in a least-squares or perhaps some other sense to be determined, with the data T_i for $i = 1, \ldots, D$.

The quantity A_{ij} is simply the total-field anomaly at observation location $\mathbf{r}_i$ caused by a unit magnetization of block j. The magnetic field of a 2-D uniformly magnetized prism has a well-known analytic formula (Blakely, 1995). The matrix elements are

$$A_{ij} = \sum_{L=1}^{N} \left[(B_x)^{L_j} \hat{x} \cdot \hat{B}_{REF} + (B_z)^{L_j} \hat{z} \cdot \hat{B}_{REF} \right], \tag{11.10}$$

where the sum indexed by L is over the edges of the j-th prism, hence in this case $N = 4$ since each prism has four edges. The formula (11.10) is valid only if the origin is set at the location $\mathbf{r}_i = (x_i, 0)$ of the i-th data point. Thus, as the matrix A is built, the coordinate origin moves along the data profile, which is in the $\hat{x}$ direction. In the previous equation, as Blakely (1995) has shown,

$$(B_x)^{L_j} = -\frac{\mu_0}{2\pi} \hat{n} \cdot \hat{B}_{REF} \left[S_x ln\left(\frac{r_2}{r_1}\right) - S_z(\theta_1 - \theta_2) \right]; \tag{11.11}$$

$$(B_z)^{L_j} = -\frac{\mu_0}{2\pi} \hat{n} \cdot \hat{B}_{REF} \left[S_z ln\left(\frac{r_2}{r_1}\right) + S_x(\theta_1 - \theta_2) \right]. \tag{11.12}$$

The vector $\hat{n}$ is the unit vector normal to the L-th edge of the j-th prism. The vector $\mathbf{S} = S_x\hat{x} + S_z\hat{z}$ is the unit vector tangential to the L-th edge of the j-th prism. The tangential unit vector of each prism edge is oriented clockwise around the prism. The quantities (r_1, θ_1) and (r_2, θ_2) are the polar coordinates of the endpoints of the L-th edge of the j-th prism.

The system of equations (11.9) is solved using the least-squares approach described in the previous section. This yields a set of optimal values for the block magnetizations, namely $\mathbf{M}^* = (M_1^*, M_2^*, \ldots, M_P^*)$. The forward response of this optimal model provides the best fit to the observed total-field anomaly measurements $(T_1, T_2, \ldots, T_D)$. At this point the inverse problem is solved and it then becomes a task of geological interpretation to relate the optimal distribution of near-surface magnetization to the objectives of the geophysical investigation. In subsequent sections of this chapter we explore modifications to the least-squares approach in the case that the data contain observation errors. We also explore, in detail, the effects of adding constraints on the model parameters and we provide some comments about the stability of the model-parameter reconstruction.

11.5 Example: deconvolution

Recall from Chapter 2 that a measured geophysical response can be regarded as the convolution of an idealized impulse response with the actual source function. The impulse response is of fundamental theoretical importance because it reveals how the Earth behaves to an idealized input signal without having to take into account potentially confounding effects of a complicated source distribution. The near-surface seismologist, for example, is often more interested to know how the Earth would have responded to an idealized short, sharp mechanical impulse rather than how it actually did respond to the drawn-out, complicated signature of an actual sledgehammer, shotgun, or explosive source.

Deconvolution is the process of extracting the impulse response from the measured geophysical response. In other words, deconvolution addresses the question: "given the measured response to the actual source function, how would the Earth have responded had the source function been an ideal impulse?" Viewed in this way, it is clear that the objective of deconvolution is to undo the effects of a complicated source signature.

Deconvolution in its most straightforward guise can be posed as a linear least-squares inverse problem of the standard form $\mathbf{d} = A\mathbf{m} + \Delta$. In this case, the "forward problem" consists of the convolution of a known source function discretized as $\mathbf{s} = (s_1, s_2, \ldots, s_K)$ with an unknown filter that represents the impulse response of the subsurface. The model-parameter vector $\mathbf{m} = (m_1, m_2, \ldots, m_P)$ of length P consists of the sought-after filter coefficients. The data vector $\mathbf{d} = (d_1, d_2, \ldots, d_D)$ of length D consists of the measured geophysical response. The $D \times P$ "physics matrix" A describes, as usual, the mechanism that connects the model to the predicted response vector $\mathbf{u} = A\mathbf{m}$, which in this case is simply the convolution

$$u_k = \sum_{n=1}^{P} s_{k-n} m_n, \quad k = 1, \ldots, D. \tag{11.13}$$

The elements of matrix A are easily seen to be given by $A_{ij} = s_{i-j}$. The least-squares optimal filter $\mathbf{m}^*$, defined as the filter that minimizes $\Delta^T\Delta$ where $\Delta = \mathbf{d} - \mathbf{u}$, is then calculated by the standard formula $\mathbf{m}^* = (A^TA)^{-1}A^T\mathbf{d}$, as prescribed by Equation (11.3). The optimal filter $\mathbf{m}^*$ computed in this manner is sometimes called a Wiener filter. It can be readily appreciated from the form of Equation (11.13) that the matrix A^TA involves the autocorrelation of the sequence $\mathbf{s}$ while matrix $A^T\mathbf{d}$ involves the cross-correlation of the sequences $\mathbf{s}$ and $\mathbf{d}$. The interested reader is referred to Berkhout (1977) for further analysis of least-squares inverse filters, and a discussion of their limitations for practical deconvolution of seismic data. The paper by Irving and Knight (2003) explains how removal of the effects due to frequency-dependent attenuation from GPR data can also be regarded as a deconvolution problem.

11.6 Data covariance

It is always important to take full consideration of the errors in any scientific experiment. As mentioned above, the geophysicist must distinguish between observation errors and modeling errors. Observation errors are those that affect the actual measurements made in the field, or those that occur during the subsequent analysis of field recordings, and typically include some combination of instrument error, reading error, systematic or random background noise and, possibly, coherent signal-generated noise. Modeling error is caused by unrecognized or oversimplified physics and other approximations used in conceptualizing the forward problem and is not related to the measurement or data-analysis processes. In our notation, observation errors are described by the error vector **e** while modeling errors are described by the vector Δ. In this section, only observation errors are discussed.

Observation errors should be treated as *random variables* since their actual values cannot be determined. Accordingly, only statistical quantities can be found such as the mean μ_x and variance σ_x^2 of the random variable x, and the covariance of pairs of random variables x, y. The covariance C_{xy} is defined by

$$C_{xy} = \frac{1}{N-1}\sum_{k=1}^{N}\left(x^{(k)} - \mu_x\right)\left(y^{(k)} - \mu_y\right); \tag{11.14}$$

where $x^{(k)}$, $y^{(k)}$ is the k-th of N realizations of the pair x, y. The quantities (μ_x, σ_x^2, C_{xy}) are most easily found simply by averaging a number of repeated measurements. However, if repeat measurements are not available, as is commonly the case in geophysics, values for (μ_x, σ_x^2, C_{xy}) must be assigned based on theoretical considerations combined with the experience and judgment of the geophysicist. Error estimation is one of the most challenging tasks faced by near-surface applied geophysicists.

The covariance of two independent random variables x and y is zero since any change in the first variable is unrelated to a change in the second variable. The covariance between x and y, however, is non-zero if the two random variables are correlated. In the multivariate case, the covariances amongst pairs of elements of a vector of D observations $\mathrm{d} = (d_1, d_2, \ldots, d_D)$, where each observation is viewed as a random variable, is a matrix C whose i, j-th entry is

$$C_{ij} = C(d_i, d_j) = \frac{1}{D-1}\sum_{k=1}^{D}\left(d_i^{(k)} - \mu_{d_i}\right)\left(d_j^{(k)} - \mu_{d_j}\right). \tag{11.15}$$

If all D observations are independent, which is often assumed to be the case in geophysics, the covariance matrix takes the diagonal form $C = \mathrm{diag}(\sigma_{d_1}^2, \sigma_{d_2}^2, \ldots, \sigma_{d_D}^2)$. A non-diagonal covariance matrix typically develops when the observations exhibit serial correlation or when they have been rotated or otherwise transformed or combined.

Recall that the normal equations $A^T A\mathbf{m} = A^T\mathbf{d}$ derived earlier provide the least-squares solution to the linear inverse problem. The formulation however did not take the data error vector **e** into explicit account. An improved estimate of the model vector **m** is obtained if the covariance matrix C associated with the data vector **d** is considered. In such a case, the normal equations become

$$(A^T C^{-1} A)\mathbf{m} = A^T C^{-1}\mathbf{d}, \tag{11.16}$$

with solution

$$\mathbf{m}^* = (A^T C^{-1} A)^{-1} A^T C^{-1} \mathbf{d}. \tag{11.17}$$

This equation is derived in Dean (1988) and Gubbins (2004). The solution (11.17) is the model that minimizes the weighted misfit function $\Delta^T C^{-1} \Delta$.

11.7 The null space

The under-determined linear inverse problem in which $D < P$ is commonly formulated by geophysicists. Here, the number of model parameters exceeds the number of data that constrain them. Hence, the model is *over-parameterized* with respect to the comparatively small number of constraints imposed on it by a sparse dataset. It is likely that many different combinations of model parameters could be found to fit the paucity of constraints.

In the under-determined case ($D < P$) there are fewer linearly independent constraints on the model than there are unknown model parameters. Since the problem is linear, we can determine only up to D independent linear combinations of the model parameters uniquely. In the idealized case without errors, these D linear combinations would *exactly* satisfy the constraint equations, i.e. completely explain all the data. This means that the remaining $N = P - D$ linear combinations of model parameters are completely unresolved and, in fact, have no effect on the response vector $\mathbf{u}$.

The above considerations imply the existence of an N-dimensional *null space*. Following Gubbins (2004), let us select a particular model $\mathbf{m}_0$ from the null space. Since the model vector $\mathbf{m}_0$, as mentioned, has no effect on the data it must satisfy $A\mathbf{m}_0 = 0$. Thus, we see that $A(\mathbf{m} + \alpha\mathbf{m}_0) = \mathbf{d}$, where α is an arbitrary scalar. Therefore $\mathbf{m}^*$ and $\mathbf{m}^* + \alpha\mathbf{m}_0$ for any choice of α are equivalent solutions to the inverse problem. Hence, the under-determined inverse problem has infinitely many solutions since α is arbitrary. It must be mentioned that both $\mathbf{m}$ and $\mathbf{m}_0$ are P-dimensional vectors but $\mathbf{m}$ has only D linearly independent components while $\alpha\mathbf{m}_0$ has N linearly independent components, with $D + N = P$. Notice that adding another data point, i.e. increasing $D \to D + 1$, simply decreases by one the dimension of the null space but does not significantly mitigate the non-uniqueness of the inverse problem.

11.8 The minimum-norm solution

The previous section showed that there are infinitely many different models $\mathbf{m} + \alpha\mathbf{m}_0$ that fit the data equally well. We might wish to choose one from amongst them to report as the best solution to the under-determined inverse problem. This ultimately becomes a subjective choice. A popular option is to select the model with the minimum norm, $\mathbf{m}^T\mathbf{m}$. The *minimum-norm* solution, being the smallest solution, presumably contains no superfluous information. This property helps to justify its popularity. Notice that any model $\mathbf{m}_0$ that lies in the null space contains no information at all because, as mentioned earlier, $\mathbf{m}_0$ has no effect on the response vector **u** since $A\mathbf{m}_0 = 0$. The minimum-norm solution is designed to eliminate contributions from the null space. Put another way, if the minimum-norm model did contain a part that lies in the null space, it would not be the smallest possible solution.

How is the minimum-norm solution constructed? The most efficient way is to pose a *constrained optimization* problem: find the model $\mathbf{m}^*$ that minimzes $\mathbf{m}^T\mathbf{m}$ subject to the constraints $A\mathbf{m} = \mathbf{d}$. Mathematically, this statement can be written

$$\mathbf{m}^* = \arg\left[\min_{A\mathbf{m}=\mathbf{d}} (\mathbf{m}^T\mathbf{m})\right]. \tag{11.18}$$

Notice that we are trying to minimize a quadratic function $\mathbf{m}^T\mathbf{m}$ subject to a set of linear constraints. The technique of Lagrange multipliers provides the standard approach to solve linearly constrained quadratic optimization problems. In this technique, we start by constructing the scalar *Lagrangian* function

$$L(\mathbf{m}, \lambda) = \mathbf{m}^T\mathbf{m} + \lambda^T(A\mathbf{m} - \mathbf{d}), \tag{11.19}$$

where $\boldsymbol{\lambda}$ is a D-dimensional vector of Lagrange multipliers. It is clear that minimizing the Lagrangian function $L(\mathbf{m}, \boldsymbol{\lambda})$ with respect to each multiplier λ_i for $i = 1, \ldots, D$ is equivalent to solving the equations $\partial L/\partial\lambda_i = 0$. This implies that the constraints $A\mathbf{m} = \mathbf{d}$ shall automatically be satisfied, as shown below:

$$\frac{\partial L}{\partial \lambda_i} = \frac{\partial}{\partial \lambda_i}\left\{\mathbf{m}^T\mathbf{m} + \sum_{j=1}^{D}\lambda_j\left[\sum_{k=1}^{P}A_{jk}m_k - d_j\right]\right\} = \sum_{k=1}^{P}A_{ik}m_k - d_i = 0.$$

Thus, the minimization of $L(\mathbf{m}, \boldsymbol{\lambda})$ with respect to the i-th Lagrange multiplier λ_i ensures that the i-th constraint is upheld. Accordingly, minimization of $L(\mathbf{m}, \boldsymbol{\lambda})$ with respect to all the Lagrange multipliers ensures that all the constraints $A\mathbf{m} = \mathbf{d}$ are upheld. Mathematically, we can write this as $\partial L(\mathbf{m}, \boldsymbol{\lambda})/\partial\boldsymbol{\lambda} = 0 \Rightarrow A\mathbf{m} = \mathbf{d}$.

The minimum-norm solution $\mathbf{m}^*$ is the *unconstrained* minimum of the Lagrangian,

$$(\mathbf{m}^*, \lambda^*) = \arg\left[\min L(\mathbf{m}, \lambda)\right], \tag{11.20}$$

and it is found by enforcing the necessary condition $\partial L(\mathbf{m}, \lambda)/\partial m_k = 0$ on each model parameter $k = 1, \ldots, P$ along with $\partial L(\mathbf{m}, \lambda)/\partial\lambda_j = 0$ on each Lagrange multiplier $j = 1, \ldots, D$. The first condition minimizes $\mathbf{m}^T\mathbf{m}$, as required, while the second condition ensures satisfaction of the contraints $A\mathbf{m} = \mathbf{d}$. The two necessary conditions are referred to as the *augmented normal equations*. After some algebra the solution of the augmented normal equations reduces to (Gubbins, 2004)

$$\mathbf{m}^* = A^T(AA^T)^{-1}\mathbf{d}. \tag{11.21}$$

The matrix $A^\dagger = A^T(AA^T)^{-1}$ is often called the *generalized inverse* of matrix A.

Notice that the construction of the minimum-norm solution involves the conversion of a P-dimensional constrained optimization problem into a larger $D + P$-dimensional unconstrained optimization problem. It is then required to find optimal values of the Lagrange multipliers simultaneously with finding optimal values of the model parameters. The optimal Lagrange multipliers $\boldsymbol{\lambda}^*$, however, have no physical significance.

11.9 The trade-off curve

There are other choices for the auxiliary constraint on the model parameters that one could make besides minimizing the norm $\mathbf{m}^T\mathbf{m}$ that produces the smallest model. For example, we could instead minimize the quantity $|\mathbf{m} - \mathbf{m}_P|$ that measures the departure of a model $\mathbf{m}$ from a reference, or prior model, $\mathbf{m}_P$ for which we might have some physical reason to suspect is a good model. Or, we could minimize $\mathbf{m}^T W\mathbf{m}$ where W is a symmetric positive-definite matrix to be determined. These choices permit some of the model parameters to be larger than others, without penalizing them, and also allows correlations between model parameters. The matrix W is generally chosen to reflect some auxiliary, or a-priori information about the model. In general, $\mathbf{m}^T W\mathbf{m}$ can be regarded as a measure of the model complexity since any choice other than $W = I$, the identity matrix, introduces structure into the model.

A general solution to an under-determined linear inverse problem is one that minimizes the function $\mathbf{m}^T W\mathbf{m}$ subject to satisfaction of the linear constraints $A\mathbf{m} = \mathbf{d}$. In the practical case, the linear constraints cannot be exactly satisfied and some amount of error $\Delta = A\mathbf{m} - \mathbf{d}$ must be tolerated. As noted by Gubbins (2004), a flexible criterion that admits a range of possible solutions involves unconstrained minimization of the function

$$T(\theta) = \Delta^T\Delta + \theta^2(\mathbf{m}^T W\mathbf{m}), \tag{11.22}$$

where θ is a *trade-off* parameter whose value establishes the relative importance of $\Delta^T\Delta$ and $\mathbf{m}^T W\mathbf{m}$ in the optimal solution. Notice that $\Delta^T\Delta$, $\mathbf{m}^T W\mathbf{m}$, and θ^2 are all positive numbers so that $T(\theta) > 0$. The minimization of $T(\theta)$ is a compromise between fitting the data and generating a model that has a desired structure. As $\theta \rightarrow 0$, the problem reduces to the conventional least-squares solution of minimizing $\Delta^T\Delta$. As $\theta \rightarrow \infty$, the problem reduces to the unconstrained minimization of $\mathbf{m}^T W\mathbf{m}$ which has trivial solution $\mathbf{m} = 0$, as would be expected since the data are ignored in that case. A typical trade-off curve that is obtained by varying the parameter θ and repeatedly minimizing $T(\theta)$ for each value of θ is portrayed in Figure 11.4.

Recall that the least-squares equation in the presence of data errors, found by minimizing the quantity $\Delta^T C^{-1}\Delta$, is $\mathbf{m}^* = (A^T C^{-1} A)^{-1} A^T C^{-1}\mathbf{d}$ where C is the data covariance matrix. If the quantity $T(\theta)$ is minimized in the same fashion, the result is

$$\mathbf{m}^* = (A^T C^{-1} A + \theta^2 W)^{-1} A^T C^{-1}\mathbf{d} \tag{11.23}$$

in which W, if it is a diagonally dominant matrix, acts to stabilize the computations. The *damped-least-squares* solution is the special case in which $W = I$.

To summarize, the trade-off curve $\mathbf{m}^*(\theta)$ is constructed by minimizing Equation (11.22) repeatedly for different values of the trade-off parameter θ and then plotting the resulting values of $\Delta^T\Delta$ and $\mathbf{m}^T W\mathbf{m}$. At small values of θ, the misfit $\Delta^T\Delta$ (or $\Delta^T C^{-1}\Delta$ in the case of weighted data) is small and the model norm $\mathbf{m}^T W\mathbf{m}$ is large. Increasing the value of θ has the effect of reducing the model norm $\mathbf{m}^T W\mathbf{m}$, that is, reducing extraneous model structure, but at the cost of increasing the misfit. At large values of θ, the model norm is small but the attendant simplicity in the model comes at the expense

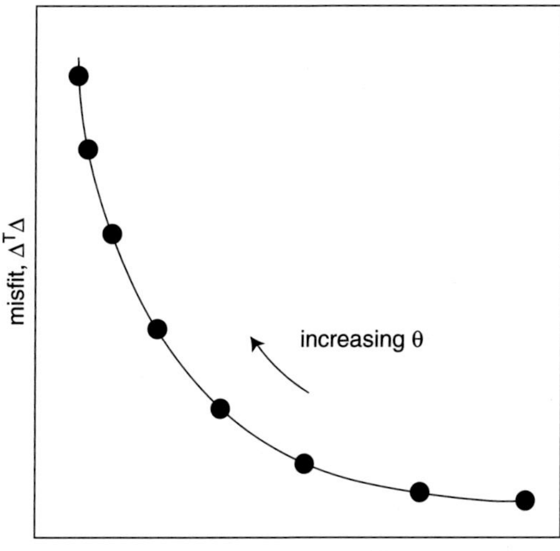

Figure 11.4 Trade-off between misfit and model complexity. Each dot represents an optimal model $\mathbf{m}^*(\theta)$.

of the misfit, which is large. Any model $\mathbf{m}^*(\theta)$ along the trade-off curve is a *bona fide* optimal solution to the inverse problem. The geophysicist can either present the entire suite of optimal models, as in Figure 11.4, and leave it at that or, better, he or she can use auxiliary information, experience, and judgment to select and justify a preferred optimal model or a preferred subset of the optimal models from those located along the $\mathbf{m}^*(\theta)$ trade-off curve.

11.10 Regularization

The strategy of regularized inversion is to find a model whose response fits the data to a given tolerance but at the same time is optimal in some other pre-determined attribute such as *smoothness*, or *adherence* to a prior model that is based on previous knowledge. In this discussion we restrict focus to the smoothness criterion. The regularized inverse problem can then be stated as: *find the smoothest model that fits the data to a given tolerance.* The search for the least-squares minimum-misfit model is therefore abandoned, since such a model invariably overfits the data and becomes excessively rough. The solution to a smoothness-constrained inverse problem is attractive since the optimal model contains no more structure than is strictly demanded by the data.

The quest for such simple solutions to scientific problems has a distinguished historical precedent. The fourteenth-century philosopher William of Ockham stated, in effect: *it is vain to do with more that which can be done with fewer*; and in so doing provided inspiration for the doctrine that has become known as Occam's razor. We apply the razor in regularized inverse problems to "cut away" all unnecessary structure from the model.

We now provide an algorithm for constructing smooth solutions to linear inverse problems of the familiar type $\mathbf{d} = A\mathbf{m} + \Delta$ where $\mathbf{d} = (d_1, d_2, \ldots, d_D)$ represents the data, $\mathbf{m} = (m_1, m_2, \ldots, m_P)$ represents the model, A is the $D \times P$ matrix that contains the physics of the forward problem, and $\Delta = (\Delta_1, \Delta_2, \ldots, \Delta_D)$ is the modeling error vector. The simple case of a 1-D inverse problem is herein considered in which case the model parameters $\{m_i\}$ for $i = 1, \ldots, P$ are physical properties such as layer resistivities or prism magnetizations in the DC-resistivity and magnetics problems, respectively. The extension of the algorithm to multi-dimensional inverse problems is straightforward but not considered here.

The method of linear regularization is very similar to that of damped-least-squares, which we have already analyzed. The damped-least-squares problem is solved by minimizing the function $T(\theta)$ given by Equation (11.22).

Recall that a trade-off exists between misfit $\Delta^T\Delta$ and model complexity $\mathbf{m}^T W\mathbf{m}$. A smooth regularized inversion seeks a trade-off between the misfit and model roughness defined as $\mathbf{m}^T H\mathbf{m}$, where $W = H$ is now some kind of *roughening matrix*, one variant of which is described below.

We may consider model roughness $R^2[\mathbf{m}]$ to be a scalar quantity that minimizes the sum of the squared differences between adjacent model parameters,

$$\begin{aligned} R^2[\mathrm{m}] &= \sum_{i=1}^{P-1} [m_i - m_{i+1}]^2 \\ &= [m_1 - m_2]^2 + [m_2 - m_3]^2 + \cdots + [m_{P-1} - m_P]^2. \end{aligned} \tag{11.24}$$

Expanding the right-hand side and collecting terms, we arrive at

$$R^2[\mathrm{m}] = (m_1, m_2, \ldots, m_P)^T H (m_1, m_2, \ldots, m_P) = \mathrm{m}^T H \mathrm{m} \tag{11.25}$$

where H is the tridiagonal matrix (for the case $P = 6$),

$$H = \begin{pmatrix} 1 & -1 & 0 & 0 & 0 & 0 \\ -1 & 2 & -1 & 0 & 0 & 0 \\ 0 & -1 & 2 & -1 & 0 & 0 \\ 0 & 0 & -1 & 2 & -1 & 0 \\ 0 & 0 & 0 & -1 & 2 & -1 \\ 0 & 0 & 0 & 0 & -1 & 1 \end{pmatrix}. \tag{11.26}$$

With this choice for H, the linear-regularization problem becomes: find the unconstrained minimum of the function $T(\theta) = \Delta^T\Delta + \theta^2\mathbf{m}^T H\mathbf{m}$ where now $\theta^2 > 0$ is termed the regularization parameter. The minimization is performed in the usual way, that is, by solving the augmented normal equations. The result is

$$\mathbf{m}^* = (A^T A + \theta^2 H)^{-1} A^T \mathbf{d} \tag{11.27}$$

which compares closely to the conventional damped-least-squares solution $\mathbf{m}^* = (A^T A + \theta^2 I)^{-1} A^T \mathrm{d}$. The function $T(\theta)$ can be minimized for different values of the regularization parameter θ^2. Large values of θ^2 correspond to very smooth models with high misfit, whereas small values of θ^2 generate rough models with low misfits.

A subjective trade-off must be made in the choice between misfit and roughness. An important feature of linear regularization is that the term $\theta^2 H$ in Equation (11.27) effectively stabilizes the inversion of matrix $(A^T A + \theta^2 H)$ since the roughening matrix H is diagonally dominant.

11.11 Example: EM loop–loop sounding

In the previous section we found that the construction of smooth solutions to linear inverse problems in geophysics involves minimization of the function $T(\theta) = \Delta^T\Delta + \theta^2\mathbf{m}^T H\mathbf{m}$, where $\Delta^T\Delta$ is the misfit, θ^2 is the regularization parameter, and $\mathbf{m}^T H\mathbf{m}$ is the model roughness characterized by the roughening matrix H. In this section, we explore an example involving the determination of smooth electrical-conductivity depth profiles $\sigma(z)$ from loop–loop electromagnetic-response data.

In the frequency-domain loop–loop EM prospecting method, as discussed earlier in Chapter 8 of this book, a transmitter (TX) loop is energized with a current oscillating at ~ 1–100 kHz. The TX–RX separation distance is typically L ~ 10–100 m. The oscillatory nature of the current through the TX loop generates a primary magnetic field $\mathbf{B}_P$ which, in the absence of the conductive Earth, has a simple analytic form. The oscillating flux Φ_P through a RX coil, due to the primary field, by Faraday's law generates a voltage in the RX coil that is proportional to $\partial\Phi_P/\partial t$.

The conductive Earth, in accordance with Faraday's law of induction, responds to the primary-field excitation by establishment of *eddy currents* flowing in the subsurface. The steady-state distribution of eddy currents in the ground depends on the conductivity profile $\sigma(z)$. The eddy-current loops act as a secondary source of magnetic field, $\mathbf{B}_S$. Thus, the RX loop senses a secondary magnetic flux $\Phi_S \sim 10^{-6}\Phi_P$. The eddy currents are deeper if the Earth is resistive and closer to the surface if the ground is conductive, according to the *skin effect*. In a homogeneous ground, the eddy currents are concentrated within a skin depth δ [m]

$$\delta = \sqrt{\frac{2}{\mu_0 \sigma \omega}} \tag{11.28}$$

of the surface.

According to the skin effect, the secondary magnetic flux Φ_S through the RX loop is larger if the ground is conductive, since the secondary source is close to the surface. It can be shown that, if the condition $\delta << L$ holds and the ground is homogeneous, then the flux ratio Φ_S/Φ_P is linearly proportional to the ground conductivity σ. The following Fredholm integral equations can then be derived (Borchers *et al.*, 1997):

$$m^H(h) = \int_0^\infty \varphi^H(z+h)\sigma(z)dz; \tag{11.29}$$

$$m^V(h) = \int_0^\infty \varphi^V(z+h)\sigma(z)dz. \tag{11.30}$$

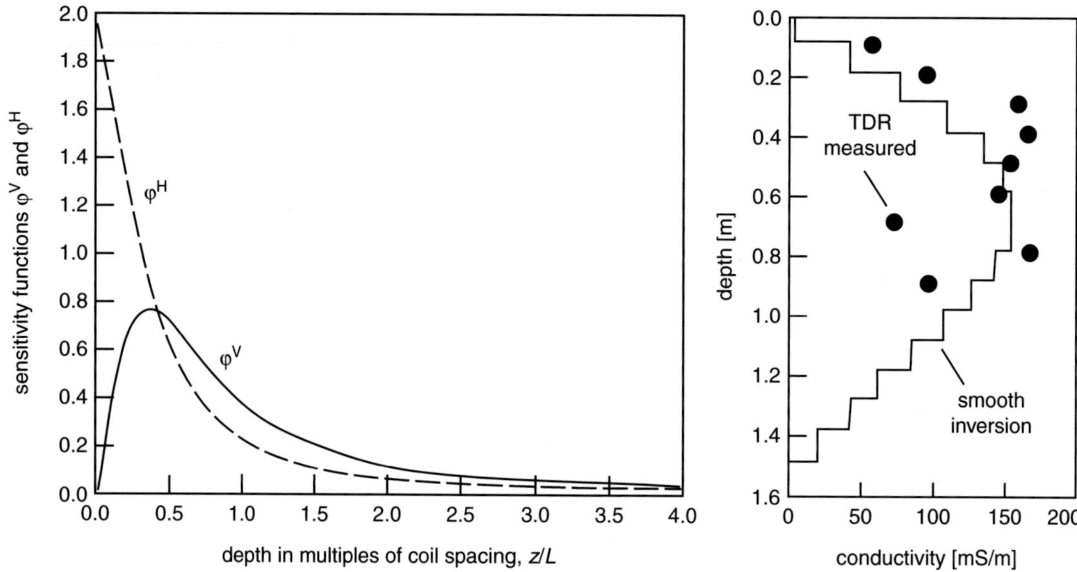

Figure 11.5 (Left) The sensitivity functions $\varphi^H(z)$ and $\varphi^V(z)$ for EM loop–loop prospecting. (Right) A field example of smooth inversion of EM-38 apparent-conductivity data. After Borchers *et al.* (1997)

In the above equations, $m^{H/V}(h)$ represents the measured flux ratio Φ_S/Φ_P as a function of the TX loop height h above the ground. The vertical distribution of electrical conductivity $\sigma(z)$ can be probed by taking measurements of the apparent resistivity with the TX and RX pair positioned at different heights above the ground. The superscripts H and V correspond to two different configurations of the TX and RX loops, oriented as coaxial horizontal and coplanar vertical magnetic dipoles, respectively.

The sensitivity functions $\varphi^H(z)$ and $\varphi^V(z)$ have the following forms (McNeill, 1980):

$$\varphi^H(z) = 2 - \frac{4(z/L)}{\sqrt{4(z/L)^2 + 1}}; \tag{11.31}$$

$$\varphi^V(z) = \frac{4(z/L)}{\left[4(z/L)^2 + 1\right]^{3/2}}, \tag{11.32}$$

where depth z is conveniently scaled by the TX–RX separation L. Plots of the sensitivity functions $\varphi^H(z)$ and $\varphi^V(z)$, shown in Figure 11.5, left, reveal that the coaxial horizontal configuration is very sensitive to the surface conductivity whereas the coplanar vertical configuration responds most strongly to conductivity in the ground at depth $z \sim 0.5L$.

The inverse problem is then to determine the conductivity profile $\sigma(z)$ from measurements of $m^H(h)$ and $m^V(h)$ for various values of TX–RX loop height h above the ground. The problem can be converted into standard linear form $\mathbf{d} = A\mathbf{m}$ where $\mathbf{d} = (m_1^{H/V}, m_2^{H/V}, \ldots, m_D^{H/V})^T$ represents the data, $\mathbf{m} = (\sigma_1, \sigma_2, \ldots, \sigma_P)$ represents the model, namely the conductivities of a number of fixed layers, and A is the $D \times P$ matrix that contains the physics of the forward problem. In this case, following Borchers *et al.* (1997) we can write

$$A_{ij} = \varphi^{H/V}(\tilde{z}_j + h_i)(l_j - u_j), \text{ for } i = 1, \ldots, D; \; j = 1, \ldots, P. \quad (11.33)$$

In the above equation, $\tilde{z}_j$ is the average depth of the j-th layer; h_i is the TX–RX height associated with the i-th data point; and l_j and u_j are, respectively, the depth to the bottom and top of the j-th layer. The superscript "H/V" means either H or V depending on the orientation of the coils for that measurement.

The inverse problem is then solved by minimizing the function $T(\theta) = \Delta^T\Delta + \theta^2\mathbf{m}^T H\mathbf{m}$. It is common practice to over-parameterize the conductivity profile such that the smoothness constraint (minimization of $\mathbf{m}^T H\mathbf{m}$) will generate, in a numerically stable manner, a smoothly varying depth profile $\sigma(z)$ that presumably should be readily interpretable from a geological perspective.

A field example from Borchers *et al.* (1997) demonstrates smooth inversion of loop–loop electromagnetic data using the Geonics EM-38 instrument (frequency 14.6 kHz; TX–RX separation 1.0 m). The smooth conductivity profile they found is shown in Figure 11.5, right. Both horizontal and vertical coil orientations were used with measurements taken at different heights up to 1.2 m above the surface. The field site in New Mexico is in clay–loam soil of the Rio Grande floodplain. The points labeled "TDR measured" are *in situ* electrical-conductivity readings taken using a time-domain reflectometer probe (Nadler *et al.*, 1991).

11.12 Singular-value decomposition

The stability of damped-least-squares or smoothness-constrained inversion is easily understood for the special cases of $W = I$, the identity matrix, in which case $\mathbf{m}^* = (A^T A + \theta^2 I)^{-1} A^T\mathbf{d}$ or $W = H$, the roughening matrix, in which case $\mathbf{m}^* = (A^T A + \theta^2 H)^{-1} A^T\mathbf{d}$. The matrix inversions in these expressions are stable since both the identity and roughening matrices are diagonally dominant. If the parameter θ^2 is large enough, the matrix to be inverted becomes positive definite and hence invertible.

In this section we develop an alternate method for stable inversion of geophysical data that are modeled by a set of linear equations $A\mathbf{m} = \mathbf{d}$. The method (Hansen, 1987), is based on a truncated singular-value decomposition (SVD) of the A matrix. Recall that an *eigenvalue* λ of a square matrix B satisfies the equation $B\mathbf{v} = \lambda\mathbf{v}$, where $\mathbf{v}$ is an *eigenvector*. We shall be interested in the eigenvalues of the square matrices $A^T A$ and AA^T.

Recall that matrix $A^T A$ is $P \times P$, symmetric, and positive semi-definite. Accordingly, there exists a set of mutually orthogonal eigenvectors $\mathbf{u}^{(i)}$ for $i = 1, \ldots, P$ of this matrix which spans P-dimensional model space. By definition, the eigenvectors satisfy $A^T A\mathbf{u}^{(i)} = \lambda_i\mathbf{u}^{(i)}$ where the scalars $\lambda_i \geq 0$ are the eigenvalues. Since the eigenvectors span model space, *any* model vector $\mathbf{m} = (m_1, m_2, \ldots, m_P)^T$ can be expressed as a linear combination of them,

$$\mathbf{m} = \sum_{i=1}^{P} m_i \mathbf{u}^{(i)}. \tag{11.34}$$

Inserting Equation (11.34) into the normal equations $A^T A\mathbf{m} = A^T\mathbf{d}$ and recognizing the orthogonality of the eigenvectors $\mathbf{u}^{(i)T}\mathbf{u}^{(j)} = \delta_{ij}$ results in

$$A^T A\left[\sum_{i=1}^{P} m_i \mathbf{u}^{(i)}\right] = A^T\mathbf{d} \tag{11.35}$$

$$\sum_{i=1}^{P} m_i \lambda_i \mathbf{u}^{(i)} = A^T\mathbf{d}$$

$$\sum_{i=1}^{P} m_i \lambda_i \mathbf{u}^{(j)T}\mathbf{u}^{(i)} = \mathbf{u}^{(j)T}A^T\mathbf{d}$$

$$\sum_{i=1}^{P} m_i \lambda_i \delta_{ij} = \mathbf{u}^{(j)T}A^T\mathbf{d}$$

$$m_j = \frac{\mathbf{u}^{(j)T}A^T\mathbf{d}}{\lambda_j}$$

as the formal solution for the j-th model parameter. Notice that this expression is unstable or ill-defined in the case of a zero eigenvalue, $\lambda_j = 0$. If an eigenvalue λ_j is non-zero but sufficiently small, the resulting estimate of m_j from Equation (11.35) tends to be unstable since it involves a division by a small number. In general, the eigenvectors corresponding to small eigenvalues contribute comparatively little to the data but tend to de-stabilize the model reconstruction.

A *ranking* of eigenvalues according to size permits us to identify and remove these non-contributing eigenvectors from the expansion (11.34). This procedure, known as *winnowing*, greatly stabilizes the inverse problem. The threshold value below which eigenvalues are selected for removal remains a subjective choice. This forms the conceptual basis for the SVD approach to linear inversion, described below.

A fundamental result of linear algebra (Golub and van Loan, 1996) is that a rectangular matrix A has the following SVD

$$A_{D\times P} = U_{D\times D}\Lambda_{D\times P}V^T_{P\times P}; \tag{11.36}$$

where U and V are *unitary matrices* such that $U^T U = I$ and $V^T V = I$. We focus here on the under-determined case for which $D < P$. The $D \times P$ matrix Λ has the structure of a $D \times D$ diagonal submatrix augmented by a non-square $D \times N$ null submatrix consisting entirely of zero entries,

$$\Lambda_{D\times P} = (\text{diag}\{\Lambda_1, \Lambda_2, \ldots, \Lambda_D\}|0_{D\times N}), \tag{11.37}$$

where $N = P - D$. The matrices in the SVD decomposition (11.36) have some interesting properties that are relevant for linear inverse problems. To explore further, we can form the $A^T A$ matrix,

$$A^T A = V\Lambda^T U^T U\Lambda V^T = V\Lambda^T \Lambda V^T = (\Lambda^T \Lambda)VV^T; \tag{11.38}$$

where $\Lambda^T\Lambda = \mathrm{diag}\{\Lambda_1^2, \Lambda_2^2, \ldots, \Lambda_D^2, 0, \ldots, 0\}$ is a $P \times P$ diagonal matrix. Post-multiplication of Equation (11.38) by the unitary matrix V yields the matrix equation

$$(A^T A)V = (\Lambda^T \Lambda)V. \tag{11.39}$$

Thus, it is evident that the entries of the diagonal matrix $\Lambda^T\Lambda$ are eigenvalues of matrix A^TA and that the columns of V are eigenvectors of matrix A^TA. For example, the i-th column of the matrix equation (11.39) is $(A^TA)\mathbf{v}^{(i)} = \Lambda_i^2\mathbf{v}^{(i)}$ where $\mathbf{v}^{(i)}$ is the eigenvector of matrix V corresponding to the eigenvalue Λ_i^2. By a similar calculation it is evident that the columns of U are eigenvectors of matrix AA^T, that is, $(AA^T)\mathbf{u}^{(i)} = \Lambda_i^2\mathbf{u}^{(i)}$.

From Equation (11.36) we have

$$A^{-1} = (U\Lambda V^T)^{-1} = (V^T)^{-1}\Lambda^{-1}U^{-1}. \tag{11.40}$$

The latter expression can be simplified since $U^TU = I$ implies $U^{-1} = U^T$ and $V^TV = I$ implies $(V^T)^{-1} = V^T$. The result is

$$A^{-1} = V\Lambda^{-1}U^T, \tag{11.41}$$

where $\Lambda^{-1} = \mathrm{diag}\{1/\Lambda_1, 1/\Lambda_2, \ldots, 1/\Lambda_D\}$. Note that small eigenvalues cause the corresponding entries of Λ^{-1} to become large and hence the inversion formula (11.40) to become unstable. To avoid the small eigenvalues, we use only the Q largest eigenvalues $\{\Lambda_1, \Lambda_2, \ldots, \Lambda_Q\}$, where $Q < D$, in the inversion formula (11.41). This means that only the matrices U_Q and V_Q formed from the first Q columns of U and V, respectively, are included in the inversion. Methods to select optimal or reasonable values of Q have been explored in the literature but this topic is beyond the present scope.

Note that the formal solution to the geophysical inverse problem is $\mathbf{m}^* = A^\dagger\mathbf{d}$, where $A^\dagger$ is a generalized inverse of matrix A. The generalized inverse

$$A^\dagger = V_Q\Lambda_Q^{-1}U_Q^T, \tag{11.42}$$

where $\Lambda_Q^{-1} = \mathrm{diag}\{1/\Lambda_1, 1/\Lambda_2, \ldots, 1/\Lambda_Q\}$, represents the truncated SVD solution to the linear inverse problem. The optimal model is

$$\mathbf{m}^* = A^\dagger\mathbf{d} = V_Q\Lambda_Q^{-1}U_Q^T\mathbf{d}. \tag{11.43}$$

To review the truncated SVD algorithm, the essentials steps are: (a) find the eigenvalues and eigenvectors of the A^TA matrix; (b) find the eigenvalues and eigenvectors of the AA^T matrix; (c) choose a truncation level Q; and (d) construct the generalized inverse $A^\dagger$ and then calculate the optimal model $\mathbf{m}^*$.

The basic truncated SVD method, along with various extensions of it, has been widely used to solve linear geophysical inverse problems or as a step in the solution of a non-linear geophysical inverse problem. The latter are inverse problems in which each iterative step typically requires the solution of a linear inverse subproblem. Examples of the use of truncated SVD include Huang and Palacky (1991) in airborne electromagnetics, Song and Zhang (1999) in seismic traveltime tomography, and Xu (1998) in the gravity technique.

Problems

1. Derive the normal equations for both unweighted and weighted data.
2. Show that the solution to a well-determined linear inverse problem reduces from $\mathbf{m}^* = (A^TA)^{-1}A^T\,\mathbf{d}$ to $\mathbf{m}^* = A^{-1}\mathbf{d}$.
3. Derive the minimum-norm solution $\mathbf{m}^*$ in Equation (11.20). Explain why the minimum-norm solution is more stable than the minimum-misfit solution.
4. Show that entries of $\Lambda^2 = \text{diag}\{\Lambda_1^2, \Lambda_2^2, \ldots, \Lambda_P^2\}$ are also eigenvalues of matrix AA^T and the columns of U are eigenvectors of matrix AA^T.
5. **Computer assignment**. Consider the total-field magnetic anomaly dataset "magfrance.dat" (http://geoweb.tamu.edu/everettresearch/) that is in the form $T_i(x)$ for $i = 1, \ldots, D$. The data were collected along a south-to-north profile (in the $+\hat{x}$ direction) in Normandy, France in June, 2004. The station spacing is 0.2 m. The elevation of the sensor above ground level is 1.0 m. The model geometry shall consist of a row of two-dimensional magnetized prisms with strike in the $+\hat{y}$ direction. The width and height of each prism is 2.0 m. The prisms are touching each other and have vertical sides. The top of the prisms is located at depth 3.0 m below the surface. There are a total of P prisms. The row of prisms is centered beneath the data profile. You are to find the prism magnetizations M_j for $j = 1, \ldots, P$. It is assumed that the prisms are magnetized in the direction of the ambient geomagnetic field, i.e. there is no remanence. (a) Choose a number P of prisms, arrange them in a row beneath the data profile, and then construct the least-squares model $\mathbf{m}^* = (A^TA)^{-1}A^T\mathbf{d}$. Investigate how the optimal model $\mathbf{m}^*$ changes as you change the number of prisms and the horizontal position of the row of prisms beneath the data profile. What happens when the row of prisms extends beyond one or both ends of the data profile? (b) Repeat the exercise using the minimum-norm model $\mathbf{m}^* = A^T(AA^T)^{-1}\mathbf{d}$. (c) Construct the trade-off curve $T(\lambda)$. Display the models for various choices of λ along the trade-off curve and, for each model, report the misfit $\Delta^T\Delta$ and the model norm $\mathbf{m}^T\mathbf{m}$. Comment on how the misfit and model norm change along the trade-off curve. (d). Construct the SVD solution $\mathbf{m}^* = V_Q\Lambda_Q^{-1}U_Q^T\mathbf{d}$ to the inverse problem. Examine and explain how your solution changes as a function of the number Q of eigenvalues that are kept.

12 Non-linear inversion: local methods

In the previous chapter we examined linear inverse problems in which a model parameter vector $\mathbf{m} = (m_1, m_2, \ldots, m_P)$ is sought that is related to the observations by a linear system of equations $\mathbf{d} = A\mathbf{m} + \Delta$. The vector $\mathbf{d} = (d_1, d_2, \ldots, d_D)$ represents the data while the matrix A presumably captures the essential physics of the forward problem. The matrix–vector product $\mathbf{u} = A\mathbf{m}$ represents the model prediction, or response vector $\mathbf{u}$ given model $\mathbf{m}$. The error vector $\Delta = (\Delta_1, \Delta_2, \ldots, \Delta_D)$ represents the inability of model vector $\mathbf{m}$ to explain the data vector $\mathbf{d}$. The techniques that we looked at discovered models that minimized the sum of squares of the errors $\Delta^T\Delta$ or $\Delta^T C^{-1}\Delta$, or various model norms of the type $\mathbf{m}^T W\mathbf{m}$, or a linear combination of them.

12.1 Introduction

A non-linear inverse problem occurs when the response vector $\mathbf{u}$ cannot be written as a linear function of the model parameters. Instead, we write $\mathbf{d} = \mathbf{F}[\mathbf{m}] + \Delta$, where $\mathbf{F}[\mathbf{m}]$ is not a matrix–vector product $A\mathbf{m}$ but rather a *non-linear* mapping, $\mathbf{F}$: $R^P \rightarrow R^D$ from P-dimensional model space into D-dimensional response space. Since geophysical inverse problems are ill-posed and do not possess unique and stable solutions, their best solution almost always centers on the optimization of a carefully defined objective function $\varphi[\mathbf{m}]$. Excellent introductions to the methods of numerical optimization, including most of those that will be discussed in this chapter, are provided by Gill *et al.* (1982) and Nocedal and Wright (2006).

A non-linear inverse problem, solved using either local or global methods, almost always involves some kind of search or exploration of model space. As such, the geophysicist does well to keep in mind the *curse of dimensionality* (Bellman, 1957). That is, it is not possible to systematically explore a high-dimensional model space in appreciable detail. For example, if $P \sim 160$ (a moderate-sized inverse problem) then $2^P \sim 10^{48}$ forward calculations are required *just to sample the vertices at the edges of model space*. To put this number into perspective, it is estimated that the Earth contains about that many atoms. Thus, any practical inverse algorithm can visit only a vanishingly small fraction of the points contained in the immensely large model space. Excellent models that lie within vast uncharted regions of model space by necessity are left undiscovered. Scientists are actively researching means of mitigating the curse of dimensionality but it seems highly unlikely that a satisfactory remedy will be found in the near future.

12.2 Steepest-descent method

As mentioned above, the solution of a non-linear inverse problem often requires the minimization of a scalar objective function $\varphi[\mathbf{m}]$: $R^P \rightarrow R^1$ of P variables. The simplest formulation again involves the sum of squares of errors, for which we write $\varphi[\mathbf{m}] = \Delta^T\Delta = (\mathbf{d} - \mathbf{F}[\mathbf{m}])^T(\mathbf{d} - \mathbf{F}[\mathbf{m}])$. We assume that the function $\varphi[\mathbf{m}]$ is differentiable, i.e. without kinks or discontinuities. The *steepest-descent* method is the first and most elementary of the non-linear optimization techniques that we shall examine in this chapter.

A local minimization of $\varphi[\mathbf{m}]$ normally involves construction of the P-dimensional *gradient vector* defined by

$$\mathbf{g}[\mathbf{m}] = \nabla\varphi[\mathbf{m}] = \left(\frac{\partial\varphi}{\partial m_1}, \frac{\partial\varphi}{\partial m_2}, \ldots, \frac{\partial\varphi}{\partial m_P}\right)^T : R^P \rightarrow R^P. \tag{12.1}$$

At a given point in model space, say $\mathbf{m}^{(0)}$, the negative gradient vector

$$-\mathbf{g}^{(0)} = -\nabla\varphi[\mathbf{m}]|_{\mathbf{m}=\mathbf{m}^{(0)}} \tag{12.2}$$

points in the direction of the most rapid decrease of the function $\varphi[\mathbf{m}]$, i.e. $-\ \mathbf{g}^{(0)}[\mathbf{m}^{(0)}]$ points *downhill* in model space. Since $-\ \mathbf{g}^{(0)}$ points downhill toward lower values of $\varphi[\mathbf{m}]$, the search in model space could usefully proceed in that direction. Suppose indeed that a step of length t is taken in the direction of $-\ \mathbf{g}^{(0)}$; the search would then arrive at the model

$$\mathbf{m}(t) = \mathbf{m}^{(0)} - t\mathbf{g}^{(0)}. \tag{12.3}$$

Since the negative gradient vector always points downhill, it is clear that some step size $t > 0$ may be found that satisfies

$$\varphi[\mathbf{m}(t)] \le \varphi[\mathbf{m}^{(0)}]. \tag{12.4}$$

Equation (12.4) holds because the search proceeds in a *descent* direction, in fact, the direction of most rapid descent. As t increases from zero, we go further along the line of steepest descent. Eventually, however, we could reach a minimum of $\varphi[\mathbf{m}]$ along the line and $\varphi[\mathbf{m}]$ will start to increase. The step-size parameter t could usefully be chosen such that $\varphi[\mathbf{m}]$ is minimized along the line of steepest descent, in other words an *optimal* step size would satisfy

$$t_0^* = \arg\min \Phi_0(t) = \arg\min \varphi[\mathbf{m}^{(0)} - t\mathbf{g}^{(0)}]. \tag{12.5}$$

The updated model will be $\mathbf{m}^{(1)} = \mathbf{m}^{(0)} - t_0^*\mathbf{g}^{(0)}$. At the new location $\mathbf{m}^{(1)}$, we calculate the next direction of steepest descent, $-\mathbf{g}^{(1)} = -\nabla\varphi[\mathbf{m}^{(1)}]$, and set off in that direction to determine the next model $\mathbf{m}^{(2)}$ of the sequence. In this way, a sequence of models of the form

$$\mathbf{m}^{(k)} = \mathbf{m}^{(k-1)} - t_{k-1}^*\mathbf{g}^{(k-1)}, \tag{12.6}$$

for $k = 1, 2, \ldots$ is constructed, for which it is guaranteed that $\varphi[\mathbf{m}^{(k)}] \le \varphi[\mathbf{m}^{(k-1)}]$.

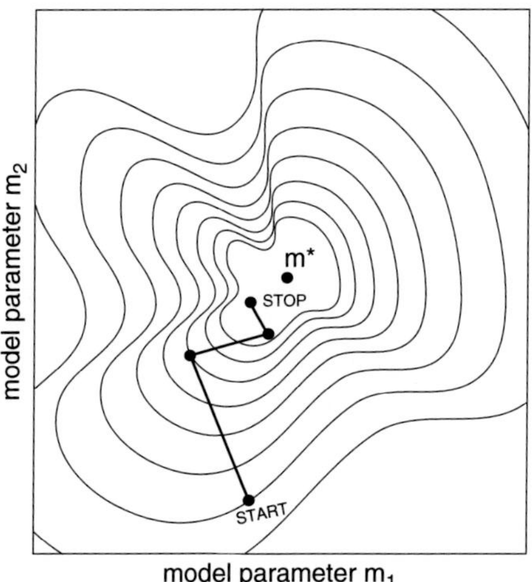

Figure 12.1 First three steps in a steepest-descent optimization of $\varphi[\mathbf{m}]$, for the case $P = 2$.

The steepest-descent method terminates when the step does not generate an appreciable reduction in the value of the function $\varphi[\mathbf{m}]$, that is, when $\varphi[\mathbf{m}^{(k)}] - \varphi[\mathbf{m}^{(k+1)}] < \varepsilon$, where ε is some pre-set tolerance. Figure 12.1 is a schematic representation of the first three steps of a steepest-descent optimization of the function $\varphi[\mathbf{m}]$.

There are a number of situations in which convergence of the steepest-descent method can fail to occur. First, a saddle point of the function could be reached at which the function $\varphi[\mathbf{m}]$ is locally flat. Second, the boundary of model space can be reached. This situation is dealt with below. In both cases, the steepest-descent method could be re-started with a different initial guess $\mathbf{m}^{(0)}$. In fact, the method can be run many times with different starting models. In general, the steepest-descent path through model space, and hence quite likely the final model, is different for each choice of starting model. The geophysicist must then try to evaluate, discriminate, and interpret the range of final models so attained.

A word now on the determination of the optimal step size. There are many classical one-dimensional optimizations methods that can be used to solve Equation (12.5) and hence find the optimal step size t^* at each iteration of the method. Herein we describe a simple bisection algorithm that is not difficult to program. The algorithm proceeds as follows. Suppose the admissible model space has been chosen such that the model parameters are bounded by $\mathbf{m}_i^- \leq \mathbf{m}_i \leq \mathbf{m}_i^+$, for each $i = 1, \ldots, P$. A *ray* emanating from the current model $\mathbf{m}^{(k)}$ and pointing in the direction of steepest descent must eventually intersect the boundary of model space. Let the value of the ray parameter be denoted by $t = 0$ at the origin of the ray and by $t = t_{BDY}$ at its point of intersection with the boundary of model space.

Next, divide the line segment $t \in [0, t_{BDY}]$ into N equal intervals such that $t_0 = 0$ and $t_N = t_{BDY}$. The choice $N \sim 10$ works well in most cases. Then, evaluate the function $\Phi_k(t_i) = F[\mathbf{m}^{(k)} - t_i\mathbf{g}^{(k)}]$ on each "node" t_i for $i = 1, \ldots, N$. We are trying to minimize the function $\Phi_k(t)$, so determine the two "nodes" along the steepest-descent ray which bracket the minimum $t_k^* = \arg\min_{t\in[0,\, t_{BDY}]} \Phi_k(t)$. Label these nodes as t_A and t_B, respectively, with

$t_A \le t_k^* \le t_B$. The sought-after minimum of $\Phi_k(t)$ is located somewhere in the interval $[t_A, t_B]$. This interval is then subdivided into N equal parts, again, and the evaluation procedure described above is iterated to generate a smaller bracketing interval $[t_A, t_B]$. The idea is to repeat the subdivision until the bracketing interval $[t_A, t_B]$ is less than some pre-set *minimum step size* δ.

Once $[t_A, t_B]$ is sufficiently small such that $|t_B - t_A| < \delta$, then set

$$t^* = t_A \text{ if } \Phi_k(t_A) < \Phi_k(t_B) \text{ or } t^* = t_B \text{ if } \Phi_k(t_B) < \Phi_k(t_A) \tag{12.7}$$

and exit. Several problems could arise. Note that if $t_A = t_0$ the method has *stalled.* In this case, increase N or decrease δ, or just accept the solution $\mathbf{m}^{(k)} - t_0\mathbf{g}^{(k)}$. If $t_B = t_{BDY}$, an error exit has occurred, since the boundary of model space has been reached. This should not happen unless a point on the boundary is a feasible point, i.e. its response fits the data to some pre-set tolerance Δ_{TOL}. In this case, either reduce the tolerance Δ_{TOL} and enlarge the model space, or just accept the solution $\mathbf{m}^{(k)} - t_{BDY}\mathbf{g}^{(k)}$.

The steepest-descent method is simple in concept but there are major limitations to its practical implementation for solving large-scale three-dimensional (3-D) inverse problems in which $P = \dim[\mathbf{m}]$ is measured in hundreds or thousands. For example, the computation of the gradient vector $\nabla\varphi[\mathbf{m}]$ with $\varphi[\mathbf{m}] = \Delta^T\Delta = (\mathbf{d} - \mathbf{F}[\mathbf{m}])^T(\mathbf{d} - \mathbf{F}[\mathbf{m}])$ at each iteration requires the solution of $\sim 2DP$ forward problems. The cost rapidly becomes prohibitive if advanced numerical simulations are required to solve the forward problem. The steepest-descent method is oftentimes slowly convergent, as discussed later in this chapter.

12.3 Non-linear least-squares method

In this section we consider the non-linear least-squares method. Jahns (1966) used this approach to solve an inverse hydraulic problem in which the data are transient pressure responses within boreholes and the model parameters are gridded values of subsurface fluid reservoir properties such as transmissivity and storativity. Here we follow the general methodology of Jahns (1966) but develop the equations in a general setting that could be applied to a wide range of non-linear inverse problems in near-surface applied geophysics. An excellent introduction to least-squares inversion of geophysical data is provided by Lines and Treitel (1984).

The objective is to minimize the weighted misfit, assuming independent data errors,

$$\Delta^T\Delta = \sum_{i=1}^{D} \left[\frac{d_i - F_i[\mathbf{m}]}{e_i}\right]^2 \tag{12.8}$$

using the classical Gauss–Newton method of non-linear least squares. This method is sometimes known as non-linear multiple regression. The optimization problem is written formally as

$$\mathbf{m}^* = \arg\min_{\mathbf{m}\in M} \sum_{i=1}^{D} \left[\frac{d_i - F_i[\mathbf{m}]}{e_i}\right]^2. \tag{12.9}$$

The normal equations are readily constructed by simple differentiation,

$$\frac{\partial \Delta^T \Delta}{\partial m_j} = -2 \sum_{i=1}^{D} \left[\frac{d_i - F_i[\mathbf{m}]}{e_i} \right] \frac{\partial F_i}{\partial m_j} = 0, \text{for } j = 1, 2, \ldots, P. \tag{12.10}$$

These are a set of P non-linear equations. Consider a starting model $\mathbf{m}^{(0)}$, which we hope is a good approximation of the optimal model $\mathbf{m}^*$. We can write the first two terms of a multivariate Taylor series

$$\boldsymbol{F}[\mathbf{m}^*] \approx \boldsymbol{F}[\boldsymbol{m}^{(0)}] + \sum_{k=1}^{P} \frac{\partial \boldsymbol{F}[\mathbf{m}]}{\partial m_k} \delta m_k \tag{12.11}$$

where $\delta\mathbf{m} = \mathbf{m}^* - \mathbf{m}^{(0)}$. Inserting Equation (12.11) into the normal Equations (12.10) results in

$$\sum_{i=1}^{D} \frac{1}{e_i^2} \left[d_i - F_i[\mathbf{m}^{(0)}] - \sum_{k=1}^{P} \frac{\partial F_i}{\partial m_k} \delta \mathbf{m}_k \right] \frac{\partial F_i}{\partial m_j} = 0, \text{for } j = 1, 2, \ldots, P. \tag{12.12}$$

Re-arranging the above expression gives

$$\sum_{i=1}^{D} \frac{1}{e_i^2} \frac{\partial F_i}{\partial m_j} \sum_{k=1}^{P} \frac{\partial F_i}{\partial m_k} \delta \mathbf{m}_k = \sum_{i=1}^{D} \frac{1}{e_i^2} [d_i - F_i[\mathbf{m}^{(0)}]] \frac{\partial F_i}{\partial m_j}, \text{for } j = 1, 2, \ldots, P \tag{12.13}$$

which can be written in matrix notation as

$$A \delta \mathbf{m} = \mathbf{b} \tag{12.14}$$

with matrix elements

$$A_{jk} = \sum_{i=1}^{D} \frac{1}{e_i^2} \frac{\partial F_i}{\partial m_j} \frac{\partial F_i}{\partial m_k}, \quad \text{for } j, k = 1, 2, \ldots, P \tag{12.15}$$

and right-side vector elements

$$b_j = \sum_{i=1}^{D} \frac{1}{e_i^2} [d_i - F_i[\mathbf{m}^{(0)}]] \frac{\partial F_i}{\partial m_j}, \quad \text{for } j = 1, 2, \ldots, P. \tag{12.16}$$

The solution to Equation (12.14) provides an expression for the model perturbation vector $\delta\mathbf{m}$. The vector $\delta\mathbf{m}$ in the non-linear least-squares method plays an equivalent role to the negative gradient vector in the steepest-descent method. Both $\delta\mathbf{m}$ and $-\mathbf{g}$ provide a *search direction* in model space.

Note that the term $\partial F_i/\partial m_j$ in Equation (12.15) expresses the change in the i-th component of the response vector $\mathbf{u} = \mathbf{F}[\mathbf{m}]$ with respect to a small change in the j-th model parameter m_j. In other words, $\partial F_i/\partial m_j$ is a measure of the *sensitivity* of the model response to a small perturbation of the model. A $D \times P$ sensitivity matrix, termed the *Jacobian matrix* may be constructed in which each element of the Jacobian J is a sensitivity,

$$J_{ij} = \frac{\partial F_i}{\partial m_j}. \tag{12.17}$$

In terms of the Jacobian matrix, Equation (12.14) can be re- written as

$$(J^T J)\delta\mathbf{m} = J^T \Delta, \tag{12.18}$$

where for notational convenience we have suppressed the weighting by data errors. The basic Gauss–Newton solution to the non-linear least-squares problem is therefore

$$\delta\mathbf{m} = (J^T J)^{-1} J^T \Delta, \tag{12.19}$$

which has the same form as the optimal solution $\mathbf{m}^*$ to the linear least-squares problem. In fact, it is easy to see that Equation (12.19) reduces to $\mathbf{m}^* = (A^T A)^{-1} A^T \mathbf{d}$ in the special case that $\mathbf{F}[\mathbf{m}] = A\mathbf{m}$.

In the Gauss–Newton method, an iterative sequence of models is defined by analogy with Equation (12.6) as

$$\mathbf{m}^{(k)}(t) = \mathbf{m}^{(k-1)} + t_{k-1}\delta\mathbf{m}^{(k)}, \tag{12.20}$$

where t is the step size is model space. The optimal step size t_{k-1} is found by a one-dimensional line-search minimization of the objective function $\varphi[\mathbf{m}]$, as in the steepest-descent method, that is,

$$t^*_{k-1} = \arg\min \varphi[\mathbf{m}^{(k-1)} + t_{k-1}\delta\mathbf{m}^{(k)}]. \tag{12.21}$$

The method iterates until $\varphi[\mathbf{m}^{(k)}] - \varphi[\mathbf{m}^{(k+1)}] < \varepsilon$, as in the steepest-descent method.

12.4 Levenberg–Marquardt method

The Gauss–Newton solution for $\delta\mathbf{m}$ (Equation (12.19)) involves an inversion of the matrix $J^T J$. This matrix is often ill-conditioned or singular, especially for large P, in which case the Gauss–Newton solution can become unstable. A well-known method to stabilize the inversion was developed by Marquardt (1963). The resulting Levenberg–Marquardt method, described in detail by Seber and Wild (2003), generalizes the Gauss–Newton solution (12.19) to

$$\delta\mathbf{m} = (J^T J + \eta I)^{-1} J^T \Delta \tag{12.22}$$

where η is a stabilization parameter and I is the $P \times P$ identity matrix. For $\eta > 0$, the matrix $J^T J + \eta I$ is better conditioned than the matrix $J^T J$ since positive-definite weight is added to the diagonal.

It is easy to see from Equation (12.22) that the Gauss–Newton solution is recovered as $\eta \to 0$. Conversely, the steepest descent solution $\delta\mathbf{m} \to J^T \Delta$ is recovered as η becomes large.

As clearly described in the tutorial article by Motulsky and Ransnas (1987), the steepest-descent method often starts off well but later iterations soon begin to zigzag through model space, either stagnating or making very slow progress toward finding the minimum of $\Delta^T\Delta$.

It is well known that the Gauss–Newton method is effective in the vicinity of a local minimum for which the $\Delta^T\Delta$ surface is well approximated by a P-dimensional hyper-ellipsoid. A strategy due to Levenberg–Marquardt then suggests itself that has proven successful on a range of non-linear problems. The idea is to initiate an iterative sequence with some starting value of the stabilization parameter, say $\eta^{(0)} \sim 0.01$. As progress slows toward reducing the value of $\Delta^T\Delta$ due to stagnation or zig-zagging in model space, the stabilization parameter is steadily decreased, for example in the manner of $\eta^{(k+1)} = \eta^{(k)}/10$. This strategy ensures that the optimization becomes less like the steepest-descent and increasingly like the Gauss–Newton method as the minimum is approached.

12.5 Quasi-Newton methods

In the steepest-descent method, the search direction at the k-th iteration is chosen to be $-\mathbf{g}^{(k)} = -\nabla\varphi[\mathbf{m}^{(k)}]$. Thus, only first-derivative information is used in the calculation of the search direction. It should not be surprising to learn that a more efficient optimization scheme can be derived if more general information about $\varphi[\mathbf{m}]$ is incorporated into the calculation of the search direction.

The family of optimization methods discussed in this section use second-derivative information. Recall that the function to be minimized is $\varphi[\mathbf{m}]$ with $\mathbf{m} \in R^P$. Suppose we have already found a model $\mathbf{m}^{(k)}$ as part of an iterative sequence of models that we hope converges to a minimizer $\mathbf{m}^*$ of $\varphi[\mathbf{m}]$. It is now desired to find a model update vector $\delta\mathbf{m}$ such that the new model $\mathbf{m}^{(k+1)} = \mathbf{m}^{(k)} + \delta\mathbf{m}$ generates a lower value of the objective function. To this end, consider the multivariate Taylor series expansion

$$\varphi[\mathbf{m}^{(k)} + \delta\mathbf{m}] \approx \varphi[\mathbf{m}^{(k)}] + \nabla\varphi[\mathbf{m}^{(k)}]^T\delta\mathbf{m} + \frac{1}{2}\delta\mathbf{m}^T B\delta\mathbf{m} \qquad (12.23)$$

where B marks the *Hessian matrix* whose elements are defined by the second-order derivatives in model space,

$$B_{ij} = \frac{\partial^2\varphi}{\partial m_i \partial m_j}, \text{ for } i,j = 1,2,\ldots,P. \qquad (12.24)$$

Taking the gradient in model space of Equation (12.23) gives

$$\nabla\varphi[\mathbf{m}^{(k)} + \delta\mathbf{m}] = \nabla\varphi[\mathbf{m}^{(k)}] + B\delta\mathbf{m} + \text{higher order terms}.$$

A necessary condition for the function $\varphi[\mathbf{m}]$ to achieve a minimum at the location $\mathbf{m}^{(k)} + \delta\mathbf{m}$ is the vanishing of its gradient there. Accordingly, setting the above equation to zero, we obtain the basic Newton model-update formula

$$\delta\mathbf{m} = -B^{-1}\nabla\varphi[\mathbf{m}^{(k)}]. \tag{12.25}$$

The Newton formula is simple and elegant, yet impractical. It is prohibitively costly at each step of an iterative search through model space to calculate the Hessian matrix and then determine its inverse. The *quasi-Newton methods* constitute an efficient alternative; they are based on updating at each step an approximate Hessian found in the previous step. A number of Hessian-updating methods have been proposed in the literature. A popular choice has long been the Broyden–Fletcher–Goldfarb–Shanno (BFGS) update, named after its four independent discoverers, given by

$$B^{(k+1)} = B^{(k)} + \frac{\mathbf{y}^{(k)}\mathbf{y}^{(k)T}}{\mathbf{y}^{(k)T}\delta\mathbf{m}} - \frac{B^{(k)}\delta\mathbf{m}\delta\mathbf{m}^T B^{(k)}}{\delta\mathbf{m}^T B^{(k)}\delta\mathbf{m}} \tag{12.26}$$

where

$$\mathbf{y}^{(k)} = \nabla\varphi[\mathbf{m}^{(k)} + \delta\mathbf{m}] - \nabla\varphi[\mathbf{m}^{(k)}]. \tag{12.27}$$

The quasi-Newton algorithm with the standard BFGS update is then

1. choose starting model $\mathbf{m}^{(0)}$ and approximate Hessian $B^{(0)}=I$
2. set $k = 1$
3. compute $\delta\mathbf{m} = -[B^{(k-1)}]^{-1}\nabla\varphi[\mathbf{m}^{(k-1)}]$
4. find step size t by univariate line search, as in steepest-descent method
5. compute $\mathbf{m}^{(k)} = \mathbf{m}^{(k-1)} + t\delta\mathbf{m}$
6. exit if a stopping criterion is satisfied
7. compute $B^{(k)}$ using BFGS update formula
8. set $k \to k + 1$ and go to 3.

Romdhane *et al.* (2011) used a quasi-Newton algorithm incorporating an efficient, limited-memory BFGS update strategy in their 2-D full-waveform inversion of synthetic elastic wavefields for the purpose of imaging near-surface seismic velocity structure in the presence of complex topography. A limited-memory BFGS update (Nocedal and Wright, 2006) is one in which the approximate inverse Hessian $[B^{(k-1)}]^{-1}$ is estimated from the $m \sim 10$ most recent model and gradient vectors.

Romdhane *et al.* (2011) simulated the elastic wavefield, including surface waves and extreme topography, that would have been acquired if explosive sources and horizontal and vertical geophones were arrayed over the Super Sauze earthflow in the French Alps. Figure 12.2a, b show reconstructed S-wave and P-wave velocity models of the earthflow region obtained using the quasi-Newton inversion procedure. One of the synthetic shot gathers from the true model is shown in Figure 12.2c. Figure 12.2d gives an indication of the goodness of fit between the former and a synthetic shot gather computed using the reconstructed velocity models. Romdhane *et al.* (2011) conclude for this problem that the quasi-Newton approach outperforms a preconditioned conjugate-gradient method, a technique to which we now turn attention.

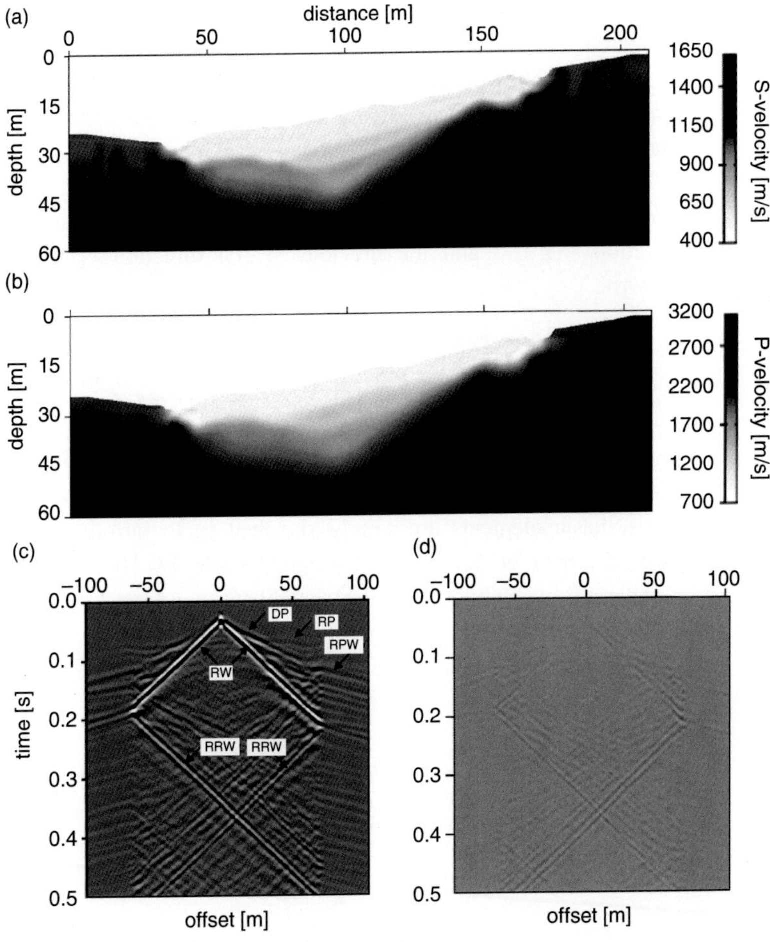

Figure 12.2 Quasi-Newton inversion of synthetic 2-D elastic waveforms for a realistic landslide scenario: (a) V_S and (b)V_P reconstructed models; (c) vertical-component synthetic shot gather acquired in center of profile (DP = direct, RP = refracted, RPW = reflected P-waves, RW = Rayleigh waves, RRW = back-propagated Rayleigh waves); (d) difference between (c) and shot gather calculated from the reconstructed models. After Romdhane *et al.* (2011).

12.6 Conjugate-gradient method

The non-linear inversion methods that we have considered so far are classical optimization techniques used to navigate through P-dimensional model space and hopefully arrive at a local minimum $\mathbf{m}^*$ of the objective function $\varphi[\mathbf{m}]$. These methods are characterized by a first stage that involves calculation of a *search direction* $\delta\mathbf{m}$ from the current location $\mathbf{m}^{(k)}$ in model space, and a second stage of finding an optimal step size t along the current search direction in order to advance to an improved location $\mathbf{m}^{(k+1)}$ in model space.

A powerful optimization method can be obtained if all P axes of model space are used as search directions. The generic method of *alternating directions* uses cyclically the P-coordinate axes of model space. However, a better method is one which chooses its sequence of directions adaptively, taking into account accumulated knowledge about the shape of $\varphi[\mathbf{m}]$.

The *conjugate-gradient* method is one such approach. In this method, the current search direction $\delta\mathbf{m}^{(k+1)}$ is chosen to be a linear combination of the steepest-descent direction $-\ \mathbf{g}^{(k+1)}$ and the previous search directions $\{\delta\mathbf{m}^{(0)}, \delta\mathbf{m}^{(1)}, \ldots, \delta\mathbf{m}^{(k)}\}$. Thus we write

$$\delta\mathbf{m}^{(k+1)} = -\mathbf{g}^{(k+1)} + \sum_{j=0}^{k} \beta_{jk}\delta\mathbf{m}^{(j)}. \tag{12.28}$$

The task is to determine optimal values of the β_{jk} coefficients. A complete and rigorous derivation of the conjugate-gradient method is beyond the scope of this book; herein some of the basic elements are simply sketched in to introduce the method. An advanced treatment may be found in the original paper by Hestenes and Steifel (1952) or the monograph by Meurant (2006). A useful starting point to learn about conjugate-gradient techniques for optimization is the book by Kowalik and Osborne (1968).

Following Fletcher and Reeves (1964), we shall require

$$\mathbf{g}^{(k+1)T}\delta\mathbf{m}^{(k+1)} = 0. \tag{12.29}$$

Moreover, suppose that the shape of the non-linear function $\varphi[\mathbf{m}]$ is nearly quadratic in the vicinity of location $\mathbf{m}^{(k)}$. In that case, we can write the approximation

$$\varphi[\mathbf{m}^{(k+1)}] \approx \varphi[\mathbf{m}^{(k)}] + \frac{1}{2}[\mathbf{m}^{(k+1)} - \mathbf{m}^{(k)}]^T G^{(k)}[\mathbf{m}^{(k+1)} - \mathbf{m}^{(k)}], \tag{12.30}$$

where $G^{(k)}$ is the $P \times P$ symmetric Hessian matrix whose elements are given by

$$G_{ij}^{(k)} = \frac{\partial^2 \varphi}{\partial m_i^{(k)} \partial m_j^{(k)}}, \tag{12.31}$$

such that $G = G^T$. Taking the gradient of Equation (12.30) with respect to $\mathbf{m}^{(k+1)}$ gives

$$\mathbf{g}^{(k+1)} = G^{(k)}\delta\mathbf{m}^{(k)}. \tag{12.32}$$

Combining Equations (12.29) and (12.32) results in

$$\delta\mathbf{m}^{(k+1)T}G^{(k)}\delta\mathbf{m}^{(k)} = 0. \tag{12.33}$$

In other words, sequential search directions in model space are found naturally to be *G-conjugate* to each other. Hestenes and Steifel (1952) demonstrate that a powerful exploration of model space is enabled if all the search directions are mutually G-conjugate, namely,

$$\delta\mathbf{m}^{(i)T}G\delta\mathbf{m}^{(j)} = 0 \text{ for } i \neq j. \tag{12.34}$$

The coefficients β_{jk} in Equation (12.28) should then be selected in such a manner that the conditions (12.29) and (12.34) are satisfied. From (12.28) and (12.29) it follows that

$$\mathbf{g}^{(k+1)T}\mathbf{g}^{(k+1)} = \sum_{j=0}^{k} \beta_{jk}\mathbf{g}^{(k+1)T}\delta\mathbf{m}^{(j)}. \tag{12.35}$$

Making use of Equation (12.32) and the G-conjugacy requirement (12.34) reduces the above equation to

$$\mathbf{g}^{(k+1)T}\mathbf{g}^{(k+1)} = \beta_{k,k}\delta\mathbf{m}^{(k)T}G^{(k)}\delta\mathbf{m}^{(k)}, \tag{12.36}$$

which shows that only the $j = k$ term in Equation (12.28) survives, so that

$$\beta_k = \frac{\mathbf{g}^{(k+1)T}\mathbf{g}^{(k+1)}}{\delta\mathbf{m}^{(k)T}G^{(k)}\delta\mathbf{m}^{(k)}}. \tag{12.37}$$

A few steps of algebra show that the denominator can be simplified as

$$\delta\mathbf{m}^{(k)T}G^{(k)}\delta\mathbf{m}^{(k)} = \mathbf{g}^{(k)T}\mathbf{g}^{(k)} \tag{12.38}$$

so that the final result is

$$\beta_k = \frac{\mathbf{g}^{(k+1)T}\mathbf{g}^{(k+1)}}{\mathbf{g}^{(k)T}\mathbf{g}^{(k)}}. \tag{12.39}$$

We now have all the pieces in place for the basic conjugate-gradient algorithm,

1. choose starting model $\mathbf{m}^{(0)}$
2. choose initial search direction $\delta\mathbf{m}^{(0)} = -\,\mathbf{g}^{(0)}$
3. find $\mathbf{m}^{(k+1)}$ by minimizing $\varphi[\mathbf{m}]$ along direction $\delta\mathbf{m}^{(k)}$ (line search)
4. evaluate the gradient $\mathbf{g}^{(k+1)}$ at the new point $\mathbf{m}^{(k+1)}$
5. set $\beta_k = [\mathbf{g}^{(k+1)T}\mathbf{g}^{(k+1)}]/[\mathbf{g}^{(k)T}\mathbf{g}^{(k)}]$
6. choose new search direction $\delta\mathbf{m}^{(k+1)} = -\,\mathbf{g}^{(k+1)} + \beta_k\delta\mathbf{m}^{(k)}$
7. repeat from step 3 until the method converges.

A practical implementation of the conjugate-gradient algorithm is quite straightforward, which has led to its popularity for inverse problems in which P is large. The conjugate-gradient method derives much of its theoretical justification from the fact that it can be shown to converge in exactly P steps if $\varphi[\mathbf{m}]$ is a *quadratic function*. The idea is that an algorithm that works exactly on quadratic functions should perform well for general non-linear problems. The rationale for this optimism is that, in the vicinity of a minimum, a non-linear function can often be well approximated by a quadratic function. Furthermore if $\varphi[\mathbf{m}]$ is the misfit of a linear forward problem, then $\varphi[\mathbf{m}]$ is a quadratic function. Finally, note that the conjugate-gradient algorithm operates only with vectors and does not require a large matrix to be stored.

Polak (1971) has introduced, and justified on theoretical grounds, the following modification to Equation (12.39)

$$\beta_k = \frac{\mathbf{g}^{(k+1)T}\left[\mathbf{g}^{(k+1)} - \mathbf{g}^{(k)}\right]}{\mathbf{g}^{(k)T}\mathbf{g}^{(k)}} \tag{12.40}$$

and it has been found to improve the convergence rate of the conjugate-gradient method for non-linear functions. The replacement of Equation (12.39) by (12.40) is termed the Polak–Ribiere correction. An excellent example of its employment in a non-linear conjugate-gradient inversion of geophysical data may be found in Rodi and Mackie (2001).

12.7 Example: seismic traveltime tomography

This section describes the use of the conjugate-gradient method to solve a *linear* inverse problem of the form $\mathbf{d} = A\mathbf{m} + \Delta$ that occurs in seismic traveltime tomography, following the approach outlined by Scales (1987). The matrix A is of dimension $D \times P$ matrix and describes approximately the physics of the forward problem; while as usual $\mathbf{m}$ is the P-vector of model parameters and $\mathbf{d}$ is the D-dimensional data vector.

The goal of seismic traveltime tomography is to invert observed seismic traveltimes for an underlying subsurface velocity model, $V(\mathbf{r})$. It is presumed that the traveltime τ between a seismic source and receiver is related to the velocity structure $V(\mathbf{r})$ by the linear equation

$$\tau = \int\limits_{l(s)} s(\mathrm{r})dl; \tag{12.41}$$

where $s(\mathbf{r}) = 1/V(\mathbf{r})$ is the wave *slowness* and $l(s)$ is the ray path from the source to the receiver. Notice that parameterizing the subsurface by the slowness $s(\mathbf{r})$ instead of the velocity $V(\mathbf{r})$ is done since the traveltime is a non-linear function of the velocity structure. However, Equation (12.41) does not yet describe a linear function since the traveltime depends on a potentially complicated path $l(s)$.

To cast the seismic traveltime tomography inverse problem into linear form, we begin by discretizing a vertical section of the Earth into two-dimensional cells. The problem we consider is idealized in the sense that the seismic sources and receivers may be distributed in an arbitrary manner over the boundary of the discretized region and the ray paths through the model are assumed to be straight lines. A straight seismic ray i then connects the i-th transmitter–receiver pair. Let Ψ_{ij} be the distance travelled by seismic ray i in cell j, as shown in Figure 12.3. In practical ground surveys, TX and RX locations would be restricted to the upper boundary of the solution domain. In borehole-to-borehole (cross-well) seismic surveys, TX and RX pairs can be located on the side boundaries of the solution domain, as shown in the figure.

We write the slowness $s(\mathbf{r})$ in the Earth as

$$\mathbf{s}(\mathbf{r}) = \mathbf{s}^{(0)}(\mathbf{r}) + \delta\mathbf{s}(\mathbf{r}) \tag{12.42}$$

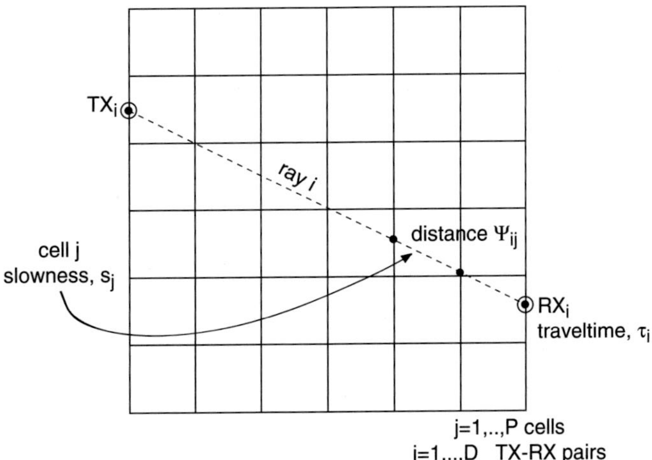

Figure 12.3 Seismic ray *i* for the idealized traveltime tomography problem.

where $s^{(0)}(\mathbf{r})$ is a known reference model and $\delta s(\mathbf{r})$ is the unknown perturbation about the reference model that is to be determined by the geophysical inversion procedure. The first step is to draw straight-line rays through the reference model $s^{(0)}(\mathbf{r})$ to find reference traveltimes $\tau_i^{(0)}$ for each of $i = 1, \ldots, D$ rays. A basic *assumption* here is that the perturbation $\delta s(\mathbf{r})$ is sufficiently small that the set of rays evaluated for the reference model $s^{(0)}(\mathbf{r})$ is approximately the same as would be evaluated for the actual structure $s(\mathbf{r})$ to be determined, and that this set of rays consists of straight lines. In a more realistic tomographic reconstruction, the seismic rays would bend around or into slowness heterogeneities and would need to be re-computed at each iterative step of the inversion.

In the simple case of straight-line rays, we have

$$\tau_i^{(0)} = \sum_{j=1}^{P} \Psi_{ij} s_j^{(0)} \tag{12.43}$$

and, since the same set of rays is used for all models,

$$\delta\tau_i = \sum_{j=1}^{P} \Psi_{ij} \delta s_j. \tag{12.44}$$

This relation is of the standard linear form $\mathbf{u} = A\mathbf{m}$ with matrix $\mathrm{A}_{ij} = \Psi_{ij}$, model vector $\mathbf{m} = \delta\mathbf{s}$ and model prediction vector $\mathbf{u} = \delta\boldsymbol{\tau}$. We could use the least-squares method to find the solution to this inverse problem in one step, in which case

$$\delta\mathbf{s}^* = (\Psi^T\Psi)^{-1}\Psi^T\delta\boldsymbol{\tau}. \tag{12.45}$$

The optimal model $\delta\mathbf{s}^*$ is, as usual, a minimum of the misfit function $\varphi[\delta\mathbf{s}] = \Delta^T\Delta$.

To develop a conjugate-gradient solution to the same optimization problem, note that the function $\varphi[\delta\mathbf{s}] = \Delta^T\Delta$ to be minimized can be expanded as

$$\varphi[\delta\mathbf{s}] = -\delta\boldsymbol{\tau}^T\Psi\delta\mathbf{s} + \frac{1}{2}\delta\mathbf{s}^T\Psi^T\Psi\delta\mathbf{s}. \quad (12.46)$$

Notice that $\varphi[\delta\mathbf{s}]$ is quadratic in the model vector $\delta\mathbf{s}$. It is easy to show that

$$\mathbf{g} = \nabla\varphi[\delta\mathbf{s}] = -\Psi^T\delta\boldsymbol{\tau} + \Psi^T\Psi\delta\mathbf{s}. \quad (12.47)$$

Thus, the determination of the search direction is analytic. In the general non-linear case, the gradient vector $\mathbf{g}$ must be found numerically using, for example, a finite-difference approximation. We are now ready to apply the conjugate-gradient algorithm to minimize the function $\varphi[\delta\mathbf{s}]$.

The basic conjugate-gradient algorithm can be applied to find the optimal model $\delta\mathbf{s}$*. The remaining difficult step is the line search. Recall that the optimal step size satisfies

$$t^* = \arg\min \varphi[\delta\mathbf{s}^{(k)} + t\delta\mathbf{m}^{(k)}], \quad (12.48)$$

where $\delta\mathbf{m}^{(k)}$ is the change in the model $\delta\mathbf{s}$ at step k, that is $\delta\mathbf{m}^{(k)} = \delta[\delta\mathbf{s}]^{(k)}$. The optimal step size is then found by solving the necessary condition for a minimum,

$$\frac{\partial}{\partial t}\{\varphi[\delta\mathbf{s}^{(k)} + t\delta\mathbf{m}^{(k)}]\} = 0. \quad (12.49)$$

A brief calculation yields an analytic solution to the above equation

$$t^* = \alpha^{(k+1)} = \frac{\mathbf{g}^{(k)T}\mathbf{g}^{(k)}}{\mathbf{Q}^{(k)T}\mathbf{Q}^{(k)}}, \quad (12.50)$$

where $\mathbf{Q}^{(k)} = \Psi\delta\mathbf{m}^{(k)}$. The principal simplification of the conjugate-gradient algorithm applied to the least-squares linear inverse problem, as opposed to a general non-linear problem, is the availability of the analytic solution (12.50) for the optimal step size t* at each iteration. In the general non-linear conjugate-gradient method, the optimal step size must be found numerically using a line search algorithm.

Putting everything together, the conjugate-gradient algorithm for solving the seismic tomographic linear inverse problem is

1. choose starting slowness perturbation $\delta\mathbf{s}^{(0)}$
2. choose initial search direction $\delta\mathbf{m}^{(0)} = \mathbf{g}^{(0)} = \Psi^{\mathrm{T}}(\delta\boldsymbol{\tau} - \Psi\delta\mathbf{s}^{(0)})$
3. define vector $\mathbf{Q}^{(0)} = \Psi\delta\mathbf{m}^{(0)}$ iterate from $k = 0$ to $k = P$
4. evaluate optimal step size $\alpha^{(k+1)} = [\mathbf{g}^{(k)T}\mathbf{g}^{(k)}]/[\mathbf{Q}^{(k)T}\mathbf{Q}^{(k)}]$
5. evaluate the new model $\delta\mathbf{s}^{(k+1)} = \delta\mathbf{s}^{(k)} + \alpha^{(k+1)}\delta\mathbf{m}^{(k)}$
6. evaluate the new gradient $\mathbf{g}^{(k+1)} = \mathbf{g}^{(k)} - \alpha^{(k+1)}\Psi^T\mathbf{Q}^{(k)}$
7. set $\beta_k = [\mathbf{g}^{(k+1)T}\mathbf{g}^{(k+1)}]/[\mathbf{g}^{(k)T}\mathbf{g}^{(k)}]$
8. choose new search direction $\delta\mathbf{m}^{(k+1)} = \mathbf{g}^{(k+1)} + \beta_k\delta\mathbf{m}^{(k)}$
9. evaluate new vector $\mathbf{Q}^{(k+1)} = \Psi\delta\mathbf{m}^{(k+1)}$.

The steps (1–9) are equivalent to Equations (5a–5g) in the paper by Scales (1987). The derivation of the expression for $\mathbf{g}^{(k+1)}$ in step 6 of the above conjugate-gradient algorithm is left as an exercise for the reader. Notice that the matrix Ψ does not need to be stored since the only times it appears in the algorithm is in the context of matrix–vector multiplications such as $\Psi^T\mathbf{Q}^{(k)}$ or $\Psi\delta\mathbf{m}^{(k)}$.

12.8 Bayesian inversion

To this point, we have considered only deterministic linear and non-linear inversion of geophysical data. It is also possible to develop stochastic inversion methods, based on an entirely different philosophical stance. In a Bayesian analysis, for example, the physical property of the Earth under investigation is presumed to be a spatially random function. The actual spatial distribution of the property inside the Earth is envisioned to be a single realization of that random function. A central objective of Bayesian analysis is then to establish the *most probable* spatial distribution of the physical property.

The general approach that a geophysicist might take to conduct a Bayesian inversion can be divided crudely into the following two steps: (a) one builds an admissible set of models M and assigns a *prior* probability $P(\mathbf{m})$ to each of the models in M based on one's theoretical knowledge, scientific intuition, consultations with other experts, and personal experience; (b) one then uses Bayes' rule to develop a *posterior* probability distribution $P(\mathbf{m}|\mathbf{d})$ that essentially sharpens the prior distribution in light of the new evidence contained in the geophysical dataset $\mathbf{d}$. An informal introduction to the foundations of Bayesian reasoning is provided by Jaynes (1986) while Scales and Snieder (1997) broadly discuss the topic in the context of geophysical inversion. The comprehensive paper by von Toussaint (2011) treats Bayesian inference from the point of view of a physicist.

Bayes' rule (Mackay, 1992) for combining probabilities may be stated as

$$P(\mathbf{m}|\mathbf{d}) = \frac{P(\mathbf{d}|\mathbf{m})P(\mathbf{m})}{P(\mathbf{d})} \tag{12.51}$$

where $P(\mathbf{m}|\mathbf{d})$ is the *posterior* probability that model $\mathbf{m}$ is indeed responsible for the observed dataset $\mathbf{d}$; while $P(\mathbf{d}|\mathbf{m})$ is the theoretically determined *likelihood* that model $\mathbf{m}$ is capable of generating dataset $\mathbf{d}$; with $P(\mathbf{m})$ being the probability that model $\mathbf{m}$ is the actual model, as determined *prior* to any consideration of the dataset $\mathbf{d}$; and finally $P(\mathbf{d})$ is the probability of observing the particular dataset $\mathbf{d}$ from amongst all the possible datasets that could have been observed.

It is straightforward to develop a standard non-linear regularized inverse problem within a Bayesian context. The following example is taken from Ulrych *et al.* (2001). Suppose the non-linear inverse problem is $\mathbf{d} = \mathbf{F}[\mathbf{m}] + \Delta$ where $\mathbf{m} = (m_1, m_2, \ldots, m_P)$ is the model parameter vector and $\mathbf{d} = (d_1, d_2, \ldots, d_D)$ is the data vector. The D random variables in the data vector are assumed to be independent, such that the data covariance matrix C assumes the familiar form $C = \mathrm{diag}(\sigma^2_{d_1}, \sigma^2_{d_2}, \ldots, \sigma^2_{d_D})$. The likelihood function $P(\mathbf{d}|\mathbf{m})$ is assumed to be a Gaussian probability function

$$P(\mathbf{d}|\mathbf{m}) = \frac{1}{\sqrt{(2\pi)^P|C|^{1/2}}}\exp\left[-\frac{1}{2}\Delta^T C^{-1}\Delta\right], \tag{12.52}$$

with the diagonal covariance matrix C. The form of Equation (12.52) is appropriate in the case of uncorrelated, Gaussian-distributed errors, and assigns higher likelihood to models whose response generates a low value of the misfit $\Delta^T C^{-1} \Delta$.

It is also required to specify an appropriate prior probability distribution. In the case that smoothly varying models are preferable to rough models, such a prior distribution could be written as

$$P(\mathbf{m}) = \left(\frac{\eta}{2\pi}\right)^{\frac{D-1}{2}} \exp\left[-\frac{\eta}{2}\mathbf{m}^T D^T D\mathbf{m}\right] \tag{12.53}$$

where it will be seen that η plays the role of a regularization parameter. The denominator $P(\mathbf{d})$ in Equation (12.51) can be regarded as a constant normalizing factor insofar as the inversion is concerned since it does not depend on the model vector $\mathbf{m}$. With these assignments, the Bayes posterior probability function emerges as

$$P(\mathbf{m}|\mathbf{d}) = \frac{1}{\sqrt{(2\pi)^P |C|^{1/2}}} \left(\frac{\eta}{2\pi}\right)^{\frac{D-1}{2}} \exp\left[-\frac{1}{2}\Phi(\mathbf{m}, \eta)\right] \tag{12.54}$$

with

$$\Phi(\mathbf{m}, \eta) = \Delta^T C^{-1} \Delta + \eta \mathbf{m}^T D^T D\mathbf{m}. \tag{12.55}$$

The role of η as a regularization parameter should now be apparent. The most probable model $\mathbf{m}^*$ is defined as the one that maximizes the posterior probability $P(\mathbf{m}|\mathbf{d})$ or, equivalently, minimizes the function $\Phi(\mathbf{m},\eta)$ for an appropriate choice of regularization parameter η. The minimization of $\Phi(\mathbf{m},\eta)$ can be carried out using non-linear least-squares, conjugate gradients, or some other optimization method.

Bayes inversion is seen by this example to be computationally equivalent to the deterministic non-linear regularized inversion provided a multi-dimensional Gaussian is chosen for the likelihood function and a smooth prior is assumed. Clearly, the form of the prior model distribution has a great impact on the posterior model distribution. Oftentimes however, due to factors including Earth's complexity across a wide range of spatial scales, prior model statistics are exceedingly difficult to define or estimate (Scales and Snieder, 1997). The prior model distribution then becomes effectively a tool at the disposal of the geophysicist to regularize or shape the posterior distribution.

The solution to the inverse problem, in the Bayesian case, has a probabilistic interpretation. The posterior distribution $P(\mathbf{m}|\mathbf{d})$ describes the probability that model $\mathbf{m}$ is the actual Earth model given the evidence $\mathbf{d}$. This interpretation is made even though the geophysicist is fully aware that the true Earth $\mathbf{m}_T$ lies outside the admissible set M. Following the advice of Jaynes (1986), the Bayesian geophysicist has little choice but to proceed anyway *as if* the true Earth $\mathbf{m}_T$ were inside the admissible set M.

Some authors have pointed out that Bayes' rule contains basic assumptions that may not be appropriate in all situations. For example, it is clear from Equation (12.51) that Bayesian analysts treat the prior information and the new information in a non-symmetric way. Bayes' rule also contains the fundamental assumption that the probabilities of a hypothesis and its negation must sum to unity. In response to this, a number of other rules for combining probabilities have been explored. The Dempster–Shafer theory of evidence (Shafer, 1976), for example, holds that all the available information, prior and new, should be treated on equal footing. Moreover, a person should assign one portion of his belief to a hypothesis and a second portion to its negation; however, the two portions need not sum to unity if the person finds no cause to assign his entire belief to either the hypothesis or its negation. Dempster–Shafer reasoning, or other alternatives to Bayesian reasoning, may see increasing use in near-surface applied geophysics as a framework for combining information from disparate types of datasets.

12.9 Auxiliary sensitivity analysis

A major difficulty in solving a non-linear inverse problem $\mathbf{d} = \mathbf{F}[\mathbf{m}] + \Delta$ is to compute the sensitivities, or partial derivatives, of the data prediction vector $\mathbf{u} = \mathbf{F}[\mathbf{m}]$ with respect to the entries of the model parameter vector $\mathbf{m}$ (see Equation (12.17)). The sensitivities are used to refine an initial model so that an improved fit to the observed data can be obtained.

There are two standard numerical approaches for computing the sensitivities: the *sensitivity-equation* approach and the *adjoint-equation* approach. In the former, the governing differential equation of the forward problem is differentiated with respect to a model parameter and the resulting new boundary-value problem is solved. This technique requires P forward solutions if there are P model parameters so that it is not very efficient if P is large. In the adjoint approach, an auxiliary equation (to be defined later) is formulated and solved. The cost of solving the auxiliary equation is typically comparable to the cost of solving the forward problem. However, only D auxiliary solutions are required, where D is the number of data points. Thus, the auxiliary-equation approach offers a considerable advantage over the sensitivity-equation approach if $D < P$, which occurs in the common case of an *over-parameterized* inverse problem.

The auxiliary approach can be derived formally, using rigorous mathematics, via the development of an adjoint differential operator along with appropriate boundary conditions. The formal derivation takes advantage of the *reciprocity* properties of geophysical data. However, a simple derivation of the auxiliary sensitivity equations has been developed by McGillivray *et al.* (1994) that does not require the construction of an adjoint operator and does not appeal to reciprocity. Herein we follow their approach.

We illustrate the method in the context of a continuous 2-D forward problem written as a second-order partial differential equation

$$\nabla^2 u + \alpha(\mathbf{r})u = f, \tag{12.56}$$

with

$$\nabla^2 = \frac{\partial^2}{\partial x^2} + \frac{\partial^2}{\partial z^2}. \tag{12.57}$$

Equation (12.56) is general enough to describe a wide range of geophysical forward problems including seismic wave propagation, electromagnetic induction, heat flow, and ground-penetrating radar wave propagation. The spatially varying physical property of the Earth to be determined by inversion is $\alpha(\mathbf{r})$. Measurements of the field $u(\mathbf{r})$ are presumed to be made on the boundary $\partial\Omega$ of the solution domain Ω. For simplicity it is assumed that the source term is a negative unit impulse at some transmitter location $\mathbf{r}_{TX}$, that is,

$$f(\mathbf{r}) = -\delta(\mathbf{r} - \mathbf{r}_{TX}). \tag{12.58}$$

First, discretize the solution domain Ω into a 2-D grid of cells such that the physical property in the j-th cell Ω_j is a constant $\alpha = \alpha_j$ for each $j = 1, \ldots, P$. The continuous function $\alpha(\mathbf{r})$ may then be decomposed as

$$\alpha(\mathbf{r}) = \sum_{j=1}^{P} \alpha_j \psi_j(\mathbf{r}) \tag{12.59}$$

where $\psi_j(\mathbf{r})$ is a boxcar function which is unity in cell j and vanishes elsewhere,

$$\psi_j(\mathbf{r}) = \begin{cases} 1 & \text{for } \mathbf{r} \in \Omega_j \\ 0 & \text{otherwise} \end{cases}. \tag{12.60}$$

Differentiating Equation (12.56) with respect to the k-th model parameter α_k gives

$$\nabla^2 \frac{\partial u}{\partial \alpha_k} + \psi_k u + \alpha \frac{\partial u}{\partial \alpha_k} = 0. \tag{12.61}$$

Next, introduce the auxiliary forward problem

$$\nabla^2 v + \alpha(\mathbf{r}) v = g, \tag{12.62}$$

with source function $g(\mathbf{r})$ to be determined later. It is assumed that Equation (12.62) can be solved for the auxiliary field $v(\mathbf{r})$ using the same forward algorithm as used to solve the original forward Equation (12.56). The source $g(\mathbf{r})$, as we will see below, is not a physical source but rather a convenient fictional source introduced to facilitate an efficient sensitivity analysis.

Pre-multiply Equation (12.61) by the auxiliary field v to obtain

$$v\nabla^2 \frac{\partial u}{\partial \alpha_k} + \psi_k v u + \alpha v \frac{\partial u}{\partial \alpha_k} = 0. \tag{12.63}$$

Similarly, pre-multiply the auxiliary Equation (12.62) by $\partial u/\partial \alpha_k$ to obtain

$$\frac{\partial u}{\partial \alpha_k} \nabla^2 v + \alpha v \frac{\partial u}{\partial \alpha_k} - \frac{\partial u}{\partial \alpha_k} g = 0. \tag{12.64}$$

Subtraction of Equations (12.63) and (12.64) gives

$$v\nabla^2 \frac{\partial u}{\partial \alpha_k} - \frac{\partial u}{\partial \alpha_k} \nabla^2 v + \psi_k v u + \frac{\partial u}{\partial \alpha_k} g = 0. \tag{12.65}$$

Now, Equation (12.65) can be integrated over the solution domain Ω to give

$$\int_{\Omega} \left[v\nabla^2 \frac{\partial u}{\partial \alpha_k} - \frac{\partial u}{\partial \alpha_k} \nabla^2 v \right] d\Omega + \int_{\Omega} \left[\psi_k v u + \frac{\partial u}{\partial \alpha_k} g \right] d\Omega = 0. \tag{12.66}$$

The first term can be simplified with the aid of Green's identity

$$\int_{\Omega} a\nabla^2 b d\Omega = -\int_{\Omega} \nabla a \cdot \nabla b d\Omega + \int_{\partial\Omega} a\,\hat{n}\cdot\nabla b dS, \tag{12.67}$$

where $a(\mathbf{r})$ and $b(\mathbf{r})$ are arbitrary once-differentiable functions defined on Ω and $\hat{n}$ is the outward normal of the surface $\partial\Omega$ bounding the solution domain Ω. Using (12.67) in Equation (12.66) yields

$$\int_{\Omega}\left[-\nabla v\cdot\nabla\frac{\partial u}{\partial\alpha_k} + \nabla\frac{\partial u}{\partial\alpha_k}\cdot\nabla v\right]d\Omega + \int_{\Omega}\left[\psi_k vu + \frac{\partial u}{\partial\alpha_k}g\right]d\Omega = 0. \tag{12.68}$$

The surface integral in Equation (12.67) vanishes if the solution domain Ω is large enough that the field $u(\mathbf{r})$ on the bounding surface $\partial\Omega$ due to the source f at the TX location $\mathbf{r}_{TX}$ vanishes. It is readily seen also that the first term in Equation (12.68) vanishes. This leaves

$$\int_{\Omega}\frac{\partial u}{\partial\alpha_k}g d\Omega = -\int_{\Omega}\psi_k vu d\Omega. \tag{12.69}$$

Now we can specify a convenient auxiliary source,

$$g(\mathbf{r}) = -\delta(\mathbf{r}-\mathbf{r}_{RX}); \tag{12.70}$$

it is a negative unit impulse function deployed at the RX location $\mathbf{r}_{RX}$. For this choice, Equation (12.69) simplifies to

$$-\int_{\Omega}\frac{\partial u}{\partial\alpha_k}\delta(\mathbf{r}-\mathbf{r}_{RX})d\Omega = -\int_{\Omega}\psi_k vu d\Omega \tag{12.71}$$

which leads immediately to the desired result

$$\left.\frac{\partial u}{\partial\alpha_k}\right|_{\mathbf{r}=\mathbf{r}_{RX}} = \int_{\Omega_k} vu d\Omega_k \,(\text{auxiliary sensitivity analysis}). \tag{12.72}$$

Thus, the sensitivity of the field u at receiver location $\mathbf{r}_{RX}$ to a perturbation in the k-th model parameter α_k is given by an integration of the primary field u and the auxiliary field v over the k-th cell volume Ω_k. The primary field u is found by solving the forward problem for a negative unit impulse source located at the TX position. The auxiliary field v is found by solving the forward problem for a negative unit impulse source located at the RX position.

The total number of forward solves for computing the full sensitivity matrix, using the auxiliary approach, is equal to the number of receivers $N_{RX}+1$, which is the same as

$D+1$ if each receiver measures one datum. The total number of required forward solutions using the traditional approach is $2P$, since

$$\left.\frac{\partial u}{\partial \alpha_k}\right|_{\mathbf{r}=\mathbf{r}_{RX}} \approx \frac{u(\alpha_k + \Delta\alpha_k) - u(\alpha_k - \Delta\alpha_k)}{2\Delta\alpha_k} \text{(traditional analysis)}. \tag{12.73}$$

It is easy to see, therefore, that the auxiliary sensitivity analysis is much more efficient than the traditional approach in the case of an over-parameterized inverse problem for which the number of model parameters greatly exceeds the number of data points, $P >> D$.

Problems

1. Derive the expression for $\mathbf{g}^{(k+1)}$ in step 6 of the conjugate-gradient algorithm for seismic tomography.
2. **Computer assignment**. The aim of the assignment is to develop a program for tomographic 2-D reconstruction of a single, uniform block of anomalous P-wave slowness embedded in a uniform background. Use the linear conjugate-gradient algorithm developed in this chapter, based on the paper by Scales (1987), to solve the seismic traveltime inverse problem. Develop your computer program according to the following steps: (a) build a 2-D regular grid of rectangular cells; (b) assign a uniform background slowness to each cell; (c) define a number of TX locations distributed around the boundary on all four sides of the grid; (d) ditto for RX locations; (e) calculate the background traveltimes for each TX–RX pair based on straight rays; (f) calculate the distance traveled by each ray in each cell; (g) define an anomalous block embedded somewhere near the center of your grid; (h) assign an anomalous slowness value to the block; (i) calculate the anomalous TX–RX traveltimes with the block present; (j) add a small component of random noise to the anomalous traveltimes to simulate actual data; (k) construct the linear system for the unknown traveltime perturbations; (l) solve the linear system using the conjugate-gradient method; (m) calculate and report the error between your inversion result and the actual anomalous slowness; (n) once your program is working, repeat the analysis for one or more different values in each of: amount of random noise, TX and RX configuration, size of anomalous block, anomalous slowness, and number of cells in the original grid.

13 Non-linear inversion: global methods

In the previous chapter we examined descent methods for non-linear geophysical inverse problems in which each step of the search through the admissible model space M is designed to lower an objective function $\varphi[\mathbf{m}]$ such that $\varphi[\mathbf{m}^{(k+1)}] \leq \varphi[\mathbf{m}^{(k)}]$. The descent methods converge toward a *local* minimum

$$\mathbf{m}^* = \arg\min_{\mathbf{m}\in M} \varphi[\mathbf{m}] \tag{13.1}$$

in model space and involve evaluations of the local gradient $\nabla\varphi[\mathbf{m}]$, plus sometimes also second-derivative information.

However, for practical geophysical inverse problems characterized by non-linear forward mappings $\mathbf{F}[\mathbf{m}]$, the objective function $\varphi[\mathbf{m}]$ almost always contains many local minima in P-dimensional model space. Methods of *global optimization* are sought that, unlike the local methods, can jump out of local minima. This behavior is shown schematically in Figure 13.1. Global optimization methods are typically not gradient-based, rather they are most often based on direct evaluations of the objective function $\varphi[\mathbf{m}]$.

In this chapter just a brief introduction is presented to a few of the global optimization methods that are currently in routine use in geophysical inversion. The techniques are largely *heuristic* in the sense that there exist no firm rules leading to the guaranteed discovery of a globally optimal solution for any given inverse problem. Instead, geophysicists are more interested in the average or expected performance of a method as it is applied to a large-scale inverse problem. A useful overview of global optimization methods in geophysical inversion is provided by Sen and Stoffa (2013).

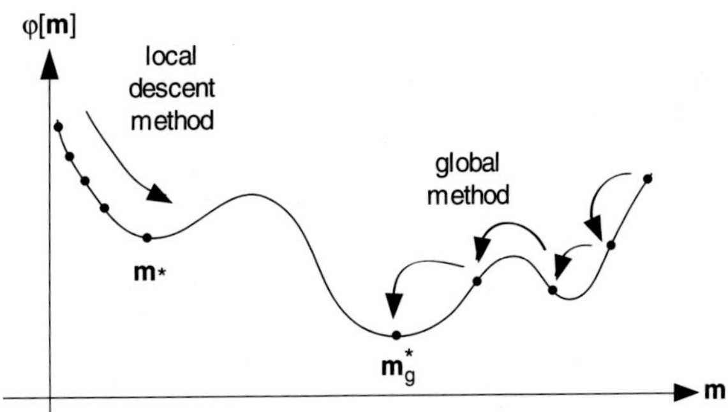

Figure 13.1 Local descent methods find only a local minimum m* of the non-linear objective function ϕ[m]. Global methods can jump out of local minima and find the global minimum m_g*.

13.1 Markov chain Monte Carlo (MCMC) method

Perhaps the simplest approach to global optimization is the Monte Carlo (MC) search. Here, models from the admissible set M are visited in a pure random walk (Doyle and Snell, 1984) through model space. The objective function $\varphi[\mathbf{m}]$ is evaluated at each step of the random walk. The MC algorithm proceeds indefinitely until time or computational resources are exhausted. The inversion result is the model $\mathbf{m}^*$ that happened to generate the lowest value of the objective function $\varphi[\mathbf{m}]$. A strict MC search is impractical, however, since it is not feasible to explore P-dimensional model space in appreciable detail, as noted in the previous chapter with the discussion about the curse of dimensionality.

A type of algorithm that outperforms the strict random walk involves the construction of an efficient Markov chain of models $\{\mathbf{m}^{(0)}, \mathbf{m}^{(1)}, \mathbf{m}^{(2)}, \ldots\}$. In a Markov chain, each successive model $\mathbf{m}^{(k+1)}$ depends only on the previous one $\mathbf{m}^{(k)}$. The resulting family of optimization methods are called Markov chain Monte Carlo (MCMC) methods.

An enduring MCMC algorithm was developed by Metropolis *et al.* (1953) and Hastings (1970). Formulated in the context of Bayesian inference, the algorithm consists of three steps: (i) an initial "burn-in" phase during which the random walker is methodically guided toward the region of model space that contains the most-probable model $\mathbf{m}^*$, i.e. the peak of the posterior probability density function (pdf); (ii) the drawing of a candidate model $\mathbf{m}'$ from a pdf $q(\mathbf{m}|\mathbf{m}')$ that approximates the posterior pdf; (iii) with acceptance probability α, the addition of the candidate $\mathbf{m}'$ to the Markov chain. The algorithm repeatedly executes steps (i)–(iii). A good introduction to the Metropolis–Hastings algorithm is provided by Chib and Greenberg (1995).

In geophysics, Malinverno (2002) has developed a Metropolis–Hastings type of MCMC algorithm for inversion of DC apparent resistivity data acquired over a layered Earth using a Schlumberger electrode array (for details about the forward problem, see Chapter 4). The forward mapping $\mathbf{F}[\mathbf{m}]$ is non-linear but it can be calculated using an analytic formula. Here we have space to provide only a rough qualitative sketch of the Malinverno (2002) approach. Before proceeding, the reader might wish to briefly review the section on Bayesian inversion from the previous chapter.

Recall that the Bayesian solution to an inverse problem is the most-probable model $\mathbf{m}^*$, i.e. the one that maximizes the posterior pdf $P(\mathbf{m}|\mathbf{d})$. Normally also reported is an estimate of statistical properties of the distribution $P(\mathbf{m}|\mathbf{d})$ such as its mean and variance. In a one-dimensional (1-D) resistivity inversion, it is conventional to fix the number of layers k and to parameterize the thickness and resistivity of each layer. Malinverno (2002), however, treats the number of layers k as an adjustable model parameter. His model vector is then $\mathbf{m} = (k, \mathbf{z}, \boldsymbol{\rho})$ where $\mathbf{z}$ is a $(k-1)$-dimensional vector containing logarithms of the depths to the resistivity interfaces, while $\boldsymbol{\rho}$ is a k-dimensional vector containing logarithms of resistivity. Log-depths are chosen to reflect the exponential loss of resolution with depth while log-resistivities are chosen to ensure positivity of the modeled resistivities and to account for the wide dynamic range in resistivity that is found in geomaterials.

For a fixed number of layers k, the prior pdf considered by Malinverno (2002) is the product

$$P(\mathbf{m}|k) = P(\mathbf{z}|k)P(\boldsymbol{\rho}|k), \tag{13.2}$$

where $P(\mathbf{z}|k)$ is a prior pdf describing layers that have a prescribed minimum thickness h_{min} and whose interfaces are distributed with equal probability over the admissible depth range $z \in [z_{min}, z_{max}]$. The other prior pdf $P(\boldsymbol{\rho}|k)$ appearing in Equation (13.2) requires that the log-resistivities obey a multivariate Gaussian distribution that specifies the most-probable values and the variance of the layer resistivities.

The likelihood function $P(\mathbf{d}|\mathbf{z},\boldsymbol{\rho})$ is a standard multivariate Gaussian pdf (as used in Ulrych *et al.*, 2001) that describes zero-mean, uncorrelated components of the modeling error vector Δ. Malinverno (2002) then motivates and derives a second likelihood function $P(k|\mathbf{d})$ that assigns greater probability to models with fewer layers. The use of this likelihood function acts as a type of regularization that penalizes the appearance of models with a large number of layers. Putting everything together, the posterior pdf $P(\mathbf{m}|\mathbf{d})$ to be maximized in the Bayesian inversion follows from the chain rule

$$P(\mathbf{m}|\mathbf{d}) = P(k|\mathbf{d})P(z|k,\mathbf{d})P(\boldsymbol{\rho}|k,\mathbf{d}). \tag{13.3}$$

The heuristic MCMC algorithm of Malinverno (2002) involves the following steps:

(i) choose an arbitrary starting model $\mathbf{m}^{(0)}$;
(ii) add, delete, or move a layer interface according to prescribed probabilities;
(iii) construct a candidate model $\mathbf{m}'$ by selecting layer resistivities from a pdf $q(\mathbf{m}|\mathbf{m}')$ that approximates $P(\mathbf{m}|\mathbf{d})$;
(iv) compute the acceptance probability α;
(v) select a random number $r \in [0, 1]$ and accept $\mathbf{m}'$ if $r < \alpha$;
(vi) if the burn-in period is over (see below), add $\mathbf{m}'$ to the Markov chain;
(vii) return to step (*ii*) until a stopping criterion is satisfied. Methods to compute $q(\mathbf{m}|\mathbf{m}')$ and the acceptance probability α are explained in Chib and Greenberg (1995).

The burn-in period is ended once the objective function $\varphi[\mathbf{m}]$ has been reduced to a prescribed acceptable value, in which case the current model is presumed to be in a favorable region of model space. The MCMC algorithm terminates when the Markov chain is of sufficient length such that the statistical properties of the posterior pdf $P(\mathbf{m}|\mathbf{d})$ no longer change appreciably with the addition of a new model to the chain.

Some of the results of the Bayesian MCMC inversion on synthetic 1-D apparent resistivity data are given in Figure 13.2. At left is shown a three-layer geoelectrical model and its synthetic response with error bars. A superposition of the models sampled by the Markov chain is shown at the right of the figure. The ensemble of models provides a pictorial representation of the posterior pdf. Notice, for example, that the low-resistivity values are more tightly constrained than the intervening zone of higher-resistivity ones (as highlighted by the dashed white line). The reader should also observe that there is a moderate uncertainty associated with the layer-thickness estimates.

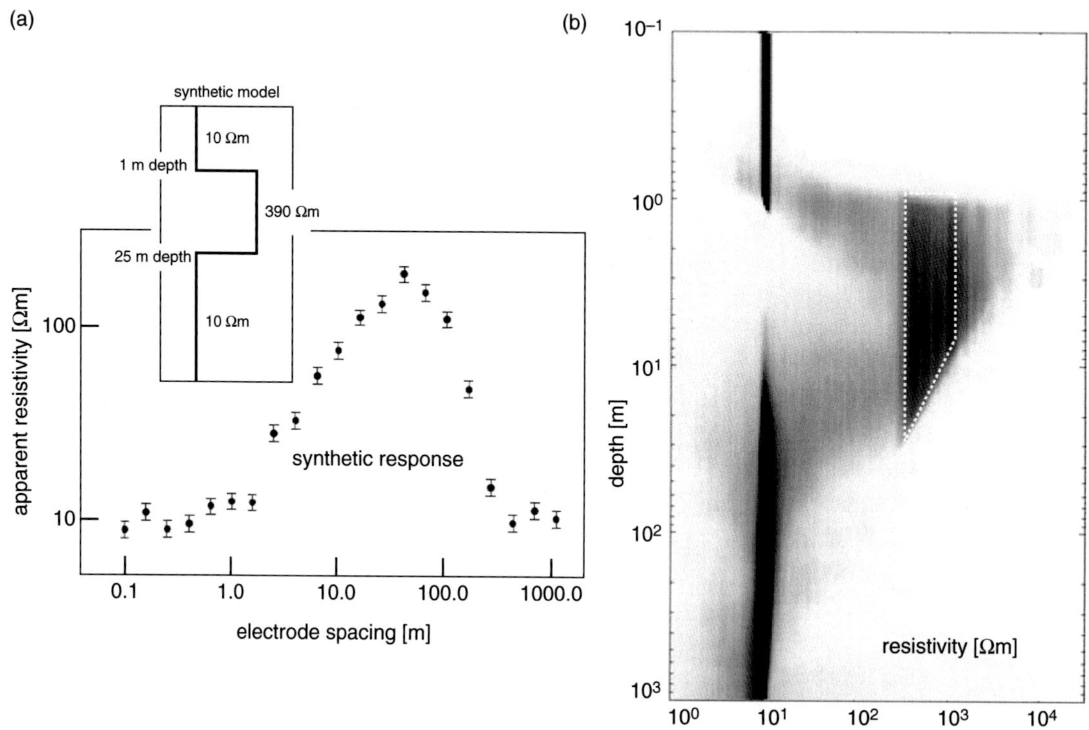

Figure 13.2 Synthetic test of the Bayesian MCMC inversion algorithm: (a) three-layer model (inset) and its synthetic Schlumberger-array response curve; (b) superposition of models sampled by the Markov chain. After Malinverno (2002). © 2002 RAS.

The results of the Bayesian MCMC inversion on actual data measured in central Australia are given in Figure 13.3. The superposition of the models sampled by the Markov chain is shown in the figure inset. The most-probable value of the resistivity–depth profile is indicated by the solid white line in Figure 13.3 while the width of the shaded region indicates the variance of the model parameters.

13.2 Simulated-annealing (SA) method

A number of global optimization approaches to geophysical inversion are based on the computer simulation of a physical, chemical, or biological process found in nature. The simulated-annealing (SA) method, for example, mimics the statistical mechanical process of the condensation, or *annealing*, of a liquid into a regular crystalline solid (Kirkpatrick *et al.*, 1983; Sen and Stoffa, 2013). The physical process may be described briefly as follows. Consider an ensemble of freely moving atoms at some initial temperature T_0. As the temperature is lowered to the freezing point, the atoms become less mobile and the ensemble eventually condenses into the solid state. If the cooling is performed slowly and

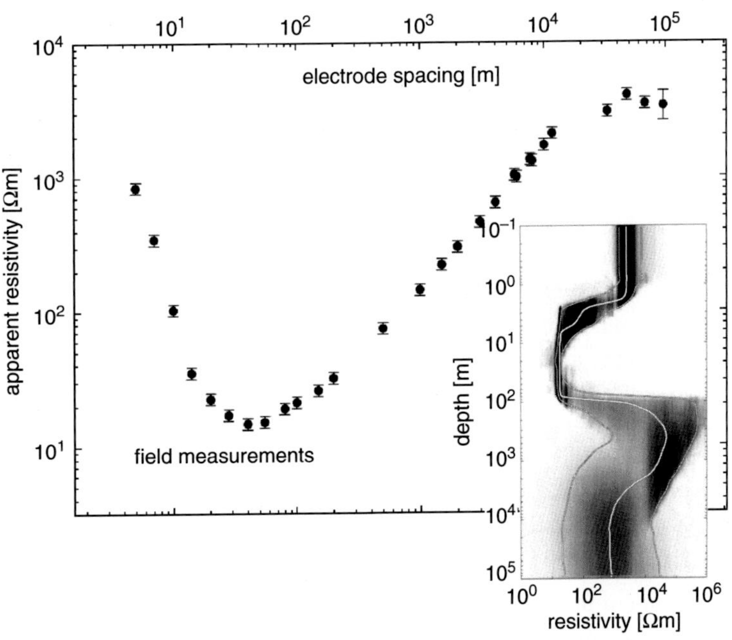

Figure 13.3 Field example of the Bayesian MCMC inversion algorithm: symbols with error bars indicate Schlumberger-array responses from central Australia; (inset) superposition of models sampled by the Markov chain, the central solid white line shows the most-probable value of the model ensemble. After Malinverno (2002). © 2002 RAS.

carefully in stages, the ensemble is afforded the chance to evolve into a rare, but important, ground-state configuration. The ground state is the sought-after state of global minimum energy and, in nature, would correspond to a perfect crystal of regular geometry. If the system is cooled too fast, or quenched, it will develop either a large number of crystal defects or a non-crystalline glassy texture. Either of these states is only locally optimal in terms of its energy.

At each stage of the cooling process, the system of atoms is permitted to come into thermal equilibrium with a heat bath of temperature T. The probability of the system being found in any configuration k that has energy E_k is given by the Gibbs distribution

$$P(E_k) \sim \exp\left(-\frac{E_k}{k_B T}\right) \tag{13.4}$$

where k_B is Boltzmann's constant. The minimum energy, or ground state, is attained in the limit $T \rightarrow 0$.

In geophysical inverse problems solved by the SA method, the energy E is represented by the objective function $\varphi[\mathbf{m}]$ while different model vectors $\mathbf{m}$ are analogous to different atomic configurations. The probability associated with a specific model $\mathbf{m}_k$ is then given by

$$P(\mathbf{m}_k) \sim \exp\left(-\frac{\varphi[\mathbf{m}_k]}{T}\right), \tag{13.5}$$

where T is now interpreted as an effective temperature.

A slow cooling is essential for the SA method to possibly jump out of local minima of the objective function. Local optimization methods that necessarily lower the objective function $\varphi[\mathbf{m}]$ at each step act instead like a rapid quenching process, rather than a slow annealing process, and consequently they arrive only at a locally optimal or metastable energy state.

A variant of the Markov-chain algorithm due to Metropolis *et al.* (1953) is capable of simulating a slow annealing process. The algorithm proceeds step-wise as follows. A random perturbation $\Delta\mathbf{m}$ to the current model $\mathbf{m}$ is made. The resulting change in the objective function is $\Delta\varphi = \varphi[\mathbf{m} + \Delta\mathbf{m}] - \varphi[\mathbf{m}]$. If $\Delta\varphi \leq 0$, the objective function is lowered, the model perturbation is definitely accepted, and the new model $\mathbf{m} + \Delta\mathbf{m}$ becomes the next member of the Markov chain. If $\Delta\varphi > 0$, the objective function is raised, but the model perturbation may still be accepted with a probability

$$P(\Delta\varphi[\mathbf{m}]) \sim \exp\left(-\frac{\Delta\varphi[\mathbf{m}]}{T}\right). \tag{13.6}$$

The acceptance or rejection of $\Delta\mathbf{m}$ in this case is determined by generating a random number on the interval [0, 1] and comparing it to $P(\Delta\varphi[\mathbf{m}])$ in Equation (13.6). Repeated application of the Metropolis algorithm drives the function $\varphi[\mathbf{m}]$ toward its global minimum $\mathbf{m}^*$. Transitions out of local minima are possible since perturbations that raise the value of the objective function are allowed. Generally, experience with the method on different types of problems has found that large-scale features of the eventual globally optimal model $\mathbf{m}^*$ tend to appear early in the Markov chain, with the finer-scale details emerging at lower temperatures.

The effective temperature T in the simulated annealing process acts as a control parameter that is reduced in stages. At each T, there must be enough Metropolis steps to ensure that the system has properly equilibrated to that value. In this manner, a *cooling schedule* should be defined consisting of the temperature increments and the number of Metropolis steps at each increment. It can be shown (Geman and Geman, 1984) that a schedule of the form

$$T(k) = \frac{T_0}{lnk}, \tag{13.7}$$

with k being the iteration number, guarantees the eventual convergence of the SA method. However, the logarithmic form for $T(k)$ leads to a very slow algorithm.

Improvements to the rate of convergence of the SA algorithm have been made by a number of authors. Szu and Hartley (1987) demonstrated a fast algorithm (FSA) that uses a cooling schedule of the form $T(k) = T_0/k$. Their method selects the model perturbations $\Delta\mathbf{m}$ that are to be tested by the Metropolis criterion according to a Cauchy-type distribution

$$f(\Delta\mathbf{m}) \sim \frac{T}{\sqrt{\Delta\mathbf{m}^2 + T^2}}. \tag{13.8}$$

A very fast algorithm termed VFSA has also been developed, initially by Ingber (1989). This algorithm admits separate perturbations to each model parameter m_i for $i = 1, \ldots, P$, along with a number of other significant improvements. A detailed description of the FSA and VFSA algorithms may be found in Sen and Stoffa (2013).

In a near-surface geophysical application, Ryden and Park (2006) used the FSA algorithm to invert dispersive seismic surface wave in terms of thickness and stiffness properties of pavement layers. Their field setup involves a single receiver (RX) at a fixed location and a hammer (TX) source deployed at several locations along a profile. The resulting seismic wavefield $u(x, t)$ is compiled as a function of the TX–RX offset x. A Fourier transform then yields $U(x,\omega)$ where ω is the angular frequency. As described in the section on the multichannel analysis of surface waves (MASW) technique in Chapter 7, a slant-stack of the form

$$S(\omega, c_T) = \int \exp\left(-i\left[\frac{\omega}{c_T}\right]x\right)U(x,\omega)dx \tag{13.9}$$

is constructed for a range of values of phase velocity c_T. A maximum in the measured phase-velocity spectrum $S_{meas}(\omega, c_T)$ occurs for the value of c_T equal to the true phase velocity of each frequency component. Dispersion curves are then identified as the high-amplitude loci on plots of the two-dimensional phase-velocity spectrum. Given a model **m** of the pavement layer properties, an analytic solution $S_{anal}(\mathbf{m};\omega, c_T)$ can be derived starting from the basic *Haskell–Thomson* theory described in Chapter 7. The goal of the inversion is then to find a layered model whose corresponding phase-velocity spectrum best matches the observed spectrum.

A test of the FSA algorithm on synthetic data is demonstrated by Ryden and Park (2006). A three-layer pavement model is considered for which the parameters that are allowed to vary are assembled into a $P = 6$-dimensional model vector $\mathbf{m} = (h_1, h_2, V_{S1}, V_{S2}, V_{S3}, \nu_2)^T$, where h_i is the thickness of the *i-th* layer, V_{Si} is the shear velocity of each layer and that of the underlying halfspace, and ν_2 is Poisson's ratio of the second layer. The data vector **d** contains the modulus of the measured phase-velocity spectrum evaluated on a discrete 2-D set of points, i.e. $d_{jk} = |S_{meas}(\omega_j, c_{Tk})|$. Similarly, the model prediction vector $\mathbf{u}[\mathbf{m}]$ contains the discrete values $u_{jk}[\mathbf{m}] = |\, S_{anal}(\mathbf{m};\omega_j, c_{Tk})|$. The objective function $\varphi[\mathbf{m}]$ is defined as

$$\varphi[\mathbf{m}] = 1 - \frac{\mathbf{d}\cdot\mathbf{u}[\mathbf{m}]}{|\mathbf{d}|^2}, \tag{13.10}$$

which vanishes in the case of $\mathbf{d} = \mathbf{u}[\mathbf{m}]$, i.e. a perfect fit of theory to the data. Synthetic data are generated for a range of TX–RX offsets $0.05 \le x \le 5.0$ m at a station spacing of 0.05 m. The source excitation is a unit vertical point load.

After some trial and error, a geometric FSA cooling schedule of the form $T = T_0 a^k$ was set up with $T_0 = 30$ and $a = 0.99$. At each stage of the inversion, the temperature was decreased to the next-lower stage after $n = 5$ model perturbations were accepted by the Metropolis criterion. The inversion was terminated after 457 of 9095 random model perturbations were accepted, with a final value of the objective function $\varphi[\mathbf{m}] = 0.327\%$. The results are shown in Figure 13.4.

The true model that generated the synthetic data is $\mathbf{m}_T = (0.22$ m, 0.40 m, 1400.0 m/s, 300.0 m/s, 100.0 m/s, 0.35). The best-matching model found by the inversion is very similar, $\mathbf{m}^* = (0.224$ m, 0.404 m, 1401.5 m/s, 303.0 m/s, 103.4 m/s, 0.349), indicating that

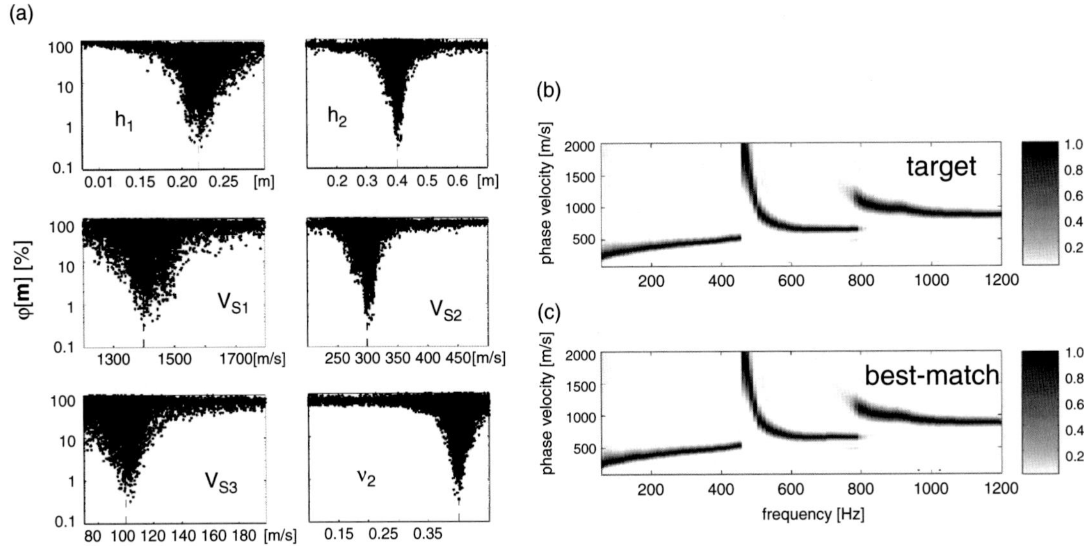

Figure 13.4 Synthetic FSA inversion of multichannel seismic surface wave data: (a) Pavement models examined and their corresponding objective function values, with the true parameter values indicated by vertical dashed lines; (b) the phase-velocity spectrum of the true pavement model; (c) the spectrum of the optimal pavement model. After Ryden and Park (2006).

a successful convergence was achieved. Ryden and Park (2006) further demonstrated a field test of this inversion method from a pavement construction site in Malmö, Sweden.

13.3 Genetic-algorithm (GA) method

The genetic-algorithm (GA) optimization method can be regarded as a computer implementation of the "survival of the fittest" principle from evolutionary biology. The fitness of a model **m** in this context is defined by the value of the objective function $\varphi[\mathbf{m}]$. The GA method does not involve the construction of a Markov chain but nevertheless it can be regarded as a directed Monte Carlo search through model space. The method operates on a group of Q models simultaneously. The starting group, or population, is selected at random with each model coded as a binary bit sequence. The Q binary sequences are then manipulated, as if they were chromosomes, according to rules that mimic the biological processes of reproduction, gene crossover, and gene mutation. These manipulations result in a new generation of Q models. As the algorithm proceeds through a number of iterations, the fittest models tend to be preserved in later generations while the less-fit models die out. The GA method has become a widely used global optimization technique in many fields of science and engineering; its earliest appearance is attributed to Holland (1975).

The defining characteristic of the GA method for geophysical inversion is the encoding of the Earth model parameter vector **m** into a concatenated sequence of 0 and 1 values with each model parameter m_i (for $i = 1, \ldots, P$) replaced by its corresponding value in the binary numeral system. For example, the familiar base-10 value $m_i = 53$ has an equivalent base-2 value of 110101 using a six-bit string. A model vector of $\mathbf{m} = (53,22,61)^T$ would then be encoded in this format as 110101101100111101. As described below, the encoded bit sequences are manipulated to mimic the way that certain biological systems evolve toward an optimal configuration. Gallagher *et al.* (1991), who considered a 1-D inversion of seismic waveforms, were amongst the first authors to use GA for geophysical inversion.

It is conventional to construct the admissible set M by assigning bounds to the model parameters of the form

$$\mathbf{m}_i^{(-)} \leq \mathbf{m}_i \leq \mathbf{m}_i^{(+)} \tag{13.11}$$

for each $i = 1,2, \ldots, P$. The model *granularity* d_i should also be specified. The quantity d_i defines the smallest admissible perturbation of model parameter m_i. The model bounds and granularity together determine the number of possible discrete values of a model parameter. This, in turn, determines the number of bits that should be used in the binary representations.

Following Gallagher *et al.* (1991), each iteration of a basic GA algorithm consists of three stages. First is the reproduction stage, in which a new generation of Q models is selected from the previous generation of Q models according to a probability distribution $P_R(\mathbf{m})$. The probability $P_R(\mathbf{m})$ of a given model $\mathbf{m}$ being selected for the next generation is determined by the value of $\varphi[\mathbf{m}]$, such that the fittest models are likely to be drawn, perhaps more than once, for inclusion in the new generation while the least-fitting models are not likely to be selected for reproduction.

The second stage of the basic GA algorithm is the crossover step, which is designed to shuffle the existing genetic information amongst the current population of Q models. In this step, the models are paired off at random to produce $Q/2$ couples. According to a crossover probability P_C, the two members of a couple interchange a randomly chosen contiguous, bit substring. This creates two new models with slightly different genetic sequences than the original two members. With probability $1 - P_C$, no such crossover occurs and the couple remains unchanged.

The final stage of the basic GA algorithm is the mutation step, which introduces entirely new genetic information to the current generation of Q models. Recall that each model consists of a binary sequence of bits with values of 0 or 1. According to a small mutation probability P_M, the polarity of each bit within each model sequence is reversed. The mutation probability is kept small so that the search remains biased to the favorable region of model space toward which the current generation of Q models is evolving. If there is too much mutation the GA algorithm is at risk to devolve into a Monte Carlo-like search and would not converge.

It should be kept in mind that a GA algorithm, like many of the global optimization methods, is a heuristic technique that is not guaranteed to converge in a finite number of steps. The GA method also contains several tuning parameters, such as the population size Q; the probabilities $P_R(\mathbf{m})$, P_C, and P_M; the model bounds $\mathrm{m}_i^{(\pm)}$; and the model granularity d_i. These parameters are set by trial and error or arbitrarily prescribed.

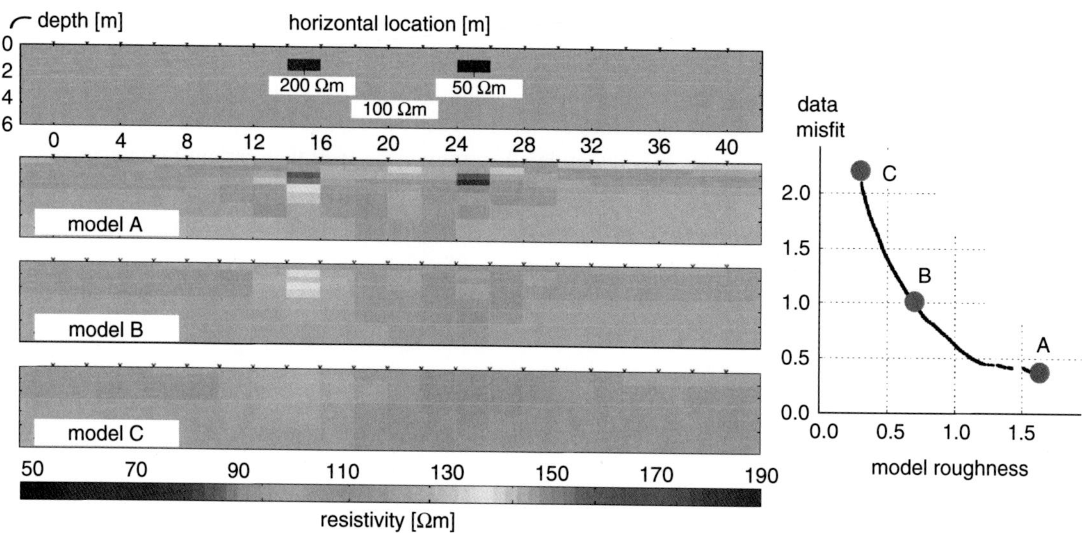

Figure 13.5 Genetic algorithm inversion of 2-D resistivity data: (left, top) synthetic model; (left, bottom three panels) inversion results; (right) the locations of the three models A, B, C on the misfit–roughness trade-off curve. After Schwarzbach *et al.* (2005). See plate section for color version.

A number of GA inversion studies have appeared in the geophysical literature including Everett and Schultz (1993) and Boschetti *et al.* (1997).

An advanced genetic algorithm has recently been implemented within a parallel computational environment for 2-D regularized inversion of DC apparent resistivity data by Schwarzbach *et al.* (2005). The model parameters are piecewise-constant log-resistivities of the blocks of a 2-D finite difference grid that is used to solve the forward problem based on Poisson's equation. The GA was directed to construct a family of models that lie along the trade-off curve (see Chapter 12) between data misfit and model roughness. Synthetic and field datasets consisted of $D = 114$ dipole–dipole measurements made using an array of 21 electrodes at 2.0-m spacing. The modeling domain was discretized into $P = 120$ grid blocks with each resistivity value constrained to lie in the range 10–10^4 Ωm. Model parameters were encoded as eight-bit binary strings. The model search proved costly: a single run of 2048 generations with population size $Q = 8192$ on a 32-node parallel computer took ~ 2 months of CPU time. An example of the inversion results for a synthetic dataset is shown in Figure 13.5. The best reconstruction of the blocky true model is apparently model A, as the smoother models B and C fail to resolve the blocks.

13.4 Neural-network (NN) methods

An ordinary laptop computer is excellent at performing fast arithmetic and carrying out the precisely programmed logical steps of a serial algorithm. It is less proficient, however, at more complex tasks such as processing noisy data, tolerating faults, recognizing patterns, or

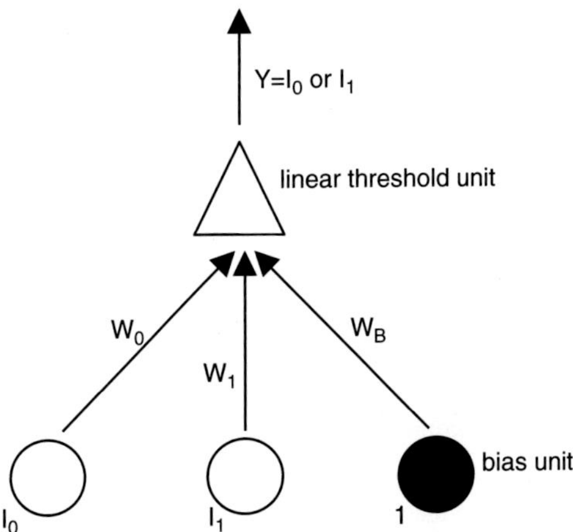

Figure 13.6 A single-unit adaptive network. After Smith (2001).

adapting to changing circumstances. Artificial neural networks (NNs) have been developed to solve these types of problems and others that are characterized by high levels of uncertainty. The latter include geophysical inverse problems for which the true Earth structure $\mathbf{m}_T$ is unknown but we have access to sparsely sampled, noisy realizations of its response $\mathbf{d}$. An introduction to the use of NNs in geophysics is given by van der Baan and Jutten (2000).

An NN, whether it is biological or artificial, consists of processing elements (neurons) that are highly interconnected and capable of passing messages to each other. A biological neuron located in a brain may have as many as 10 000 different inputs, and may send its output (the presence or absence of a short-duration spike) to a like number of other neurons. The biological NN found in the human brain remains far more complex than any of the artificial NNs that is has inspired.

An NN must be trained before it can be used to solve problems. Following the example of Smith (2001), let us look at the process of training a simple artificial NN. The network shown in Figure 13.6 has two binary inputs, I_0 and I_1. The network is assigned the task of learning a basic OR operation, defined by the following: namely, the desired output is $D = 1$ if $I_0 = 1$ and/or $I_1 = 1$. The network is exposed to a sequence of input/output pairs and it adapts by changing its weights $\{W_0, W_1, W_B\}$ according to the difference between the desired output D and the actual output Y.

The binary output Y is a weighted sum of the inputs,

$$Y = \begin{cases} 1 & \text{if } W_0 I_0 + W_1 I_1 + W_B > 0 \\ 0 & \text{if } W_0 I_0 + W_1 I_1 + W_B \le 0 \end{cases}. \tag{13.12}$$

The adjustment of the weights is controlled by the elementary learning rule

$$\Delta W_i = \eta (D - Y) I_i \tag{13.13}$$

for $i = \{0,1, B\}$, where $I_B = 1$ is fixed, and acts as a bias, while η is the *learning rate*.

To explore how the network learns, suppose an untrained network is exposed to the following set of four input patterns $\{I_0, I_1\}$ that define the OR operation:

I_0	I_1	D
0	0	0
0	1	1
1	0	1
1	1	1

The actual output of the network is desired to be $Y = D$, as tabulated above for all four input patterns. The goal of the training is to achieve this result. We simplify the process by fixing the learning rate to be $\eta = 0.5$. It is also convenient to start with an initial guess of zero weights $W_0 = 0$ and $W_1 = 0$ and $W_B = 0$. After exposure to the four patterns, as the reader can verify it is readily found that the weights, updated according to the learning rule (13.13), are changed to $W_0 = 0$ and $W_1 = 0.5$ and $W_B = 0.5$. The network returns the correct output in response to the final three test patterns but using these weights it predicts an incorrect response, $Y = 1$, when exposed to the first pattern. After two more exposures to the four patterns, the correct weights $W_0 = 0.5$ and $W_1 = 0.5$ and $W_B = 0$ are found. At this point the network is finished learning since $(D - Y) = 0$ for all patterns. Note that if the starting weights are set to $W_0 = W_1 = W_B = 0.2$ the network again requires three exposures to the test patterns to learn the OR operation but this time it returns a different correct set of weights $W_0 = 0.7$ and $W_1 = 0.7$ and $W_B = -0.3$.

The simple network described above admits only a binary classification of the input pattern since only two output states exist, $Y = 0$ or $Y = 1$. More advanced networks can learn arbitrary input/output mappings based on real or complex input and output vectors. Neural networks used in geophysical applications can be trained, for example, on arbitrary $\{\mathbf{m}; \mathbf{u}[\mathbf{m}]\}$ pairs. These represent Earth models $\mathbf{m}$ and their predicted responses $\mathbf{u}[\mathbf{m}]$. The training of a network is called *supervised* when a teacher, who is aware of the physics connecting the input and the output, informs the network what the desired output D should be to a given input vector. The training is *unsupervised* if the network learns the mapping adaptively, by itself, simply from the input/output pairs to which it has been exposed and without regard to any physics that might connect them.

Interpolation based on supervised training is the philosophy behind many of the published NN inversions of geophysical data. The network is first trained on a set of known geophysical responses to known Earth models, as taught by the supervisor. Then, following its exposure to a new geophysical response, the network is prompted to provide its best guess of the underlying, unknown Earth model.

Feed-forward network architecture and back-propagation. A typical network architecture, shown in Figure 13.7, consists of a layer of input units at the bottom, one or more intermediate layers of *hidden* units, and a layer of output units at the top. The units are always connected upwards, in a *feed-forward* configuration, such that information is fed from the bottom to the top of the network. The contents of the units in the hidden layer are not examined explicitly although they do encode important information about the input/output (I/O) relationship.

The goal, as before, is to iteratively adjust the weights of the NN connections so as to minimize the difference between the actual and the desired output vectors. To this end, we describe the popular *back-propagation algorithm* introduced by Rumelhart *et al.* (1986).

Suppose an untrained feed-forward network is assigned the task of learning an inverse geophysical mapping $\mathbf{u}^{-1}[\mathbf{m}]$: $R^D \rightarrow R^P$. An input to the network is a response vector $\mathbf{u} = (u_1, u_2, \ldots, u_D)^T$. The desired output of the network is a model vector

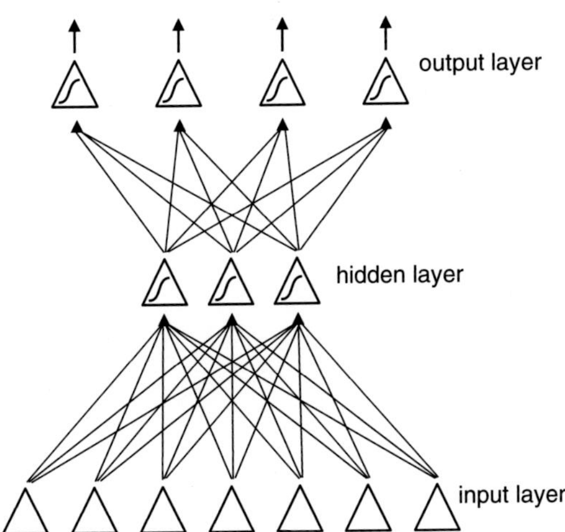

Figure 13.7 Typical feed-forward network architecture. After Smith (2001).

$\mathbf{m} = (m_1, m_2, \ldots, m_P)^T$ that could have generated the response $\mathbf{u}$. The learning is supervised in the sense that the desired output is known in advance; it is determined by the pair $\{\mathbf{m}, \mathbf{u}\}$ that solves the forward-modeling problem. Suppose the output of the untrained or partially trained network (before the weights are optimized) is not $\mathbf{m}$ but rather the vector $\mathbf{y} = (y_1, y_2, \ldots, y_P)^T$. The error in the case that there are Q training pairs $\{\mathbf{m}^{(q)}, \mathbf{u}^{(q)}\}$ for $q = 1,2, \ldots, Q$ may then be defined as

$$E = \frac{1}{2}\sum_{q=1}^{Q}\sum_{j=1}^{P}[y_j^{(q)} - m_j^{(q)}]^2. \tag{13.14}$$

For simplicity in the following formulas, and without losing the essential aspects of the back-propagation algorithm, we hereinafter suppress the first summation over training pairs. The essential idea of the back-propagation algorithm is first to determine the sensitivity $\partial E/\partial W_{ij}$ of the error E in Equation (13.14) with respect to the weights W_{ij} of each network layer, working sequentially downward from the output layer. These sensitivities are then used to update the weights.

A specific input vector $\mathbf{u} = (u_1, u_2, \ldots, u_D)^T$ is presented to the network by setting the D bottom-layer input units to these values, as shown in Figure 13.8. The values of the other units in the network are then determined by a *forward pass* through the network, working upward from the bottom layer. As indicated in the figure, the total input x_i to a given unit i located in the hidden layer is computed according to the rule

$$x_i = \sum_{h=1}^{D} W_{hi}y_h + W_{hb}b, \tag{13.15}$$

where y_h is the output from unit h and W_{hi} is the weight of the connection $h \to i$. The total input x_i to a given unit i has an added *bias* $b = 1.0$, weighted by W_{hb} (not shown in Figure 13.8). Similarly, working further up the network, the total input x_j into unit j of the output layer is given by

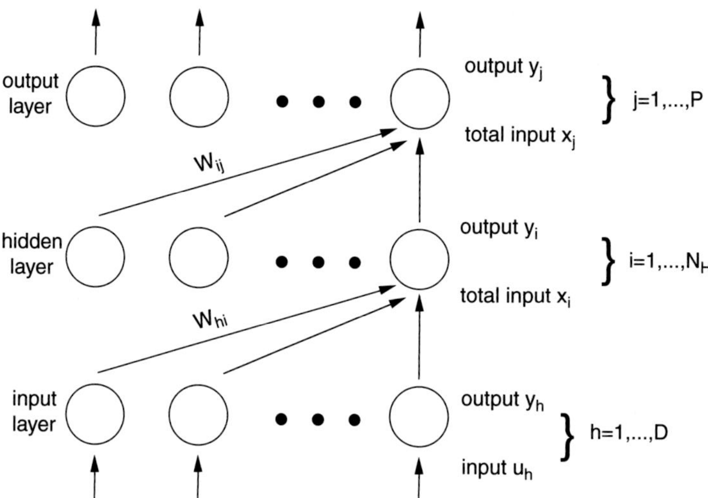

Figure 13.8 Nomenclature for the back-propagation algorithm.

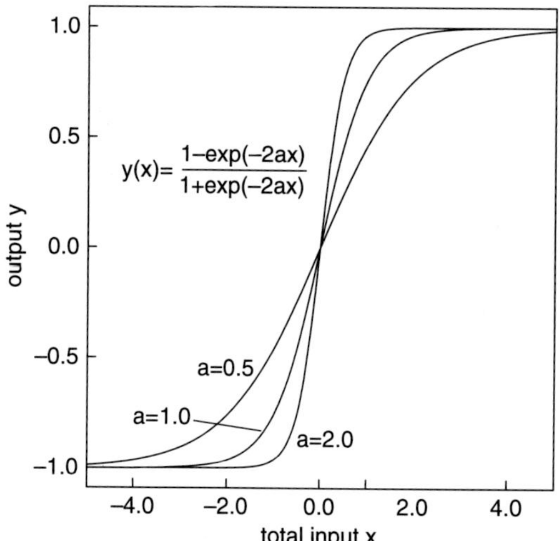

Figure 13.9 The sigmoid, or unit activation, function.

$$x_j = \sum_{i=1}^{N_H} W_{ij} y_i + W_{ib} b, \tag{13.16}$$

where N_H is the number of units in the hidden layer.

The output y of any given unit depends on its total input x. A suitable I/O relationship that describes $y(x)$ is the sigmoid function shown in Figure 13.9; in the present context it is called the *unit activation function*. The family of functions shown in the figure is given by

$$y(x) = \frac{1 - \exp(-2ax)}{1 + \exp(-2ax)}. \tag{13.17}$$

This unit activation function is continuous and bounded, with the properties that $y \rightarrow 1$ as $x \rightarrow +\infty$; $y = 0$ when $x = 0$ and $y \rightarrow -1$ as $x \rightarrow -\infty$. This choice of the unit activation function is not the only one possible; many other functional forms have been successfully applied in NNs. A main requirement is that $y(x)$ is bounded for any value of x; this is necessary to ensure the stability of the back-propagation algorithm.

The goal of network learning is to reduce the total error E. This is achieved, as earlier mentioned, by adjusting the weights of the unit connections. It is desirable, therefore, to analyze how small perturbations to the weights will affect the total error E. Thus we need to compute the partial derivatives $\partial E/\partial w_{ji}$. This is accomplished by a *backward pass* through the network from the top layer to the bottom.

The back-propagation algorithm starts with

$$\frac{\partial E}{\partial y_j} = y_j - m_j, \tag{13.18}$$

by differentiating Equation (13.14) with $Q = 1$, as explained earlier. Then, according to the chain rule we have

$$\frac{\partial E}{\partial x_j} = \frac{\partial E}{\partial y_j}\frac{\partial y_j}{\partial x_j}. \tag{13.19}$$

We use an approximate form of (13.17), namely $y(x) \sim [1 + \exp(-x)]^{-1}$, we can write

$$\frac{\partial y_j}{\partial x_j} \sim \frac{\partial}{\partial x_j}\left\{\frac{1}{1 + \exp(-x_j)}\right\} = \frac{\exp(-x_j)}{[1 + \exp(-x_j)]^2} = y_j(1 - y_j) \tag{13.20}$$

and hence

$$\frac{\partial E}{\partial x_j} = \frac{\partial E}{\partial y_j} y_j(1 - y_j). \tag{13.21}$$

The quantity $\partial E/\partial x_j$ in Equation (13.21) is a measure of how a change in the total input x_j to the output unit j affects the total error E. Now, since x_j is a linear function of the weights of connections from units in the hidden layer, the chain rule gives

$$\frac{\partial E}{\partial W_{ij}} = \frac{\partial E}{\partial x_j}\frac{\partial x_j}{\partial W_{ij}} = \frac{\partial E}{\partial x_j} y_i \tag{13.22}$$

where y_i is the output from the connected unit i in the hidden layer. The expression (13.22) suggests a weight-modification rule of the form

$$\Delta W_{ij} = \varepsilon \frac{\partial E}{\partial W_{ij}} \tag{13.23}$$

where ε is a constant to be determined, perhaps by trial and error. Once the weights have been adjusted according to Equation (13.23), the quantity $\partial E/\partial y_i$ for the hidden layer is computed as

$$\frac{\partial E}{\partial y_i} = \sum_{j=1}^{P} \frac{\partial E}{\partial x_j}\frac{\partial x_j}{\partial y_i} = \sum_{j=1}^{P} \frac{\partial E}{\partial x_j} W_{ij}, \tag{13.24}$$

where $\partial E/\partial x_j$ is given by Equation (13.21). The weights connecting the input layer to the hidden layer are adjusted by repeating the process. This completes the essential ideas of the error back-propagation and weight-updating. To summarize, the back-propagation algorithm is

step 0: run the network on the training set;
step 1: compute $\partial E/\partial y_j = y_j - m_j$ for the top layer;
step 2: compute $\partial E/\partial x_j = (\partial E/\partial y_j)y_j\,(1 - y_j)$ for the top layer;
step 3: compute $\partial E/\partial W_{ij} = (\partial E/\partial x_j)y_i$ where y_i is the output from the next lower layer;
step 4: adjust weights $\Delta W_{ij} = \varepsilon\,\partial E/\partial W_{ij}$ of the connections between the top and the next-lower layers;
step 5: compute $\partial E/\partial y_i$ with Equation (13.24) for the next-lower layer using the newly adjusted weights;
step 6: go back to step 2 and repeat until the bottom of the network;
step 7: re-run the network with the adjusted weights, and repeat the sequence of forward pass (step 0) and backward pass (steps 1–6) until convergence.

An NN inversion of Schlumberger DC resistivity sounding data, using error back-propagation for connection weight adjustment, is described by Calderon-Macias *et al.* (2000). The inverse problem is to determine a layered Earth resistivity model of the form $\rho(z)$ from an observed apparent-resistivity sounding curve $\rho_a(AB/2)$ such as the one shown by the open circles in Figure 13.10, where $AB/2$ is one-half of the spacing between the current injection and withdrawal electrodes. The forward problem has the analytic form derived in Chapter 4. The geology of the area in which the data are acquired is a sedimentary sequence consisting of a thin alluvial cover underlain by alternating sand and clay.

The NN inversion strategy starts by training the network on a number of synthetic $\rho_a(AB/2)$ sounding curves, along with the layered models that generated them. After training is completed, the network is then exposed to the observed $\rho_a(AB/2)$ sounding curve. The corresponding output of the trained network is the solution to the inverse problem.

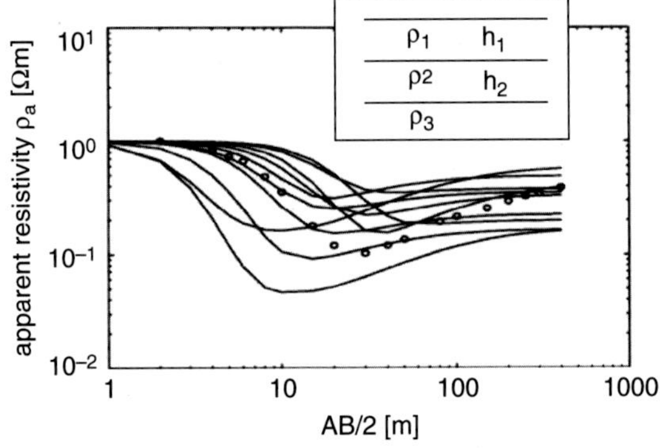

Figure 13.10 Synthetic apparent-resistivity curves generated from a random ensemble of models; the open circles represent the dataset to be inverted. After Calderon-Macias *et al.* (2000).

Table 13.1 Model parameter bounds for neural-network resistivity inversion

Parameter	Minimum value	Maximum value
ρ_1	1.00 S/m	1.00 S/m
ρ_2	0.03 S/m	0.20 S/m
ρ_3	0.15 S/m	0.60 S/m
h_1	1.0 m	10.0 m
h_2	3.0 m	20.0 m

The input vector **u** to the network is a D-dimensional set of computed responses, in the form of apparent-resistivity values

$$\mathbf{u} = (\rho_{a1}, \rho_{a2}, \ldots, \rho_{aD})^T; \tag{13.25}$$

where ρ_{ai} is the apparent resistivity measured at the i-th electrode separation distance. The output vector **m** of length $P = 2p - 1$ consists of the model parameters

$$\mathbf{m} = (\rho_1, \rho_2, \ldots, \rho_p; h_1, h_2, \ldots, h_{p-1})^T \tag{13.26}$$

where ρ_i and h_i are, respectively, the resistivity and thickness of the i-th of p Earth layers. The training set consists of a total of Q pairs of $\{\mathbf{u}, \mathbf{m}\}$ input/output vectors. The synthetic curves $\rho_a(AB/2)$ in Figure 13.10 are generated from a random suite of three-layer models. The model parameters are selected within the bounds shown in Table 13.1, below, with the constraint $\rho_1 > \rho_2$ and $\rho_3 > \rho_2$. Note that the parameter $\rho_1 = 1.00$ S/m is fixed to a value that is representative of the alluvial overburden. The fixing of the upper-layer resistivity is found to help stabilize the network weight adjustments.

A feed-forward network containing one hidden layer with $N_H = 8$ neurons was designed. The network was trained on $Q = 250$ examples of synthetic curves $\rho_a(AB/2)$. The performance of the trained network was then tested on a further sample of 100 synthetic curves.

In general, the training error E was found to increase as the size Q of the training set increases, while the subsequent testing error decreases as Q increases. The reason for this is straightforward (van der Baan and Jutten, 2000). An NN with few training examples (low Q) but a large number of connections is equivalent to an under-determined inverse problem. Such a network fits the training data accurately but it is unable to predict correct answers from new data. In effect, the NN has simply memorized the training examples but cannot generalize. The testing error therefore is large. Conversely, an NN trained on many examples (large Q) does not necessarily fit every one closely, but it should be able to generalize and consequently it has good predictive value. The testing error in that case is small.

After training and testing were successfully completed, the performance of the trained NN was evaluated on three new synthetic data examples and one field example. The results are shown in Figure 13.11a. The dashed sounding curves represent the NN input vectors. The solid lines in the insets show the resistivity models that generated the input vectors

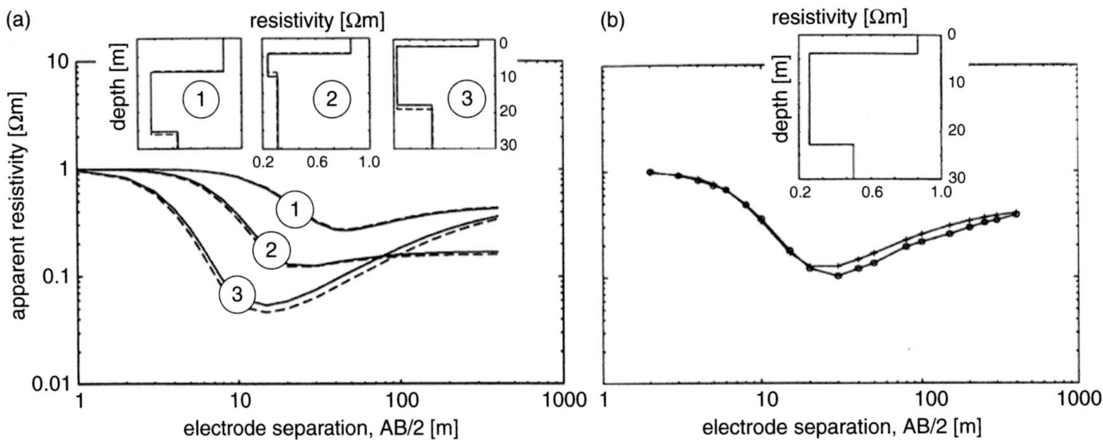

Figure 13.11 (a): Performance of the trained fast-forward NN on new three-layer synthetic curves; (b): neural-network inversion of the field resistivity data. After Calderon-Macias *et al.* (2000).

while the dashed lines in the insets show the NN-output resistivity models. The solid sounding curves are the computed responses of the output models. The NN inversion of the field data is shown in Figure 13.11b. The inset shows the NN-output resistivity model. The open symbols show the field sounding curve while the *crosses* show the sounding curve computed from the NN-output model.

13.5 The self-organizing map (SOM)

As described in the previous section, the feed-forward NN architecture with error back-propagation provides an important framework for *supervised* machine learning. As we have seen, the input–output mapping is determined by iterative adjustment of the unit connection weights. An entirely different network architecture can be envisioned consisting of a 2-D grid of cells, which may or may not explicitly interact with each other. The cell weights configure themselves to become efficient at recognizing specific input signal patterns. The *self-organizing map* (SOM) (Kohonen, 1990) is such an example of an *unsupervised* machine-learning algorithm.

In an elementary SOM, the cells are arrayed as a single grid-like structure, although hierarchical layers of such maps are also possible. Once trained, only a single cell, or a localized group of cells, should respond to a given input signal. An active response might consist, for example, of the turning on of a light at the active cell locations. The locations of the actively responding cells gives information about key defining *features* of the input signal. Ideally, the presence or absence of an active response at a given cell location enables the geophysicist to make a meaningful *interpretation* of the input signal.

The SOM draws its inspiration from studies of the brain. It has long been recognized that different regions of the brain are dedicated to perform different tasks. Early electrophysiological measurements, for example, identified specific regions of the brain that respond to certain input stimuli consisting of electrical impulse trains. Advanced imaging techniques have refined our knowledge of how various areas of the brain respond to complex stimuli such as familiar faces, speech, or handwriting.

A basic concept underlying the training of a simple SOM is *competitive learning*, otherwise known as "winner-take-all." Suppose that a given sequence of input vectors $\{\mathbf{x}\}=\{\mathbf{x}(t_1), \mathbf{x}(t_2), \ldots, \mathbf{x}(t_n)\}$ is presented to the network. This means that, at each time, or *epoch* t_i, the network is exposed a different training vector $\mathbf{x}(t_i)$. Further suppose that there exists a set of reference vectors $\{\mathbf{m}_1, \mathbf{m}_2, \ldots, \mathbf{m}_k\}$ known as the *codebook*. The codebook, which ultimately becomes the key to the pattern recognition, evolves with time t as the network is being trained.

In the training phase, the current input vector $\mathbf{x}(t)$ is compared to each of the reference vectors in the current codebook. The *best-matching* reference vector, say $\mathbf{m}_C$, is found. That vector is then *updated*, according to some pre-defined rule, so that it matches even more closely the input vector $\mathbf{x}(t)$. All of the other codebook vectors are left unchanged.

In this way, after several input vectors $\mathbf{x}(t_1), \mathbf{x}(t_2), \ldots, \mathbf{x}(t_n)$ are presented to the network, the codebook vectors become specifically *tuned* to different classes of input signal. A class is defined as a subset of the input vectors whose members closely resemble each other in some sense. After a large number of input signals are processed, it can be shown that the pdf $p(\mathbf{x})$ of the input signals becomes approximated by the statistical distribution of the codebook vectors. In other words, the set of codebook vectors acquires the same statistical properties as the sequence of input vectors, but with fewer degrees of freedom. In this sense, the SOM can be regarded as a dimensional reduction, data compression, or coding algorithm.

Recall that the index C is associated with the best-matching codebook vector $\mathbf{m}_C$ for the current input vector $\mathbf{x}(t)$. Thus, the following must hold at each epoch:

$$\left\|\mathbf{x} - \mathbf{m}_C\right\| = \min_{i \in [1,k]} \left\|\mathbf{x} - \mathbf{m}_i\right\|. \tag{13.27}$$

The optimal codebook can be defined as the set of vectors $\{\mathbf{m}_1, \mathbf{m}_2, \ldots, \mathbf{m}_k\}$ which minimizes the error measure

$$E = \int \|\mathbf{x} - \mathbf{m}_C\|^2 p(\mathbf{x}) d\mathbf{x}; \tag{13.28}$$

such that each input vector $\mathbf{x}(t)$ has, as far as possible, a closely matching reference vector in the codebook.

The determination of the optimal codebook, i.e. the minimization of the error E, must be performed iteratively. There is no known direct solution to this problem. In a classical technique known as *vector quantization*, the codebook updating formula is

$$\mathbf{m}_C(t+1) = \mathbf{m}_C(t) + \alpha(t)[\mathbf{x}(t) - \mathbf{m}_C(t)]; \tag{13.29}$$

with $\mathbf{m}_i(t+1) = \mathbf{m}_i(t)$ for $i \neq C$. In other words, rule (13.29) states that only the best-matching codebook vector is updated while the others remain unchanged.

In the above equation, the parameter $\alpha(t)$ is a monotonically decreasing function of time with $0 < \alpha(t) < 1$. The function $\alpha(t)$ should decrease at each successive epoch so that, as the training proceeds, the codebook updates become smaller and smaller to reflect the fact that the codebook has already learned many of the essential features of the input sequence. After the updating Equation (13.29) at epoch t is complete, the next input vector $\mathbf{x}(t+1)$ is presented to the network and the codebook is further updated. The trained codebook may be regarded as a set of *feature-sensitive detectors*.

The codebook is related to the SOM topology in a simple way. Consider the simplest type of map, a two-dimensional grid of non-interacting cells. Each cell has its own codebook vector $\mathbf{m}_i$. As described above, an input vector $\mathbf{x}$ is presented to the network and the best-match, or winning, cell C is determined. The codebook, or weight, vector $\mathbf{m}_C$ of cell C is then updated using Equation (13.29), while all other weights are unaffected. In this way, the various cells within the network are somewhat independent of each other. This is the main idea of the winner-take-all strategy.

In a more sophisticated and powerful algorithm, which may be called *shared rewards*, a neighbourhood N_C of cells surrounding the winning cell C can be defined. In this case, all cells within the neighbourhood are updated while the others outside N_C are left unchanged. It has been found experimentally that best results are obtained when N_C is large at the start of the training process and is made smaller as additional input vectors are introduced. After a certain amount of training, the winner-take-all strategy $N_C = C$ can be used.

Example. Unexploded ordnance (UXO) discrimination.

A metal detector based on electromagnetic induction (EM induction, or EMI) principles (see Chapter 8) measures the eddy-current response of buried conductive targets, such as unexploded ordnance (UXO). Detection and proper identification of hazardous UXO is important for remediation of former battlefields or military areas reassigned to civilian use. The main drawback of metal detectors is the high false-alarm rate that increases the cost of clean-up operations.

A main challenge to improve EMI performance is to develop algorithms that *discriminate* between intact UXO, clutter such as scrap metal or ordnance fragments, and geological background (small signals that are measureable but there is no target present). *Supervised* training of feed-forward neural networks, based on a forward model of the UXO as an idealized spheroid or dipole, has achieved a certain level of success. An alternative approach is based on *unsupervised* learning using a SOM.

The goal is to train a computer to recognize the distinct patterns in the EMI responses that arise from different targets. Then, the patterns that arise from UXO are assigned to certain key locations on the map. Clutter and null responses would be assigned to other locations on the map. The geophysicist then looks for an active response to occur at one of the key map locations; this becomes the vital cue that a UXO target has been detected. In essence, the SOM simply provides a method to project the high-dimensional input vectors, namely the EMI responses, into a lower-dimensional space, namely the 2-D grid of cells. This is accomplished by *clustering* input signal patterns that resemble one other. The SOM

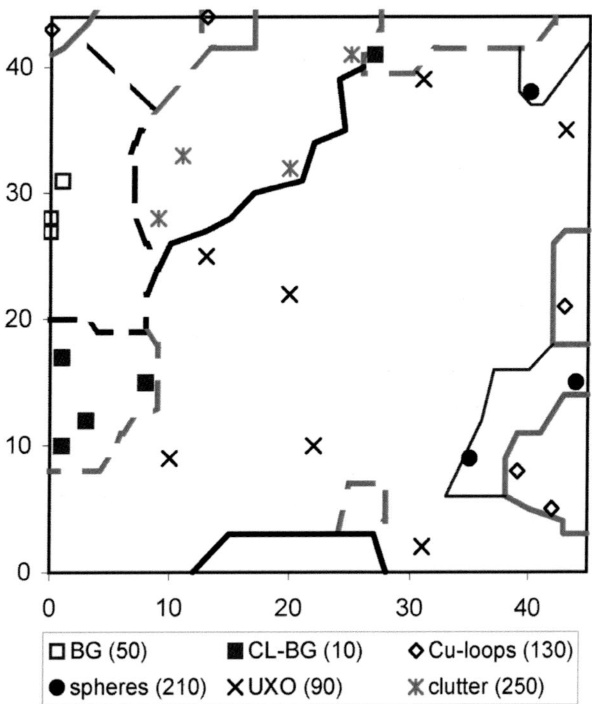

Figure 13.12 The cluster boundaries of a 45 × 45 self-organizing map used as a UXO discriminator based on EMI geophysical responses. BG = background response with no target present; CL-BG = clutter response with background response subtracted. After Benavides *et al.* (2009).

learning algorithm described above, with *neighborhood* N_C, has been developed for UXO discrimination by Benavides *et al.* (2009).

Suppose an input data vector set contains more than one class of signal pattern. For example, EMI responses from UXO, clutter, and background would constitute three different signal classes. A trained SOM algorithm would then cluster these signals into different areas of the map. In this way, the SOM is used as a pattern classifier. Cluster boundaries may be defined on the map. Then, if the active response of an input vector falls within a given cluster boundary, it can be assigned to the corresponding class. The SOM is thereby *partitioned* into UXO, clutter and background regions separated by cluster boundaries.

How are the cluster boundaries determined? Herein, we sketch an elementary method. Further details and more advanced techniques appear in Ultsch (2005). Assign a number U_i for each cell i, characterized by its codebook vector $\mathbf{m}_i$, using the formula

$$U_i = \sum_{j \in N_C} \|\mathbf{m}_i - \mathbf{m}_j\|^2, \tag{13.30}$$

such that the summation is over the cells in the neighborhood N_C of cell i. The quantity U_i is then a measure of how much the codebook vector $\mathbf{m}_i$ varies across the neighborhood N_C.

When the quantity U_i is plotted in map form, it generally presents as a smooth surface containing valleys and ridges. The valleys, corresponding to low U values, contain cells that have similar codebook vectors. The ridges (high U values) contain cells that have distinctive codebook vectors, much different from their neighbors. This basic procedure can be refined (Benavides *et al.*, 2009) to generate a U^* map which further flattens the valleys (cluster regions) and sharpens the ridges (cluster boundaries.) A U^* map corresponding to the classification of UXO-related EMI responses is shown in Figure 13.12. Note that different classes of buried objects generate active responses, as shown by the symbols, in different regions of the map. For example, as shown by the diamond symbols, the copper loops (from which 130 different EMI responses were measured in different test configurations) cluster toward the lower-right boundary of the map.

If a-posteriori labeling of the winning cell $\mathbf{m}_C$ is done in the training stage, the U^* map becomes a useful UXO discriminator. The different cluster regions can be assigned a color based on the class label. Thus, a UXO signal would produce an active response within a region of a specific color, while a clutter signal would activate a different region that has a different color. One advantage of color-coding the different regions of the map is that signal classification can then easily be performed in near-real-time by an unskilled operator while the geophysical data are being acquired.

Problem

1. Write a short computer program that trains a neural network to respond as: (i) an OR logic gate; (ii) an AND logic gate; (iii) a XOR logic gate.

Appendix A Shannon sampling theorem

In this appendix we prove the *Shannon sampling theorem* that states: no information is lost by regular sampling provided that the sampling frequency is greater than twice the highest frequency component in the function $f(t)$ being sampled. Several fundamental properties of Fourier transforms will be required to prove this theorem; these properties will be stated without proof as they are required.

In order to analyze the effects of sampling, consider the *comb function* comb(t) which consists of an infinite set of regularly spaced unit impulse functions (Figure 2.1). The formula for the comb function is

$$\mathrm{comb}(t) = \sum_{n=-\infty}^{+\infty} \delta(t - n\Delta t). \tag{A–1}$$

Suppose we are sampling a band-limited function $f(t)$ whose Fourier spectrum $F(\omega)$ is zero outside the frequency band $|\omega| \geq \omega_N$ (Figure A.1). The sampled function $f_S(t)$ may be regarded as the multiplication of the continuous function $f(t)$ by the comb function,

$$f_S(t) = f(t) \sum_{n=-\infty}^{+\infty} \delta(t - n\Delta t). \tag{A–2}$$

It is necessary to analyze the Fourier spectrum $F_S(\omega)$ of the sampled function $f_S(t)$. Thus, we take a Fourier transform of Equation (A–2),

$$F_S(\omega) = \mathrm{F}\{f_S(t)\} = \mathrm{F}\left\{f(t) \sum_{n=-\infty}^{+\infty} \delta(t - n\Delta t)\right\} \tag{A–3}$$

where F{} denotes the Fourier transform of the argument. Note that a multiplication of two functions in the time domain is equivalent to a convolution in the frequency domain.

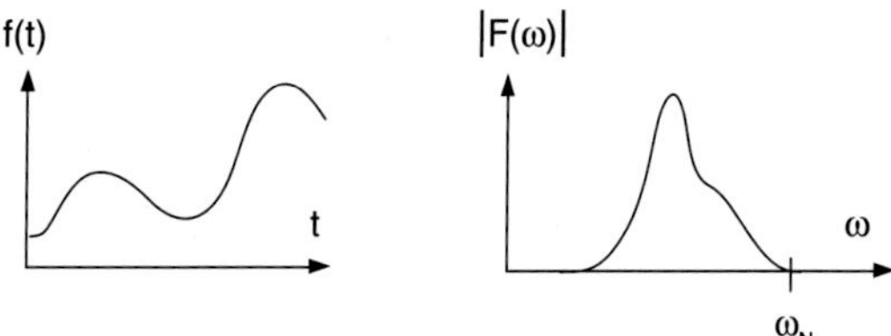

Figure A.1 A band-limited function $f(t)$ and its Fourier amplitude spectrum $|F(\omega)|$.

Specifically, the Fourier transform of an product of two arbitrary signals $f(t)g(t)$ is $\mathrm{F}\{f(t)g(t)\} = F(\omega) * G(\omega)/2\pi$, where $F(\omega)$ and $G(\omega)$ are the Fourier transforms of, respectively, $f(t)$ and $g(t)$. Thus, Equation (A–3) becomes

$$F_S(\omega) = \frac{1}{2\pi}\mathrm{F}\{f(t)\} * \mathrm{F}\left\{\sum_{n=-\infty}^{+\infty} \delta(t - n\Delta t)\right\}. \tag{A–4}$$

Next, we use the fact that the Fourier transform of the comb function is also a comb function. Accordingly, the Fourier transforms in Equation (A–4) reduce to

$$F_S(\omega) = \frac{1}{2\pi}F(\omega) * \frac{2\pi}{\Delta t}\sum_{n=-\infty}^{+\infty} \delta(\omega - n\Delta\omega) = \frac{1}{\Delta t}\sum_{n=-\infty}^{+\infty} F(\omega) * \delta(\omega - n\Delta\omega). \tag{A–5}$$

In Chapter 2 we examined the convolution of two data sequences $\{h\}$ and $\{s\}$. The corresponding formula for the convolution of two continuous functions $h(t)$ and $s(t)$ is

$$h(t) * s(t) = \int_{-\infty}^{\infty} h(\tau)s(t - \tau)d\tau \tag{A–6}$$

Applying the definition in Equation (A–6) to the right-side of Equation (A–5) gives

$$F_S(\omega) = \frac{1}{\Delta t}\sum_{n=-\infty}^{+\infty} \int_{\infty}^{+\infty} F(x)\delta(\omega - n\Delta\omega - x)dx. \tag{A–7}$$

This integration can be done analytically using the integral property of the Dirac delta function,

$$\int_{-\infty}^{+\infty} f(x)\delta(x - x_0)dx = f(x_0). \tag{A–8}$$

The result is

$$F_S(\omega) = \frac{1}{\Delta t}\sum_{n=-\infty}^{+\infty} F(\omega - n\Delta\omega), \tag{A–9}$$

which is equivalent under a shift of the index $n \to -n$ to

$$F_S(\omega) = \frac{1}{\Delta t}\sum_{n=-\infty}^{+\infty} F(\omega + n\Delta\omega) \tag{A–10}$$

Equating (A–9) and (A–10) shows that $F_S(\omega - n\Delta\omega) = F_S(\omega + n\Delta\omega)$, which implies that the spectrum $F_S(\omega)$ of the sampled function is *periodic* with period $\Delta\omega = 2\pi/\Delta t$. A representative sketch of the spectrum $|F_S(\omega)|$ is given in Figure A.2.

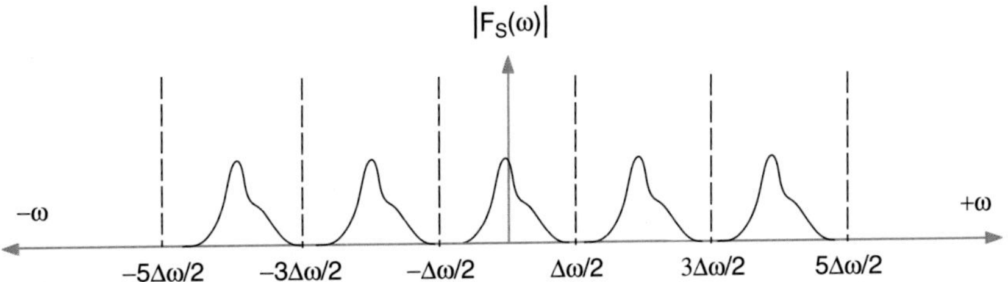

Figure A.2 The Fourier amplitude spectrum $|F_S(\omega)|$ of a regularly sampled signal.

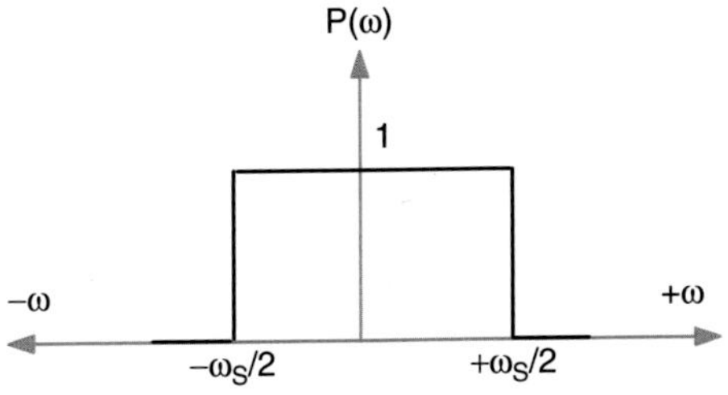

Figure A.3 The unit boxcar function $P(\omega)$.

The function $F_S(\omega)$ consists of periodic repetitions of the spectrum $F_S(\omega)$ of the underlying continuous function $f(t)$. Note that we can write $F(\omega)$ as the product of $F_S(\omega)$ and the *boxcar* function $P(\omega)$, shown in Figure A.3 and defined as

$$P(\omega) = \begin{cases} 1 & \text{if } |\omega| \leq \Delta\omega/2 \\ 0 & \text{otherwise} \end{cases}. \tag{A–11}$$

Thus we have

$$F(\omega) = P(\omega)\Delta t F_S(\omega). \tag{A–12}$$

This is the desired equation which relates the spectra of the original function $f(t)$ and the sampled function $f_S(t)$. Now we can take the inverse Fourier transform of Equation (A–12), which leads to

$$\begin{aligned} f(t) &= \mathrm{F}^{-1}\{F(\omega)\} = \mathrm{F}^{-1}\{P(\omega)\Delta t\, F_S(\omega)\} \\ &= \Delta t\, \mathrm{F}^{-1}\left\{P(\omega)\sum_{n=-\infty}^{+\infty} f(n\Delta t)\exp(-in\omega\Delta t)\right\} \\ &= \Delta t \sum_{n=-\infty}^{+\infty} f(n\Delta t)\mathrm{F}^{-1}\{P(\omega)\exp(-in\omega\Delta t)\}. \end{aligned} \tag{A–13}$$

The last two equations follow from the equation

$$F_S(\omega) = \sum_{n=-\infty}^{+\infty} f(n\Delta t)\mathrm{F}\{\delta(t - n\Delta t)\} \tag{A–14}$$

and the Fourier transform of the Dirac delta function

$$\mathrm{F}\{\delta(t-a)\} = \exp(-i\omega a). \tag{A–15}$$

Finally, we use the *shift property* of the Fourier transform

$$\mathrm{F}^{-1}\{F(\omega)\exp(-i\omega a)\} = f(t-a) \tag{A–16}$$

and the inverse Fourier transform of the boxcar function

$$\mathrm{F}^{-1}\{P(\omega)\} = \frac{\sin(\Delta\omega/2)}{\pi t} \tag{A–17}$$

to get the final result

$$f(t) = \Delta t \sum_{n=-\infty}^{+\infty} f(n\Delta t) \frac{\sin[\Delta\omega(t-n\Delta t)/2]}{\pi(t-n\Delta t)}. \tag{A–18}$$

This equation shows that the band-limited function $f(t)$ is completely determined by its sampled values $f(n\Delta t)$ for $n = -\infty, \ldots, +\infty$.

B Appendix B Solution of Laplace's equation in spherical coordinates

We wish to find the general solution in spherical coordinates (r, θ, ϕ) for Laplace's equation

$$\nabla^2 U = 0, \tag{B–1}$$

for a potential field $U(r, \theta, \phi)$. Expressing (B–1) in spherical coordinates gives

$$\frac{1}{r^2}\frac{\partial}{\partial r}\left(r^2\frac{\partial U}{\partial r}\right) + \frac{1}{r^2 \sin\theta}\frac{\partial}{\partial \theta}\left(\sin\theta\frac{\partial U}{\partial \theta}\right) + \frac{1}{r^2 \sin^2\theta}\frac{\partial^2 U}{\partial \phi^2} = 0. \tag{B–2}$$

This appears to be a difficult equation to solve in terms of elementary functions. Fortunately, it has been studied for hundreds of years and standard mathematical techniques have been developed. The most widely used solution technique is termed *separation of variables* which begins by postulating that the solution $U(r, \theta, \phi)$ can be written in the form

$$U(r, \theta, \phi) = R(r)T(\theta)P(\phi), \tag{B–3}$$

that is, the product of a radial function $R(r)$, an azimuthal function $T(\theta)$, and a polar function $P(\phi)$. The best way to determine whether Laplace's equation can be solved by separation of variables is simply to plug (B–3) into (B–2) and see where it leads.

Doing this, Equation (B–2) becomes

$$\frac{1}{r^2}\frac{\partial}{\partial r}\left(r^2\frac{\partial[RTP]}{\partial r}\right) + \frac{1}{r^2 \sin\theta}\frac{\partial}{\partial \theta}\left(\sin\theta\frac{\partial[RTP]}{\partial \theta}\right) + \frac{1}{r^2 \sin^2\theta}\frac{\partial^2[RTP]}{\partial \phi^2} = 0. \tag{B–4}$$

Dividing through by RTP yields

$$\frac{1}{Rr^2}\frac{\partial}{\partial r}\left(r^2\frac{\partial R}{\partial r}\right) + \frac{1}{Tr^2 \sin\theta}\frac{\partial}{\partial \theta}\left(\sin\theta\frac{\partial T}{\partial \theta}\right) + \frac{1}{Pr^2 \sin^2\theta}\frac{\partial^2 P}{\partial \phi^2} = 0, \tag{B–5}$$

and now multiply through by $r^2 \sin^2\theta$ to get

$$\frac{\sin^2\theta}{R}\frac{\partial}{\partial r}\left(r^2\frac{\partial R}{\partial r}\right) + \frac{\sin\theta}{T}\frac{\partial}{\partial \theta}\left(\sin\theta\frac{\partial T}{\partial \theta}\right) = -\frac{1}{P}\frac{\partial^2 P}{\partial \phi^2}. \tag{B–6}$$

The left side is a function of the variables (r, θ) while the right side is a function of the variable ϕ. The only way for both functions to be equal for all values of (r, θ, ϕ) is for both sides to be equal to a constant, c. Thus, we have separated out two equations,

$$\frac{\sin^2\theta}{R}\frac{\partial}{\partial r}\left(r^2\frac{\partial R}{\partial r}\right)+\frac{\sin\theta}{T}\frac{\partial}{\partial\theta}\left(\sin\theta\frac{\partial T}{\partial\theta}\right)=c; \tag{B–7}$$

$$-\frac{1}{P}\frac{\partial^2 P}{\partial\phi^2}=c. \tag{B–8}$$

Rearranging the last equation results in

$$\frac{\partial^2 P}{\partial\phi^2}+cP=0. \tag{B–9}$$

Since we require the function $U(r,\theta,\phi)$ to be continuous, this means that $P(\phi)$ must be periodic with period 2π, in other words, $P(\phi)=P(\phi+2\pi)$. Functions which have this property are $\sin m\phi$ and $\cos m\phi$, with m an integer. The trigonometric functions $(\sin m\phi, \cos m\phi)$ satisfy the differential equation

$$\frac{\partial^2 P}{\partial\phi^2}+m^2 P=0. \tag{B–10}$$

Thus, comparing Equations (B–9) and (B–10) we see that the separation constant must be of the form $c=m^2$. The general form of the azimuthal function $P(\phi)$ is then a linear combination

$$P(\phi)=\sum_{m=0}^{\infty}a_m\sin m\phi+b_m\cos m\phi. \tag{B–11}$$

The radial/polar equation (B–7) becomes

$$\frac{\sin^2\theta}{R}\frac{\partial}{\partial r}\left(r^2\frac{\partial R}{\partial r}\right)+\frac{\sin\theta}{T}\frac{\partial}{\partial\theta}\left(\sin\theta\frac{\partial T}{\partial\theta}\right)=m^2. \tag{B–12}$$

Divding through by $\sin^2\theta$ and re-arranging, there results

$$\frac{1}{R}\frac{\partial}{\partial r}\left(r^2\frac{\partial R}{\partial r}\right)=-\frac{1}{T\sin\theta}\frac{\partial}{\partial\theta}\left(\sin\theta\frac{\partial T}{\partial\theta}\right)+\frac{m^2}{\sin^2\theta}. \tag{B–13}$$

The left side is a function of the variable r while the right side is a function of the variable θ. The only way for both functions to be equal for all values of (r,θ) is for both sides to be equal to a constant, c. Thus we have two equations,

$$\frac{1}{R}\frac{\partial}{\partial r}\left(r^2\frac{\partial R}{\partial r}\right)=c; \tag{B–14}$$

$$-\frac{1}{T\sin\theta}\frac{\partial}{\partial\theta}\left(\sin\theta\frac{\partial T}{\partial\theta}\right)+\frac{m^2}{\sin^2\theta}=c. \tag{B–15}$$

Rearranging the last equation results in

$$\frac{1}{\sin\theta}\frac{\partial}{\partial\theta}\left(\sin\theta\frac{\partial T}{\partial\theta}\right)-\frac{m^2 T}{\sin^2\theta}+cT=0. \tag{B–16}$$

Differential equations of this type were studied in the nineteenth century by Legendre. For a general non-integer constant c, the equation does not have a simple solution. However, Legendre discovered that the special case

$$\frac{1}{\sin\theta}\frac{\partial}{\partial\theta}\left(\sin\theta\frac{\partial T}{\partial\theta}\right)-\frac{m^2T}{\sin^2\theta}+n(n+1)T=0, \tag{B–17}$$

for particular choice of constant $c = n(n+1)$, with n an integer and $n \geq m$, has simple polynomial solutions denoted as $P_{nm}(\theta)$. These are called the associated Legendre functions. The functions $P_{nm}(\theta)$ are continuous and finite at the poles, $\theta = 0$ and $\theta = \pi$, as expected on physical grounds. Legendre showed that the second independent solution of Equation (B–17), denoted as as $Q_{nm}(\theta)$, is singular at the north pole, $\theta = 0$, and hence unphysical. Thus, the polar equation for our purposes has the general solution

$$T(\theta)=\sum_{n=0}^{\infty}\sum_{m=0}^{n}c_{nm}P_{nm}(\theta). \tag{B–18}$$

We have now separated out the polar and azimuthal equations and found general expressions for $P(\phi)$ and $T(\theta)$. Their product is sometimes called a *surface harmonic expansion* and is given by

$$S(\theta,\phi)=T(\theta)P(\phi)=\sum_{n=0}^{\infty}\sum_{m=0}^{n}(d_{nm}\sin m\phi+e_{nm}\cos m\phi)P_{nm}(\theta). \tag{B–19}$$

It remains now to solve the radial equation

$$\frac{1}{R}\frac{\partial}{\partial r}\left(r^2\frac{\partial R}{\partial r}\right)=c=n(n+1), \tag{B–20}$$

which can be re-arranged to give

$$\frac{\partial}{\partial r}\left(r^2\frac{\partial R}{\partial r}\right)-n(n+1)R=0. \tag{B–21}$$

This equation is readily solved by the functions r^n and r^{-n-1}. Hence, the general solution of the radial equation is

$$R(r)=\sum_{n=0}^{\infty}f_nr^n+g_nr^{-n-1}. \tag{B–22}$$

Now we have found all three functions R, T, and P. The general solution of the Laplace's equation $\nabla^2U = 0$ is their product,

$$U(r,\theta,\phi)=\sum_{n=0}^{\infty}\sum_{m=0}^{n}(f_nr^n+g_nr^{-n-1})(d_{nm}\sin m\phi+e_{nm}\cos m\phi)P_{nm}(\theta). \tag{B–23}$$

Now consider a sphere of radius a. Note that the function r^n blows up as $r \to \infty$ outside the sphere, while the function r^{-n-1} goes to zero. Therefore, outside the sphere we must have $f_n = 0$ to keep the potential finite. Hence,

$$U(r,\theta,\phi) = \sum_{n=0}^{\infty}\sum_{m=0}^{n}\left(\frac{r}{a}\right)^{-n-1}(h_{nm}\sin m\phi + i_{nm}\cos m\phi)P_{nm}(\theta), \text{ for } r \geq a \qquad \text{(B–24)}$$

Similarly, the function r^{-n-1} blows up as $r \rightarrow 0$ inside the sphere while r^n goes to zero. Thus, to keep the potential finite inside the sphere we require $g_n = 0$ so that

$$U(r,\theta,\phi) = \sum_{n=0}^{\infty}\sum_{m=0}^{n}\left(\frac{r}{a}\right)^{n}(j_{nm}\sin m\phi + k_{nm}\cos m\phi)P_{nm}(\theta), \text{ for } r \leq a \qquad \text{(B–25)}$$

In Equations (B–24) and (B–25) we have introduced the normalization $r \rightarrow r/a$. The coefficients $\{h_{nm}, i_{nm}\}$ and $\{j_{nm}, k_{nm}\}$ are called the *spherical harmonic coefficients* of potential U.

C

Appendix C The linear τ–p transformation of seismic data

The linear τ–p transformation of direct, reflected, and refracted seismic waves in a multi-layer medium is derived in this appendix. Consider a seismic wave propagating as a straight ray in a uniform medium characterized by velocity V. Suppose the ray makes an incident angle i with the vertical axis. After a time interval Δt, the distance traveled by the seismic wave is $V\Delta t$. If x is the horizontal coordinate and z is the vertical coordinate, then

$$\Delta x = V\Delta t \sin i; \Delta z = V\Delta t \cos i. \tag{C–1}$$

Re-arrangement of the foregoing equation results in

$$2\Delta t = \frac{\Delta x}{V\sin i} + \frac{\Delta z}{V\cos i}. \tag{C–2}$$

Multiplication of the right side of the above equation by unity, in the form $\sin^2 i + \cos^2 i$, and simplifying, results in

$$\Delta t = \frac{\Delta x \sin i}{V} + \frac{\Delta z \cos i}{V}. \tag{C–3}$$

Recall from the discussion of Snell's law in Chapter 6 that the ray parameter p is constant along the entire ray, regardless of layering. We can identify p as a horizontal slowness [s/m] and q as a vertical slowness [s/m] with the following assignments

$$p = \frac{\sin i}{V}; q = \frac{\cos i}{V}; \tag{C–4}$$

where the path slowness U [s/m] is defined as

$$U = \sqrt{p^2 + q^2} = \frac{1}{V}. \tag{C–5}$$

Note that the traveltime of a ray is equal to the product of the distance traveled times the slowness. Accordingly, with reference to Figure C.1 and Equation (C–3), the traveltime for both the reflected and refracted waves is

$$T = px + 2\sum_{j=1}^{N} q_j h_j, \tag{C–6}$$

where h_j is the thickness of the j-th layer and q_j is the vertical slowness of the j-th layer. The TX–RX separation distance is x. Note that q_j in each layer is different for reflected and refracted waves, since the ray geometry and hence the incident angle i is different for each

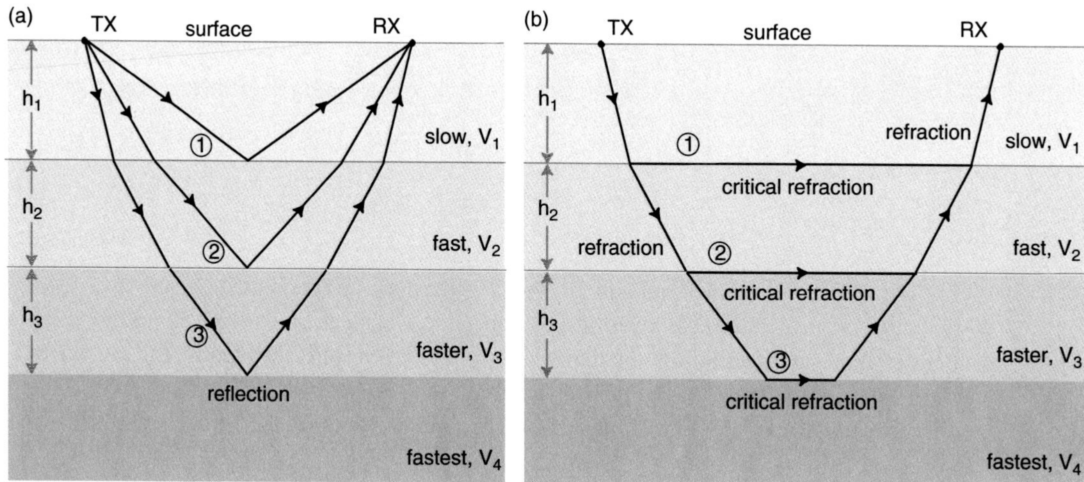

Figure C.1 Ray paths for a three-layer velocity model; (a) reflected waves; (b) refracted waves.

type of wave. The factor 2 in Equation (C–6) is due to the up-going and down-going contributions to the traveltime.

Now we can define the parameter τ as

$$\tau = 2\sum_{j=1}^{N} q_j h_j \tag{C–7}$$

such that $\tau = T - px$. From Equations (C–5) and (C–7), we can see that the contribution to τ from the j-th layer is just

$$\tau_j = 2h_j\sqrt{U_j^2 - p^2}. \tag{C–8}$$

Re-arranging the terms in Equation (C–8) results in the equation of an ellipse in the (τ, p) domain,

$$\frac{\tau_j^2}{4h_j^2 U_j^2} + \frac{p^2}{U_j^2} = 1 \tag{C–9}$$

with semi-axes $2h_jU_j$ and U_j.

Let us now examine how the direct wave and the various reflected and refracted waves in a multi-layer medium are mapped into the (τ, p) domain. The direct wave traveling horizontally from the TX to the RX location has traveltime $T = px$, i.e. the product of horizontal slowness and the TX–RX separation. Since $\tau = T - px$, it follows that the direct wave collapses to the single point D (shown in Figure C.2) with coordinates $(\tau, p) = (0, U_1)$.

It is easy to see from Equation (C–9) that the primary reflection ray path marked as 1 on Figure C.1a maps into an ellipse in the (τ, p) domain with semi-axes $2h_1U_1$ and U_1. This elliptic curve is the top one shown in Figure C.2. The refraction ray path marked 1 in

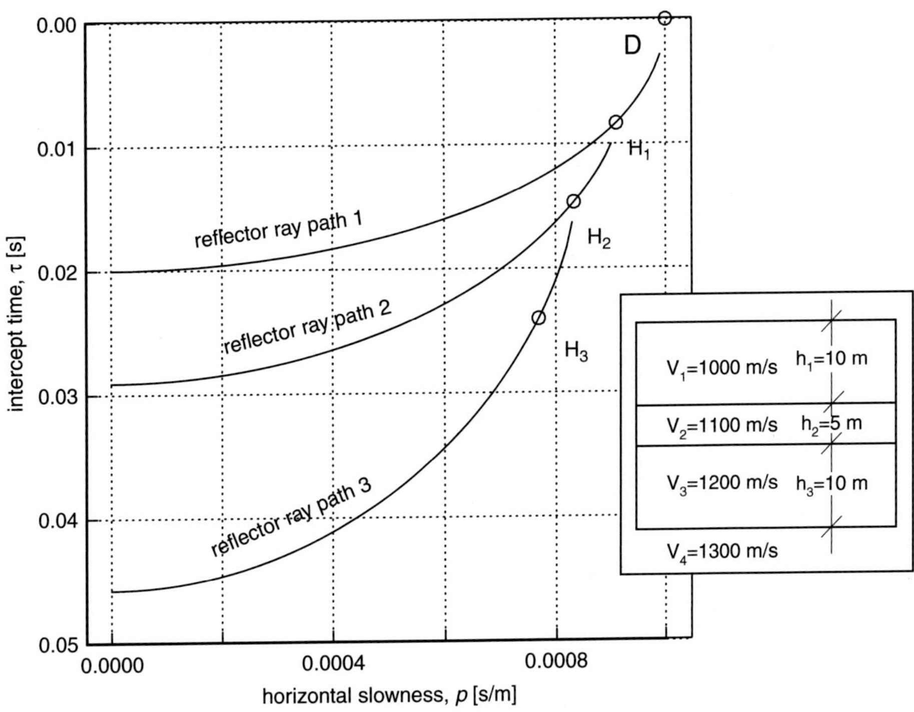

Figure C.2 Reflected (elliptic curves) and direct or refracted (open circles) ray paths, mapped onto the (τ, p) domain for the three-layer velocity model shown in the inset.

Figure C.1b can also be mapped into the (τ, p) domain. Note that the ray parameter $p = U_2$ for this ray path, since the interface ray is horizontal and the wave propagates at the velocity of the underlying medium, which in this case is V_2. The traveltime of this ray path is $T = px + 2q_1h_1$ and hence, using the formula $\tau = T - px$, we find $\tau_1 = 2h_1\sqrt{U_1^2 - U_2^2}$. Thus, refraction ray path 1 maps into the single point located at $(\tau, p) = (2h_1\sqrt{U_1^2 - U_2^2}, U_2)$ and it is marked as H_1 on Figure C.2.

In a similar fashion, the other reflection and refraction ray paths in Figure C.1 can be mapped onto the (τ, p) domain. For example, the elliptic curve associated with reflected ray path 2 is given by $\tau = 2h_1\sqrt{U_1^2 - p^2} + 2h_2\sqrt{U_2^2 - p^2}$. This curve is the middle one plotted in Figure C–2. The refraction ray path 2 maps onto the single point $(\tau, p) = (2h_1\sqrt{U_1^2 - U_3^2} + 2h_2\sqrt{U_2^2 - U_3^2}, U_3)$; this is marked as H_2 on the figure. The reflection and refraction ray paths 3 are mapped onto the (τ, p) domain in a similar fashion.

D Appendix D Horizontal loop over a conducting halfspace

We wish to determine the electromagnetic response of a uniformly conducting Earth of electrical conductivity σ occupying the halfspace $z \geq 0$ (so that $\hat{\mathbf{z}}$ is directed downward) due to its excitation by a harmonic current flow of the form $I(t) = I_0 \exp(i\omega t)$ in a horizontal loop transmitter of radius a situated at height h above the surface $z = 0$.

The symmetry of the problem lends itself to cylindrical coordinates (ρ, φ, z), which are defined relative to the Cartesian coordinates by

$$x = \rho \cos\varphi; \quad y = \rho \sin\varphi; \quad z = z. \tag{D–1}$$

It is readily shown that the inverse coordinate relationships are

$$\rho = \sqrt{x^2 + y^2}; \quad \varphi = \tan^{-1}\frac{y}{x}; \quad z = z. \tag{D–2}$$

An arbitrary vector field $\mathbf{A}(\rho, \varphi, z)$ in cylindrical coordinates has the general representation

$$\mathbf{A} = A_\rho(\rho, \varphi, z)\,\hat{\boldsymbol{\rho}} + A_\varphi(\rho, \varphi, z)\,\hat{\boldsymbol{\varphi}} + A_z(\rho, \varphi, z)\,\hat{\mathbf{z}} \tag{D–3}$$

with the curl of the vector $\mathbf{A}$ being given by

$$\nabla \times \mathbf{A} = \hat{\boldsymbol{\rho}}\left[\frac{1}{\rho}\frac{\partial A_z}{\partial \varphi} - \frac{\partial A_\varphi}{\partial z}\right] + \hat{\boldsymbol{\varphi}}\left[\frac{\partial A_\rho}{\partial z} - \frac{\partial A_z}{\partial \rho}\right] + \hat{\mathbf{z}}\left[\frac{1}{\rho}\frac{\partial}{\partial \rho}(\rho A_\varphi) - \frac{1}{\rho}\frac{\partial A_\rho}{\partial \varphi}\right]. \tag{D–4}$$

Again by the symmetry of the problem, we anticipate that the induced electric currents in the subsurface flow only horizontally. Also, we anticipate that the induced Earth currents should not vary with azimuth, φ. Thus, the induced current density $\mathbf{j}$ reduces to

$$\mathbf{j} = j_\rho(\rho, z)\,\hat{\boldsymbol{\rho}} + j_\varphi(\rho, z)\,\hat{\boldsymbol{\varphi}}. \tag{D–5}$$

Finally, we anticipate that there will be zero radial current flow, $j_\rho = 0$, since that would imply a source or sink of electric current exists somewhere inside the Earth. This would run contrary to the principle of conservation of charge. Thus, the current density is simply

$$\mathbf{j} = j_\varphi(\rho, z)\,\hat{\boldsymbol{\varphi}} \tag{D–6}$$

and since by Ohm's law $\mathbf{j} = \sigma\mathbf{e}$ where $\mathbf{e}$ is the electric field vector, we deduce that the electric-field distribution everywhere inside the Earth has the simple representation

$$\mathbf{e} = e(\rho, z)\,\hat{\boldsymbol{\varphi}}. \tag{D–7}$$

The electric field vector $\mathbf{e}$ satisfies the vector diffusion equation

$$\nabla \times \nabla \times \mathbf{e} + i\mu_0\sigma\omega\,\mathbf{e} = -i\mu_0\omega\mathbf{j}_S, \tag{D–8}$$

in which $\mathbf{j}_S$ [A/m^2] is the transmitter current density.

Inserting the simplified form $\mathbf{e} = e(\rho, z)\,\hat{\boldsymbol{\varphi}}$ into equation (D–8) yields a scalar diffusion equation for the electric field e

$$\frac{\partial}{\partial \rho}\left[\frac{1}{\rho}\frac{\partial}{\partial \rho}(\rho e)\right] + \frac{\partial^2 e}{\partial z^2} + \alpha^2 e = i\mu_0 \omega j_S \tag{D–9}$$

where we have defined

$$\alpha = \sqrt{-i\mu_0 \omega \sigma} \tag{D–10}$$

as the characteristic wavenumber of the conducting halfspace. In obtaining Equation (D–9) we have used the representation $\mathbf{j}_S = j_S\,\hat{\boldsymbol{\varphi}}$ that is appropriate for the horizontal-loop transmitter.

The scalar diffusion equation (D–9) is similar to the diffusion equation governing plane-wave excitation problems, except for the appearance of the first term on the left side. This term reflects the fact that the electric field $\mathbf{e}$ decays with horizontal distance ρ away from the transmitter loop. In the vertically incident plane-wave problem there is zero horizontal dependence ($\partial/\partial\rho \to 0$) to the electromagnetic field, so that the first term is absent. Most of the mathematical techniques we require to solve the loop problem are necessary to handle this horizontal decay of the electric field away from the loop. We shall need to examine the following three concepts:

(i) the Dirac delta function in cylindrical coordinates;
(ii) the source current density $\mathbf{j}_S$ appropriate for a horizontal loop source;
(iii) Hankel transforms that can handle the ρ-dependence of the electric field.

Dirac delta function in cylindrical coordinates

In a Cartesian plane-wave excitation problem, the current source is presumed to be a horizontal sheet of infinite extent located in the air at some height h above Earth's surface $z = 0$. The current sheet is infinitesimal in the vertical $\hat{\mathbf{z}}$ direction and of infinite extent in the two horizontal directions, $\hat{\mathbf{x}}$ and $\hat{\mathbf{y}}$. Thus, the current density can be cast in terms of a one-dimensional delta function $\delta(z+h)$. The horizontal loop source of radius a, by contrast, is infinitesimally thin in two directions $\hat{\boldsymbol{\rho}}$ and $\hat{\mathbf{z}}$, but occupies all values of azimuth φ. We therefore describe the loop source current density $\mathbf{j}_S$ in terms of the delta functions $\delta(\rho - a)$ and $\delta(z + h)$. To see how these delta functions combine we need to look in general at three-dimensional delta functions. Let us start with Cartesian coordinates. A function that is zero everywhere except infinitely large at position $\mathbf{r}_0 = x_0\,\hat{\mathbf{x}} + y_0\,\hat{\mathbf{y}} + z_0\,\hat{\mathbf{z}}$ is denoted by

$$\delta(\mathbf{r} - \mathbf{r}_0) = \delta(x - x_0)\delta(y - y_0)\delta(z - z_0) \tag{D–11}$$

and has SI units [1/m^3], along with the integral property

$$\int_R \delta(\mathbf{r} - \mathbf{r}_0)\,dx\,dy\,dz = 1 \tag{D–12}$$

where R is any region of space containing the point $\mathbf{r}_0$. The function $\delta(\mathbf{r}-\mathbf{r}_0)$ with these properties is the three-dimensional Dirac delta function. We now seek an expression for $\delta(\mathbf{r}-\mathbf{r}_0)$ in cylindrical coordinates. We require that the following condition is upheld:

$$\int_R \delta(\mathbf{r}-\mathbf{r}_0)d\tau = 1, \quad \text{(D–13)}$$

where the volume element in cylindrical coordinates is

$$d\tau = \rho\, d\rho\, d\varphi\, dz. \quad \text{(D–14)}$$

Hence we require that

$$\int_R \delta(\mathbf{r}-\mathbf{r}_0)\,\rho\, d\rho\, d\varphi\, dz = 1 \quad \text{(D–15)}$$

which suggests that the correct expression for $\delta(\mathbf{r}-\mathbf{r}_0)$ in cylindrical coordinates is

$$\delta(\mathbf{r}-\mathbf{r}_0) = \frac{1}{\rho}\delta(\rho-\rho_0)\delta(\varphi-\varphi_0)\delta(z-z_0). \quad \text{(D–16)}$$

Source current density for a horizontal loop

Consider now a current element $Id\mathbf{s}$ of the horizontal circular loop source. The current element is directed azimuthally such that

$$Id\mathbf{s} = Iad\varphi\, \hat{\boldsymbol{\varphi}}. \quad \text{(D–17)}$$

The current density for a hypothetical transmitter that is constructed out of just a small element of the entire circular loop would therefore be

$$d\mathbf{j}_S = Iad\varphi\, \hat{\boldsymbol{\varphi}}\, \delta(r-r_0) \quad \text{(D–18)}$$

where $\mathbf{r}_0 = a\,\hat{\mathbf{r}} + \varphi_0\,\hat{\boldsymbol{\varphi}} - h\hat{z}$. The angle φ_0 represents the azimuth of the small current element. Thus, the Dirac delta function in Equation (D–18) becomes

$$\delta(\mathbf{r}-\mathbf{r}_0) = \frac{1}{\rho}\delta(\rho-a)\delta(\varphi-\varphi_0)\delta(z+h). \quad \text{(D–19)}$$

However, the circular loop consists of the superposition of all the small, azimuthally oriented current elements. Hence, the loop source current density is found by integrating the contributions of each element, namely

$$\mathbf{j}_S = \int_0^{2\pi} \frac{Ia}{\rho} d\varphi\, \hat{\boldsymbol{\varphi}}\, \delta(\rho-a)\delta(\varphi-\varphi_0)\delta(z+h). \quad \text{(D–20)}$$

The integration over azimuth is trivial and results in

$$\mathbf{j}_S = \frac{Ia}{\rho}\, \hat{\boldsymbol{\varphi}}\, \delta(\rho-a)\delta(z+h), \quad \text{(D–21)}$$

which has units [A/m^2], as required. Note that the parameter φ_0 is arbitrary and we have used the identity

$$\int_0^{2\pi} \delta(\varphi - \varphi_0) d\varphi = 1. \tag{D–22}$$

Hankel transforms

Recall that we are trying to solve the scalar diffusion equation (D–9) for the electric field $e(\rho, z)$ in cylindrical coordinates where the source current density is given by Equation (D–21). First, we will assume that the loop is situated in a wholespace of conductivity σ. Later, we will specialize to the case of a conducting halfspace ($z \geq 0$) with the loop source located in the air at height h above the air–Earth interface. We define the Hankel transform $\tilde{e}(\lambda, z)$ of an axisymmetric function $e(\rho, z)$ in cylindrical coordinates as

$$e(\rho, z) = \int_0^\infty \tilde{e}(\lambda, z) J_1(\lambda\rho)\lambda d\lambda. \tag{D–23}$$

We shall also need the Fourier transform $\tilde{\varepsilon}(\lambda, k)$ of the function $\tilde{e}(\lambda, z)$, which is

$$\tilde{e}(\lambda, z) = \frac{1}{2\pi}\int_0^\infty \tilde{\varepsilon}(\lambda, k)\exp(+ikz)dk \tag{D–24}$$

such that the Hankel–Fourier transform now reads

$$e(\rho, z) = \frac{1}{2\pi}\int_{-\infty}^\infty dk \int_0^\infty \lambda d\lambda \tilde{\varepsilon}(\lambda, k)\exp(+ikz)J_1(\lambda\rho). \tag{D–25}$$

We now insert (D–25) into the governing equation (D–9). The resulting equation greatly simplifies since we can make use of the identities

$$\frac{\partial}{\partial\rho}\left[\frac{1}{\rho}\frac{\partial}{\partial\rho}\{\rho J_1(\lambda\rho)\}\right] = -\lambda^2 J_1(\lambda\rho); \tag{D–26}$$

$$\frac{\partial^2}{\partial z^2}\exp(+ikz) = -k^2\exp(+ikz). \tag{D–27}$$

Using equations (D–26) and (D–27), the scalar diffusion equation in the Hankel–Fourier domain becomes

$$(-\lambda^2 - k^2 + \alpha^2)\tilde{\varepsilon}(\lambda, k) = i\mu_0\omega\tilde{\xi}_S(\lambda, k), \tag{D–28}$$

where $\tilde{\xi}_S(\lambda, k)$ is the Hankel–Fourier current density satisfying

$$\begin{aligned} j_S(\rho, z) &= \frac{1}{2\pi}\int_{-\infty}^\infty dk \int_0^\infty \lambda d\lambda \tilde{\xi}_S(\lambda, k)\exp(+ikz)J_1(\lambda\rho) \\ &= \frac{Ia}{\rho}\delta(\rho - a)\delta(z + h). \end{aligned} \tag{D–29}$$

Using the delta function identities

$$\delta(\rho - a) = \rho \int_0^\infty J_1(\lambda a) J_1(\lambda \rho) \lambda d\lambda; \tag{D–30}$$

$$\delta(z + h) = \frac{1}{2\pi} \int_{-\infty}^{\infty} \exp(+ikz) \exp(+ikh) dk, \tag{D–31}$$

we can see that Equation (D–29) can be written as

$$j_S(\rho, z) = \frac{Ia}{2\pi} \int_{-\infty}^{\infty} dk \, \exp(+ikz) \exp(+ikh) \int_0^\infty \lambda d\lambda \, J_1(\lambda a) J_1(\lambda \rho). \tag{D–32}$$

Comparing Equation (D–32) with Equation (D–29) it is clear that

$$\tilde{\xi}_S(\lambda, k) = Ia \, \exp(+ikh) J_1(\lambda a). \tag{D–33}$$

This is the expression for the current density in the Hankel–Fourier domain. Notice that, in this domain, the Dirac delta functions do not appear and are replaced with the well-behaved exponential and Bessel functions. Inserting this equation into (D–28) yields the scalar diffusion equation in the Hankel–Fourier domain

$$(-\lambda^2 - k^2 + \alpha^2) \tilde{\varepsilon}(\lambda, k) = i\mu_0 \omega Ia \, \exp(+ikh) J_1(\lambda a). \tag{D–34}$$

This equation is algebraic and easy to solve; the result is

$$\tilde{\varepsilon}(\lambda, k) = \frac{i\mu_0 \omega Ia \, \exp(+ikh) J_1(\lambda a)}{[\alpha^2 - (\lambda^2 + k^2)]}. \tag{D–35}$$

The ease of solving Equation (D–34) is the principal reason that we work in the Hankel–Fourier domain. Essentially, we have now found the solution to the scalar diffusion equation with a loop-source term in cylindrical coordinates, for a conducting wholespace.

For a loop source in free space, in which $\sigma = 0$ and hence $\alpha = 0$, Equation (D–35) further simplifies to

$$\tilde{\varepsilon}(\lambda, k) = -\frac{i\mu_0 \omega Ia \, \exp(+ikh) J_1(\lambda a)}{\lambda^2 + k^2}. \tag{D–36}$$

In the Hankel domain, expressing the electric field in the vertical spatial domain, we have, by inserting (D–36) into (D–24),

$$\tilde{e}(\lambda, z) = -\frac{i\mu_0 \omega Ia}{2\pi} J_1(\lambda a) \int_{-\infty}^{\infty} \frac{\exp[+ik(z + h)]}{\lambda^2 + k^2} dk. \tag{D–37}$$

The exponential in the integrand is expanded as $\exp(+ik[z + h]) = \cos(k[z + h]) + i\sin(k[z + h])$. The function $\cos(k[z + h])$ is *even* with respect to k whereas the function $\sin(k[z + h])$ is *odd* with respect to k. An odd function integrated over symmetric limits is zero. An even function integrated over symmetric limits is twice the integral over the positive half of the limits, thus

$$\tilde{e}(\lambda,z) = -\frac{i\mu_0\omega I a}{\pi}J_1(\lambda a)\int_0^\infty \frac{\cos[k(z+h)]}{\lambda^2+k^2}dk. \quad \text{(D–38)}$$

The integral in (D–38) is analytic. From tables, we have

$$\tilde{e}(\lambda,z) = -\frac{i\mu_0\omega I a}{\pi}J_1(\lambda a)\left[\frac{\pi}{2\lambda}\exp(-\lambda[z+h])\right], \text{for } z+h \geq 0. \quad \text{(D–39)}$$

Equation (D–39) is valid in free space beneath the loop. It is simplified to

$$\tilde{e}(\lambda,z) = -\frac{i\mu_0\omega I a}{2\lambda}J_1(\lambda a)\exp(-\lambda[z+h]). \quad \text{(D–40)}$$

In the presence of the conducting Earth, there should be a reflection coefficient R along with an "up-going signal" of the form $\exp(\lambda[z-h])$, such that for $z<0$,

$$\tilde{e}(\lambda,z) = -\frac{i\mu_0\omega I a}{2\lambda}J_1(\lambda a)\{\exp(-\lambda[z+h]) + R\exp(\lambda[z-h])\}. \quad \text{(D–41)}$$

As we shall see, the reflection coefficient R is determined by matching the fundamental electromagnetic boundary conditions at the air–Earth interface $z=0$.

In the Earth region $z \geq 0$, there is no source but there is a non-zero conductivity so that the governing scalar diffusion equation there becomes

$$\frac{\partial}{\partial\rho}\left[\frac{1}{\rho}\frac{\partial}{\partial\rho}(\rho e)\right] + \frac{\partial^2 e}{\partial z^2} + \alpha^2 e = 0. \quad \text{(D–42)}$$

This equation is readily solved using a Hankel transform without the need for a Fourier transform. Taking a Hankel transform of Equation (D–42),

$$-\lambda^2\tilde{e}(\lambda,z) + \frac{\partial^2\tilde{e}(\lambda,z)}{\partial z^2} + \alpha^2\tilde{e}(\lambda,z) = 0 \quad \text{(D–43)}$$

$$\frac{\partial^2\tilde{e}(\lambda,z)}{\partial z^2} + [\alpha^2-\lambda^2]\tilde{e}(\lambda,z) = 0. \quad \text{(D–44)}$$

Defining $\gamma^2 = \alpha^2 - \lambda^2$ we have the general solution

$$\tilde{e}(\lambda,z) = A\exp(+i\gamma z) + B\exp(-i\gamma z). \quad \text{(D–45)}$$

Since we expect that the electric field $\tilde{e}(\lambda,z)$ should vanish as $z\to+\infty$, we require $A=0$ so that

$$\tilde{e}(\lambda,z) = B\exp(-i\gamma z), \quad \text{(D–46)}$$

for $z>0$. Equations (D–41) and (D–46) represent two equations in two unknowns, R and B. To find these two unknowns, we impose continuity of the tangential electric- and magnetic-field components at the air–Earth interface.

The continuity of $\tilde{e}(\lambda,z)$ at $z=0$ provides the constraint

$$-\frac{i\mu_0\omega I a}{2\lambda}J_1(\lambda a)(1+R)\exp(-\lambda h) = B \quad \text{(D–47)}$$

while the continuity of $\partial\tilde{e}(\lambda,z)/\partial z$ at $z=0$ provides the other constraint

$$\frac{i\mu_0\omega Ia}{2}J_1(\lambda a)(1-R)\exp(-\lambda h) = -i\gamma B. \tag{D–48}$$

Solving these two equations by eliminating B gives the reflection coefficient

$$R = \frac{\lambda - i\gamma}{\lambda + i\gamma}, \tag{D–49}$$

which the reader can easily verify. This value for R is inserted into the expression for the electric field in Equation (D–41) to obtain the electric field in the Hankel domain in the air region beneath the transmitter loop. We can then insert the resulting expression for $\tilde{e}(\lambda, z)$ from (D–41) into the Hankel transform equation (D–23) to get the sought-after result

$$e(\rho, z) = -\frac{i\mu_0\omega Ia}{2}\int_0^\infty \{e^{-\lambda[z+h]} + Re^{\lambda[z-h]}\}J_1(\lambda a)J_1(\lambda\rho)d\lambda. \tag{D–50}$$

We can now consider special cases, such as the case in which both transmitter and receiver are located on the surface of the Earth. This situation is computed by setting $z = 0$ and $h = 0$ in the above equation. The result is

$$e(\rho) = -\frac{i\mu_0\omega Ia}{2}\int_0^\infty [1 + R]J_1(\lambda a)J_1(\lambda\rho)d\lambda, \tag{D–51}$$

but since $1 + R = 2\lambda/(\lambda + i\gamma)$ we find

$$e(\rho) = -i\mu_0\omega Ia\int_0^\infty \frac{\lambda}{\lambda + i\gamma}J_1(\lambda a)J_1(\lambda\rho)d\lambda. \tag{D–52}$$

It is common for the receiver to be a horizontal coil measuring the vertical magnetic flux. The vertical magnetic field at Earth's surface as a function of horizontal distance ρ from the transmitter loop is therefore of practical importance. From Faraday's law $\nabla \times \mathbf{e} = -i\omega\mathbf{b}$ we find that

$$b_z(\rho) = \frac{i}{\omega\rho}\frac{\partial}{\partial\rho}[\rho e(\rho)], \tag{D–53}$$

which becomes, on insertion of Equation (D–52),

$$b_z(\rho) = \frac{\mu_0 Ia}{\rho}\int_0^\infty \frac{\lambda}{\lambda + i\gamma}J_1(\lambda a)\frac{\partial}{\partial\rho}[\rho J_1(\lambda\rho)]d\lambda. \tag{D–54}$$

The Bessel function identity

$$\frac{1}{\rho}\frac{\partial}{\partial\rho}[\rho J_1(\lambda\rho)] = \lambda J_0(\lambda\rho) \tag{D–55}$$

allows us to write the final result as

$$b_z(\rho) = \mu_0 Ia\int_0^\infty \frac{\lambda^2}{\lambda + i\gamma}J_1(\lambda a)J_0(\lambda\rho)d\lambda. \tag{D–56}$$

E Appendix E Radar TE waveguide mode equations

Consider a simple planar waveguide formed by a low-radar-velocity layer v_1 overlain by a higher-radar-velocity medium v_0 and underlain by a second high-radar-velocity medium v_2. Let the thickness of the low-velocity layer be h and assume that it extends from $z = -h/2$ to $z = +h/2$, with $\hat{\mathbf{z}}$ positive downward. The GPR dipole antenna is assumed to be oriented in the $\hat{\mathbf{x}}$ direction, orthogonal to the direction $\hat{\mathbf{y}}$ of wave propagation such that predominantly the TE mode is energized.

The TE mode electric field in the $\hat{\mathbf{x}}$ (out-of-plane) direction obeys the wave equation

$$\frac{\partial^2 E_x}{\partial y^2} + \frac{\partial^2 E_x}{\partial z^2} - \mu_0 \varepsilon \frac{\partial^2 E_x}{\partial t^2} = 0, \tag{E–1}$$

which in the wavenumber–frequency (k_y, ω) domain, presuming that E_x propagates in the $\hat{\mathbf{y}}$ direction as a plane wave of the form $E_x \sim \exp[i(k_y y - \omega t)]$, becomes

$$\frac{\partial^2 E_x}{\partial z^2} - (k_y^2 - \mu_0 \varepsilon \omega^2) E_x = 0. \tag{E–2}$$

An *even* solution to Equation (E–2) in the low-velocity layer or slab is

$$E_x(k_y, \omega, z) = B \cos(\alpha_1 z + \psi), \quad -\frac{h}{2} < z < +\frac{h}{2}. \tag{E–3}$$

where $\alpha_1^2 = \mu_0 \varepsilon_1 \omega^2 - k_y^2$. The presence of the phase shift ψ accounts for the asymmetry in the $\hat{\mathbf{z}}$-direction of the layered medium. Outside the slab, the electric fields decay exponentially with vertical distance from the interfaces,

$$E_x(k_y, \omega, z) = C \exp(-\alpha_2 z), \quad z > +\frac{h}{2} \tag{E–4}$$

$$E_x(k_y, \omega, z) = D \exp(\alpha_0 z), \quad z < -\frac{h}{2} \tag{E–5}$$

where $\alpha_0^2 = \mu_0 \varepsilon_0 \omega^2 - k_y^2$ and $\alpha_2^2 = \mu_0 \varepsilon_2 \omega^2 - k_y^2$. The unknown constants $\{B, C, D, \psi\}$ are determined by imposing the two fundamental electromagnetic boundary conditions on the two interfaces, that is, continuity of tangential electric- (E_x) and tangential magnetic- $(\sim \partial E_x/\partial z)$ field components. This results in a set of four linear equations in four unknowns.

Eliminating $\{B, C, D\}$ from the four equations results in a pair of equations

$$\frac{\alpha_1}{\alpha_2} \tan\left(\frac{\alpha_1 h}{2} + \psi\right) = 1; \tag{E–6}$$

$$-\frac{\alpha_1}{\alpha_0}\tan\left(-\frac{\alpha_1 h}{2}+\psi\right)=1. \tag{E–7}$$

Recognizing that $\tan x = -\tan(-x)$ and the periodicity $\tan x = \tan(x + m\pi/2)$ where $m = 0, 2, 4, \ldots$ is an arbitrary even integer, we can write

$$-\frac{\alpha_2}{\alpha_1}=\tan\left(-\frac{\alpha_1 h}{2}-\psi+\frac{m\pi}{2}\right); \tag{E–8}$$

$$-\frac{\alpha_0}{\alpha_1}=\tan\left(-\frac{\alpha_1 h}{2}+\psi+\frac{m\pi}{2}\right); \tag{E–9}$$

Taking the inverse tangents of both sides,

$$-\frac{\alpha_1 h}{2}-\psi+\frac{m\pi}{2}=\tan^{-1}\left(-\frac{\alpha_2}{\alpha_1}\right); \tag{E–10}$$

$$-\frac{\alpha_1 h}{2}+\psi+\frac{m\pi}{2}=\tan^{-1}\left(-\frac{\alpha_0}{\alpha_1}\right); \tag{E–11}$$

and adding Equations (E–10) and (E–11) together results in

$$-\alpha_1 h+m\pi=\tan^{-1}\left(-\frac{\alpha_2}{\alpha_1}\right)+\tan^{-1}\left(-\frac{\alpha_0}{\alpha_1}\right). \tag{E–12}$$

Finally using the identity $\tan^{-1}(-x) = -\tan^{-1}x$ we obtain the desired *eigenvalue*, or *mode equation*

$$\alpha_1 h=\tan^{-1}\left(\frac{\alpha_2}{\alpha_1}\right)+\tan^{-1}\left(\frac{\alpha_0}{\alpha_1}\right)+m\pi;\, m=0,2,4,\ldots\text{ (even)}. \tag{E–13}$$

References

Ahmad, J., Schmitt, D. R., Rokosh, C. D., and Pawlowicz, J. G. (2009). High-resolution seismic and resistivity profiling of a buried Quaternary subglacial valley: Northern Alberta, Canada. *Geological Society of America Bulletin*, **121**, 1570–1583.

Ajo-Franklin, J. B., Geller, J. T., and Harris, J. M. (2007). Ultrasonic properties of granular media saturated with DNAPL/water mixtures. *Geophysical Research Letters*, **34**, 2006GL029200.

Aki, K. (1957). Space and time spectra of stationary stochastic waves, with special reference to microtremors. *Bulletin of the Earthquake Research Instititute, University of Tokyo*, **35**, 415–456.

al-Garni, M. and Everett, M. E. (2003). The paradox of anisotropy in electromagnetic loop–loop responses over a uniaxial halfspace. *Geophysics*, **68**, 892–899.

Alnes, H., Eiken, O., and Stenvold, T. (2008). Monitoring gas production and CO_2 injection at the Sleipner field using time-lapse gravimetry. *Geophysics*, **73**, WA155–WA161.

Anderson, W. L. (1979). Computer program: numerical integration of related Hankel transforms of orders 0 and 1 by adaptive digital filtering. *Geophysics*, **44**, 1287–1305.

Andrew, E. R. (2007). Magnetic resonance imaging: a historical overview. In *Encyclopedia of Magnetic Resonance: Historical Perspectives*. New York: John Wiley & Sons.

Annan, A. P. (2009). Electromagnetic principles of ground-penetrating radar. In *Ground Penetrating Radar: Theory and Applications*, ed. J. M. Jol. Amsterdam: Elsevier, pp. 3–40.

Annan, A. P. and Davis, J. L. (1997). Ground penetrating radar – coming of age at last!! In *Proceedings of Exploration 97: 4th Decennial International Conference on Mineral Exploration*, ed. A. G. Gubins. pp. 515–522.

Archie, G. E. (1942). The electrical resistivity log as an aid in determining some reservoir characteristics. *Transactions of the American Institute of Mining, Metallurgical and Petroleum Engineers*, **146**, 54–61.

Arfken, G., Weber, H., and Harris, F.E. (2012). *Mathematical Methods for Physicists*, 7th edn. Waltham, MA: Academic Press.

Arms, R. J. and Hama, F. R. (1965). Localized-induction concept on a curved vortex and motion of an elliptic vortex ring. *Physics of Fluids*, **8**, 553–559.

Asquith G. B. (1995). Determining carbonate pore types from petrophysical logs. In *Carbonate Facies and Sequence Stratigraphy: Practical Applications of Carbonate Models*, ed. P. H. Pause and M. P. Candelaria. Tulsa, OK: Society of Economic and Paleontological Mineralogists – Permian Basin Section, pp. 69–80.

Aster, R. C., Borchers, B., and Thurber, C. H. (2012). *Parameter Estimation and Inverse Problems*, 2nd ed. San Diego, CA: Academic Press.

ASTM. (2011). *Standard Guide for Selecting Surface Geophysical Methods*. ASTM D6429–99.

Auken, E., Pellerin, L., Christensen, N. B., and Sørensen, K. (2006). A survey of current trends in near-surface electrical and electromagnetic methods. *Geophysics*, **71**, G249–G260.

Auken, E., Violette, S., d'Ozouville, N., *et al.* (2009). An integrated study of the hydrogeology of volcanic islands using helicopter borne transient electromagnetic: application in the Galápagos Archipelago. *Comptes Rendus Geoscience*, **341**, 899–907.

Aziz, Z., van Geen, A., Stute, M., *et al.* (2008). Impact of local recharge on arsenic concentrations in shallow aquifers inferred from the electromagnetic conductivity of soils in Araihazar, Bangladesh. *Water Resources Research*, **44**, 2007WR006000.

Baker, G. S. (1999). Processing near-surface seismic-reflection data: a primer. In *Course Notes Series* **9**, ed. R. A. Young. Tulsa, OK, Society of Exploration Geophysicists Press.

Baker, G. S., Steeples, D. W., Schmeissner, C., and Spikes, K. T. (2000). Source-dependent frequency content of ultrashallow seismic reflection data. *Bulletin of the Seismological Society of America*, **90**, 494–499.

Baker, G. S., Strasser, J. C., Evenson, E. B., *et al.* (2003). Near-surface seismic reflection profiling of the Matanuska Glacier, Alaska. *Geophysics*, **68**, 147–156.

Bard, A. J. and Faulkner, L. R. (2001). *Electrochemical Methods: Fundamentals and Applications*, 2nd edition. New York, NY: John Wiley & Sons.

Baumgartner, F. and Christensen, N. B. (1998). Analysis and application of a non-conventional underwater geoelectrical method in Lake Geneva, Switzerland. *Geophysical Prospecting*, **46**, 527–541.

Beamish, D. (2000). Quantitative 2-D VLF data interpretation. *Journal of Applied Geophysics*, **45**, 33–47.

Beamish, D. (2011). Low induction number, ground conductivity meters: a correction procedure in the absence of magnetic effects. *Journal of Applied Geophysics*, **75**, 244–253.

Becker, E. D., Fisk, C. L., and Khetrapal, C. L. (2007). Development of NMR: from the early beginnings to the early 1990s. In *Encyclopedia of Magnetic Resonance: Historical Perspectives*. New York, NY: John Wiley & Sons.

Bellman, R. E. (1957). *Dynamic Programming*. Princeton, NJ: Princeton University Press.

Benavides, A. and Everett, M. E. (2005). Target signal enhancement in near-surface controlled source electromagnetic data. *Geophysics*, **70**, G59–67.

Benavides, A., Everett, M. E., and Pierce, C. (2009). Unexploded ordnance discrimination using time-domain electromagnetic induction and self-organizing maps. *Stochastic Environmental Research and Risk Assessment*, **23**, 169–179.

Berkhout, A. J. (1977). Least-squares inverse filtering and wavelet deconvolution. *Geophysics*, **42**, 1369–1383.

Berkowitz, B. and Scher, H. (2005). Quantification of non-Fickian transport in fractured formations. In *Dynamics of Fluids and Transport in Fractured Rock*, eds. B. Faybishenko, P. A. Witherspoon and J. Bale. AGU Geophysical Monograph **162**, 23–31.

Bexfield, C. E., McBride, J. H., Pugin, A. J. M., *et al.* (2006). Integration of P- and SH-wave high-resolution seismic reflection and micro-gravity techniques to improve interpretation of shallow subsurface structure: new Madrid seismic zone. *Tectonophysics*, **420**, 5–21.

Bhattacharya, P. K. and Patra, H. P. (1968). *Direct Current Geoelectric Sounding*. Amsterdam: Elsevier.

Blakely, R. C. (1995). *Potential Theory in Gravity and Magnetic Applications*. Cambridge: Cambridge University Press.

Bloom, A. L. (1962). Principles of operation of the rubidium vapor magnetometer. *Applied Optics*, **1**, 61–68.

Boadu, F. K., Gyamfi, J., and Owusu, E. (2005). Determining subsurface fracture characteristics from azimuthal resistivity surveys: a case study at Nsawam, Ghana. *Geophysics*, **70**, B35–B42.

Bonnet, E., Bour, O., Odling, N. E., *et al.* (2001). Scaling of fracture systems in geological media. *Reviews of Geophysics*, **39**, 347–383.

Borchers, B., Uram, T., and Henrickx, J. M. H. (1997). Tikhonov regularization of electrical conductivity depth profiles in field soils. *Soil Science Society of America Journal*, **61**, 1004–1009.

Boschetti, F., Dentith, M., and List, R. (1997). Inversion of potential field data by genetic algorithms. *Geophysical Prospecting*, **45**, 461–478.

Boudreault, J. P., Dube, J. S., Chouteau, M., Winiarski, T., and Hardy, E. (2010). Geophysical characterization of contaminated urban fills. *Engineering Geology*, **116**, 196–206.

Bracewell, R. N. (2000). *The Fourier Transform and its Applications*, 3rd edn. New York. NY: McGraw-Hill.

Bradford, J. H. and Deeds, J. C. (2006). Ground-penetrating radar theory and application of thin-bed offset-dependent reflectivity. *Geophysics*, **71**, K47–K57.

Branston, M. W. and Styles, P. (2003). The application of time-lapse micro gravity for the investigation and monitoring of susidence at Northwich, Chesire. *Quarterly Journal of Engineering Geology and Hydrogeology*, **36**, 231–244.

Brigham, E. O. (1988). *The Fast Fourier Transform and its Applications*. Upper Saddle River, NJ: Prentice-Hall.

Budker, D. and Romalis, M. (2007). Optical magnetometry. *Nature Physics*, **3**, 227–234.

Butler, D. K. (2003). Implications of magnetic backgrounds for unexploded ordnance detection, *Journal of Applied Geophysics*, **54**, 111–125.

Butler, D. W. (2005). What is near-surface geophysics? In *Near-Surface Geophysics,* ed. D. W. Butler. Tulsa, OK: Society of Exploration Geophysicists Press, pp. 1–6.

Butler, K. E. (2009). Trends in waterborne electrical and EM induction methods for high resolution sub-bottom imaging. *Near Surface Geophysics*, **7**, 241–246.

Calderon-Macias, C., Sen, M. K., and Stoffa, P. L. (2000). Artificial neural networks for parameter estimation in geophysics. *Geophysical Prospecting*, **48**, 21–47.

Campbell, W. H. (2003). *Introduction to Geomagnetic Fields,* 2nd edn. Cambridge: Cambridge University Press.

Candansayar, M. E. and Tezkan, B. (2008). Two-dimensional joint inversion of radiomagnetotelluric and direct current resistivity data. *Geophysical Prospecting*, **56**, 737–749.

Cassidy, N. J. (2009). Ground penetrating radar data processing, modelling and analysis. In *Ground Penetrating Radar: Theory and Applications*, ed. H. M. Jol. Amsterdam: Elsevier, pp. 141–176.

Castagna, J. P. (1993). AVO analysis – tutorial and review. In *Offset-Dependent Reflectivity – Theory and Practice of AVO Analysis,* ed. J. P. Castagna and M. M. Backus. Tulsa, OK, Society of Exploration Geophysicists, pp. 3–36.

Cavinato, G. P., Di Luzio, E., Moscatelli, M., *et al.* (2006). The new Col di Tenda tunnel between Italy and France: integrated geological investigations and geophysical prospections for preliminary studies on the Italian side. *Engineering Geology*, **88**, 90–109.

Chambers, J. E., Wilkinson, P. B., Kuras, O., *et al.* (2011). Three-dimensional geophysical anatomy of an active landslide in Lias Group mudrocks, Cleveland basin, UK. *Geomorphology*, **125**, 472–484.

Chapman, C. H. (2004). *Fundamentals of Seismic Wave Propagation*. Cambridge: Cambridge University Press.

Chave, A. D. (1983). Numerical integration of related Hankel transforms by quadrature and continued fraction expansion. *Geophysics*, **48**, 1671–1686.

Chave, A. D. and Jones, A. G. (2012). *The Magnetotelluric Method: Theory and Practice*. Cambridge: Cambridge University Press.

Chelidze, T. L. and Gueguen, Y. (1999). Electrical spectroscopy of porous rocks: a review – I. Theoretical models. *Geophysical Journal International*, **137**, 1–15.

Chen, Q. F., Liu, L. B., Wang, W. J., and Rohrbach, E. (2009). Site effects on earthquake ground motion based on microtremor measurements for metropolitain Beijing. *Chinese Science Bulletin*, **54**, 280–287.

Cheo, B. R. S. (1965). A reciprocity theorem for electromagnetic fields with general time dependence. *IEEE Transactions on Antennas and Propagation*, **13**, 278–284.

Chib, S. and Greenberg, E. (1995). Understanding the Metropolis–Hastings algorithm. *American Statistician*, **49**, 327–335.

Ciminale, M. and D. Gallo (2008). High-resolution magnetic survey in a quasi-urban environment. *Near Surface Geophysics*, **9**, 97–103.

Clark, D. A. (1997). Magnetic petrophysics and magnetic petrology: aids to geological interpretation of magnetic surveys. *Journal of Australian Geology and Geophysics*, **17**, 83–103.

Cole, K. S. and Cole, R. H. (1941). Dispersion and absorption in dielectrics. I. Alternating current characteristics. *Journal of Chemical Physics*, **9**, 341–351.

Collins, J. L., Everett, M. E., and Johnson, B. (2006). Detection of near-surface horizontal anisotropy in a weathered metamorphic schist at Llano Uplift (Texas) by transient electromagnetic induction. *Physics of the Earth and Planetary Interiors*, **158**, 159–173.

Conway, B. E. and Barradas, R. G. (1966). *Chemical Physics of Ionic Solutions*, New York, NY: John Wiley & Sons.

Conyers, L. B. (2011). Discovery, mapping and interpretation of buried cultural resources non-invasively with ground-penetrating radar. *Journal of Geophysical Engineering*, **8**, S13–S22.

Cooper, G. R. J. and Cowan D. R., (2011). A generalized derivative operator for potential field data. *Geophysical Prospecting*, **59**, 188–194.

Corrington, M. S. and Kidd, M. C. (1951). Amplitude and phase measurements on loud-speaker cones. *Proceedings of the Institute of Radio Engineers*, **39**, 1021–1026.

Cullity, B. D. and Graham, C. D. (2009). *Introduction to Magnetic Materials*, 2nd edn. Hoboken, NJ: Wiley–IEEE Press.

Dahlin, T. (2001). The development of DC resistivity imaging techniques. *Computers and Geosciences*, **27**, 1019–1029.

Daily, W. and Owen, E. (1991). Cross-borehole resistivity tomography. *Geophysics*, **56**, 1228–1235.

Davis, J. L. and Annan, A. P. (1989). Ground-penetrating radar for high-resolution mapping of soil and rock stratigraphy. *Geophysical Prospecting*, **37**, 531–551.

Day-Lewis, F. D., White, E. A., Johnson, C. D., and Lane, J. W., Jr., (2006). Continuous resistivity profiling to delineate submarine groundwater discharge – examples and limitations. *The Leading Edge*, **25**, 724–728.

Dean, E. B. (1988). Linear least squares for correlated data. Proceedings, 10th Annual Conference of the International Society of Parametric Analysts, Brighton, UK.

Dekkers, M. J. (1997). Environmental magnetism: an introduction. *Geologie en Mijnbouw*, **76**, 163–182.

Delgado, A. V., Gonzalez-Caballero, F., Hunter, R. J., Koopal, L. K., and Lyklema, J. (2005). Measurement and interpretation of electrokinetic phenomena. *Pure and Applied Chemistry*, **77**, 1753–1805.

Deparis, J. and Garambois, S. (2009). On the use of dispersive APVO GPR curves for thin-bed properties estimation: theory and application to fracture characterization. *Geophysics*, **74**, J1–J12.

Deregowski, S. M. (1986). What is DMO? *First Break*, **4**, 7–24.

Dey, A. and Ward, S. H. (1970). Inductive sounding of a layered Earth with a horizontal magnetic dipole. *Geophysics*, **35**, 660–703.

Dias, C. A. (2000). Developments in a model to describe low-frequency electrical polarization of rocks. *Geophysics*, **65**, 437–451.

Diebold, J. B. and Stoffa, P. L. (1981). The traveltime equation, tau–*p* mapping, and inversion of common midpoint data, *Geophysics,* **46**, 238–254.

Docherty, P. and Kappius, R. (1993). A workstation implementation of 3-D refraction statics. *SEG Expanded Abstracts, 63rd Annual Meeting*, Washington, DC, pp. 1166–1169.

Doherty, R., Kulessa, B., Ferguson, A. S., *et al.* (2010). A microbial fuel cell in contaminated ground delineated by electrical self-potential and normalized induced polarization data. *Journal of Geophysical Research*, **115**, 2009JG001131.

Doll, W. E., Miller, R. D., and Bradford, J. (2012). The emergence and future of near-surface geophysics. *The Leading Edge*, **31**, 684–692.

Domenico, P. A. and Schwartz, F. W. (1998). *Physical and Chemical Hydrogeology*, 2nd edition. Chichester: Wiley.

Doyle, P. G. and Snell, J. L. (1984). Random walks and electric networks. *Carus Mathematical Monographs*, **22**.

Dupuis, J. C., Butler, K. E., and Kepic, A. W. (2007). Seismoelectric imaging of the vadose zone of a sand aquifer. *Geophysics*, **72**, A81–A85.

Duque, C., Calvache, M. L., Pedrera, A., Martin-Rosales, W., and López-Chicano, M. (2008). Combined time domain electromagnetic soundings and gravimetry to determine marine intrusion in a detrital coastal aquifer (southern Spain). *Journal of Hydrology*, **349**, 536–547.

Eke, A., Herman, P., Bassingthwaighte, J. B., *et al.* (2000). Physiological time series: distinguishing fractal noises from motions. *Pflugers Archiv European Journal of Physiology*, **439**, 403–415.

Eliseevnin, V. A. (1965). Analysis of waves propagating in an inhomogeneous medium. *Soviet Physics Acoustics*, **10**, 242–245.

Engheta, N., Papas, C. H., and Elachi, C. (1982). Radiation patterns of interfacial dipole antennas. *Radio Science*, **17**, 1557–1566.

Eskola, L., Puranen, R., and Soininen, H. (1999). Measurements of magnetic properties of steel sheets. *Geophysical Prospecting*, **47**, 593–662.

Etgen, J., Gray, S. H., and Zhang, Y. (2009). An overview of depth imaging in exploration geophysics. *Geophysics*, **74**, WCA5–WCA17.

Everett, M. E. (2005). What do electromagnetic induction responses measure? *The Leading Edge*, **24**, 154–157.

Everett, M. E. (2012). Theoretical developments in electromagnetic induction geophysics with selected applications in the near surface. *Surveys in Geophysics*, **33**, 29–63.

Everett, M. E. and Constable, S. (1999). Electric dipole fields over an anisotropic seafloor: theory and application to the structure of 40 Ma Pacific Ocean lithosphere. *Geophysical Journal International*, **136**, 41–56.

Everett, M. E. and Meju, M. A. (2005). Near-surface controlled-source electromagnetic induction: background and recent advances. In *Hydrogeophysics*, ed. Y. Rubin and S. S. Hubbard. New York, NY: Springer, pp. 157–183.

Everett, M. E. and Schultz, A. (1993). Two-dimensional nonlinear magnetotelluric inversion using a genetic algorithm. *Journal of Geomagnetism and Geoelectricity*, **45**, 1013–1026.

Everett, M. E. and Weiss, C. J. (2002). Geological noise in near-surface electromagnetic induction data. *Geophysical Research Letters*, **29**, 2001GL014049.

Everett, M. E., Pierce, C. J., Save, N., *et al.* (2006). Geophysical investigation of the June 6, 1944 D-Day invasion site at Pointe du Hoc, Normandy, France. *Near Surface Geophysics*, **4**, 289–304.

Fassbinder, J. W. E., Stanjek, H., and Vali, H. (1990). Occurrence of magnetic bacteria in soil. *Nature*, **343**, 161–163.

Finizola, A., Revil, A., Rizzo, E., *et al.* (2006). Hydrogeological insights at Stromboli volcano (Italy) from geoelectrical, temperature, and CO_2 soil degassing investigations. *Geophysical Research Letters*, **33**, 2006GL026842.

Finlay, C. C., Maus, S., Beggan, C. D., *et al.* (2010). International Geomagnetic Reference Field: the eleventh generation. *Geophysical Journal International*, **183**, 1216–1230.

Fischer, T. and Guest, A. (2011). Shear and tensile earthquakes caused by fluid injection. *Geophysical Research Letters*, **38**, 2010GL045447.

Fitterman, D. V. and Anderson, W. L. (1987). Effect of transmitter turn-off time on transient soundings. *Geoexploration*, **24**, 131–146.

Fitterman, D. V. and Labson, V. F. (2005). Electromagnetic induction methods for environmental problems. In *Near-Surface Geophysics*, ed. D. K. Butler. Tulsa, OK: Society of Exploration Geophysicists, pp. 301–356.

Fletcher, R. and Reeves, C. M. (1964). Function minimization by conjugate gradients. *Computer Journal*, **7**, 149–154.

Florsch, N., Llubes, M., Tereygeol, F., Ghorbani, A., and Roblet, P. (2011). Quantification of slag heap volumes and masses through the use of induced polarization: application to the Castel-Minier site. *Journal of Archaeological Science*, **38**, 438–451.

Focke, J. W. and Munn, D. (1987). Cementation exponents in Middle Eastern carbonate reservoirs, *Society of Petroleum Engineers Formation Evaluation*, **2**, 155–167.

Fowler, C. M. R. (2005). *The Solid Earth: An Introduction to Global Geophysics*, 2nd edn. Cambridge: Cambridge University Press.

French, A. P. (1971). *Vibrations and Waves*. New York, NY: W.W. Norton and Co.

Frigui, H., Zhang, L., Gader, P., *et al.* (2012). An evaluation of several fusion algorithms for anti-tank landmine detection and discrimination. *Infomation Fusion*, **13**, 161–174.

Frisch, U. and Sornette, D. (1997). Extreme deviations and applications. *Journal de Physique I France*, **7**, 1155–1171.

Frohlich, C., Hayward, C., Stump, B., and Potter, E. (2011). The Dallas–Fort Worth earthquake sequence: October 2008 through May 2009. *Bulletin of the Seismological Society of America*, **101**, 327–340.

Furman, A., Ferre, T. P.A., and Warrick, A. W. (2003). A sensitivity analysis of electrical resistivity tomography array types using analytical element modeling. *Vadose Zone Journal*, **2**, 416–423.

Gaffney, C. F., Gater, J. A., Linford, P., Gaffney, V. L., and White, R. (2000). Large-scale systematic fluxgate gradiometry at the Roman city of Wroxeter. *Archaeological Prospection*, **7**, 81–99.

Gal, D., Dvorkin, J., and Nur, A. (1999). Elastic-wave velocities in sandstones with non-load-bearing clay. *Geophysical Research Letters*, **26**, 939–942.

Gallagher, K., Sambridge, M., and Drijkoningen, G. (1991). Genetic algorithms: an evolution from Monte Carlo methods for strongly non-linear geophysical optimization problems. *Geophysical Research Letters*, **18**, 2177–2180.

Garambois, S. and Dietrich, M. (2001). Seismoelectric wave conversions in porous media: field measurements and transfer function analysis. *Geophysics*, **66**, 1417–1430.

Gardner, L. W. (1967). Refraction seismograph profile interpretation. In *Seismic Refraction Prospecting*, ed. A. W. Musgrave. Tulsa, OK: Society of Exploration Geophysicists Press, pp. 338–347.

Geman, S. and Geman, D. (1984). Stochastic relaxation, Gibbs distribution and the Bayesian restoration of images. *IEEE Transactions on Pattern Analysis and Machine Intelligence*, **6**, 721–741.

Gharibi, M. and Bentley, L. R. (2005), Resolution of 3-D electrical resistivity images from inversions of 2-D orthogonal lines. *Journal of Environmental and Engineering Geophysics*, **10**, 339–349.

Gill, P. E., Murray, W., and Wright, M. H. (1982). *Practical Optimization*. Bingley: Emerald Publishing Group.

Gjoystdal, H., Iversen, E., Lecomte, I., *et al.* (2007). Improved applicabilty of ray tracing in seismic acquisition, imaging, and interpretation. *Geophysics*, **72**, SM261–SM271.

Goldstein, M. A. and Strangway, D. W. (1975). Audio-frequency magnetotellurics with a grounded electric dipole source. *Geophysics*, **40**, 669–683.

Golub, G. H. and van Loan, C. F. (1996). *Matrix Computations*, 3rd edn. Baltimore, MD: Johns Hopkins University Press.

Gorham, P., Saltzberg, D., Odian, A., *et al.* (2002). Measurements of the suitability of large rock salt formations for radio detection of high-energy neutrinos. *Nuclear Instruments and Methods A*, **490**, 476–491.

Goring, D. G. and Nikora, V. I. (2002). Despiking acoustic Doppler velocimeter data. *ASCE Journal of Hydraulic Engineering*, **128**, 117–126.

Gorman, E. M., Everett, M. E., and Johnson, B. (1998). Controlled-source electromagnetic mapping of a faulted sandstone aquifer in central Texas. *Proceedings of SAGEEP*, **11**, 975–984.

Grant, F. S. and West, G. F. (1965). *Interpretation Theory in Applied Geophysics*. New York, NY: McGraw-Hill.

Grauch, V. J. S. (2002). *High Resolution Aeromagnetic Survey to Image Shallow Faults, Dixie Valley Geothermal Field, Nevada*. USGS Open-File Report, 02–0384.

Gray, S. H., Etgen, J., Dellinger, J., and Whitmore, D. (2001). Seismic migration problems and solutions. *Geophysics*, **66**, 1622–1640.

Grotzinger, J. and Jordan, T. H. (2010). *Understanding Earth*, 6th edn. New York, NY: W. H. Freeman.

Gubbins, D. (2004). *Time Series Analysis and Inverse Theory for Geophysicists*. Cambridge: Cambridge University Press.

Gudmundsson, A. (2011). *Rock Fractures in Geological Processes*. Cambridge: Cambridge University Press.

Gueguen, Y. and Palciauskas, V. (1994). VIII. Electrical conductivity. In *Introduction to the Physics of Rocks*. Princeton, NJ: Princeton University Press, pp. 182–211.

Guptasarma, D. and Singh, B. (1997). New digital linear filters for Hankel J_0 and J_1 transforms. *Geophysical Prospecting*, **45**, 745–762.

Gurer, A., Bayrak, M., Gurer, O. F., and Sahin, S. Y. (2008). Delineation of weathering in the Catalca granite quarry with the very low frequency (VLF) electromagnetic method. *Pure and Applied Geophysics*, **165**, 429–441.

Haartsen, M. W. and Pride, S. R. (1997). Electroseismic waves from point sources in layered media. *Journal of Geophysical Research*, **102**, 24745–24769.

Hansen, P. C. (1987). The truncated SVD as a method for regularization. *BIT*, **27**, 534–553.

Hansen, R. O., Racic, L., and Grauch, V. J. S. (2005). Magnetic methods in near-surface geophysics. In *Near-Surface Geophysics*, ed. D. W. Butler. Tulsa, OK: Society of Exploration Geophysicists Press, pp. 151–176.

Harris, F. J. (1978). On the use of windows for harmonic analysis with the discrete Fourier transform. *Proceedings of the Institute of Electrical and Electronics Engineers*, **66**, 51–83.

Haskell, N. A. (1953). The dispersion of surface waves on multilayered media. *Bulletin of the Seismological Society of America*, **43**, 17–34.

Hastings, W. K. (1970). Monte Carlo sampling methods using Markov chains and their applications. *Biometrika*, **57**, 97–109.

Hauser, J., Sambridge, M., and Rawlinson, N. (2008). Multiarrival wavefront tracking and its applications. *Geochemistry, Geophysics, Geosystems*, **9**, 2008GC002069.

Hertrich, M. (2008). Imaging of groundwater with nuclear magnetic resonance. *Progress in Nuclear Magnetic Resonance Spectroscopy*, **53**, 227–248.

Hestenes, M. R. and Stiefel, E. (1952). Methods of conjugate gradients for solving linear systems. *Journal Research National Bureau of Standards*, **49**, 409–436.

Hintz, K. J. (2004). SNR improvements in NIITEK ground penetrating radar. *Proceedings SPIE*, **5415**, 399–408.

Hizem, M., Budan, H., Deville, B., *et al.* (2008). Dielectric dispersion: a new wireline petrophysical measurement, *Society of Petroleum Engineers Technical Papers,* **116130**.

Holland, A. (2011). *Examination of Possibly Induced Seismicity from Hydraulic Fracturing in the Eola Field, Garvin County, Oklahoma*, Oklahoma Geological Survey Open-File Report, OF1–2011.

Holland, J. H. (1975). *Adaptation in Natural and Artificial Systems.* Ann Arbor, MI: University Michigan Press.

Hollender, F. and Tillard, S. (1998). Modeling ground-penetrating radar wave propagation and reflection with the Jonscher parameterization. *Geophysics*, **63**, 1933–1942.

Huang, H. and Palacky, G. J. (1991). Damped least-squares inversion of time-domain airborne EM data based on singular value decomposition. *Geophysical Prospecting*, **39**, 827–844.

Huang, H. and Won, I. J. (2003). Automated anomaly picking from broadband electromagnetic data in an unexploded ordnance (UXO) survey. *Geophysics*, **68**, 1870–1876.

Hughes, L. J. (2009). Mapping contaminant-transport structures in karst bedrock with ground-penetrating radar. *Geophysics*, **74**, B197–B208.

Huntley, D. (1986). Relations between permeability and electrical resistivity in granular aquifers. *Ground Water*, **24**, 466–474.

Hunziker, J., Slob, E., Fan, Y., Snieder, R., and Wapenaar, K. (2012). Two-dimensional controlled-source electromagnetic interferometry by multidimensional deconvolution: spatial sampling aspects, *Geophysical Prospecting*, **60**, 974–994.

Hurlimann, M. P. (2012). Well logging. In *Encyclopedia of Magnetic Resonance.* New York, NY: John Wiley & Sons.

Ingber, L. (1989). Very fast simulated re-annealing. *Mathematical and Computer Modeling*, **12**, 967–973.

Irving, J. D. and Knight, R. J. (2003). Removal of wavelet dispersion from ground-penetrating radar data. *Geophysics*, **68**, 960–970.

Ismail, N. and Pedersen, L. (2011). The electrical conductivity distribution of the Hallandas horst, Sweden: a controlled source radiomagnetotelluric study. *Near Surface Geophysics*, **9**, 45–54.

Jackson, J. D. (1998). *Classical Electrodynamics*, 3rd edn. New York, NY: John Wiley & Sons.

Jacob, T., Bayer, R., Chery, J., and Le Moigne, N. (2010). Time-lapse microgravity surveys reveal water storage heterogeneity of a karst aquifer. *Journal of Geophysical Research*, **115**, 2009JB006616.

Jahns, H. O. (1966). A rapid method for obtaining a two-dimensional reservoir description from well pressure response data. *Society of Petroleum Engineers Journal*, **6**, 315–327.

Jaynes, E. T. (1986). Bayesian methods: general background. In *Maximum Entropy and Bayesian Methods in Applied Statistics*, ed. J. H. Justice. Cambridge: Cambridge University Press, pp. 1–25.

Johnson, W. E. (1932). Probability: the deductive and inductive problems. *Mind*, **41**, 409–423.

Jones, D. S. (1964). *Theory of Electromagnetism*. New York, NY: Macmillan.

Jonscher, A. K. (1977). The univerzal dielectric response. *Nature*, **267**, 673–679.

Jouniaux, L., Maineult, A., Naudet, V., Pessel, M., and Sailhac, P. (2009). Review of self-potential methods in hydrogeophysics. *Comptes Rendus Geoscience*, **341**, 928–936.

Kaiser, A. E., Horstmeyer, H., Green, A. G., *et al.* (2011). Detailed images of the shallow Alpine fault zone, New Zealand, determined from narrow-azimuth 3D seismic reflection data. *Geophysics*, **76**, B19–B32.

Kaiser, G. (1994). *A Friendly Guide to Wavelets*. Cambridge, MA: Birkhauser.

Kanasewich, E. R. (1981). *Time Sequence Analysis in Geophysics*, 3rd edn. Edmonton, Alberta: University of Alberta Press.

Katz, A. J. and Thompson, A. H. (1986). Quantitative prediction of permeability in porous rock. *Physical Review B*, **34**, 8179–8181.

Keating, P. and Sailhac, P. (2004). Use of the analytic signal to identify magnetic anomalies due to kimberlite pipes. *Geophysics*, **69**, 180–190.

Keller, G. V. and Frischknecht, F. C. (1966). *Electrical Methods in Geophysical Prospecting*. New York, NY: Pergamon Press.

Keller, J. B. (1957). Diffraction by an aperture. *Journal of Applied Physics*, **28**, 426–442.

Kirkpatrick, S., Gelatt, C. D., and Vecchi, M. P. (1983). Optimization by simulated annealing. *Science*, **220**, 671–680.

Kittel, C. (2004). *Introduction to Solid State Physics*, 8th edn. New York, NY: John Wiley & Sons.

Klein, J. D. and Sill, W. R. (1982). Electrical properties of artificial clay-bearing sandstone. *Geophysics*, **47**, 1593–1605.

Kleinberg, R. L. (1996). Probing oil wells with NMR. *The Industrial Physicist*, **2**, 18–21.

Knapp, R. W. and Steeples, D. W. (1986). High-resolution common-depth-point seismic reflection profiling; field acquisition parameter design. *Geophysics*, **51**, 283–294.

Knight, R. (2001). Ground penetrating radar for environmental applications. *Annual Review of Earth and Planetary Science*, **29**, 229–255.

Knight, R. J. and Endres, A. L. (2005). An introduction to rock physics principles for near-surface geophysics. In *Near-Surface Geophysics*, ed. D. K. Butler. Tulsa, OK: Society of Exploration Geophysicists Press, pp. 31–70.

Kohonen, T. (1990). The self-organizing map. *Proceedings of the IEEE*, **78**, 1464–1480.

Kowalik, J. S. and Osborne, M. R. (1968). *Methods for Unconstrained Optimization Problems*, Amsterdam: Elsevier.

Krawczyk, C. M., Polom, U., Trabs, S., and Dahm, T. (2011). Sinkholes in the city of Hamburg – new urban shear-wave reflection seismic system enables high-resolution imaging of subrosion structures. *Journal of Applied Geophysics*, **78**, 133–143.

Krohn, C. E. (1984). Geophone ground coupling. *Geophysics*, **49**, 722–731.

Kruschwitz, S., Binley, A., Lesmes, D., and Elshenawy, A. (2010). Textural controls on low-frequency electrical spectra of porous media. *Geophysics*, **75**, WA113–WA123.

Ku, C. C. and Sharp, J. A. (1983). Werner deconvolution for automated magnetic interpretation and its refinement using Marquardt's inverse modeling. *Geophysics*, **48**, 754–774.

Kumar, P. and Foufoula-Georgiou, E. (1997). Wavelet analysis for geophysical applications. *Reviews of Geophysics*, **35**, 385–412.

LaBrecque, D. and Daily, W. (2008). Assessment of measurement errors for galvanic-resistivity electrodes of different composition. *Geophysics*, **73**, F55–F64.

Lamb, H. (1994). *Hydrodynamics*, 6th edn. Cambridge: Cambridge University Press.

Lange, A. L. and Barner, W. L. (1995). Application of the natural electrical field for detecting karst conduits on Guam. In *Karst Geohazards*, ed. B. F. Beck. Rotterdam: Balkema, pp. 425–441.

Langel, R. A. and Estes, R. H. (1982). A geomagnetic field spectrum, *Geophysical Research Letters*, **9**, 250–253.

Le Masne, D. and Vasseur, G. (1981). Electromagnetic field of sources at the surface of a homogeneous conducting halfspace with horizontal anisotropy: applications to fissured media. *Geophysical Prospecting*, **29**, 803–821.

Lee, Y. H. and Shih, Y. X. (2011). Coseismic displacement, bilateral rupture, and structural characteristics at the southern end of the 1999 Chi-Chi earthquake rupture, central Taiwan. *Journal of Geophysical Research*, **116**, 2010JB007760.

Legchenko, A. and Valla, P. (1998). Processing of surface proton magnetic resonance signals using nonlinear fitting. *Journal of Applied Geophysics*, **39**, 77–83.

Leroy, P., Revil, A., Kemna, A., Cosenza, P., and Ghorbani, A. (2008). Complex conductivity of water-saturated packs of glass beads. *Journal of Colloid and Interface Science*, **321**, 103–117.

Lesmes, D. P. and Friedman, S. P. (2005). Relationships between the electrical and hydrogeological properties of rocks and soils. In *Hydrogeophysics*, ed. Y. Rubin and S. S. Hubbard. Dordrecht: Springer, pp. 87–128.

Lesmes, D. P. and Morgan, F. D. (2001). Dielectric spectroscopy of sedimentary rocks. *Journal of Geophysical Research*, **106**, 13329–13346.

Levine, S., Marriott, J. R., Neale, G., and Epstein, N. (1975). Theory of electrokinetic flow in fine cylindrical capillaries at high zeta-potentials. *Journal of Colloid and Interface Science*, **52**, 136–149.

Lindley, D. V. (1956). On a measure of the information provided by an experiment. *The Annals of Mathematical Statistics*, **27**, 986–1005.

Lindsey, J. P. (1989). The Fresnel zone and its interpretive significance. *The Leading Edge*, **8**, 33–39.

Liner, C. L. (1999). Concepts of normal and dip moveout. *Geophysics*, **64**, 1637–1647.

Lines, L. R. and Treitel, S. (1984). A review of least-squares inversion and its application to geophysical problems. *Geophysical Prospecting*, **32**, 159–186.

Linford, N. (2006). The application of geophysical methods to archaeological prospection. *Reports on Progress in Physics*, **69**, 2205–2257.

Loewenthal, D., Lu, L., Roberson, R., and Sherwood, J. (1976). The wave equation applied to migration. *Geophysical Prospecting*, **24**, 380–399.

Loke, M. H. (1999). *Electrical Imaging Surveys for Environmental and Engineering Studies: A Practical Guide to 2-D and 3-D Surveys*. Available at www.geometrics.com.

Louie, J. N. (2001). Faster, better: shear-wave velocity to 100 meters depth from refraction microtremor arrays. *Bulletin of the Seismological Society of America.*, **91**, 347–364.

Lucia, F. J. (1983). Petrophysical parameters estimated from visual descriptions of carbonate rocks: a field classification of carbonate pore space. *Journal of Petroleum Technology*, **35**, 629–637.

MacDonald, A. M., Burleigh, J., and Burgess, W. G. (1999). Estimating transmissivity from surface resistivity soundings: an example from the Thames Gravels. *Quarterly Journal of Engineering Geology*, **32**, 199–205.

MacDonald, J., Knopman, D., Clancy, N., Mc Ever, J., and Willis, H. (2004). *Transferring Army BRAC Lands Containing Unexploded Ordnance: Lessons Learned and Future Options*. Santa Monica, CA: The RAND Corporation.

Mackay, D. J. C. (1992). Bayesian interpolation. *Neural Computation*, **4**, 415–447.

Malinverno, A. (2002). Parsimonious Bayesian Markov chain Monte Carlo inversion in a nonlinear geophysical problem. *Geophysical Journal International*, **151**, 675–688.

Mandelbrot, B. B. and van Ness, J. W. (1968). Fractional Brownian motions, fractional noises and applications. *SIAM Review*, **10**, 422–437.

Maraschini, M., Ernst, F., Foti, S., and Socco, L. V. (2010). A new misfit function for multimodal inversion of surface waves. *Geophysics*, **75**, G31–G43.

Marion, D., Nur, A., Yin, H., and Han, D. (1992). Compressional velocity and porosity in sand–clay mixtures. *Geophysics*, **57**, 554–563.

Marquardt, D. W. (1963). An algorithm for least-squares estimation of nonlinear parameters. *Journal of the Society for Industrial and Applied Mathematics*, **11**, 431–441.

Martí, D., Carbonell, R., Flecha, I., *et al.* (2008). High-resolution seismic characterization in an urban area: Subway tunnel construction in Barcelona, Spain. *Geophysics*, **73**, B41–B50.

Martin, J. S., Larson, G. D., and Scott, Jr. W. R. (2006). An investigation of surface-contacting sensors for the seismic detection of buried landmines. *Journal of the Acoustical Society of America*, **120**, 2676–2685.

Mauri, G., Williams-Jones, G., and Saracco, G. (2010). Depth determinations of shallow hydrothermal systems by self-potential and multi-scale wavelet tomography. *Journal of Volcanology and Geothermal Research*, **191**, 233–244.

McGillivray, P. R., Oldenburg, D. W., Ellis, R. G., and Habashy, T. M. (1994). Calculation of sensitivities for the frequency-domain electromagnetic problem. *Geophysical Journal International*, **116**, 1–4.

McNeill, J. D. (1980). *Electromagnetic Terrain Conductivity Measurement at Low Induction Numbers*. Mississauga, Ontario: Geonics Ltd, Technical Note TN-6.

McNeill, J. D. (1990). Use of electromagnetic methods for groundwater studies. In *Geotechnical and Environmental Geophysics*, ed. S. H. Ward. Tulsa, OK: Society of Exploration Geophysicists, pp. 191–218.

McNeill, J. D. and Labson, V. F. (1991). Geological mapping using VLF radio fields. In *Electromagnetic Methods in Applied Geophysics*, vol. 2, ed. M. N. Nabighian. Tulsa, OK: Society of Exploration Geophysicists, pp. 521–640.

Menke, W. (1984). *Geophysical Data Analysis: Discrete Inverse Theory.* San Diego, CA: Academic Press.

Metropolis, N., Rosenbluth, A.W., Rosenbluth, M. N., Teller, A. H., and Teller, E. (1953). Equation of state calculations by fast computing machines. *Journal of Chemical Physics*, **21**, 1087–1092.

Metwaly, M. (2007). Detection of metallic and plastic landmines using the GPR and 2-D resistivity techniques. *Natural Hazards and Earth Systems Science*, **7**, 755–763.

Metz, C. E. (1978). Basic principles of ROC analysis. *Seminars in Nuclear Medicine*, **8**, 283–298.

Meurant, G. (2006). *The Lanczos and Conjugate Gradient Algorithms*. Philadelphia, PA: SIAM.

Miller, G. F. and Pursey, H. (1955). On the partition of energy between elastic waves in a semi-infinite solid. *Proceedings of the Royal Society of London A*, **233**, 55–69.

Minsley, B. J., Sogade, J., and Morgan, F. D. (2007). Three-dimensional source inversion of self-potential data. *Journal of Geophysical Research*, **112**, 2006JB004262.

Moreau, F., Gibert, D., Holschneider, M., and Saracco, G. (1997). Wavelet analysis of potential fields. *Inverse Problems*, **13**, 165–178.

Moreau, F., Gibert, D., Holschneider, M., and Saracco, G. (1999). Identification of sources of potential fields with the continuous wavelet transform: basic theory. *Journal of Geophysical Research*, **104**, 5003–5013.

Morrison, H. F., Phillips, R. J., and O'Brien, D. P. (1969). Quantitative interpretation of transient electromagnetic fields over a layered half space. *Geophysical Prospecting*, **17**, 82–101.

Motulsky, H. J. and Ransnas, L. A. (1987). Fitting curves to data using nonlinear regression: a practical and nonmathematical review. *FASEB Journal*, **1**, 365–374.

Muller, T. M., Gurevich, B., and Lebedev, M. (2010). Seismic wave attenuation and dispersion resulting from wave-induced flow in porous rocks – a review. *Geophysics*, **75**, 75A147–75A164.

Muller-Petke, M. and Yaramanci, U. (2010). QT inversion-comprehensive use of the complete surface NMR data set. *Geophysics*, **75**, WA199–WA209.

Muller-Petke, M., Dlugosch, R., and Yaramanci, U. (2011). Evaluation of surface nuclear magnetic resonance-estimated subsurface water content. *New Journal of Physics*, **13**, 095002.

Nabighian, M. N. (ed.) (1988). *Electromagnetic Methods in Applied Geophysics, Volume 1, Theory.* Tulsa, OK: Society of Exploration Geophysicists.

Nabighian, M. N. (ed.) (1991). *Electromagnetic Methods in Applied Geophysics, Volume 2, Application*. Tulsa, OK: Society of Exploration Geophysicists.

Nabighian, M. N. and Macnae, J. C. (1991). Time domain electromagnetic prospecting methods. In *Electromagnetic Methods in Applied Geophysics, Volume 2*, ed. M. N. Nabighian. Tulsa, OK: Society of Exploration Geophysicists, pp. 427–450.

Nabighian, M. N., Grauch, V. J. S., Hansen, R. O., *et al.* (2005). The historical development of the magnetic method in exploration. *Geophysics*, **70**, 33ND–61ND.

Nadler, A., Dasberg, S., and Lapid, I. (1991). Time-domain reflectometry measurements of water content and electrical conductivity of layered soil columns. *Soil Science Society of America Journal*, **55**, 938–943.

Nazarian, S. and Stokoe II, K. H. (1986). Use of surface waves in pavement evaluation. *Transportation Research Record*, **1070**, 132–144.

Nazarian, S., Stokoe II, K. H., and Hudson, W. R. (1983). Use of spectral analysis of surface waves method for determination of moduli and thicknesses of pavement systems. *Transportation Research Record*, **930**, 38–45.

NCHRP. (2006). *Use of Geophysics for Transportation Projects*. NCHRP Synthesis **357**. Washington, DC: Transportation Research Board.

Neal, A. (2004). Ground-penetrating radar and its use in sedimentology: principles, problems and progress. *Earth-Science Reviews*, **66**, 261–330.

Nicolson, H., Curtis, A., Baptie, B., and Galetti, E. (2012). Seismic interferometry and ambient noise tomography in the British Isles. *Proceedings of the Geologists' Association*, **123**, 74–86.

Nobes, D. C. (1996). Troubled waters: environmental applications of electrical and electomagnetic methods. *Surveys in Geophysics*, **17**, 393–454.

Nocedal, J. and Wright, S. J. (2006). *Numerical Optimization*, 2nd edn. Berlin: Springer.

Noutchogwe, C. T., Koumetio, F., and Manguelle-Dicoum, E. (2010). Structural features of South-Adamawa (Cameroon) inferred from magnetic anomalies: Hydrogeological implications. *Comptes Rendus Geoscience*, **342**, 467–474.

Olhoeft, G. R. (1985). Low-frequency electrical properties. *Geophysics*, **50**, 2492–2503.

Olhoeft, G. R. (1986). Direct detection of hydrocarbon and organic chemicals with ground penetrating radar and complex resistivity. In *Proceedings of the NWWA/API Conference on Petroleum Hydrocarbons and Organic Chemicals in Ground Water*. Worthington, OH: National Water Well Association, pp. 284–305.

Paillet, F. L. and Ellefsen, K. J. (2005). Downhole applications of geophysics. In *Near-Surface Geophysics*, ed. D. W. Butler. Tulsa, OK: Society of Exploration Geophysicists Press, pp. 439–471.

Paine, J. G. (2003). Determining salinization extent, identifying salinity sources, and estimating chloride mass using surface, borehole, and airborne electromagnetic induction methods. *Water Resources Research*, **39**, 2001WR000710.

Painter, S. (1996). Evidence for non-Gaussian scaling behavior in heterogeneous sedimentary formations. *Water Resources Research*, **32**, 1183–1195.

Pape, H., Riepe, L., and Schopper, J. R. (1987). Interlayer conductivity of rocks – a fractal model of interface irregularities for calculating interlayer conductivity of natural porous mineral systems. *Colloids and Surfaces*, **27**, 97–122.

Parasnis, D. S. (1997). *Principles of Applied Geophysics*, 5th edn. London: Chapman and Hall.

Park, C. B., Miller, R. D., and Xia, J. (1999). Multichannel analysis of surface waves. *Geophysics*, **64**, 800–808.

Park, C. B., Miller, R. D., and Xia, J. (1998). Imaging dispersion curves of surface waves on multi-channel record. *SEG Expanded Abstracts*, 1377–1380.

Parker, R. L. (1994). *Geophysical Inverse Theory.* Princeton, NJ: Princeton University Press.

Parsons, R. (1990). Electrical double layer: recent experimental and theoretical developments. *Chemical Reviews*, **90**, 813–826.

Pasion, L. R. (2007). Inversion of time domain electromagnetic data for the detection of unexploded ordnance. PhD Thesis, University of British Columbia.

Passalacqua, H. (1983). Electromagnetic fields due to a thin resistive layer. *Geophysical Prospecting*, **31**, 945–976.

Passaro, S. (2010). Marine electrical resistivity tomography for shipwreck detection in very shallow water: a case study from Agropoli (Salerno, southern Italy). *Journal of Archaeological Science*, **37**, 1989–1998.

Paul, M. K. (1965). Direct interpretation of self-potential anomalies caused by inclined sheets of infinite horizontal extensions. *Geophysics*, **30**, 418–423.

Pellerin, L. (2002). Applications of electrical and electromagnetic methods for environmental and geotechnical investigations. *Surveys in Geophysics*, **23**, 101–132.

Pelton, W. H., Ward, S. H., Hallof, P. G., Sill, W. R., and Nelson, P. H. (1978). Mineral discrimination and removal of inductive coupling with multifrequency IP. *Geophysics*, **43**, 588–609.

Petiau, G. (2000). Second generation of lead–lead chloride electrodes for geophysical applications. *Pure and Applied Geophysics*, **157**, 357–382.

Phillips, R. J., Zuber M. T., Smrekar, S. E, Mellon, M. T., *et al.* (2008). Mars north polar deposits: stratigraphy, age and geodynamical response. *Science*, **320**, 1182–1185.

Poddar, M. (1983). A rectangular loop source of current on multilayered Earth. *Geophysics*, **48**, 107–109.

Polak, E. (1971). *Computational Methods in Optimization*. San Diego, CA: Academic Press.

Pride, S. R. (1994). Governing equations for the coupled electromagnetics and acoustics of porous media. *Physical Review B*, **50**, 15678–15696.

Pride, S. R., Berryman, J. G., and Harris, J. M. (2004). Seismic attenuation due to wave-induced flow. *Journal of Geophysical Research*, **109**, 2003JB002639.

Radzevicius, S. J. and Daniels, J. J. (2000). Ground penetrating radar polarization and scattering from cylinders. *Journal of Applied Geophysics*, **45**, 111–125.

Ravat, D. (1996). Magnetic properties of unrusted steel drums from laboratory and field-magnetic measurements. *Geophysics*, **61**, 1325–1357.

Reid, A. B., Allsop, J. M., Granser, H., Millett, A. J., and Somerton, I. W. (1990). Magnetic interpretation in three dimensions using Euler deconvolution. *Geophysics*, **55**, 80–91.

Revil, A., Karaoulis, M., Johnson, T., and Kemna, A. (2012). Some low-frequency electrical methods for subsurface characterization and monitoring in hydrogeology. *Hydrogeology Journal*, **20**, 617–658.

Reynolds, J. M. (2011). *An Introduction to Applied and Environmental Geophysics*. New York, NY: Wiley-Blackwell.

Rice, C. L. and Whitehead, R. (1965). Electrokinetic flow in a narrow cylindrical capillary. *Journal of Physical Chemistry*, **69**, 4017–4024.

Richart, F. E., Hall, J. R., and Woods, R. D. (1970). *Vibrations of Soils and Foundations*. Englewood Cliffs, NJ: Prentice-Hall Inc.

Robinson, D. A. and Friedman, S. P. (2001). Effect of particle size distribution on the effective dielectric permittivity of saturated granular media. *Water Resources Research*, **37**, 2000WR900227.

Rodi, W. and Mackie, R. L. (2001). Nonlinear conjugate gradients algorithm for 2-D magnetotelluric inversion. *Geophysics*, **66**, 174–187.

Roest, W. R., and Pilkington, M. (1993). Identifying remanent magnetization effects in magnetic data. *Geophysics*, **58**, 653–659.

Romdhane, A., Grandjean, G., Brossier, R., *et al.* (2011). Shallow-structure characterization by 2D elastic full-waveform inversion. *Geophysics*, **76**, R81–R93.

Rosenblad, B. L., Bailey, J., Csontos, R., and van Arsdale, R. (2010). Shear wave velocities of Mississippi embayment soils from low frequency surface wave measurements. *Soil Dynamics and Earthquake Engineering*, **30**, 691–701.

Rucker, D. F., Loke, M. H., Levitt, M. T., and Noonan, G. E. (2010). Electrical-resistivity characterization of an industrial site using long electrodes. *Geophysics*, **75**, WA95–WA104.

Rumelhart, D. E., Hinton, G. E., and Williams, R. J. (1986). Learning representations by back-propagating errors. *Nature*, **323**, 533–536.

Ryden, N. and Park, C. B. (2006). Fast simulated annealing inversion of surface waves on pavement using phase-velocity spectra. *Geophysics*, **71**, R49–R58.

Ryu, J., Morrison, H. F., and Ward, S. H. (1970). Electromagnetic fields about a loop source of current. *Geophysics*, **35**, 862–896.

Saccorotti, G., Chouet, B., and Dawson, P. (2003). Shallow-velocity models at the Kilauea volcano, Hawaii, determined from array analyses of tremor wavefields. *Geophysical Journal International*, **152**, 633–648.

Santamarina, J. C., Rinaldi, V. A., Fratta, D., *et al.* (2005). A survey of elastic and electromagnetic properties of near-surface soils. In *Near-Surface Geophysics*, ed. D. K. Butler. Tulsa, OK: Society of Exploration Geophysicists Press, pp. 71–88.

Sassen, D. S. (2009). GPR methods for the detection and characterization of fractures and karst features: polarimetry, attribute extraction, inverse modeling and data mining techniques. Ph.D dissertation, Texas A&M University.

Sassen, D. S. and Everett, M. E. (2005). Multi-component ground penetrating radar for improved imaging and target discrimination. *Proceedings of SAGEEP*, **18**, 11–20.

Sassen, D. S. and Everett, M. E. (2009). 3D Polarimetric GPR coherency attributes and full-waveform inversion of transmission data for characterizing of fractured rock. *Geophysics*, **74**, J23–J34.

Sato, M. and Mooney, H. M. (1960). The electrochemical mechanism of sulphide self-potentials. *Geophysics*, **25**, 226–249.

Sato, M. and Takahashi, K. (2012). ALIS deployment in Cambodia. *Proceedings SPIE*, **8357**, 83571A.

Scales, J. A. (1987). Tomographic inversion via the conjugate gradient method. *Geophysics*, **52**, 179–185.

Scales, J. A. and Snieder, R. (1997). To Bayes or not to Bayes? *Geophysics*, **62**, 1045–1046.

Schuster, G. T. (2010). *Seismic Interferometry.* Cambridge: Cambridge University Press.

Schwarzbach, C., Boerner, R. U., and Spitzer, K. (2005). Two-dimensional inversion of direct current resistivity data using a parallel, multi-objective genetic algorithm. *Geophysical Journal International*, **162**, 685–695.

Schwindt, P. D. D., Knappe, S., Shah, V., *et al.* (2004). Chip-scale atomic magnetometer. *Applied Physics Letters*, **85**, 6409–6411.

Seber, G. A. F. and Wild, C. J. (2003). *Nonlinear Regression*. Hoboken, NJ: Wiley–Interscience.

Segall, P. (1989). Earthquakes triggered by fluid extraction. *Geology*, **17**, 942–946.

Segall, P., Grasso, J. R., and Mossop, A. (1994). Poroelastic stressing and induced seismicity near the Lacq gas field, southwestern France. *Journal of Geophysical Research*, **99**, 15423–15438.

Sein, J. J. (1982). Derivation of the laws of reflection and refraction by Huygens' construction. *American Journal of Physics*, **50**, 180–181.

Selvan, K. T. (2009). A revisiting of scientific and philosophical perspectives on Maxwell's displacement current. *IEEE Antennas and Propagation Magazine*, **51**, 36–46.

Sen, M. K. and Stoffa, P. L. (2013). *Global Optimization Methods in Geophysical Inversion,* 2nd edn. Amsterdam: Elsevier.

Shafer, G. (1976). *A Mathematical Theory of Evidence*. Princeton, NJ: Princeton University Press.

Shapiro, N. M., Campillo, M., Stehly, L., and Ritzwoller, M. H. (2005). High-resolution surface wave tomography from ambient seismic noise. *Science*, **307**, 1615–1618.

Sharma, P. V. (1997). *Environmental and Engineering Geophysics*. Cambridge: Cambridge University Press.

Shuey, R. T. (1985). A simplification of the Zoeppritz equations. *Geophysics*, **50**, 609–614.

Simpson, F. and Bahr, K. (2005). *Practical Magnetotellurics*. Cambridge: Cambridge University Press.

Slater, L. (2007). Near surface electrical characterization of hydraulic conductivity: from petrophysical properties to aquifer geometries – a review. *Surveys in Geophysics*, **28**, 169–197.

Slater, L., Knight, R., Singha, K., Binley, A., and Atekwana, E. (2006a). Near-surface geophysics: A new focus group. *Eos*, **87**, 249–250.

Slater, L., Ntarlagiannis, D., and Wishart, D. (2006b). On the relationship between induced polarization and surface area in metal–sand and clay–sand mixtures. *Geophysics*, **71**, A1–A5.

Slichter, C. P. (1996). *Principles of Magnetic Resonance*, 3rd edn. Berlin: Springer.

Slob, E. and Wapenaar, K. (2008). Practical representations of electromagnetic interferometry for GPR applications: a tutorial. *Near Surface Geophysics*, **6**, 391–402.

Smith, L. (2001). An introduction to neural networks. See http://www.cs.stir.ac.uk/~lss/NNIntro/InvSlides.html.

Smythe, W. R. (1950). *Static and Dynamic Electricity*, 2nd edn. New York, NY: McGraw-Hill.

Socco, L. V. and Strobbia, C. (2004). Surface-wave method for near-surface characterization: a tutorial. *Near-Surface Geophysics*, **4**, 165–185.

Socco, L. V., Foti, S., and Boiero, D. (2010). Surface-wave analysis for building near-surface velocity models – established approaches and new perspectives. *Geophysics*, **75**, 75A83–75A102.

Sogade, J. A., Scira-Scappuzzo, F., Vichabian, Y., *et al.* (2006). Induced-polarization detection and mapping of contaminant plumes. *Geophysics*, **71**, B75–B84.

Song, L. P. and Zhang, S. Y. (1999). Singular value decomposition-based reconstruction algorithm for seismic traveltime tomography. *IEEE Transactions on Image Processing*, **8**, 1152–1154.

Soupios, P. M., Kouli, M., Vallianatos, F., Vafidis, A., and Stavroulakis, G. (2007). Estimation of aquifer hydraulic parameters from surficial geophysical methods: a case study of Keritis basin in Chania (Crete – Greece). *Journal of Hydrology*, **338**, 122–131.

Spies, B. R. and Frischknecht, F. C. (1991). Electromagnetic sounding. In *Electromagnetic Methods in Applied Geophysics, Volume 2*, ed. M. N. Nabighian. Tulsa, OK: Society of Exploration Geophysicists, pp. 285–425.

Spitzer, R., Nitsche, F. O., and Green, A. G. (2001). Reducing source-generated noise in shallow seismic data using linear and hyperbolic τ–p transformations. *Geophysics*, **66**, 1612–1621.

Spitzer, R., Nitsche, F. O., Green, A. G., and Horstmeyer, H. (2003). Efficient acquisition, processing, and interpretation strategy for shallow 3D seismic surveying. *Geophysics*, **68**, 1792–1806.

Stein, S. and M. Wysession (2003). *An Introduction to Seismology, Earthquakes, and Earth Structure*. Chichester: Wiley-Blackwell.

Stratton, J. A. (1941). *Electromagnetic Theory*. New York, NY: McGraw-Hill.

Streich, R. (2007). Accurate 3-D vector-imaging of ground-penetrating radar data based on exact-field radiation patterns. PhD thesis, ETH Zurich (Swiss Federal Institute of Technology).

Streich, R. and van der Kruk, J. (2007). Accurate imaging of multicomponent GPR data based on exact radiation patterns. *IEEE Transactions on Geoscience and Remote Sensing*, **45**, 93–103.

Streich, R., van der Kruk, J., and Green, A. G. (2007). Vector-migration of standard copolarized 3D GPR data. *Geophysics*, **72**, J65–J75.

Strobbia, C. and Cassiani, G. (2007). Multilayer ground-penetrating radar guided waves in shallow soil layers for estimating soil water content. *Geophysics*, **72**, J17–J29.

Stummer, P., Maurer, H., and Green, A. G. (2004). Experimental design: electrical resistivity data sets that provide optimum subsurface information. *Geophysics*, **69**, 120–139.

Szu, H. and Hartley, R. (1987). Fast simulated annealing. *Physics Letters A*, **122**, 157–162.

Tarantola, A. (2004). *Inverse Problem Theory and Methods for Model Parameter Estimation*. Philadelphia, PA: Society for Industrial and Applied Mathematics.

Taylor, G. I. (1958). On the dissipation of eddies. In *The Scientific Papers of Sir Geoffrey Ingram Taylor, Volume 2*, ed. G. K. Batchelor. Cambridge: Cambridge University Press, pp. 96–101.

Telford, W. M., Geldart, L. P., and Sheriff, R. E. (1990). *Applied Geophysics*, 2nd edn. Cambridge: Cambridge University Press.

Tezkan, B. (1999). A review of environmental applications of quasi-stationary electromagnetic techniques. *Surveys in Geophysics*, **20**, 279–308.

Thompson, A. H. and Gist, G. A. (1993). Geophysical applications of electrokinetic conversion. *The Leading Edge*, **12**, 1169–1173.

Thompson, D. T. (1982). EULDPH – a new technique for making computer-assisted depth estimates from magnetic data. *Geophysics*, **47**, 31–37.

Thomson, W. T. (1950). Transmission of elastic waves through a stratified solid medium. *Journal of Applied Physics*, **21**, 89–93.

Tipler, P. A. and Mosca, G. (2007). *Physics for Scientists and Engineers*, 6th edn. New York: W. H. Freeman.

Tite, M. S. and Mullins, C. E. (1971). Enhancement of the magnetic susceptibility of soils on archaeological sites. *Archaeometry*, **13**, 209–219.

Titov, K., Komarov, V., Tarasov, V., and Levitski, A. (2002). Theoretical and experimental study of time domain-induced polarization in water-saturated sands. *Journal of Applied Geophysics*, **50**, 417–433.

Toole, F. E. (1986). Loudspeaker measurements and their relationship to listener preferences. *Journal of the Audio Engineering Society*, **34**, 227–235; 323–348.

Topp, G. C., Davis, J. L., and Annan, A. P. (1980). Electromagnetic determination of soil water content: Measurements in coaxial transmission lines. *Water Resources Research*, **16**, 574–582.

Tsui, F. and Matthews, S. L. (1997). Analytic modelling of the dielectric properties of concrete for subsurface radar applications. *Construction and Building Materials*, **11**, 149–161.

Ulrych, T. J., Sacchi, M. D., and Woodbury, A. (2001). A Bayes tour of inversion: a tutorial. *Geophysics*, **66**, 55–69.

Ultsch, A. (2005). Clustering with SOM: U*C. Proceedings, Workshop on Self-Organizing Maps, Paris, pp.75–82.

Unsworth, M. J., Lu, X., and Watts, M. D. (2000). CSAMT exploration at Sellafield: characterization of a potential radioactive waste disposal site. *Geophysics*, **65**, 1070–1079.

Vallee, M. A., Keating, P., Smith, R. S., and St-Hilaire, C. (2004). Estimating depth and model type using the continuous wavelet transform of magnetic data. *Geophysics*, **69**, 191–199.

van Dam, R. L., Nichol, S. L., Augustinus, P. C., *et al.* (2003). GPR stratigraphy of a large active dune on Parengarenga Sandspit, New Zealand. *The Leading Edge*, **22**, 865–881.

van Dam R. L., Harrison, J. B. J., Hendrickx, J. M. H., *et al.* (2005). Mineralogy of magnetic soils at a UXO remediation site in Kaho'olawe Hawaii. Proceedings of the Symposium on the Application of Geophysics to Engineering and Environmental Problems 2005, Atlanta, GA.

van der Baan, M. and Jutten, C. (2000). Neural networks in geophysical applications. *Geophysics*, **65**, 1032–1047.

van der Kruk, J. (2001). Three-dimensional imaging of multi-component ground penetrating radar data. PhD thesis, Technische Universiteit, Delft.

van der Kruk, J., Vereecken, H., and Jacob, R. W. (2009). Identifying dispersive GPR signals and inverting for surface wave-guide properties. *The Leading Edge*, **28**, 1234–1239.

van Eck, T., Goutbeek, F., Haak, H., and Dost, B. (2006). Seismic hazard due to small magnitude, shallow-source, induced earthquakes in The Netherlands. *Engineering Geology*, **87**, 105–121.

Vanhala, H., Soininen, H., and Kukkonen, I. (1992). Detecting organic chemical contaminants by spectral-induced polarization method in glacial till environment. *Geophysics*, **57**, 1014–1017.

Varian, R. (1962). Ground liquid prospecting method and apparatus. *US Patent* **3019383**.

Viezzoli, A., Christiansen, A. V., Auken, E., and Sorensen, K. (2008). Quasi-3D modeling of airborne TEM data by spatially constrained inversion. *Geophysics*, **73**, F105–F113.

Vinegar, H. J. and Waxman, M. H. (1984). Induced polarization of shaly sands. *Geophysics*, **49**, 1267–1287.

von Hippel, A. R. (1954). *Dielectrics and Waves*. New York, NY: John Wiley & Sons.

von Toussaint, U. (2011). Bayesian inference in physics. *Reviews of Modern Physics*, **83**, 943–999.

Wadsworth, G. P., Robinson, E. A., Bryan, J. G., and Hurley, P. M. (1953). Detection of reflections on seismic records by linear operators. *Geophysics*, **18**, 539–586.

Wait, J. R. (1954). Mutual coupling of loops lying on the ground. *Geophysics*, **19**, 290–296.

Wang, B. (2006). 2D and 3D potential-field upward continuation using splines. *Geophysical Prospecting*, **54**, 199–209.

Wang, P. and Horwitz, M. H. (2007). Erosional and depositional characteristics of regional overwash deposits caused by multiple hurricanes. *Sedimentology*, **54**, 545–564.

Wangsness, R. K. (1986). *Electromagnetic Fields*, 2nd edn. New York, NY: John Wiley & Sons.

Wannamaker, P. E. (1997). Tensor CSAMT survey over the Sulphur Springs thermal area, Valles Caldera, New Mexico, USA. Part II. Implications for CSAMT methodology. *Geophysics*, **62**, 466–476.

Ward, S. H. and Hohmann, G. W. (1988). Electromagnetic theory for geophysical applications. In *Electromagnetic Methods in Applied Geophysics, Volume 1*, ed. M. N. Nabighian. Tulsa, OK: Society of Exploration Geophysicists, pp. 130–311.

Wasscher, J. D. (1961) Note on four-point resistivity measurements on anisotropic conductors. *Philips Research Reports*, **16**, 301–306.

Watson, K. A. and Barker, R. D. (1999). Differentiating anisotropy and lateral effects using azimuthal resistivity offset Wenner soundings. *Geophysics*, **64**, 739–745.

Waxman, M. H. and Smits, L. J. M. (1968). Electrical conductivities in oil-bearing shaly sands. *Society of Petroleum Engineers Journal*, **8**, 107–122.

Weichman P. B., Lavely, E. M., and Ritzwoller, M. H. (2000). Theory of surface nuclear magnetic resonance with applications to geophysical imaging problems. *Physical Review E*, **62**, 1290–1312.

West, G. F. and Macnae, J. C. (1991). Physics of the electromagnetic induction exploration method. In *Electromagnetic Methods in Applied Geophysics, Volume 2*, ed. M. N. Nabighian. Tulsa, OK: Society of Exploration Geophysicists, pp. 5–46.

West, L. J., Handley, K., Huang, Y., and Pokar, M. (2003). Radar frequency dielectric dispersion in sandstone: implications for determination of moisture and clay content. *Water Resources Research*, **39**, 2001WR000923.

Widess, M. B. (1973). How thin is a thin bed? *Geophysics*, **38**, 1176–1180.

Williams, K. H., Kemna, A., Wilkins, M. J., *et al.* (2009). Geophysical monitoring of coupled microbial and geochemical processes during stimulated subsurface bioremediation. *Environmental Science & Technology*, **43**, 6717–6723.

Worthington, P. F. (1993). The uses and abuses of the Archie equations, I: the formation factor–porosity relationship. *Journal of Applied Geophysics*, **30**, 215–228.

Wyllie, M. R. J., Gregory, A. R., and Gardner, G. H. F. (1958). An experimental investigation of factors affecting elastic wave velocities in porous media. *Geophysics*, **23**, 459–493.

Xu, P. (1998). Truncated SVD methods for discrete linear ill-posed problems. *Geophysical Journal International*, **135**, 505–514.

Yao, H., Beghein, C., and van der Hilst, R. D. (2008). Surface wave array tomography in SE Tibet from ambient seismic noise and two-station analysis – II. Crustal and upper mantle structure. *Geophysical Journal International*, **173**, 205–219.

Yilmaz, O. (2001). *Seismic Data Analysis: Processing, Inversion and Interpretation of Seismic Data*. SEG Investigations in Geophysics Series, vol. **10**. Tulsa, OK: Society of Exploration Geophysicists Press.

Yu, L. and Edwards, R. N. (1992). The detection of lateral anisotropy of the ocean floor by electromagnetic methods. *Geophysical Journal International*, **108**, 433–441.

Zelt, C. A. and Barton, P. J. (1998). Three-dimensional seismic refraction tomography: a comparison of two methods applied to data from the Faroe basin. *Journal of Geophysical Research*, **103**, 7187–7210.

Zelt, C. A., Azaria, A., and Levander, A. (2006). 3D seismic refraction traveltime tomography at a groundwater contamination site. *Geophysics*, **71**, H67–H78.

Zeng, H. (2009). How thin is a thin bed? An alternative perspective. *The Leading Edge*, **28**, 1192–1197.

Zhou, W., Beck, B. F., and Stephenson, J. B. (1999). Investigation of groundwater flow in karst areas using component separation of natural potential measurements. *Environmental Geology*, **37**, 19–25.

Zonge, K., Wynn, J., and Urquhart, S. (2005). Resistivity, induced polarization and complex resistivity. In *Near-Surface Geophysics*, ed. D. K. Butler. Tulsa, OK: Society of Exploration Geophysicists Press, pp. 265–300.

Zonge, K. L. and Hughes, L. J. (1991). Controlled source audio-frequency magnetotellurics. In *Electromagnetic Methods in Applied Geophysics, Volume 2*, ed. M. N. Nabighian. Tulsa, OK: Society of Exploration Geophysicists, pp. 713–810.

Index

Printed in the United States
By Bookmasters